The XXXII Conference on Electrical and Electronic Engineering

The XXXII Conference on Electrical and Electronic Engineering

Volume Editors

Walter Vargas
Jackeline Abad
Christian Tipantuña
Pablo Proaño
Jaime Cepeda

MDPI

Basel • Beijing • Wuhan • Barcelona • Belgrade • Novi Sad • Cluj • Manchester

Volume Editors
Walter Vargas
Electric Power
Escuela Politécnica Nacional
Quito
Ecuador

Jackeline Abad
Automation and Industrial Control
Escuela Politécnica Nacional
Quito
Ecuador

Christian Tipantuña
Electronic, Telecommunications and Information Networks
Escuela Politécnica Nacional
Quito
Ecuador

Pablo Proaño
Automation and Industrial Control
Escuela Politécnica Nacional
Quito
Ecuador

Jaime Cepeda
Electric Power
Escuela Politécnica Nacional
Quito
Ecuador

Editorial Office
MDPI AG
Grosspeteranlage 5
4052 Basel, Switzerland

This is a reprint of the Proceedings, published open access by the journal *Engineering Proceedings* (ISSN 2673-4591), freely accessible at: https://www.mdpi.com/2673-4591/77/1.

For citation purposes, cite each article independently as indicated on the article page online and as indicated below:

Lastname, A.A.; Lastname, B.B. Article Title. *Journal Name* **Year**, *Volume Number*, Page Range.

ISBN 978-3-7258-3519-5 (Hbk)
ISBN 978-3-7258-3520-1 (PDF)
https://doi.org/10.3390/books978-3-7258-3520-1

Contents

About the Editors

Walter Vargas

Walter Vargas received a degree in Electrical Engineering from the Escuela Superior Politécnica del Litoral in 2007 and a master's degree in Electric Power Systems from the University of Seville in 2013. Between 2010 and 2017, he worked at CELEC EP-Transelectric in the 500 kV department and the Operation and Maintenance department. From 2017 to 2022, he worked at the National Electricity Operator (CENACE) in the Research and Development division. Since 2022, he has been working as a university professor at the National Polytechnic School and other universities. His areas of interest include optimization, planning, protection, control, and the assessment of the security of electric power systems.

Jackeline Abad

Jackeline Abad received a B.S. degree from Escuela Politécnica Nacional (2008), Ecuador, and M.S. and Ph.D. degrees in Electrical Engineering from Washington State University (2012, 2014). She held a Fulbright grant in 2010. She is currently an Associate Professor within the Departamento de Automatización y Control Industrial, Facultad de Ingeniería Eléctrica y Electrónica, Escuela Politécnica Nacional, Quito. Her current research interests include structural analysis and the design of dynamical control networks with applications in sensor/vehicle networking, epidemic control, and controlling power system networks.

Christian Tipantuña

Christian Tipantuña received a bachelor's degree in Telecommunications Engineering from Escuela Politécnica Nacional, Ecuador, in 2011, an M.Sc. degree in wireless systems and related technologies from Politecnico di Torino, Turin, Italy, in 2013, and a Ph.D. degree in Network Engineering from Universitat Politècnica de Catalunya, Barcelona, Spain, in 2022. He is a member of the Wireless Networks Research Group at Escuela Politécnica Nacional. His research interests include UAV-enabled communications, wireless networks, software-defined radio, optical networks, and the application of machine learning to communications systems and networks.

Pablo Proaño

Pablo Proaño is a full-time Professor at the National Polytechnic School (EPN) in Quito, Ecuador. He holds a master's degree in Electrical Engineering with a specialization in Smart Grids and a bachelor's degree in Electronics and Control Engineering. Since 2016, he has been involved in teaching in the faculties of Technological Education and Electrical and Electronics engineering at EPN. His research interests include industrial process control, smart and conventional electrical installations, renewable energy systems, and robotics. His professional experience encompasses maintenance, electrical installations, process automation, and the IoT.

Jaime Cepeda

Jaime Cepeda received a degree in Electrical Engineering from Escuela Politécnica Nacional in 2005, a PhD degree in Electrical Engineering from National University of San Juan, Argentina in 2013, and a master's degree in Big Data from the European University Miguel de Cervantes, Spain in 2021. In 2013, he served as a WAMS specialist at CENACE. From 2014 to 2021 he was the Head of Research and Development and the General Manager of Technical Development at the Ecuadorian

Independent System Operator ISO CENACE, and from 2021 to 2022 he held the position of Executive Director of the Agency for the Regulation and Control of Energy and Non-Renewable Natural Resources. Currently, he works as a University Professor teaching master's and PhD Programs at Escuela Politécnica Nacional and other Universities, and as a consultant on issues related to smart grids, power system operations, power system security and vulnerability assessment, power system modeling and control, and the application of data science to power systems.

engineering proceedings

Editorial

Preface: XXXII Conference on Electrical and Electronic Engineering †

Walter Vargas [1,*], Jackeline Abad [2], Christian Tipantuña [3], Pablo Proaño [2] and Jaime Cepeda [1]

1 DEE, Escuela Politécnica Nacional, Quito 170143, Ecuador; jaime.cepeda@epn.edu.ec
2 DACI, Escuela Politécnica Nacional, Quito 170143, Ecuador; jackeline.abad@epn.edu.ec (J.A.); pablo.proano@epn.edu.ec (P.P.)
3 DETRI, Escuela Politécnica Nacional, Quito 170143, Ecuador; christian.tipantuna@epn.edu.ec
* Correspondence: walter.vargas@epn.edu.ec
† The proceeding papers of this volume are presented at the XXXII Conference on Electrical and Electronic Engineering, Quito, Ecuador, 13–15 November 2024.

1. Introduction

1.1. Conference Introduction

The XXXII Electrical and Electronic Engineering Conference (XXXII JIEE-2024) is an annual event organized by the Faculty of Electrical and Electronic Engineering at the Escuela Politécnica Nacional, Quito, Ecuador. This conference brings together students, professors, researchers, and professionals with the aim of disseminating their graduation projects, curricular integration initiatives, research project advancements, and engagement activities, as well as presenting academic papers.

The JIEE serves as an opportunity to promote the exchange of knowledge and experiences in various fields of expertise, establish professional connections, stay updated on the latest technologies, and acquire skills to tackle the challenges presented by the industry.

The XXXII JIEE-2024 aimed to contribute to the sustainable development of the Ecuadorian energy, telecommunications, and production sectors, including the electricity sector in its generation, transmission, and distribution stages; control, automation, and security of essential infrastructure in the country's production. In this sense, the XXXII JIEE-2024 provided the space to disseminate national and international research and development aimed at guaranteeing the future availability of energy and telecommunications services, improving the production chain in the country, while promoting a lower impact on the environment.

Citation: Vargas, W.; Abad, J.; Tipantuña, C.; Proaño, P.; Cepeda, J. Preface: XXXII Conference on Electrical and Electronic Engineering. *Eng. Proc.* **2024**, *77*, 34. https://doi.org/10.3390/engproc2024077034

Published: 11 December 2024

1.2. Organizers

- Faculty of Electrical and Electronic Engineering at the Escuela Politécnica Nacional;
- Ecuadorian Committee of the Regional Energy Integration Commission (ECUACIER).

1.3. Scopes and Topics

- Power and energy;
- Communications;
- Electronic circuits, systems, and devices;
- Artificial intelligence;
- Local and wide area networks;
- Mobile and web applications;
- Internet of things;
- Robotics and automation systems;
- Security and privacy;
- Systems and control;
- Engineering education.

1.4. Program and Schedule

The details can be found at https://jiee.epn.edu.ec/en/program/schedule (accessed on 10 December 2024).

2. Committee Members

Walter Vargas, Escuela Politécnica Nacional, Ecuador;
Jackeline Abad, Escuela Politécnica Nacional, Ecuador;
Christian Tipantuña, Escuela Politécnica Nacional, Ecuador;
Pablo Proaño, Escuela Politécnica Nacional, Ecuador;
Jaime Cepeda, Escuela Politécnica Nacional, Ecuador.

3. Keynote Speakers

Rafael Segundo, University of Zurich, Switzerland;
Juan Vásquez, University of Luxembourg, Luxembourg;
Vicente Mut, National University of San Juan, Argentina.

4. Sponsors

Ecuadorian Committee of the Regional Energy Integration Commission (ECUACIER).

5. Statement of Peer Review

In submitting the conference proceedings to *Engineering Proceedings*, the volume editors of the proceedings certify to the publisher that all papers published in this volume have been subjected to peer review administered by the volume editors. Reviews were conducted by expert referees to the professional and scientific standards expected of a proceedings journal.

- Type of peer review: single-blind;
- Conference submission management system: Susy MDPI;
- Number of submissions received: 53;
- Number of submissions sent for review: 47;
- Number of submissions accepted: 33;
- Acceptance rate (number of submissions accepted/number of submissions received): 62.3%;
- Average number of reviews per paper: 1.38;
- Total number of reviewers involved: 65.

6. Peer Review Criteria and Process

The review process was conducted in a single-blind manner, with the entire procedure administered through the MDPI Submission System. Evaluations were performed by reviewers from various universities and countries. Based on their feedback, revisions were made to each article, encompassing thorough examinations of the structure, content, and writing.

Conflicts of Interest: The authors declare no conflicts of interest.

MDPI

engineering proceedings

Proceeding Paper

Information-Theoretic Security of RIS-Aided MISO System Under *N*-Wave with Diffuse Power Fading Model [†]

José David Vega-Sánchez [1,*], Ana Zambrano [2], Ricardo Mena [2] and José Oscullo [2]

1 Colegio de Ciencias e Ingenierías "El Politécnico", Universidad San Francisco de Quito (USFQ), Diego de Robles s/n, Quito 170157, Ecuador
2 Departamento de Electrónica, Telecomunicaciones y Redes de Información, Escuela Politécnica Nacional (EPN), Ladrón de Guevara E11-253, Quito 170525, Ecuador; ana.zambrano@epn.edu.ec (A.Z.); ricardomenav@epn.edu.ec (R.M.); jose.oscullo@epn.edu.ec (J.O.)
* Correspondence: dvega@usfq.edu.ec
† Presented at the XXXII Conference on Electrical and Electronic Engineering, Quito, Ecuador, 12–15 December 2024.

Abstract: This paper aims to examine the physical layer security (PLS) performance of a reconfigurable intelligent surface (RIS)-aided wiretap multiple-input single-output (MISO) system over generalized fading conditions by assuming inherent phase shift errors at the RIS. Specifically, the procedures (i.e., the method) to conduct this research is based on learning-based approaches to model the magnitude of the end-to-end RIS channel, i.e., employing an unsupervised expectation-maximization (EM) approach via a finite mixture of Nakagami-*m* distributions. This general framework allows us to accurately approximate key practical factors in RIS's channel modeling, such as generalized fading conditions, spatial correlation, discrete phase shift, beamforming, and the presence of direct and indirect links. For the numerical results, the secrecy outage probability, the average secrecy rate, and the average secrecy loss under different setups of RIS-aided wireless systems are assessed by varying the fading parameters of the *N*-wave with a diffuse power fading channel model. The results show that the correlation between RIS elements and unfavorable channel conditions (e.g., Rayleigh) affect secrecy performance. Likewise, it was confirmed that the use of a RIS is not essential when there is a solid line-of-sight link between the transmitter and the legitimate receiver.

Keywords: generalized fading channels; secrecy performance; reconfigurable intelligent surface

Citation: Vega-Sánchez, J.D.; Zambrano, A.; Mena, R.; Oscullo, J. Information-Theoretic Security of RIS-Aided MISO System Under *N*-Wave with Diffuse Power Fading Model. *Eng. Proc.* **2024**, *77*, 1. https://doi.org/10.3390/engproc2024077001

Academic Editor: Jaime Cepeda

Published: 18 November 2024

1. Introduction

Recently, RISs have received academic and industrial attention for their promising potential to customize the wireless channel cost-effectively. A RIS element can be programmed to adjust the phases, amplitude, phase frequency, and even the polarization of the impinging signal on the RIS to overwhelm the hazardous consequences of the channel. Owing to the promising RIS's features, this technology will presumably permit PLS to finally advance as a defense technique for delivering information security to wireless systems by exploiting the wireless propagation's intrinsic randomness (e.g., fading and noise). Motivated by these reasons, some research works on RIS-aided secure systems under diverse fading channel conditions have been studied in the literature [1]. Nevertheless, an in-depth examination of RIS-related studies in PLS indicates that the typical premise is to consider Rayleigh fading to model the individual end-to-end RIS channels for the mathematical facility. However, the original intention of RIS is to provide reliable line-of-sight (LoS) links between the indirect RIS channels. Therefore, unlike the Rayleigh model, the generalized channel models are better suited to deliver these requirements thanks to their freedom to model different propagation environments. In recent years, some activities have been addressed to develop a more precise channel model through two procedures: On the one hand, envelope-aided fading models, such us [2]. On the other hand, ray-based channel models where the received wave is constructed from dominant specular components plus several scattered waves with random phases [3]. The pioneering ray-based model was proposed in [4] and called the N-wave with diffuse power (N-WDP) model. This type of channel model is suited to accommodate different wireless channel conditions encountered in upcoming networks for higher frequencies. Particularly, the N-WDP models have exhibited satisfactory exactness in characterizing the short-fading channel in mm wave bands [5]. Regarding the performance of the RIS over different fading channel models, in [6], the authors developed a unified framework to assess the outage performance of RIS-assisted communications over generalized fading channels in the presence of phase error noise by using Fox's H functions. Likewise, in [7], a comprehensive statistical, secrecy, and performance analysis of RIS-assisted multiple-antenna systems under Weibull fading was presented. In [8], the secrecy rate analysis for a RIS-aided multi-user MISO system over a Rician fading channel was explored. Recently, in [9], the secrecy performance of RIS-aided systems over the Rician channel by assuming spatially random eavesdroppers was analyzed.

In light of the aforementioned points, and encouraged by the RIS features, this paper investigates how more realistic fading channels, discrete phase shifts, and spatial correlation impact the secrecy performance of RIS-assisted networks. Specifically, our primary purpose is to conduct a good description of the role of N-WDP's parameters on the secrecy performance of RIS-assisted networks. Furthermore, some useful insights for reaching secure communications in RIS-aided systems are provided, which are naturally connected to the physical parameters captured by the generalized channel model.

Notation: In what follows, uppercase and lowercase bold letters denote matrices and vectors, respectively; $f_{(\cdot)}(\cdot)$, probability density function (PDF); $F_{(\cdot)}(\cdot)$, the cumulative density function (CDF); $\mathcal{U}[a,b]$, a uniform distribution on $[a,b]$; $(\cdot)^{\mathrm{T}}$, the transpose; $\|\cdot\|$, the Euclidean norm of a complex vector; $\Gamma(\cdot)$, the gamma function ([10], Equation (6.1.1)); $\Upsilon(\cdot,\cdot)$, the lower incomplete gamma function ([10], Equation (6.5.2)); ${}_2F_1(\cdot,\cdot;\cdot;\cdot)$, the hypergeometric function ([10], Equation (15.1.1)); $\mathbb{C}$, the set of complex numbers; $\mathbb{E}[\cdot]$, expectation; $\mathcal{B}(\cdot,\cdot)$, the Beta function ([10], Equation (6.2.2)); $\mathrm{diag}(\mathbf{x})$, a diagonal matrix whose main diagonal is given by $\mathbf{x}$; $\mathcal{CN}(\cdot,\cdot)$, the circularly symmetric Gaussian distribution; $\mathbf{I}_{\mathrm{N}}$, the identity matrix of size $N\times N$; $(\cdot)^{\mathrm{H}}$, the Hermitian transpose; $\mathrm{mod}(\cdot)$, the modulus operation; $\lfloor\cdot\rfloor$, the floor function, and $\mathrm{sinc}(w)=\sin(\pi w)/(\pi w)$, the sinc function.

2. System and Channel Models

Assume a RIS-aided wiretap model composed of a transmitter, Alice (A) provided with M antennas, one legitimate node called Bob (B), one eavesdropper denoted by Eve (E), and a RIS, which helps the secure communication between A and B nodes. Specifically, a MISO system is assumed, where A sends information through its multiple antennas to a legitimate receiver B and an eavesdropper E, both of which have a single antenna. In this sense, the RIS helps to redirect the incident signal toward the legitimate receiver B, complicating the decoding of the message by eavesdropper E, as shown in Figure 1. In short, B and E are equipped with a single antenna, while the RIS has L nearly reconfigurable units. So, the received signal at E and B can be expressed as

$$y_i=\sqrt{P}\left(\mathbf{h}_{2,i}^{\mathrm{T}}\mathbf{\Phi}\mathbf{G}+\mathbf{h}_{\mathrm{d},i}^{\mathrm{T}}\right)\mathbf{w}x+\widetilde{n}_i, \tag{1}$$

in which $i\in\{\mathrm{B},\mathrm{E}\}$ is either the eavesdropper or the legitimate channel, x is the transmitted information symbol, P is the transmit power at A, $\widetilde{n}\sim\mathcal{CN}(0,\sigma_i^2)$ denotes the additive white Gaussian noise (AWGN) with σ_i^2 power, $\mathbf{h}_{\mathrm{d},i}\in\mathbb{C}^{M\times 1}$ is the channel between A and B or A and E, $\mathbf{w}\in\mathbb{C}^{M\times 1}$ denotes the precoding vector at A, $\mathbf{G}=[\mathbf{g}_1,\ldots,\mathbf{g}_{\mathrm{M}}]\in\mathbb{C}^{L\times M}$ and $\mathbf{h}_{2,i}=[h_{2,i,1},\ldots,h_{2,i,L}]^{\mathrm{T}}\in\mathbb{C}^{L\times 1}$ are the paths between A and RIS, RIS and B, or RIS and E, respectively. Furthermore, $\mathbf{\Phi}=\mathrm{diag}(\delta_1e^{j\varphi_1},\ldots,\delta_Le^{j\varphi_L})$ is the phase-shift matrix generated by the RIS, where δ_l for $l=\{1,\ldots,L\}$ denotes the amplitude of the lth RIS element. Hereafter, it is considered that $\delta_l=1,\forall l$ in all subsequent derivations. The RIS adjusts the phase shifts for each unit to compensate for all phases created by the direct and indirect channels at the legitimate receiver, i.e., $\varphi_l = \angle(\mathbf{h}_{\mathrm{d,B}}^{\mathrm{H}}\mathbf{w}) - \angle(h_{2,\mathrm{B},l}^{\mathrm{H}}) - \angle(\mathbf{g}_l\mathbf{w})$ ([11], Equation (19)) where $h_{2,\mathrm{B},l}^{\mathrm{H}}$ is the lth element of $\mathbf{h}_{2,\mathrm{B}}^{\mathrm{H}}$ and $\mathbf{g}_l$ is the lth row vector from $\mathbf{G}$. In this configuration, both φ_l and the transmit precoding $\mathbf{w}$ are optimized together [11] to maximize the acquired signal-to-noise ratio (SNR) at B. Nevertheless, the RIS forces a residual random phase error to prevail in practice. Here, such an error is denoted by ϕ_l, so the final phase shifts of the lth RIS unit change from the optima to $\Theta_l = \varphi_l + \phi_l$ [11]. Therefore, the equivalent channel observed by the receivers can be formulated as

$$h_i = \left(\mathbf{h}_{2,i}^{\mathrm{T}}\boldsymbol{\Psi}\mathbf{G} + \mathbf{h}_{\mathrm{d},i}^{\mathrm{T}}\right)\mathbf{w}^{+}, \tag{2}$$

where $\boldsymbol{\Psi} = \mathrm{diag}(e^{j\Theta_1}, \ldots, e^{j\Theta_L})$ indicates the phase noise, and $\mathbf{w}^{+} = \frac{\left(\mathbf{h}_{2,\mathrm{B}}^{\mathrm{T}}\boldsymbol{\Psi}\mathbf{G}+\mathbf{h}_{\mathrm{d,B}}^{\mathrm{T}}\right)^{\mathrm{H}}}{\left\|\mathbf{h}_{2,\mathrm{B}}^{\mathrm{T}}\boldsymbol{\Psi}\mathbf{G}+\mathbf{h}_{\mathrm{d,B}}^{\mathrm{T}}\right\|} \in \mathbb{C}^{M\times 1}$ is the maximal ratio transmission (MRT) precoding. Here, we decided to use the joint active and passive MRT precoding strategy from [11] because it is a low-complexity algorithm obtained on alternating optimization. In particular, the transmit precoding at A and the RIS's phase shifts were optimized iteratively until meeting a pre-established criterion. It is worth pointing out that the MRT beamforming occurs when $\mathbf{w}$ is employed as a starting point in the distributed algorithm, which is updated in each iteration together with the RIS's phase shifts. Concerning the channel model, we supposed that the fading coefficients of the underlying system are constructed as a superposition of N-dominant specular components plus scattering waves (i.e., the N-WDP model); so, it follows that [12]

$$h_{\mathrm{d},i}^{p} = \sqrt{\beta_{\mathrm{d},i}}\left(\sum_{n=1}^{N_{\mathrm{d},i}} V_{\mathrm{d},i}^{(n,p)} e^{j\theta_{\mathrm{d},i}^{(n,p)}} + Z_{\mathrm{d},i}^{(p)}\right),\ h_{2,i,l} = \sqrt{\beta_{2,i}}\left(\sum_{n=1}^{N_{\mathrm{RIS},i}} V_{2,i}^{(n,l)} e^{j\theta_{2,i}^{(n,l)}} + Z_{2}^{(l)}\right) \tag{3}$$

$$g_{l,p} = \sqrt{\beta_1}\left(\sum_{n=1}^{N_{\mathrm{A,RIS}}} V_1^{(n,p,l)} e^{j\theta_1^{(n,p,l)}} + Z_1^{(p,l)}\right), \tag{4}$$

for $p = \{1, \ldots, M\}$, and $l = \{1, \ldots, L\}$. Here, $V_{(\cdot)}^{(n,\cdot)}$ symbolize the amplitude of an nth specular component, $\theta_{(\cdot)}^{(n,\cdot)} \sim \mathcal{U}[0, 2\pi]$ and $N_{(f),(u)}$ indicate the total number of dominant components of the channel between nodes f and u, and $Z_{(\cdot)}^{(\cdot)}$ follow Rayleigh with $\mathbb{E}\{|Z|^2\} = 2\sigma^2 = \Omega_0$, indicating the diffuse received waves. The terms β_1, $\beta_{2,i}$, and $\beta_{\mathrm{d},i}$ encompass the path loss for A-RIS, RIS-i, and A-i links for $i \in \{\mathrm{B}, \mathrm{E}\}$, respectively. Notice that the N-WDP model includes important stochastic fading models such as TWDP, Rician, and Rayleigh as special cases for $N_{(\cdot)} = 2, 1, 1$, respectively. Now, to evaluate spatial correlation for the RIS channels, the geometry network proposed in [13] is employed. Hence, we assumed that the RIS is a surface consisting of L elements, where L_V and L_H are the total units per row and per column, respectively. In this regard, the area of an individual RIS element is expressed by $A = d_H d_V$, with d_H and d_V denoting the vertical height and the horizontal width, respectively. Based on this, and by assuming an isotropic diffuse environment, the indirect RIS fading channels subject to correlation and the direct link can defined as $\mathbf{h}_{\mathrm{d},i} = \left[h_{\mathrm{d},i}^{1}, \ldots, h_{\mathrm{d},i}^{M}\right]^{\mathrm{T}}$, $\mathbf{h}_{2,i} = \sqrt{A}\mathbf{R}[h_{2,i,1}, \ldots, h_{2,i,L}]^{\mathrm{T}}$, $\mathbf{g}_p = \sqrt{A}\mathbf{R}[g_{1,p}, \ldots, g_{L,p}]^{\mathrm{T}}$, where $p = \{1, \ldots, M\}$, $\mathbf{R} \in \mathbb{C}^{L\times L}$ is the spatial correlation matrix for the RIS. The (a,b)th-entry of $\mathbf{R}$ is given by $[\mathbf{R}]_{a,b} = \mathrm{sinc}(2\|\mathbf{u}_a - \mathbf{u}_b\|/\lambda)\ a, b = 1, \ldots, L$ [13], where $\mathbf{u}_\zeta = [0, \mathrm{mod}(\zeta - 1, L_H)d_H, \lfloor(\zeta-1)/L_H\rfloor d_V]^T$, $\zeta \in \{a, b\}$, and λ is the wavelength of the wave. With the previous expressions, the received SNR at B or E is formulated by

$$\gamma_i = \frac{P|h_i|^2}{\sigma_i^2} = \overline{\gamma}_i|h_i|^2. \tag{5}$$

Now, $\overline{\gamma}_i = P/\sigma_i^2$ is defined as the average transmit SNR for B or E. As a first stage, the purpose is to approximate the magnitude of h_i using some statistical distribution. For this purpose, we employed the EM learning algorithm, as follows:

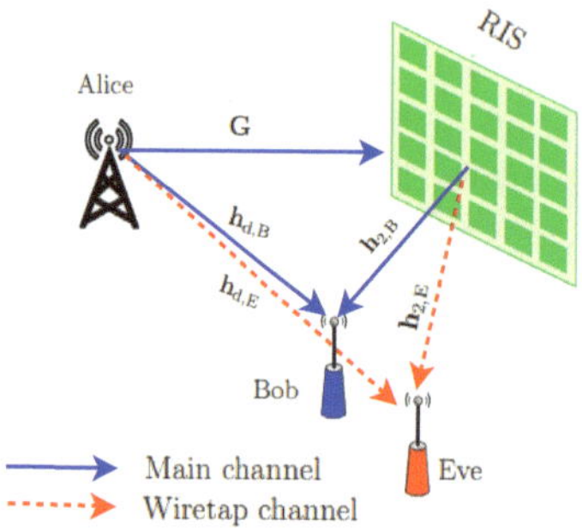

Figure 1. RIS-aided wiretap MISO wireless communication system.

3. RIS Channel Characterization

Here, using a simple mixture of Nakagami-m random variables (RVs) combined with the EM strategy, the end-to-end RIS channel was characterized in a simple manner.

Let $\mathbf{h}_i = \{h_i\}_1^t$ be a training set vector of t samples of h_i in (2); it is approximated that h_i is observed by B and E using a mixture of Nakagami-m distributions, as follows [14]:

$$f_{h_i}(r_i) \approx \sum_{z=1}^{\mathcal{C}} \omega_{z,i}\mathcal{N}_{z,i}(r_i; m_{z,i}, \Omega_{z,i}), \tag{6}$$

where $\mathcal{C}$ is the mixture size (i.e., the number of Nakagami-m components), $\mathcal{N}_{z,i}(r_i; m_{z,i}, \Omega_{z,i}) = \frac{m_{z,i}^{m_{z,i}} r_i^{2m_{z,i}} e^{-\frac{m_{z,i}r_i^2}{\Omega_{z,i}}}}{2^{-1}\Gamma(m_{z,i})\Omega_{z,i}^{m_{z,i}} r_i}$, $\omega_{z,i} \in \{1, \ldots, \mathcal{C}\}$ with $\sum_{z=1}^{\mathcal{C}} \omega_{z,i} = 1$ and $0 \leq \omega_{z,i} \leq 1$ are the mixture weights, and $\Omega_{z,i}$ and $m_{z,i}$ are the mean powers and the fading elements of the weighted PDFs, respectively. Here, we refer to the unsupervised EM method to fit the mixture parameters. Specifically, in the expectation (E)-step of the EM technique, the posterior probabilities (i.e., membership values) of the zth-weighted PDF of both B and E can be calculated as [14]

$$\tau_{zj,i}^{(k)} = \frac{\omega_{z,i}\mathcal{N}_{z,i}(h_{i,j}; m_{z,i}, \Omega_{z,i})}{\sum_{l=1}^{\mathcal{C}} \omega_{l,i}\mathcal{N}_{l,i}(h_{i,j}; m_{z,i}, \Omega_{z,i})}, \; z = 1, \ldots, \mathcal{C}, j = 1, \ldots, t, \tag{7}$$

where k represents the present iteration, t is the size of the sample set in (2), $h_{i,j}, \forall j = 1 \ldots t$ are the unlabeled samples at the receiver node $i \in \{\mathrm{B}, \mathrm{E}\}$, and z is the mixture component. Next, in the M-step of the EM algorithm, the updated parameters are fitted by maximizing the log-likelihood formulation of each RV-weighted mixture via the posterior probabilities. Hence, the updated parameters of the receive sides can be calculated as [14]

$$\Omega_{z,i}^{(k+1)} = \frac{\sum_{j=1}^{t} \tau_{zj,i}^{(k)} h_{i,j}^2}{\sum_{j=1}^{t} \tau_{zj,i}^{(k)}} \quad m_{z,i}^{(k+1)} = \frac{1 + \sqrt{1 + \frac{4\Delta_{z,i}^k}{3}}}{4\Delta_{z,i}^k}$$
$$\omega_{z,i}^{(k+1)} = \frac{\sum_{j=1}^{t} \tau_{zj,i}^{(k)}}{t}, \Delta_{z,i}^k = \frac{\sum_{j=1}^{t} \tau_{zj,i}^{(k)} \left[\log(\Omega_{z,i}) - \log(h_{i,j}^2)\right]}{\sum_{j=1}^{t} \tau_{zj,i}^{(k)}} \tag{8}$$

Algorithm 1 displays the EM Nakagami-m mixture model, where the initial mixture weights are randomly selected from $\mathcal{U}[0, 1]$, and the start values for the scaling parameters of

Nakagami-m PDFs are calculated using the maximum likelihood estimation (MLE). Also, comparable toleration practice is employed as a stopping criterion. It is worth pointing out that the EM can efficiently model any target system's wireless channel considered in (2). In fact, EM can also be utilized even with experimental measurement channel data.

Algorithm 1: Unsupervised EM machine learning algorithm for fitting ω_i, Ω_i, and m_i of Nakagami-m RVs

Input: $\mathbf{h}_i \leftarrow$ unlabeled set for $i \in \{\mathrm{B,E}\}$, $\epsilon \leftarrow 1 \times 10^{-3}$, and initial values of $\omega_{z,i}$, $\Omega_{z,i}$, and $m_{z,i}$ via MLE, for $z = 1, \dots, \mathcal{C}$;
Output: $\omega_{z,i}$, $\Omega_{z,i}$, and $m_{z,i}$, for $z = 1, \dots, \mathcal{C}$;
$k = 1,\ t = \mathrm{length}(\mathbf{h}_i)$;
while $\Lambda\Omega_{z,i}$ && $\Lambda m_{z,i} < \epsilon$ **do**
  E step:
  for $j = 1; j < t; j$++ **do**
    $\tau^k_{zj,i} = \frac{\omega^k_{z,i}\mathcal{N}_{z,i}(h_{i,j}; m^k_{z,i}, \Omega^k_{z,i})}{\sum_{l=1}^{2}\omega_{l,i}\mathcal{N}_{l,i}(h_{i,j}; m^k_{z,i}, \Omega^k_{z,i})}$, *for* $z = 1, \dots, \mathcal{C}$;
  end
  M step:
  $\Delta^k_{z,i} = \frac{\sum_{j=1}^{t}\tau^k_{zj,i}\left(\log(\Omega^k_{z,i}) - \log(h^2_{i,j})\right)}{\sum_{j=1}^{t}\tau^k_{zj,i}}$;
  $\Omega^{k+1}_{z,i} = \frac{\sum_{j=1}^{t}\tau^k_{zj,i} \times h^2_{i,j}}{\sum_{j=1}^{t}\tau^k_{zj,i}}$ $m^{k+1}_{z,i} = \frac{1+\sqrt{1+\frac{4\Delta^k_{z,i}}{3}}}{4\Delta^k_{z,i}}$;
  $\omega^{k+1}_{z,i} = \sum_{t=1}^{t}\tau^k_{zj,i}/t$, *for* $z = 1, \dots, \mathcal{C}$;
  $k++$;
end
Stop Criterion: $\Lambda m_{z,i} = |(m^{(k+1)}_{z,i} - m^{(k)}_{z,i})/m^k_{z,i}|$ $\Lambda\Omega_{z,i} = |(\Omega^{(k+1)}_{z,i} - \Omega^{(k)}_{z,i})/\Omega^k_{z,i}|$;

4. RIS Secrecy Performance

4.1. Distribution of γ_E and γ_B

The PDFs and CDFs of the received SNR in E and B, respectively, are acquired by performing a transformation of variables from (5). Thus, this yields

$$f_\mathrm{E}(\gamma_\mathrm{E}) = \sum_{z=1}^{\mathcal{C}} \frac{\gamma_\mathrm{E}^{\kappa_{z,\mathrm{E}}-1}}{(\overline{\gamma}_\mathrm{E}\theta_{z,\mathrm{E}})^{\kappa_{z,\mathrm{E}}}\Gamma(\kappa_{z,\mathrm{E}})} e^{\left(-\frac{\gamma_\mathrm{E}}{\overline{\gamma}_\mathrm{E}\theta_{z,\mathrm{E}}}\right)},\quad F_\mathrm{B}(\gamma_\mathrm{B}) = \sum_{z=1}^{\mathcal{C}} \frac{\Upsilon\left(\kappa_{z,\mathrm{B}}, \frac{\gamma_\mathrm{B}}{\overline{\gamma}_\mathrm{B}\theta_{z,\mathrm{B}}}\right)}{\Gamma(\kappa_{z,\mathrm{B}})} \tag{9}$$

where $\kappa_{z,i} = m_{z,i}$ and $\theta_{z,i} = \Omega_{z,i}/m_{z,i}$.

4.2. Secrecy Outage Probability (SOP)

This metric is useful for passive eavesdropping scenarios where Eve channel's channel state information (CSI) is not known at Alice. Thus, Alice encodes the information at an unchanging secrecy rate R_S. Based on [15], the secrecy capacity, i.e., C_S, is calculated as $C_\mathrm{S} = \max\{C_\mathrm{B} - C_\mathrm{E}, 0\}$, with $C_\mathrm{B} = \log_2(1+\gamma_\mathrm{B})$ and $C_\mathrm{E} = \log_2(1+\gamma_\mathrm{E})$ being the channel capacities at B and E, respectively. Here, secrecy is attained only when $R_\mathrm{S} \leq C_\mathrm{S}$, and it is compromised otherwise. Mathematically speaking, a lower bound of the SOP can be determined as [15] $\mathrm{SOP_L}(R_\mathrm{S}) = \int_0^\infty F_{\gamma_\mathrm{B}}(2^{R_\mathrm{S}}\gamma_\mathrm{E}) f_{\gamma_\mathrm{E}}(\gamma_\mathrm{E}) d\gamma_\mathrm{E}$. Hence, an approximation of the $\mathrm{SOP_L}$ can obtained as stated below.

Proposition 1. *The $\mathrm{SOP_L}$ for RIS-assisted MISO networks under generalized fading is approximated as*

$$\mathrm{SOP}_L(R_S) = \sum_{s=1}^{C}\sum_{z=1}^{C} \omega_{s,B}\omega_{z,E}\left(\frac{\overline{\gamma}_E\theta_{z,E}}{\overline{\gamma}_B\theta_{s,B}}\right)^{\kappa_{s,B}} \frac{2^{R_S\kappa_{s,B}}}{\kappa_{s,B}\mathcal{B}(\kappa_{s,B},\kappa_{z,E})} {}_2F_1\left(\kappa_{s,B}+\kappa_{z,E},\kappa_{s,B};1+\kappa_{s,B};-\frac{2^{R_S}\overline{\gamma}_E\theta_{z,E}}{\overline{\gamma}_B\theta_{s,B}}\right). \tag{10}$$

Proof. $\mathrm{SOP_L}$ can be easily derived from ([16], Equation (7)), with the corresponding substitutions of (9), and after some mathematical manipulations. □

4.3. Average Secrecy Rate (ASR)

This metric is relevant for the active eavesdropping case, where Alice knows the CSI of both Bob and Eve channels. Alice can guarantee secrecy by adjusting her transmission rate subject to R_S. Based on [17], the ASR can be computed as $\mathrm{ASR} = \int_0^\infty (1 - \mathrm{SOP_L}(x))dx$.

Proposition 2. *The ASR expression of the RIS underlying network is approximated by*

$$ASR \approx \sum_{c=1}^{\eta} \xi_c e^{-x_c} u_1(x_c), \tag{11}$$

where $u_1(x_c) = e^{x_c}(1 - SOP_L(x_c))$ and x_c and ξ_c are the cth zero (root) and weight of the ηth-order Laguerre polynomial ([10], Equation (22.2.13)).

Proof. Based on Gauss–Laguerre quadrature technique ([10], Equation (25.4.45)), the exact ASR can be approximated by a weighted sum of samples. □

4.4. Average Secrecy Loss (ASL)

This novel secrecy metric relates to the information leaked from A to E. Using ([17], Equation (9)), the ASL is expressed as in the following proposition.

Proposition 3. *The ASL formulation of the underlying network is expressed as*

$$ASL \approx \frac{\sum_{j=1}^{v} \rho_j e^{-x_j} u_2(x_j)}{ASR^2} - 1, \tag{12}$$

where $u_2(x_j) = 2x_j e^{x_j}(1 - SOP_L(x_j))$, x_j, and ρ_j are the jth zero (root) and weight of the vth-order Laguerre polynomial ([10], Equation (22.2.13)).

Proof. Again, the Gauss–Laguerre quadrature approach is used to approximate the integral ([17], Equation (10)), which involves the computation of the exact ASL. □

5. Numerical Results and Discussions

Here, we investigated how correlated channels over generalized fading impact the system's secrecy performance, assuming discrete phase shifts, as well as the goodness-of-fit of the approximations for the RIS channels. Monte Carlo (MC) simulations for EM-based approximations are included in all instances with markers. For the sake of comparison, the well-known moment matching method (MoM) (For informative purposes and to compare the MoM with the proposed method fairly) models the end-to-end channel in (2) through a Nakagami-m distribution. Here, the fitting parameters were obtained numerically by calculating the moments and equating them to the parameters of the Nakagami-m distribution, as indicated by the MoM procedure in ([18], Sec. III).) proposed in [18] for modeling generalized RIS channels (i.e., (2)) is included as a baseline in the secrecy analysis. Also, to estimate the parameters of the Nakagami-m mixture model with the EM method, 10^5 realizations are generated for the unlabeled training set in (2), which is required in Algorithm 1 for both Bob and Eve. Specifically, these generated channel coefficients are the observations, constituting the input in Algorithm 1. With the output of Algorithm 1 (e.g., $\omega_{z,i}$, $\Omega_{z,i}$, and $m_{z,i}$), these fitting parameters are used in (9) to be later substituted into (10)–(12); in this way, the performance of the underlying system model is evaluated. The discrete phase shift ϕ_l is built from $\mathcal{U}[-2^{-q}\pi, 2^{-q}\pi]$, where $q \geq 1$ is the amount of quantization bits utilized to encode the phase shifts errors, i.e., the discrete set of 2^q phases that can be configured on the RIS (a larger q refers to smaller phase errors on RIS's configuration).

In all figures, the following considerations are assumed: (i) a RIS-aided network system given in [13] with a frequency of 3 GHz (i.e., $\lambda = 0.1$ m), $BW = 1$ MHz of bandwidth, a transmit power of $P = 23$ dBm, and an AWGN with a power spectral density of $N_0 = -174$ dBm/Hz. Hence, $\sigma_i^2 = N_0 BW = -114$ dBm, which makes $P/\sigma_i^2 = \overline{\gamma}_i = 137$ dB, (ii) $R_S = 1$ bps/Hz for all SOP curves, and (iii) $\mathcal{C} = 2$ as the number of components of the Nakagami-m mixture.

Moreover, only Figure 2b includes both direct and indirect links. For mathematical compactness, the power ratio parameter is represented as being analogous to the Rician K factor, i.e., $K_{N_{(\cdot)}} \triangleq \frac{\Omega_{N_{(\cdot)}}}{\Omega_0}$, with $\Omega_{N_{(\cdot)}} = \sum_{n=0}^{N_{(\cdot)}} \left(V_{(\cdot)}^{(n,\cdot)}\right)^2$ being the total average power of the dominant components. Also, $K_{\text{dB}}^{(\cdot)} = 10\log_{10}\left(K_{(\cdot)} = K_{N_{(\cdot)}}\right)$ is defined. Likewise, the amplitudes of corresponding rays are depicted in terms of the amplitude of the first dominant component, i.e., $V_{(\cdot)}^{(n,\cdot)} = \alpha_{n,(\cdot)} V_{(\cdot)}^{(1,\cdot)}$ for $n = \left\{2, \ldots, N_{(\cdot)}\right\}$, with $0 < \alpha_{n,(\cdot)} < 1$. Figure 2a shows the SOP vs. $\beta_{2,\text{B}}$ for different channel setups. Here, the following cases are assumed. *Case* I: The legitimate paths are subject to Rician fading, i.e., $N_{\text{A,RIS}} = N_{\text{RIS,B}} = 1$ with $V_1^{(1,\cdot)} = V_{2,\text{B}}^{(1,\cdot)} = 1$ for $K_{\text{dB}}^{\text{A,RIS}} = K_{\text{dB}}^{\text{RIS,D}} = 12$ dB, and the eavesdropper link, i.e., RIS-to-E, follows Rayleigh fading (i.e., $N_{\text{RIS,E}} = 0$). *Case* II: All channels follow Rayleigh fading (i.e., the classical assumption). For the aforementioned cases, the remaining system parameters are $M = 5$, $L = 144$, $q = 5$, $\beta_1 = -20$ dB, $\beta_{2,\text{E}} = -30$ dB, and $d_H = d_V \in \{\lambda/2, \lambda/5\}$. Here, the impact of dealing with both Rician fading and RIS-correlated channels is explored. From all traces, we observed that a more favorable channel propagation for legitimate links (i.e., Rician fading) contributes to improving the SOP performance with respect to Rayleigh propagation's counterpart. Moreover, revealing SOP behavior is observed depending on the strength of the correlation on RIS channels. For instance, decreasing the size of RIS units (i.e., hard-correlated RIS channels) leads to lower secrecy performance.

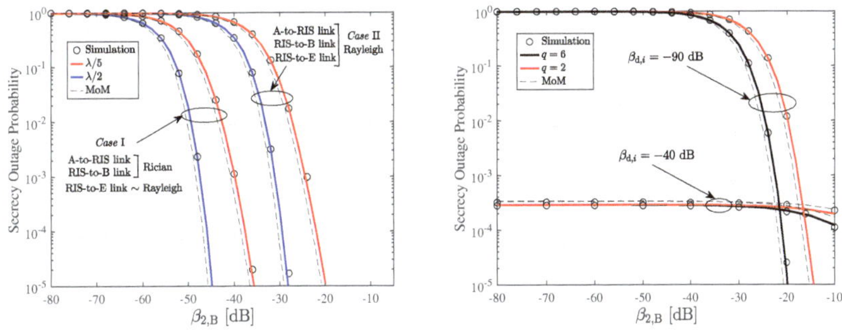

Figure 2. (**a**) SOP vs. $\beta_{2,\text{B}}$ with different channel configurations. (**b**) SOP vs. $\beta_{2,\text{B}}$ by varying both $\beta_{\text{d},i}$ and q in the presence of direct and indirect paths. The solid lines represent the proposed analytical solutions.

Figure 2b illustrates the SOP vs. $\beta_{2,\text{B}}$ when both Bob and Eve have a direct path. Here, all the indirect paths, i.e., A-to-RIS, RIS-to-B, and RIS-to-E, follow Rayleigh fading. However, the direct links, i.e., A-to-B and A-to-E, are subject to TWDP fading. Thus, $N_{\text{d,E}} = N_{\text{d,B}} = 2$ with $V_{\text{d,E}}^{(1,\cdot)} = V_{\text{d,B}}^{(1,\cdot)} = 1$, $\alpha_{2,\text{d,B}} = \alpha_{2,\text{d,E}} = 0.7$ for $K_{\text{dB}}^{\text{d,B}} = K_{\text{dB}}^{\text{d,E}} = 18$ dB. The remaining parameters are set to $M = 8$, $L = 196$, $A\beta_1 = -30$ dB, $A\beta_{2,\text{E}} = -50$ dB, and $d_H = d_V = \lambda/3$. From this figure, we observed that the combination of mild path loss (i.e., $\beta_{\mathrm{d},i} = -40$ dB) and good propagation conditions (i.e., TWDP fading) on Bob's direct link is favorable to achieve better secrecy performance. Also, for this scenario, notice that the RIS does not contribute significantly to the system's performance due to the presence of a strong LoS for Bob and Eve's links. However, the SOP's behavior changes slightly from $\beta_{2,\mathrm{B}} \approx -20$ dB since the RIS-to-B link presents better channel conditions than the direct A-to-B link. Therefore, the RIS begins to operate in this region, and a difference in the phase error's configuration is observed. Now, when direct paths are heavily attenuated (i.e., $\beta_{\mathrm{d},i} = -90$ dB), the use of the RIS makes sense in this case since different performances of the SOP are obtained by varying q.

Figure 3a presents the ASR vs. $\beta_{2,\mathrm{B}}$. In this figure, base values (i.e., baseline fading channel models) are considered to measure the system's performance by setting different propagation conditions. Also, the following scenarios are considered: *Case* III: A-to-RIS link (Rician), $N_{\mathrm{A,RIS}} = 1$ with $V_1^{(1,\cdot)} = 1$ for $K_{\mathrm{dB}}^{\mathrm{A,RIS}} = 25$ dB. RIS-to-B link (three rays), $N_{\mathrm{RIS,B}} = 3$ with $V_{2,\mathrm{B}}^{(1,\cdot)} = 1$, $\alpha_{n,\mathrm{RIS,B}} = 0.1$ for $n = \{2,3\}$, and $K_{\mathrm{dB}}^{\mathrm{RIS,B}} = 25$ dB. RIS-to-E link (TWDP), $N_{\mathrm{RIS,E}} = 2$ with $V_{2,\mathrm{E}}^{(1,\cdot)} = 1$, $\alpha_{2,\mathrm{RIS,E}} \in \{1, 0.5, 0.1\}$ for $K_{\mathrm{dB}}^{\mathrm{RIS,E}} = 25$ dB. Also, $M = 6$, $L = 196$, $q = 4$, $A\beta_1 = A\beta_{2,\mathrm{E}} = -50$ dB, and $d_H = d_V = \lambda/4$. Herein, the impact of Eve's fading TWDP parameters on the ASR performance is explored. From all traces, notice that there is a mixture of more dominant paths in B than E and that lower amplitude values associated with Eve's rays (i.e., $\alpha_{2,\mathrm{RIS,E}}$) lead to poor ASR performance. This result verifies that the amplitudes of the specular waves play relevant roles in secure communication criteria of over-generalized channel conditions.

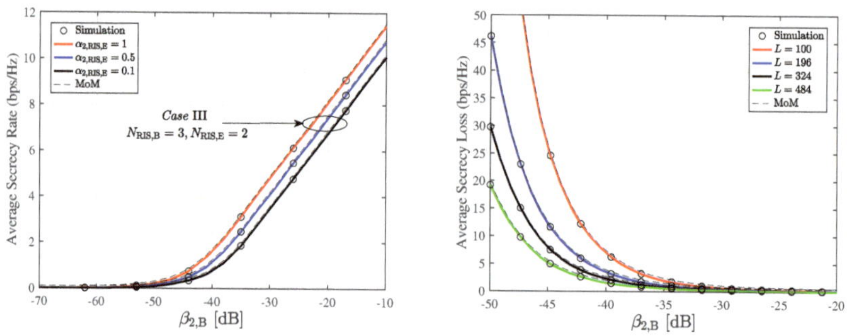

(**a**) (**b**)

Figure 3. (**a**) ASR vs. $\beta_{2,\mathrm{B}}$ with different number of specular components on the receiver sides. (**b**) ASL vs. $\beta_{2,\mathrm{B}}$ by varying the number of elements on the RIS. The solid lines represent the proposed analytical solutions.

Figure 3b depicts the ASL vs. $\beta_{2,\mathrm{B}}$ by altering the elements at the RIS. For this scenario, we considered that all indirect paths follow Rician fading, i,e., $N_{\mathrm{A,RIS}} = N_{\mathrm{RIS,B}} = N_{\mathrm{RIS,E}} = 1$ with $V_1^{(1,\cdot)} = V_{2,\mathrm{B}}^{(1,\cdot)} = V_{2,\mathrm{E}}^{(1,\cdot)} = 1$ for $K_{\mathrm{dB}}^{\mathrm{A,RIS}} = K_{\mathrm{dB}}^{\mathrm{RIS,B}} = K_{\mathrm{dB}}^{\mathrm{RIS,E}} = 20$ dB. Furthermore, $M = 4$, $q = 6$, $A\beta_1 = A\beta_{2,\mathrm{E}} = -45$ dB, and $d_H = d_V = \lambda/3$. From all instances, it was observed that information leakage was drastically reduced as the number of elements on the RIS increased. Here, the ASL shows how much $\beta_{2,\mathrm{B}}$ the system needs to drive the information leak to zero. For instance, in the proposed setup, from $\beta_{2,\mathrm{B}} \approx -35$ dB, communication between legitimate links is secure. In addition, notice that in all the analyzed curves, the proposed approximation beats the MoM in precision, mainly in the secrecy outage-based metric. Finally, it is worth mentioning that an energy consumption parameter that integrates energy collection from environmental sources with secure communication mechanisms at the physical layer is not considered. However, this topic is out of this paper's scope and will be explored in the future.

6. Conclusions

The secrecy performance of the RIS-assisted wiretap MISO system by assuming generalized fading channels and discrete phase shifts was explored. Specifically, secrecy metrics such as the SOP, ASR, and ASL were obtained in closed-form expression to evaluate the secure performance by approximating the end-to-end RIS channel through an unsupervised EM learning algorithm. Some insights reveal how different propagation conditions and RIS practical implementation (e.g., discrete phase-shift noise, spatial correlation, beamforming, presence of direct link) impact secrecy performance. For instance, when the direct link is present in the communication between Alice and Bob, using a RIS is practically unnecessary despite correctly configuring the phase error. On the contrary, when only the non-line-of-sight link between Alice and Bob is available, RIS technology is essential so that the information does not leak to the eavesdropper.

Author Contributions: Conceptualization, J.D.V.-S.; methodology, A.Z. and R.M.; writing—original draft preparation, A.Z. and J.O.; writing—review and editing, J.D.V.-S. and A.Z.; supervision, R.M. and J.O. All authors have read and agreed to the published version of the manuscript.

Funding: This research received no external funding.

Data Availability Statement: No new data were created or analyzed in this study. Data sharing is not applicable to this article.

Acknowledgments: The work of Ana Zambrano and Ricardo Mena was supported by the Escuela Politécnica Nacional, under research project PII-DETRI-2024-02, "Análisis de desempeño de comunicaciones 6G asistidas por superficies inteligentes reconfigurables o antenas fluidas".

Conflicts of Interest: The authors declare no conflicts of interest.

References

1. Zhou, G.; Pan, C.; Ren, H. Secure Wireless Communication in RIS-Aided MISO System with Hardware Impairments. *IEEE Wirel. Commun. Lett.* **2021**, *10*, 1309–1313. [CrossRef]
2. Yacoub, M.D. The κ-μ distribution and the η-μ distribution. *IEEE Antennas Propagat. Mag.* **2007**, *49*, 68–81. [CrossRef]
3. Durgin, G.D. Theory of Stochastic Local Area Channel Modeling for Wireless Communications. Ph.D. Thesis, Virginia Polytechnic Institute and State University, Blacksburg, VA, USA, 2000.
4. Chun, Y.J. A generalized fading model with multiple specular components. *arXiv* **2018**, arXiv:1810.05258.
5. Zöchmann, E.; Caban, S.; Mecklenbräuker, C.F.; Pratschner, S.; Lerch, M.; Schwarz, S.; Rupp, M. Better than Rician: Modelling millimetre wave channels as two-wave with diffuse power. *EURASIP J. Wirel. Commun. Netw.* **2019**, *2019*, 21. [CrossRef]
6. Trigui, I.; Ajib, W.; Zhu, W.-P.; Renzo, M.D. Performance Evaluation and Diversity Analysis of RIS-Assisted Communications Over Generalized Fading Channels in the Presence of Phase Noise. *IEEE Open J. Commun. Soc.* **2022**, *3*, 593–607. [CrossRef]
7. Alves Junior, M.A.; Fraidenraich, G.; Ferreira, R.C.; De Figueiredo, F.A.P.; De Lima, E.R. Multiple-Antenna Weibull-Fading Wireless Communications Enhanced by Reconfigurable Intelligent Surfaces. *IEEE Access* **2023**, *11*, 107218–107236. [CrossRef]
8. Wang, J.; Zhang, J.; Lu, J.; Wang, J.; Zhang, Q.; Wang, D. Secrecy Rate Analysis for RIS-Aided Multi-User MISO System over Rician Fading Channel. *J. Commun. Inf. Netw.* **2023**, *8*, 48–56. [CrossRef]
9. Shi, W.; Xu, J.; Xu, W.; Yuen, C.; Lee Swindlehurst, A.; Zhao, C. On Secrecy Performance of RIS-Assisted MISO Systems Over Rician Channels with Spatially Random Eavesdroppers. *IEEE Trans. Wirel. Commun.* **2024**, *23*, 8357–8371. [CrossRef]
10. Abramowitz, M.; Stegun, I.A. *Handbook of Mathematical Functions*; US Department of Commerce, National Bureau of Standards: Washington, DC, USA, 1972.
11. Wu, Q.; Zhang, R. Intelligent Reflecting Surface Enhanced Wireless Network: Joint Active and Passive Beamforming Design. In Proceedings of the 2018 IEEE Global Communications Conference (GLOBECOM), Abu Dhabi, United Arab Emirates, 9–13 December 2018; pp. 1–6. [CrossRef]
12. Durgin, G.D.; Rappaport, T.S.; De Wolf, D.A. New analytical models and probability density functions for fading in wireless communications. *IEEE Trans. Commun.* **2002**, *50*, 1005–1015. [CrossRef]
13. Björnson, E.; Sanguinetti, L. Rayleigh Fading Modeling and Channel Hardening for Reconfigurable Intelligent Surfaces. *IEEE Wirel. Commun. Lett.* **2020**, *10*, 830–834. [CrossRef]
14. Sánchez, J.D.V.; Urquiza-Aguiar, L.; Paredes, M.C.P.; López-Martínez, F.J. Expectation-Maximization Learning for Wireless Channel Modeling of Reconfigurable Intelligent Surfaces. *IEEE Wirel. Commun. Lett.* **2021**, *10*, 2051–2055. [CrossRef]
15. Bloch, M.; Barros, J.; Rodrigues, M.R.D.; McLaughlin, S.W. Wireless Information-Theoretic Security. *IEEE Trans. Inf. Theory* **2008**, *54*, 2515–2534. [CrossRef]

16. Kong, L.; Tran, H.; Kaddoum, G. Performance analysis of physical layer security over α-μ fading channel. *Electron. Lett.* **2016**, *52*, 45–47. [CrossRef]
17. Li, S.; Yang, L.; Hasna, M.O.; Alouini, M.S.; Zhang, J. Amount of Secrecy Loss: A Novel Metric for Physical Layer Security Analysis. *IEEE Commun. Lett.* **2020**, *24*, 1626–1630. [CrossRef]
18. Da Costa, D.B.; Yacoub, M.D.; Filho, J. Highly Accurate Closed-Form Approximations to the Sum of α-μ Variates and Applications. *IEEE Trans. Wirel. Commun.* **2008**, *7*, 3301–3306. [CrossRef]

engineering proceedings

MDPI

Proceeding Paper

Word Overlap Artifacts in Textual Entailment Models: A Study with Multilingual Back Translation on the SNLI Dataset †

Franklin Quilumba [1,*], Fausto Valencia [2,*] and Hugo Arcos [2,*]

1 Department of Statistics and Data Sciences and Department of Computer Science, College of Natural Sciences, University of Texas at Austin, Austin, TX 78712, USA
2 Departamento de Energía Eléctrica, Facultad de Ingeniería Eléctrica y Electrónica, Escuela Politécnica Nacional, Quito 170525, Ecuador
* Correspondence: franklin.quilumba@utexas.edu (F.Q.); fausto.valencia@epn.edu.ec (F.V.); hugo.arcos@epn.edu.ec (H.A.)
† Presented at the XXXII Conference on Electrical and Electronic Engineering, Quito, Ecuador, 12–15 November 2024.

Abstract: This paper investigates the domain of Natural Language Inference (NLI), with an emphasis on Recognizing Textual Entailment (RTE). We utilize the Stanford Natural Language Inference (SNLI) dataset, a benchmark for RTE tasks, to examine the efficacy of machine back translation and model performance in textual entailment. Our methodology employs a cost-effective approach using an open-source machine translation library like MarianMT with Helsinki-NLP/opus-mt models for back translation, applied to the comprehensive SNLI dataset. The concluding analysis demonstrates that no single model, whether back translated or augmented, consistently outperforms the reference English model in all aspects. The performance variations are particular to certain word overlap ranges and categories, suggesting that these models are essentially equivalent to the reference. This study contributes to the comprehension of machine translation's impact on textual entailment models, emphasizing the complexities in multilingual NLI tasks.

Keywords: textual entailment; natural language inference (nli); snli dataset; back translation

Citation: Quilumba, F.; Valencia, F.; Arcos, H. Word Overlap Artifacts in Textual Entailment Models: A Study with Multilingual Back Translation on the SNLI Dataset. *Eng. Proc.* **2024**, 77, 2. https://doi.org/10.3390/engproc2024077002

Academic Editor: Christian Tipantuña

Published: 18 November 2024

1. Introduction

Natural language inference (NLI) is central to numerous semantic tasks in natural language processing, such as question answering and text summarization. A key aspect of NLI is recognizing textual entailment (RTE), where the objective is to determine if a hypothesis (H) can logically be inferred from a premise (P). The Stanford Natural Language Inference (SNLI) dataset, constructed through a crowdsourcing method developed by [1], has emerged as a benchmark for RTE tasks. However, recent studies, including [2], have highlighted the prevalence of artifacts and biases in datasets like SNLI, which can lead to spurious correlations and reduced model generalizability.

Natural Language Processing (NLP) is becoming an integral component across various domains, including the electric power industry. NLP enables computers to interpret human language, offering promising opportunities to enhance operational efficiency and automate complex processes [3]. In particular, NLI can be valuable for understanding and analyzing complex systems in the electric power industry, to capture complex relationships and entailments within textual data. One notable application is ETAP 2024's AI-powered Natural Language Search, which allows engineers and operators to interact with power system analysis tools using everyday language, significantly reducing the learning curve and enhancing productivity [4]. As NLP continues to evolve, its exploration for potential applications in the power industry underscores the need to understand its capabilities for power systems starting by the fundamentals, such as those addressed in this paper.

In the context of NLI and RTE, this paper explores the potential of back translation as a method to diversify data patterns and reduce dependency on dataset artifacts in RTE classification. Back translation involves translating a text into a pivot language and then back into the original language, potentially offering a novel approach to addressing data-related issues in the SNLI dataset. By analyzing the effects of this technique, this study aims to explain its impact on model performance and data characteristics, thereby contributing to the development of more robust and reliable NLP models.

2. Related Work

RTE datasets were initially developed predominantly in English, providing greater depth in terms of sentence pairs and topical variety. This focus on linguistic aspects has spurred research into creating RTE datasets in other languages, including Italian, Spanish, German, Arabic, Greek, and Czech [5]. The translation of the SNLI and MultiNLI datasets into Turkish by [6] exemplifies efforts to broaden the linguistic scope of NLP, emphasizing the need for multilingual resources. These efforts highlight the potential differences in model performance when trained and tested on linguistically diverse data. Cross-lingual adaptation research, such as the work by [7] on English-to-Arabic RTE and the XNLI project by [8], further investigates models trained in English and evaluated on non-English datasets. These studies emphasize the significance of cross-lingual adaptability in NLP models, a notion that aligns with the concept of back translation.

In NLP, back translation is primarily used in machine translation to create paraphrased content that maintains the original meaning with diverse expressions. This approach, typically employed for data augmentation, offers potential benefits for entailment tasks by providing varied expressions of the same ideas, thereby aiding models in capturing semantic meanings beyond specific word patterns. However, the application of back translation to entailment datasets like SNLI is not as extensively documented. This study explores the potential of back translation in mitigating data biases in English NLI RTE tasks by introducing diverse expressions. This approach aims to enhance model robustness and reduce overfitting to dataset-specific idiosyncrasies. Key considerations in this approach include the quality of translation and computational costs, with a focus on preserving dataset balance to prevent introducing new biases.

3. Approach and Methodology

RTE models often struggle with the presence of artifacts in datasets like SNLI, where models may rely heavily on superficial cues, such as word overlap, rather than deeper semantic understanding. These artifacts limit model generalizability and result in overfitting to the patterns within the dataset.

To address these challenges, this study explores back translation as a potential solution for mitigating artifacts by introducing linguistic diversity and paraphrased expressions into the dataset. As native Spanish speakers, we were particularly motivated to test the effectiveness of back translation in reducing artifacts in the SNLI dataset. By translating sentences into various pivot languages and then back to English, back translation generates paraphrased versions of premise and hypothesis pairs, encouraging the model to rely more on semantic content rather than superficial lexical patterns.

The rationale behind back translation is that by using multiple languages, including Danish, German, Spanish, French, Russian, Swedish, and Chinese, we introduce a broader range of linguistic variations. This variation can expose models to different syntactic and lexical patterns, helping them generalize better and potentially reduce their reliance on artifacts like word overlap.

The following subsections detail the specific methodologies employed, including the machine translation process and the fine-tuning of ELECTRA models for textual entailment.

3.1. Machine Back Translation

The SNLI dataset comprises approximately 550k sentence pairs labeled for textual entailment in the training set, totaling about 57 million characters. Utilizing a paid translation API service for a single pass of a dataset of this magnitude would incur significant costs, especially since back translation requires a round trip. Open-source machine translation libraries like OpenNMT and MarianMT offer a cost-effective alternative, allowing local execution without API costs, albeit requiring some setup and computational resources.

The MarianMT framework [9] is a high-performance Neural Machine Translation (NMT) framework known for its efficiency, flexibility, and portability. Leveraging the Transformer architecture, it ensures accurate, fast, and memory-efficient translations. Its user-friendly API and cross-lingual adaptability make it a popular tool for researchers.

Helsinki-NLP/opus-mt [10], a collection of pre-trained open NMT models, offers several advantages for translation tasks. It provides ready-to-use NMT models for various language pairs, eliminating the need for additional training. Helsinki-NLP/opus-mt models are built upon the MarianMT framework. While newer models such as mBART and mT5 offer advanced multilingual translation capabilities, they often require significant computational resources and fine-tuning to achieve optimal performance. Given the time and resource constraints of this study, Helsinki-NLP/opus-mt was selected for its balance of translation quality, ease of use, and efficiency, allowing for high-quality back translation without the need for fine-tuning or substantial overhead.

Combining MarianMT with Helsinki-NLP/opus-mt pre-trained models through the Hugging Face Transformers library proves to be an effective approach for back translation. Sentences are translated from one language (English) to another (pivot language) and then back to the original language (English), generating paraphrased content.

In the SNLI dataset, premises were provided to crowdworkers as textual descriptions of images. The workers generated hypotheses that could be true, false, or neutral, which were then labeled as entailment, neutral, or contradiction by judges. As a result, the same premise could appear multiple times with different hypotheses. To address this, we translated only unique premises, reducing the character count from over 36 million to about 9.5 million. This reduction underscores the presence of repeated premises, which can challenge natural language inference models due to potential associations between specific premises and hypotheses.

Pivot language selection was influenced by linguistic characteristics, availability of high-quality machine translation models, and cultural context. Chosen languages included Danish, German, and Swedish from the Germanic family; Spanish and French from the Romance languages; Russian for its Slavic syntactic structure; and Chinese, offering significant syntactic and semantic variation from Indo-European languages. This diverse selection introduces a broad spectrum of linguistic variations into the SNLI dataset, enhancing the potential performance of trained models.

In assessing the quality of back-translated sentences, it is important to ensure that the core meaning and context are preserved, despite some expected deviation from the original text. This study utilizes a combination of quantitative and qualitative metrics to evaluate the back translation's fidelity and fluency: BLEU Score [11], Char-F Score [12], and BERTScore [13]. These metrics collectively offer a comprehensive view of the back translation's quality, balancing lexical and semantic fidelity with fluency and grammatical integrity. While BLEU and Char-F provide a quantitative assessment, BERTScore adds a layer of qualitative analysis, ensuring a thorough evaluation of the translation process.

3.2. Fine-Tuning ELECTRA Models for Textual Entailment

For this study, the ELECTRA model 'google/electra-small-discriminator' [14] was chosen for its efficiency in language understanding and suitability for textual entailment tasks. Its unique architecture, which differentiates between replaced and original tokens, makes it particularly effective for our study.

In this study, we conducted extensive model training, encompassing a total of 15 models to thoroughly investigate the impact of back translation on model performance. The models were trained across various datasets: the original English (en) dataset served as our baseline, and we then utilized back-translated English versions of the dataset for each target language, specifically Danish (da), German (de), Spanish (es), French (fr), Russian (ru), Swedish (sv), and Chinese (zh). In addition, we explored the effect of data augmentation by combining the original English dataset with non-repeated pairs from each of the English back-translated versions, resulting in augmented datasets labeled as en + da, en + de, en + es, en + fr, en + ru, en + sv, and en + zh.

3.2.1. Data Preparation

The preparation phase involved loading the SNLI dataset and processing both the original English and the various back-translated versions. We employed the 'datasets' library for loading and the 'AutoTokenizer' from the Transformers library for tokenization, adjusting the sequence length to a maximum of 128 tokens. The data was filtered to exclude examples without labels and then converted into a format compatible with the ELECTRA model. The total number of examples after filtering was 549,367 that were left after removing 785 examples due to having no label.

3.2.2. Training Setup

The training of the ELECTRA model ('google/electra-small-discriminator') was consistent across all datasets. A batch size of 16 was used to better utilize the available GPU resources, and a total of three training epochs were used for each model. To manage disk space effectively, model checkpoints were saved every 1000 steps, with a limit of retaining the last three checkpoints.

3.2.3. Model Training and Saving

The 'Trainer' class from the Hugging Face Transformers library was used to manage the training process. Models were trained on each dataset variant and saved on Google Drive. Due to disk space limitations, checkpoints were saved every 1000 steps, retaining only the last three. This strategy balances saving frequency and disk usage, ensuring access to well-trained model iterations while allowing the option to revert to earlier checkpoints using Google Drive's version history.

3.2.4. Reproducibility and Environment

The training was executed in a Google Colab environment, leveraging its cloud-based resources. Specifically, the training was conducted on a Google Colab Pro+ instance with a V100 16 GB GPU. This setup provided the necessary computational power while ensuring the reproducibility of the training process across different datasets.

4. Back Translation Analysis

We assessed the quality of back-translated sentences for unique premises and hypotheses across different target languages using BLEU, char-F, and BERT Score metrics. This analysis evaluates how effectively back translation preserves meaning while introducing linguistic variability.

Tables 1 and 2 present the NMT quality assessment for the pivot languages in unique premises and hypotheses, respectively, sorted based on their average ranking across all metrics. The BLEU and Char-F scores indicate that Spanish, Danish, and Swedish for unique premises, and Spanish, Danish, and German for hypotheses, performed exceptionally well in terms of lexical precision and grammatical correctness. High precision and recall in BERT scores across these languages suggest effective preservation of semantic content.

Table 1. NMT quality assessment for unique premises.

Lang	BLEU	chrF	P	R	F1
es	0.636	0.774	0.981	0.979	0.98
da	0.615	0.747	0.979	0.977	0.978
sv	0.606	0.747	0.979	0.977	0.978
de	0.583	0.737	0.979	0.976	0.977
fr	0.59	0.733	0.977	0.975	0.976
ru	0.387	0.602	0.963	0.959	0.961
zh	0.284	0.539	0.953	0.948	0.951

Table 2. NMT quality assessment for hypotheses.

Lang	BLEU	chrF	P	R	F1
es	0.632	0.78	0.981	0.979	0.98
da	0.554	0.708	0.977	0.974	0.975
de	0.533	0.696	0.976	0.973	0.975
sv	0.545	0.697	0.976	0.973	0.975
fr	0.528	0.688	0.974	0.97	0.972
ru	0.347	0.574	0.963	0.956	0.959
zh	0.262	0.538	0.953	0.948	0.950

The analysis shows that back translation in these languages is particularly effective in creating diverse training data while maintaining semantic integrity. The high precision, recall, and F1 scores across all languages suggest that despite linguistic variations, the core meaning in the back-translated sentences is well preserved. This suggests that back translation, regardless of the target language, can be a robust method for enhancing dataset quality in NLP tasks.

5. RTE Models Performance Evaluation

5.1. Evaluation Methodology

The performance of the ELECTRA models was evaluated on the filtered English SNLI validation dataset, which included 9842 examples after removing 158 without labels. Each model was assessed using key metrics: Loss, which indicates the error rate, and Accuracy, representing the proportion of correct predictions (see Table 3). Lower loss and higher accuracy values signify better performance. While Table 3 highlights Loss and Accuracy, additional metrics such as Micro F1, Macro F1, Weighted Precision, Weighted Recall, and Weighted F1 were also evaluated, with trends consistent with those seen in Accuracy, further validating the models' performance.

Table 3. Model performance evaluation on the SNLI validation dataset.

Model	Loss	Accuracy %
en + de	0.3132	90.39
en + sv	0.3136	90.33
en + da	0.3186	89.92
en + fr	0.3294	89.77
en + es	0.3235	89.76
en	0.3168	89.72
en + zh	0.3103	89.49
sv	0.3138	89.27
en + ru	0.3185	89.20
da	0.3154	89.00
es	0.3222	88.93
de	0.3198	88.90
fr	0.3273	88.75
ru	0.3392	87.84
zh	0.3387	87.58

5.2. Comparison of Model Performance

We assessed the impact of back translation by comparing models trained on the original dataset with those trained on back-translated datasets. Models using back translation are labeled with language codes (e.g., "de" for English–German–English, "en + de" for English augmented with English–German–English), while the original model is labeled "en" as described in Section 3.2. Our evaluation showed that back-translated models generally had comparable or slightly improved accuracy over the original, suggesting that back translation enhances model performance by increasing linguistic diversity and reducing reliance on dataset-specific artifacts. These results highlight back translation's potential to improve the robustness and generalizability of NLP models, though further research is needed to understand the underlying mechanisms.

6. Analyzing Word Overlap Artifacts in Multilingual Textual Entailment Models

Word overlap is a significant artifact in recognizing textual entailment (RTE) tasks in natural language inference (NLI). This issue arises when a premise and hypothesis share many words, potentially causing models to rely on lexical similarity instead of semantic understanding. For example, the premise "A group of people is sitting at a table eating food" and the hypothesis "The people are enjoying a meal together" share significant word overlap, yet the semantic relationship is not clear-cut. The hypothesis could be considered an entailment or neutral, highlighting how models might struggle with subtle semantic meanings. This challenge emphasizes the importance of addressing word overlap to improve model performance in RTE tasks.

This analysis specifically targets instances where model predictions are incorrect with a high degree of word overlap between the premise and hypothesis across multiple languages. The study involves training on the original English (en) RTE dataset and its variations through back-translated English datasets using pivot languages: Danish (da), German (de), Spanish (es), French (fr), Russian (ru), Swedish (sv), and Chinese (zh). Additionally, it includes training on augmented datasets combining original English with back-translated datasets (en + da, en + de, en + es, en + fr, en + ru, en + sv, en + zh) as explained in Section 3.2.

6.1. Methodology

The methodology to investigate this artifact is outlined as follows:

1. Calculate the Word Overlap Ratio: Compute the overlap ratio between the premise and hypothesis by tokenizing the text, removing specified punctuation, and converting it to lowercase to focus on the words.
2. Bin the Overlap Ratios: Categorize the calculated overlap ratios into bins (e.g., 0–10%, 10–20%, 90–100%) to facilitate structured analysis.
3. Count the Errors in Each Bin by Category: Count incorrect predictions within each bin, categorized by labels: 'Entailment,' 'Neutral,' and 'Contradiction,' to understand performance variations across overlap levels.
4. Calculate the Error Rate: Determine the error rate in each bin for every label category to quantify model performance across different overlap levels.
5. Visualize the Results: Plot error rates against word overlap bins on the x-axis and error rates on the y-axis, separated by label categories for comparative analysis.

This evaluation measures the proportion of incorrect predictions within various word overlap ranges, providing insights into the model's reliance on word overlap as a predictive feature. An ideally generalized model should show a uniform error rate across all overlap levels, indicating a balanced consideration of semantic content over lexical similarity.

Visual comparisons of models across word overlap bins allow us to assess their reliance on overlap for predictions and how language-specific characteristics from back translation or dataset augmentation might influence this.

6.2. Observations on Original English Model

Figure 1 shows the bar plots for the error rates across different word overlap bins for each category label (Entailment, Neutral, and Contradiction), with each bar representing a different bin. Based on the error rates across different word overlap bins for the English model in the categories of Entailment, Neutral, and Contradiction, we have:

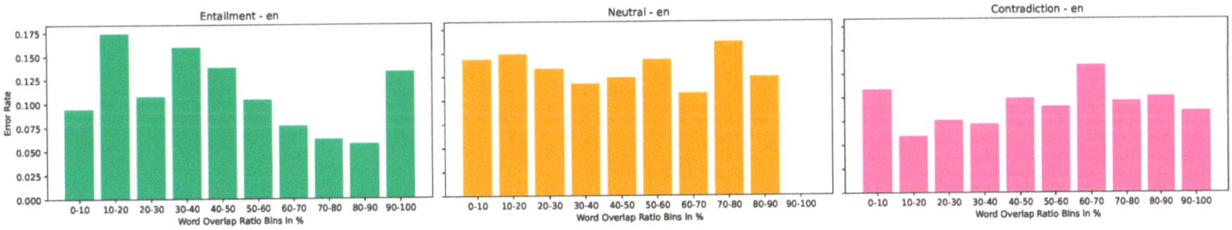

Figure 1. Error rates as a function of the word overlap ratio bins by label category using the original English model.

Entailment: The error rates start higher in the 10–20% bin, decrease slightly, and then fluctuate. The highest error rate is in the 10–20% bin (0.1744), and the lowest is in the 80–90% bin (0.0575).

Neutral: The error rates are more varied. The highest error rate is in the 70–80% bin (0.1620), which is interesting as it suggests difficulty in this range. The lowest is in the 90–100% bin, but this could be due to a smaller number of examples in this range.

Contradiction: The error rates generally decrease with a few fluctuations. The highest is in the 0–10% bin (0.1086), and the lowest is in the 10–20% bin (0.0595).

The variation in error rates suggests that the model does not consistently rely on word overlap for predictions. Notably, the peak in the 10–20% bin for Entailment, and the high error rate in the 70–80% bin for Neutral indicate potential areas of struggle, possibly due to specific linguistic features. Lower error rates in the Contradiction category across most bins suggest the model is more effective at identifying contradictions than entailments or neutral relationships involving word overlap.

6.3. Observations on Back-Translated and Augmented English + Back-Translated Models

Figure 2 shows the line plots for the error rates across different word overlap bins for each category label (Entailment, Neutral, and Contradiction), with each line representing a different model. Similarly, Figure 3 shows the line plots for the error rates for each line corresponding to one of the augmented models.

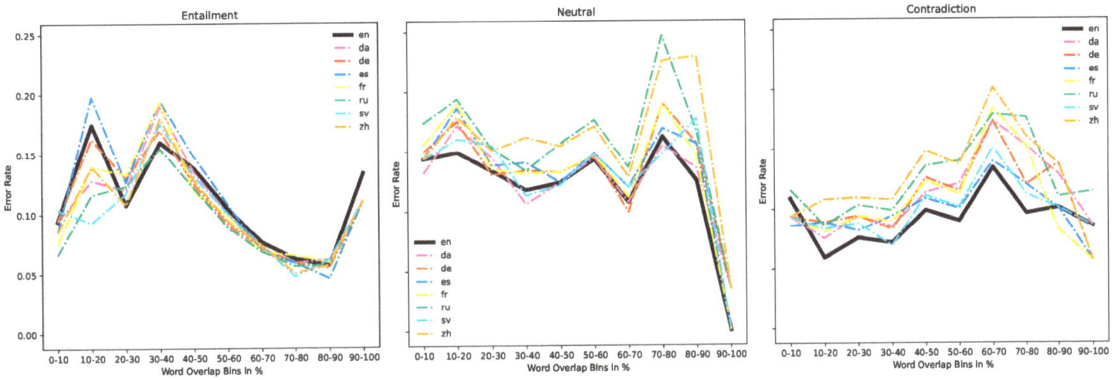

Figure 2. Error rates as a function of the word overlap ratio bins by label category using the back-translated models vs. the original English model.

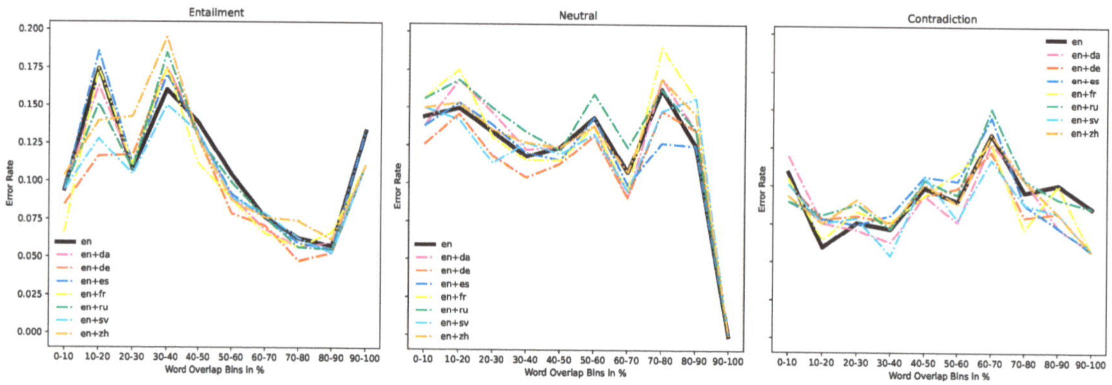

Figure 3. Error rates as a function of the word overlap ratio bins by label category using the augmented English + back-translated models vs. the original English model.

Table 4 provides a summary of observations for both back-translated and augmented models, highlighting the deviations from the English counterpart across different word overlap bins. This table offers a concise comparison of how each model performs in the Entailment, Neutral, and Contradiction categories, allowing for a clearer understanding of the impact of linguistic variations introduced through back translation. By consolidating the observations into a single table, we can more easily identify patterns and insights that may not be immediately evident from the figures alone.

Table 4. Observations for back-translated and augmented models across word overlap bins as compared to its English counterpart.

Model	Entailment	Neutral	Contradiction
da	Similar trend to English, slight deviations	Generally close to English, minor differences	Comparable to English
de	Notable deviations in certain bins	Fluctuates, some significant differences	Mostly similar to English, some exceptions
es	Follows English, some deviations	Noticeable differences in certain bins	Significant deviations from English
fr	Larger deviations in several bins	Varied performance, especially mid-range	Similar to English, with differences
ru	Considerable differences, especially higher bins	Varied performance with significant divergence	Close to English, few exceptions
sv	Significant deviations, mid to high range	Some alignment, others diverge	Similar in lower bins, diverges in higher
zh	Similar in lower bins, diverges in higher	Some variability, particularly middle bins	Similar pattern to English, some deviations
en+da	Minor deviations, similar to English	Slight differences, close to English	Some deviations, generally follows English
en+de	Noticeable differences in certain bins	Fluctuates, significant variation	Fairly close to English, minor deviations
en+es	Follows English, small deviations	Consistent with English	Some variance, generally aligns with English
en+fr	Larger deviations, different trend	Varied performance, distinct handling	Mostly follows English, some divergence
en+ru	Notable differences in some bins	Similar in several bins, few deviations	Close to English trend, some divergence
en+sv	Significant differences, distinct trend	Generally follows English, some exceptions	Similar in lower bins, diverges in others
en+zh	Similar in lower bins, diverges in higher	Similar trend, minor variations	Aligns with English, noticeable differences

6.4. Visualizing Artifact Mitigation and Distribution Changes

To further investigate the effects of back translation on artifact reduction and the introduction of new artifacts, we present a statistical analysis of word correlations in the SNLI dataset using the method outlined in [15], which explores spurious feature correlations through a z-statistic for each token. The goal of this analysis is to determine whether individual words in back-translated datasets exhibit detectable correlations with specific class labels, as seen in the original English dataset.

In Figures 4 and 5, we present the artifact statistics for the SNLI dataset across the various back-translated and augmented datasets. As highlighted in [15], certain annotated words (e.g., "nobody", "outdoors", "competition") function as artifacts in the SNLI dataset that were also reported by [2,16]. These words are labeled in the figures to track their behavior across languages and dataset combinations.

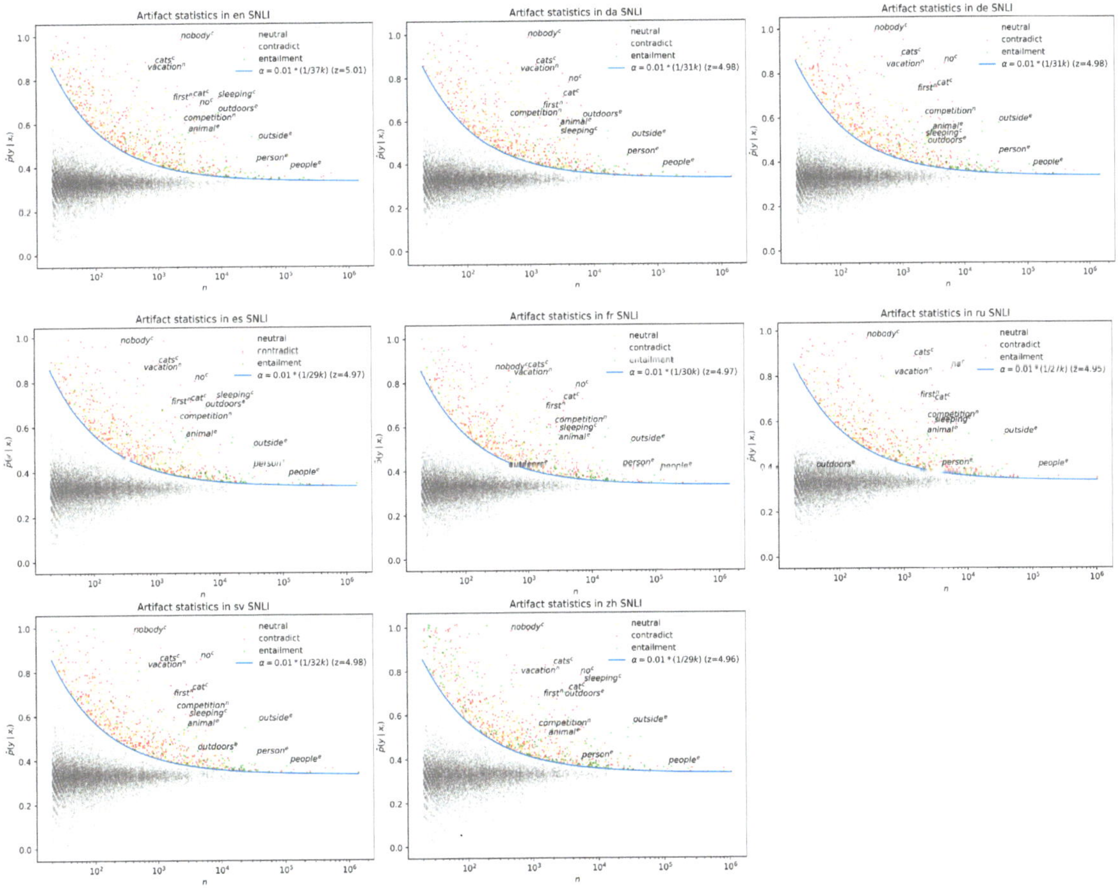

Figure 4. Artifact statistics across different language versions of the SNLI dataset. (Top row) English (en), Danish (da), German (de); (Middle row) Spanish (es), French (fr), Russian (ru); (Bottom row) Swedish (sv), Chinese (zh). Points above the blue line indicate words with statistically significant correlations to a class label, representing potential artifacts. The class labels are color-coded: neutral (orange), contradiction (red), and entailment (green), with annotated artifact words labeled with superscripts n, c, and e for neutral, contradiction, and entailment, respectively.

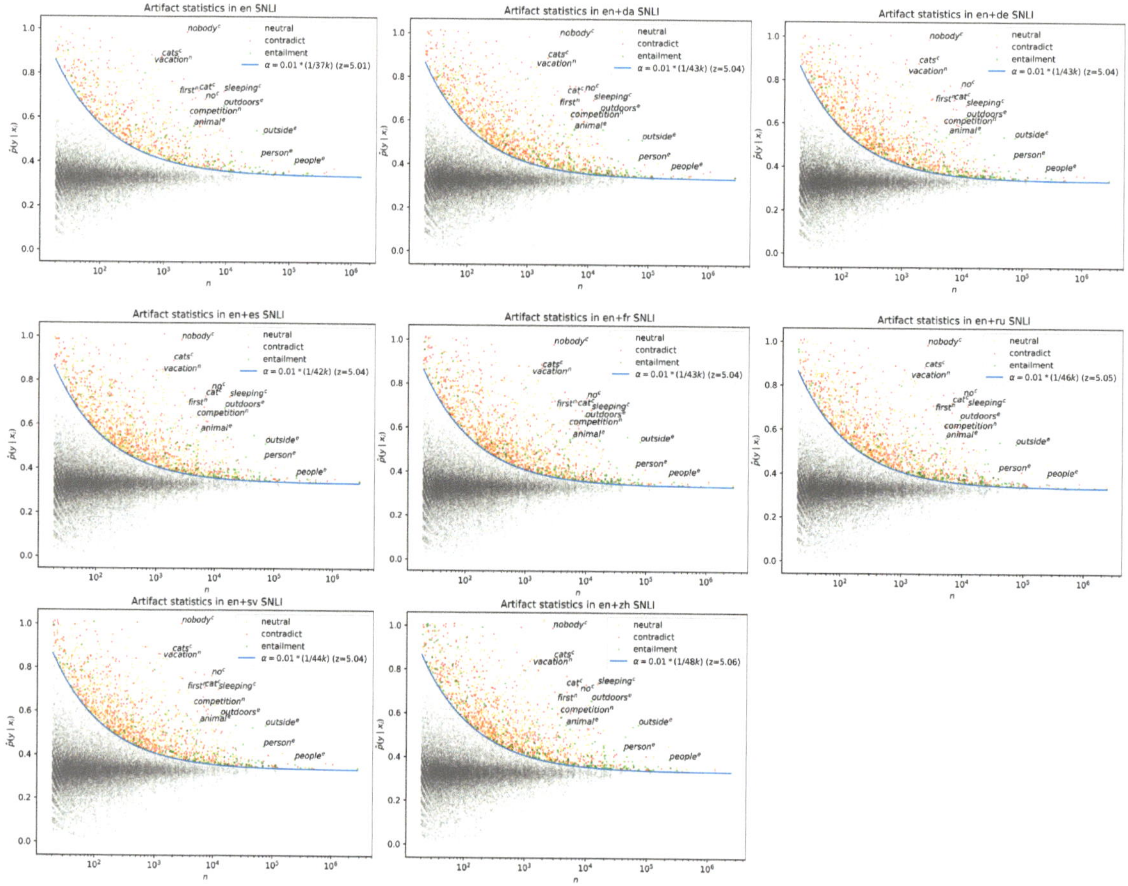

Figure 5. Artifact statistics for augmented datasets (original English combined with back-translated data). (Top row) English (en), English+Danish (en+da), English+German (en+de); (Middle row) English+Spanish (en+es), English+French (en+fr), English+Russian (en+ru); (Bottom row) English+Swedish (en+sv), English+Chinese (en+zh). The original English plot is included for reference. Points above the blue line indicate words with statistically significant correlations to a class label, highlighting potential artifacts. The class labels are color-coded: neutral (orange), contradiction (red), and entailment (green), with annotated artifact words labeled with superscripts n, c, and e for neutral, contradiction, and entailment, respectively.

In Figure 4, which focuses on the back-translated datasets (da, de, es, fr, ru, sv, zh), we observe notable shifts in the positions of these annotated words across languages. For instance, the word "outdoors", which is a strong artifact in the English dataset, gradually moves closer to the significance threshold (blue line) in the French back-translated dataset, and in the Russian dataset, it falls below the blue line, indicating a reduction in its correlation with class labels. However, not all artifacts exhibit such changes. Words like "nobody" and "competition" remain above the significance threshold across most languages, maintaining their artifact-like behavior.

In Figure 5, which compares the artifact distribution in augmented datasets (e.g., en+da, en+de, en+es), we observe that the overall artifact distribution remains quite similar to the English baseline. The density of points above the blue line is largely unchanged, suggesting that augmentation with back-translated data does not drastically alter the artifact distribution. However, there are slight shifts in the positions of the annotated words compared to their single back-translated counterparts. The shifts in annotated words are less substantial in the augmented datasets, and the overall artifact density remains similar to the English reference.

7. Conclusions

The analysis showed that no single model (neither back translated nor augmented) consistently outperformed the reference English model across all word overlap bins and categories. While some models did better in specific areas, this improvement was not uniform across the entire range of overlap bins. Therefore, it can be said that the models are broadly equivalent to the reference, with variations specific to certain word overlap ranges and categories. The analysis of models trained on back-translated and augmented datasets showed similar trends, indicating that these variations in training data did not significantly alter the models' approach to word overlap.

In particular, back translation affected the distribution of some artifacts, such as reducing the correlation of certain annotated words (e.g., "outdoors") with class labels in languages like Russian. However, in many cases, these artifacts remained largely unchanged or shifted only slightly, as seen in the augmented datasets (e.g., en+da, en+de, en+ru), where the overall artifact distribution remained similar to the English reference.

While back translation and data augmentation introduce linguistic diversity and can mitigate certain artifacts, they are not sufficient to completely eliminate the problem. The model's approach to word overlap appears balanced, incorporating it as one of several factors rather than relying solely on it as a heuristic. However, further techniques and more sophisticated preprocessing strategies may be necessary to fully address the artifact issues in NLI datasets.

Author Contributions: Conceptualization, F.Q.; methodology, F.Q.; formal analysis, F.Q.; writing—original draft preparation, F.Q.; writing—review and editing, F.V. and H.A.; supervision, F.V.; funding acquisition, H.A. All authors have read and agreed to the published version of the manuscript.

Funding: This research received no external funding.

Institutional Review Board Statement: Not applicable.

Informed Consent Statement: Not applicable.

Data Availability Statement: The original data presented in this study are openly available in the Stanford Natural Language Inference (SNLI) Corpus at https://nlp.stanford.edu/projects/snli or via https://huggingface.co/datasets/stanfordnlp/snli.

Conflicts of Interest: The authors declare no conflicts of interest.

References

1. Bowman, S.R.; Angeli, G.; Potts, C.; Manning, C.D. A Large Annotated Corpus for Learning Natural Language Inference. In Proceedings of the 2015 Conference on Empirical Methods in Natural Language Processing, Lisbon, Portugal, 17–21 September 2015; pp. 632–642.
2. Gururangan, S.; Swayamdipta, S.; Levy, O.; Schwartz, R.; Bowman, S.; Smith, N.A. Annotation Artifacts in Natural Language Inference Data. In Proceedings of the 2018 Conference of the North American Chapter of the Association for Computational Linguistics: Human Language Technologies, Volume 2 (Short Papers), New Orleans, LA, USA, 1–6 June 2018; pp. 107–112.
3. Natural Language Processing and Its Application in the Utility Industry. Available online: https://www.epri.com/research/products/000000003002017321 (accessed on 1 August 2024).
4. Meet Your AI Copilot in ETAP 2024—Webinar. Available online: https://etap.com/events/etap-university-courses/meet-your-ai-copilot-in-etap-2024 (accessed on 1 August 2024).
5. Putra, I.M.S.; Siahaan, D.; Saikhu, A. Recognizing Textual Entailment: A Review of Resources, Approaches, Applications, and Challenges. *ICT Express* **2024**, *10*, 132–155. [CrossRef]
6. Budur, E.; Özçelik, R.; Gungor, T.; Potts, C. Data and Representation for Turkish Natural Language Inference. In Proceedings of the 2020 Conference on Empirical Methods in Natural Language Processing (EMNLP), Online, 16–20 November 2020; pp. 8253–8267.
7. Agić, Ž.; Schluter, N. Baselines and Test Data for Cross-Lingual Inference. In Proceedings of the Eleventh International Conference on Language Resources and Evaluation (LREC 2018), Miyazaki, Japan, 7–12 May 2018.

8. Conneau, A.; Rinott, R.; Lample, G.; Williams, A.; Bowman, S.; Schwenk, H.; Stoyanov, V. XNLI: Evaluating Cross-Lingual Sentence Representations. In Proceedings of the 2018 Conference on Empirical Methods in Natural Language Processing, Brussels, Belgium, 31 October–4 November 2018; pp. 2475–2485.
9. Junczys-Dowmunt, M.; Grundkiewicz, R.; Dwojak, T.; Hoang, H.; Heafield, K.; Neckermann, T.; Seide, F.; Germann, U.; Aji, A.F.; Bogoychev, N.; et al. Marian: Fast Neural Machine Translation in C++. In Proceedings of the ACL 2018, System Demonstrations, Melbourne, Australia, 5–10 July 2018; pp. 116–121.
10. Tiedemann, J.; Thottingal, S. OPUS-MTBuilding Open Translation Services for the World. In Proceedings of the 22nd Annual Conference of the European Association for Machine Translation, Lisboa, Portugal, 3–5 November 2020; pp. 479–480.
11. Papineni, K.; Roukos, S.; Ward, T.; Zhu, W.-J. BLEU: A Method for Automatic Evaluation of Machine Translation. In Proceedings of the 40th Annual Meeting on Association for Computational Linguistics, Philadelphia, PA, USA, 6–12 July 2002; pp. 311–318.
12. Popović, M. chrF: Character n-Gram F-Score for Automatic MT Evaluation. In Proceedings of the Tenth Workshop on Statistical Machine Translation, Lisbon, Portugal, 17–18 September 2015; pp. 392–395.
13. Zhang, T.; Kishore, V.; Wu, F.; Weinberger, K.Q.; Artzi, Y. BERTScore: Evaluating Text Generation with BERT. *arXiv* **2020**, arXiv:1904.09675v3.
14. Clark, K.; Luong, M.-T.; Le, Q.V.; Manning, C.D. ELECTRA: Pre-Training Text Encoders as Discriminators Rather Than Generators. *arXiv* **2020**, arXiv:2003.10555v1.
15. Gardner, M.; Merrill, W.; Dodge, J.; Peters, M.; Ross, A.; Singh, S.; Smith, N.A. Competency Problems: On Finding and Removing Artifacts in Language Data. In Proceedings of the 2021 Conference on Empirical Methods in Natural Language Processing, Online/Punta Cana, Dominican Republic, 7–11 November 2021; pp. 1801–1813.
16. Wallace, E.; Feng, S.; Kandpal, N.; Gardner, M.; Singh, S. Universal Adversarial Triggers for Attacking and Analyzing NLP. In Proceedings of the 2019 Conference on Empirical Methods in Natural Language Processing and the 9th International Joint Conference on Natural Language Processing (EMNLP-IJCNLP), Hong Kong, China, 3–7 November 2019; pp. 2153–2162.

engineering proceedings

MDPI

Proceeding Paper

Storage System for Automated Object Classification Based on Color Analysis †

Michael Guerra ‡, Santiago Proaño ‡, Angélica Quito *,‡ and Samuel Peña-Bautista ‡

Facultad de Ciencias Técnicas, Universidad Internacional del Ecuador UIDE, Quito 170411, Ecuador; guerramichael27@gmail.com (M.G.); saproanopa@gmail.com (S.P.); spena@uide.edu.ec (S.P.-B.)

* Correspondence: anquitoca@uide.edu.ec

† Presented at the XXXII Conference on Electrical and Electronic Engineering, Quito, Ecuador, 12–15 November 2024.

‡ These authors contributed equally to this work.

Abstract: This article introduces a compelling study on the design and implementation of a storage system for automated object classification using color analysis. Despite notable advancements in robotic systems, the precise and reliable color-based classification in uncontrolled environments remains an ongoing challenge. The primary aim of this research is to develop an economical prototype that integrates color sensor and robotic arm to significantly enhance the efficiency of object-sorting processes across a wide range of industries. This study showcases the system's impressive high accuracy rates of 97% for red, 96% for blue, and 98% for white objects in controlled settings. However, it also highlights the drop in accuracy to 76% for red, 64% for blue, and 68% for white objects in uncontrolled environments. These findings underscore the importance of maintaining consistent environmental conditions to optimize system performance, while also pointing to promising opportunities for modernization and semi-automation within the pharmaceuticals, food, and manufacturing sectors.

Keywords: object classification; color analysis; color sensor; robotic arm; semi-automation

1. Introduction

Currently, object classification holds substantial importance across diverse fields, which require an autonomous process to be able to segment different elements of interest. In modern times, by utilizing techniques such as neural networks [1], machine learning algorithms like K-nearest neighbor [2], and image processing with computer vision [3], favorable results can be obtained to make an automatic classification machine, depending on the application you want to provide. Moreover, in robotic applications, automated color-based object classification systems can improve tasks like grasp form detection, contributing to the efficiency and reliability of robotic hand systems [4]. Overall, integrating automated object classification based on color analysis streamlines processes, saves time, and enhances accuracy in different domains.

Thus, taking into account that artificial intelligence is on the rise, offering powerful capabilities for color detection by taking advantage of machine learning models and convolutional neural networks to recognize subtle color changes imperceptible to the human eye [5–7], other simpler methods can be used, such as color sensors, as these can provide cost-effective color detection solutions by efficiently converting sensor signals to RGB colors [8]. While AI excels at complex color analysis and adaptability, a color sensor may be better suited for simpler applications where real-time processing speed and cost-effectiveness are prioritized [9].

Therefore, automated classification systems based on color analysis with sensors play a crucial role in various industries by enhancing efficiency, reducing labor costs, and improving accuracy. These systems enable the classification of products [10], such as objects [11], fruits [12], vegetables [13], and fish [14], based on their color attributes. By utilizing RGB

Citation: Guerra, M.; Proaño, S.; Quito, A.; Peña-Bautista, S. Storage System for Automated Object Classification Based on Color Analysis. *Eng. Proc.* **2024**, 77, 3. https://doi.org/10.3390/engproc2024077003

Academic Editor: Jackeline Abad

Published: 18 November 2024

sensors and color analysis techniques, these systems can accurately categorize items into different classes, detect defects, assess freshness, and sort products accordingly. The automation of this process not only increases productivity and quality assessment but also minimizes human error, speeds up decision-making processes, and facilitates remote monitoring and management through IoT connectivity. In general, automated classification systems based on color analysis with sensors are instrumental in streamlining operations, ensuring product quality, and optimizing resource utilization in various sectors [15].

2. Related Works

For this analysis, several references were explored, each adopting different approaches and methodological techniques to address research problems in the field of mechatronics. These studies employ methodologies such as experiments, controlled environment tests, and simulations, all aimed at solving specific problems.

In the context of developing classification stations, several notable concepts have emerged from the literature. The design and development of a delta robot for object classification presents a unique approach compared with conventional Cartesian robots. This application focuses on picking up and placing objects in stacking stations, and this paper demonstrates the accuracy achieved using artificial vision. However, it emphasizes that the shape of objects and the controlled environment are critical factors for success [16]. Robotic arms, irrespective of their type, showcase impressive capabilities in modern robotics. Their high degrees of flexibility enable them to maneuver in complex areas and assume challenging positions for effective object grasping. The versatility of robotic arms opens up various possibilities for industrial applications [17].

Implementing color sensors provides an affordable solution for identifying the color of objects during the classification process. Controlled by microcontrollers, these sensors enable the robotic arm to accurately determine colors using sophisticated analog readers. This low-cost development significantly enhances the efficiency and reliability of the object-picking process [18]. The integration of transportation bands or conveyor belts is essential for certain robotic arm manipulators. These bands facilitate the movement of objects to specific areas where the arm can easily grasp and relocate them to their designated stations. Integrating transportation bands with robotic arms optimizes the overall efficiency of the system [19]. The development of a human–machine interface (HMI) plays a crucial role in effectively controlling the programmable logic controller (PLC) and the entire robotic system. The HMI enables an interactive mode of operation within the station, allowing operators to monitor and control various parameters, ensuring seamless operation and precise control over the entire process [20].

In addition to the aforementioned ideas, other aspects have been considered in the literature. The use of a Cartesian arm, positioned as a square station on top of the working area, offers advantages in the classification process. It allows students to develop algorithms and control strategies more easily, enhancing their understanding and proficiency in working with robotic arms for effective object manipulation [21]. Furthermore, the accurate identification of object position is crucial in the classification process. The use of artificial vision to identify the color, shape, and position of the pieces has demonstrated 100% accuracy in all inspections conducted [22]. Open-source solutions, such as using a Raspberry Pi for color identification, have been explored. These solutions involve utilizing a camera to capture images, which are then analyzed pixel by pixel to identify RGB colors. Additionally, shape analysis is performed to determine the object's position. This approach achieves accurate color readings and employs two microcontrollers, including an Arduino for robotic arm control [23].

3. Main Contributions

As the preceding section illustrates, storage systems are essential to automated color-based item classification. Nevertheless, several of these systems' features have not been fully investigated yet. Thus, the following contributions are intended to fill in these gaps in this work:

- The machine uses color sensors to categorize objects based on their color, enhancing the efficiency of sorting processes.
- The system incorporates a cartesian arm that moves along the X and Z axes, combined with pneumatic actuators for precise object manipulation.
- This article explores a closed-loop control algorithm to regulate motor functions, ensuring accurate and reliable operation.
- The prototype integrates an HMI screen for monitoring and controlling the entire process, providing a user-friendly interface for operators.

This paper is structured as follows: Section 2 presents the methodology used to create the machine, detailing the components used for the mechanical, electronic, and computer parts, together with information about the operation of the machine. Furthermore, Section 3 shows the system in its final state and the results obtained in the experimental part with respect to the implementation of the color classification method, accompanied by the respective discussion. Finally, in Section 4, the conclusions obtained corresponding to the system implemented in this paper are presented.

4. Methodology

4.1. Mechatronic Design

Mechatronic design integrates mechanical, electrical, and information processing components to create innovative, functional, and streamlined systems [24–26]. Understanding this concept, the next section presents the mechatronic design used to develop the prototype, detailing its components and principal features for its correct operation.

4.1.1. Mechanical Design

This section describes the mechanism used to move the robotic arm in the three planes (X, Y, Z). First, structures are made in the upper part of the machine where the discs are placed for detection. Likewise, structures are incorporated to support the Cartesian arm mechanism in the same part of the machine (see Figure 1).

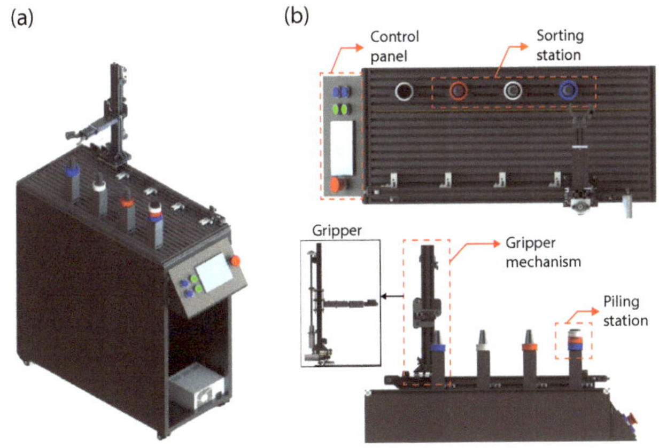

Figure 1. (**a**) CAD model of the classification system based on color analysis. (**b**) Components of the upper part of the machine in conjunction with the mechanism for classifying objects.

For the station, disc-shaped pieces were developed that were placed on the main upper structure to be classified. In this case, they were made up of two main components: The first is the lower disc that had a smaller internal diameter and the second component had a larger internal diameter (see Figure 2), which were joined under pressure. Its composition was made in this way for ease of manufacturing and for saving materials. The colors used for classification were white, blue, and red.

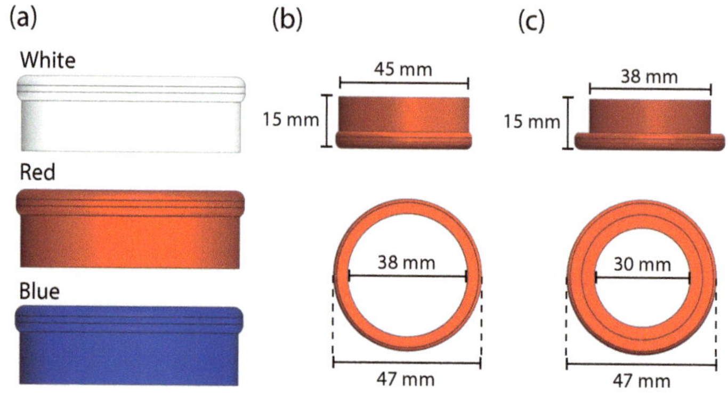

Figure 2. (**a**) Discs for classification process. (**b**) Female piece dimensions. (**c**) Male piece dimensions.

For the movement of the Cartesian arm, a base with wheels and a belt was used to move the arm horizontally from the left to the right. For movement along the Y axis, a pneumatic actuator operating at 0–10 PSI with a stroke of 50 mm was employed. This actuator, which uses compressed air, allowed the arm gripper to move forward and backward. Movement along the Z axis was achieved using a power worm mechanism with a corresponding nut, which was mounted on a support with wheels attached to the Cartesian arm. For holding objects, a parallel air gripper, operating at 14.5–101.5 PSI and with a diameter range of 20 mm (see Figure 3), was utilized for precise classification. This system ensures accurate and efficient movement and handling of the items.

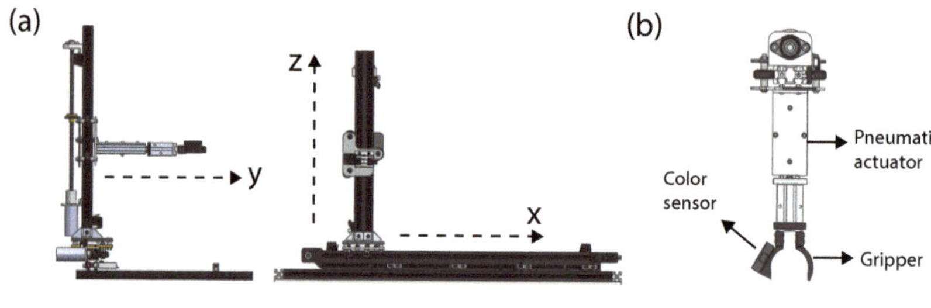

Figure 3. (**a**) Cartesian arm movement. (**b**) Gripper for holding objects.

4.1.2. Electronic Design

Referring to the electronic section of the machine, it is essential to highlight that an Xinje XC3-24RT-E PLC was employed to control various machine processes, powered by a 110 V electrical supply. Regarding the microcontrollers, two ESP32 units were utilized. The first microcontroller processes data from the TCS3472 color sensor via I2C communication for accurate color detection. Based on the detected hue, it sends signals to a corresponding relay module, facilitating the necessary voltage conversion from 5 V to 12 V, which is then transmitted to the PLC inputs. The second microcontroller is tasked with managing the DC motors in conjunction with an L298N H-bridge module. It activates the motors in response to the PLC outputs, which are also routed through relay modules to reduce voltage as needed.

Additionally, TCRT5000 proximity sensors are deployed to ascertain the position of the Cartesian arm base, while limit switches are installed on other axes to determine maximum positions along the X and Y axes. These signals are directed to the PLC inputs and, for controlling the flow of compressed air, two solenoid valves are activated by the controller outputs.

An HMI screen was integrated for process control, maintaining a constant interaction with the PLC. Moreover, power and emergency stop buttons, complete with their respective pilot lights, were installed. The electronic block diagram is depicted in Figure 4.

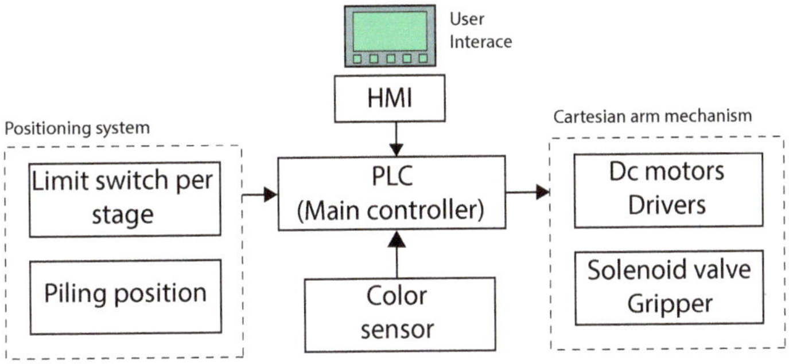

Figure 4. Block diagram representing the electronic process of the machine.

4.1.3. Software Design

Regarding the programming section, XCP-Pro V3.3 software was used for the PLC program and Arduino IDE V1.8.18 software for the microcontroller ESP32 using Ladder and C languages, respectively.

To perform the control and select the direction of the motors under the command of the ESP32, the process that can be observed in Algorithm 1 was established. It was used along with sequences based on timers from PLC programming, where each one had a certain time and, according to it each DC motor, was activated by the microcontroller. Therefore, first the Cartesian arm moves from the top to the bottom on the Z axis and then it moves from the left to the right on the X axis to place the piece in the structure of the respective color, thus ending the sequence. To move the motors and give them their respective direction, microcontroller pins connected to the H-bridge were established so that they rotated in one direction if one of the pins was "HIGH" and the other was "LOW", and vice versa for the opposite direction.

Algorithm 1 Motor control.

```
Set variables
while Motor is on do
  if First, PLC timer equals 1 then
    Move axis X rightwards
  else if Second, PLC timer equals 1 then
    Move axis X leftwards
  else if Third PLC timer equals 1 then
    Move axis Z upwards
  else if Fourth PLC timer equals 1 then
    Move axis Z downwards
  end if
end while
```

On the other hand, for the case of color-based detection and classification, the process is outlined in Algorithm 2. In this scenario, involving a maximum of three discs in the initial structure, variables were defined to determine the position of each disc. Consequently, if one of the three predefined colors was detected in the first position, the Cartesian arm would lower to that position and transfer the disc to the corresponding structure following the previously described algorithm. Upon completing this sequence, the process would repeat for each remaining object position.

Algorithm 2 Color detection ESP32.

```
Set variables
while Sensor is on do
  if Piece in position 1 then
    Activate position variable 1
    Take the piece
    if Piece in position 1 is color blue then
      Move the piece to a blue structure
    else if Piece in position 1 is color white then
      Move the piece to a white structure
    else if Piece in position 1 is color red then
      Move the piece to a red structure
    end if
  else if Piece in position 2 then
    Activate position variable 2
    Take the piece
    if Piece in position 2 is color blue then
      Move the piece to a blue structure
    else if Piece in position 2 is color white then
      Move the piece to a white structure
    else if Piece in position 2 is color red then
      Move the piece to a red structure
    end if
  else if Piece in position 3 then
    Activate position variable 3
    Take the piece
    if Piece in position 3 is color blue then
      Move the piece to a blue structure
    else if Piece in position 3 is color white then
      Move the piece to a white structure
    else if Piece in position 3 is color red then
      Move the piece to a red structure
    end if
  end if
end while
```

Finally, a graphical HMI interface was developed within the same software to control the PLC, featuring three main functions: automatic, manual, and counting. In the case of automatic classification, there is an option to position the Cartesian arm at the starting point and initiate classification according to the previously explained programming. For manual classification, a panel with icons representing the directions the Cartesian arm can move, as well as object gripping actions, was developed. Lastly, the counting option allows visualization of the number of objects classified according to their color.

5. Results and Discussion

5.1. Machine Fabrication and Integration

The machine's structure was developed and assembled internally, along with the cartesian arm system for sorting. The entire device can be seen in detail in Figure 5. As observed, the sorting structure is located at the top of the machine. Additionally, the control panel is located at the front of the structure, featuring various power and emergency stop buttons, as well as pilot lights to indicate if the machine is powered on (green) and if pieces are available (yellow). The HMI touch screen is located in the same section, allowing us to control the machine's different processes, either manually or automatically. Primarily, the process for using the machine involves first turning it on, placing the pieces to be sorted at the starting point, and selecting the sorting mode. For a better understanding of the aforementioned process, the machine's functionality can be visually appreciated at the following link, which shows the sorting process: https://youtu.be/IA7_wQ3IJ_M (accessed on 26 August 2024).

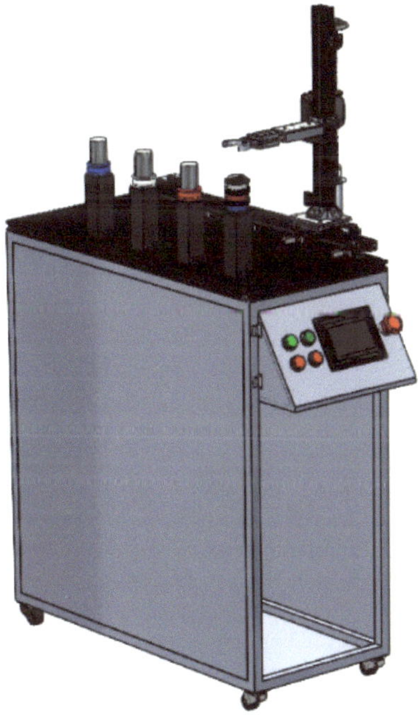

Figure 5. Classification machine.

Regarding the mechanism and mobility of the Cartesian arm, no failures were observed in its movements across the three axes, achieving smoothness in the bases equipped with wheels and proper operation of the pneumatic actuator. Likewise, the electronic components ensure the correct functioning of the machine, including the motors and sensors used to determine the maximum and minimum positions of the mechanism. Additionally, the microprocessor and the PLC controller operated optimally to determine each of the sequences and processes for sorting the objects by their respective colors.

5.2. Objects Classification

For this section, tests of the sorting system were carried out in a controlled and uncontrolled environment, where 100 discs of each color were used for testing. Therefore, the lighting conditions were modified with a flashlight included in the gripper to lighten the colors of the parts. Thus, for the uncontrolled environment, different ambient light spectra were used to determine whether the system continued to function correctly. Regarding the graphics, the green bar indicates the iterations where the color was detected, while the red bar denotes the iterations where the color was not detected. Row 1 corresponds to the color red, row 2 to blue, and row 3 to white, as shown in Figure 6.

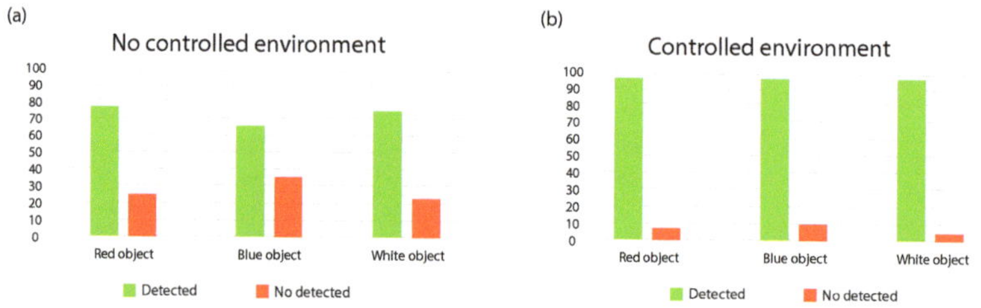

Figure 6. Results obtained for the following: (**a**) Color detection—uncontrolled environment. (**b**) Color detection—controlled environment.

The color-based object-sorting machine demonstrated high accuracy in a controlled environment, with detection rates of 97% for red, 96% for blue, and 98% for white (see Figure 6a). However, in an uncontrolled environment, the accuracy significantly decreased, achieving only 76% for red, 64% for blue, and 68% for white (see Figure 6b).

The decreased system precision is due to uncontrolled lighting conditions during the storage system operation tests. To avoid this type of uncertainty, it is recommended to analyze the environment in which the storage operations will be executed based on the color of the objects. This would reduce problems in the classification process.

These results underscore the importance of maintaining consistent environmental conditions to ensure the system's effectiveness.

6. Conclusions

In the research, it was possible to observe teams that had made similar prototypes. However, metrics still needed to be presented to validate the precision and accuracy of the system. In the present work, being able to use communication between different types of controllers has allowed for color classification, a human–machine interface, and their classification to be carried out, being an economic prototype applicable to the pharmaceutical, metalworking, mass consumption, and food sectors, among others. In these sectors, new technologies have been increasingly implemented in the form of modernization and semi-automation of initially manual processes, especially in SMEs. The color-based object-sorting machine demonstrated high accuracy in a controlled environment, with detection rates of 97% for red, 96% for blue, and 98% for white. However, in an uncontrolled environment, the accuracy significantly decreased, achieving only 76% for red, 64% for blue, and 68% for white.

Author Contributions: Conceptualization, M.G.; methodology, S.P.; software, S.P.; validation, S.P.-B.; investigation, M.G. and S.P.; resources, S.P.; data curation, S.P.; writing—original draft preparation, M.G.; writing—review and editing, M.G. and A.Q.; visualization, M.G. and S.P.-B.; project administration, S.P. All authors have read and agreed to the published version of the manuscript.

Funding: This research received no external funding.

Institutional Review Board Statement: Not applicable.

Informed Consent Statement: Not applicable.

Data Availability Statement: The data are unavailable due to privacy restrictions.

Conflicts of Interest: The authors declare no conflicts of interest.

References

1. Goyal, A.; Sharma, J.K.; Anand, D.; Gupta, M. Temperature-and Color-Based SDSS Stellar Spectral Classification Using Automated Scheme. In *Proceedings of the Intelligent Communication, Control and Devices: Proceedings of ICICCD 2017*; Springer: Singapore, 2018; pp. 1415–1425. [CrossRef]
2. Kumar, N.S.; Maheswari, S.U.; Pramila, P.; Khilar, R.; Kumar, A. Colour based Object Classification using KNN Algorithm for Industrial Applications. In Proceedings of the 2022 International Conference on Automation, Computing and Renewable Systems (ICACRS), Pudukkottai, India, 13–15 December 2022; pp. 1110–1115. [CrossRef]
3. Fadhil, A.; Abbar, K.; Qusay, A. Computer Vision-Based System for Classification and Sorting Color Objects. In Proceedings of the IOP Conference Series: Materials Science and Engineering, Baghdad, Iraq, 16–17 December 2019; Volume 745, p. 012030. [CrossRef]
4. Yacoub, K.M.; Elbialy, A. Automated Object Detection and Grasp Form Classification. In Proceedings of the 2023 3rd International Conference on Electronic Engineering (ICEEM), Menouf, Egypt, 7–8 October 2023; pp. 1–5. [CrossRef]
5. Kapgate, M.; Hatwar, K.; Jaipurkar, P.B. Color and Shape Detection Using Artificial Intelligence. *Int. Res. J. Mod. Eng. Technol. Sci.* **2023**, *5*, 1805–1808. [CrossRef]
6. Feng, F.; Ou, Z.; Zhang, F.; Chen, J.; Huang, J.; Wang, J.; Zuo, H.; Zeng, J. Artificial intelligence-assisted colorimetry for urine glucose detection towards enhanced sensitivity, accuracy, resolution, and anti-illuminating capability. *Nano Res.* **2023**, *16*, 12084–12091. [CrossRef]
7. Wu, Y.; Arwa, A.H.; Farhan, Z.A.; Alkhalifah, T.; Alturise, F.; Ali, H.E. Enhanced artificial intelligence for electrochemical sensors in monitoring and removing of azo dyes and food colorant substances. *Food Chem. Toxicol.* **2022**, *169*, 113398. [CrossRef] [PubMed]
8. Koch, G. The Color Sensor. In *Learn Engineering with LEGO: A Practical Introduction to Engineering Concepts*; Springer: Cham, Switzerland, 2023; pp. 333–383. [CrossRef]
9. Li, N.; Okmi, A.; Jabegu, T.; Zheng, H.; Chen, K.; Lomashvili, A.; Williams, W.; Maraba, D.; Kravchenko, I.; Xiao, K.; et al. van der Waals semiconductor empowered vertical color sensor. *ACS Nano* **2022**, *16*, 8619–8629. [CrossRef] [PubMed]
10. Raikar, M.M.; Meena, S.; Hubballi, S.; Kulkarni, A.; Merawade, V.; Deshpande, Y. Color based Classification of Products Using Internet of Things. In Proceedings of the International Conference on Communication, Devices and Computing, Haldia, India, 6–18 August 2023; Springer: Singapore, 2023; pp. 665–675. [CrossRef]
11. Kanjanavasoontara, P.; Suppitaksakul, C. Color shade classification using RGB sensor. In Proceedings of the 2023 8th International STEM Education Conference (iSTEM-Ed), Ayutthaya, Thailand, 20–22 September 2023; pp. 1–4. [CrossRef]
12. Elwakeel, A.E.; Mazrou, Y.S.; Tantawy, A.A.; Okasha, A.M.; Elmetwalli, A.H.; Elsayed, S.; Makhlouf, A.H. Designing, optimizing, and validating a low-cost, multi-purpose, automatic system-based RGB color sensor for sorting fruits. *Agriculture* **2023**, *13*, 1824. [CrossRef]
13. Alaya, M.A.; Tóth, Z., Géczy, A. Applied color sensor based solution for sorting in food industry processing. *Period. Polytech. Electr. Eng. Comput. Sci.* **2019**, *63*, 16–22. [CrossRef]
14. Cengizler, C. Fish spoilage classification based on color distribution analysis of eye images. *Mar. Sci. Technol. Bull.* **2023**, *12*, 63–69. [CrossRef]
15. Sinha, S.; Suman, S.K.; Kumar, A. Color Sensor-Based Object Sorting Robotic Arm. In *Proceedings of the Computing Algorithms with Applications in Engineering: Proceedings of ICCAEEE 2019*; Springer: Singapore, 2020; pp. 169–180. [CrossRef]
16. Cong, V.D.; Phuong, L.H. Design and development of a delta robot system to classify objects using image processing. *Int. J. Electr. Comput. Eng.* **2023**, *13*. ISSN 2088-8708. [CrossRef]
17. Thakare, R.D.; Thakare, M.A. Development of Dual Arm Telerobotic System Based on Color Identification and RFID Tag. *Int. J. Res. Appl. Sci. Eng. Technol.* **2018**, *8*, 1206–1213. [CrossRef]
18. Ahmad, T.; Chandan, S. Sensor Based Color Identification Robot For Type Casting. *Int. J. Eng. Res. Technol.* **2016**, *9*, 83–88.
19. Thike, A.; San, Z.M.; Oo, Z.M. Design and development of an automatic color sorting machine on belt conveyor. *Int. J. Sci. Eng. Appl.* **2019**, *8*, 176–179. [CrossRef]
20. Luna-Puente, R.; Peréz-Chimal, R.J.; Hernández-Mosqueda, C.; Muñoz-Minjarez, J.U. Control de brazo robótico clasificador mediante HMI y servidor Web Control of robotic arm classifier using HMI and Web server. *Cómputo Apl.* **2019**, 1, 8. [CrossRef]
21. Barrionuevo Freire, J.A.; Gallo Chasi, N.A. Validación de la Cinemática Directa del Robot Cartesiano Clasificador de Objetos Ferromagnéticos por Peso, Mediante un Sistema de Análisis Automático y en Tiempo Real de Movimientos por Fotogrametría. Bachelor's Thesis, Escuela Politécnica Nacional, Quito, Ecuador, 2021.
22. Mariappan, M.; Ming, T.C.T.; Nadarajan, M. Automated visual inspection: Position identification of object for industrial robot application based on color and shape. *Int. J. Intell. Syst. Appl.* **2016**, *8*, 9. [CrossRef]
23. Fredriksson, T.; Ström, S. Color Sorting Robot: Sorting Algorithm by Color Identification. Bacherlor's Thesis, KTH Royal Institute of Technology, Stockholm, Sweden, 2016.
24. Jiang, B.; Tang, L. Exploration on teaching reform of Mechatronics System Design. In Proceedings of the 2023 3rd International Conference on Modern Educational Technology and Social Sciences (ICMETSS 2023), Kuala Lumpur, Malaysia, 25–27 August 2023; Atlantis Press: Dordrecht, The Netherlands, 2023; pp. 151–157. [CrossRef]

25. Cashin, J.L. Designs and Specification of Mechatronic Systems. In *Springer Handbooks*; Springer: Cham, Switzerland, 2022; pp. 287–313. [CrossRef]
26. Murali, V.N.; Starvin, M.; Prasanth, P.; Prasad, N. *Introduction to Mechatronics and Systems*; IIP Series, Volume 3; Selfypage Developers Pvt Ltd.: Karnataka, India, 2024 ; pp. 1–10. [CrossRef]

engineering proceedings

MDPI

Proceeding Paper

On the Performance Comparison of Intelligent Control Strategies for Lithium Battery Chargers

Pablo Rivadeneira [1], William Chamorro [1], Jorge Medina [1,*], Juan Ramírez [2], Daniel Orbe [2] and Luis Salazar [2]

1 Departamento de Automatización y Control Industrial, Facultad de Ingeniería Eléctrica y Electrónica, Escuela Politécnica Nacional, Quito 170525, Ecuador; pablo.rivadeneira01@epn.edu.ec (P.R.); william.chamorro@epn.edu.ec (W.C.)

2 Departamento de Energía Eléctrica, Facultad de Ingeniería Eléctrica y Electrónica, Escuela Politécnica Nacional, Quito 170525, Ecuador; juan.ramirezd@epn.edu.ec (J.R.); daniel.orbej@epn.edu.ec (D.O.); luis.salazar@epn.edu.ec (L.S.)

* Correspondence: jorge.medinaj@epn.edu.ec

† Presented at the XXXII Conference on Electrical and Electronic Engineering, Quito, Ecuador, 12–15 November 2024.

Abstract: Lithium-ion batteries have become a beacon in modern energy storage, powering from small electronic devices to electric vehicles (EVs) and critical medical equipment. Since their commercial introduction in the 1990s, significant advancements in materials science and engineering have enhanced battery capacity, safety, and lifespan. However, the complexity of lithium-ion battery dynamics has necessitated the development of advanced charging and control strategies to optimize performance, safety, and longevity. This work proposes a comparative analysis of three advanced control methods for lithium-ion battery charging: reinforcement learning, fuzzy logic, and classic proportional–integral–derivative (PID) control. Traditional charging methods often fail to address the complexities of battery dynamics, leading to suboptimal performance. Our study evaluates these intelligent control strategies using MATLAB-Simulink simulations to enhance charging efficiency, speed, and battery lifespan. The findings indicate that reinforcement learning offers superior adaptability, fuzzy logic provides robust handling of nonlinearity, and PID control ensures reliable performance with minimal computational resources.

Keywords: lithium-ion battery; fuzzy logic; reinforcement learning; PID; charger; MATLAB-Simulink

Citation: Rivadeneira, P.; Chamorro, W.; Medina, J.; Ramírez, J.; Orbe, D.; Salazar, L. On the Performance Comparison of Intelligent Control Strategies for Lithium Battery Chargers. *Eng. Proc.* **2024**, *77*, 4. https://doi.org/10.3390/engproc2024077004

Academic Editor: Jackeline Abad

Published: 18 November 2024

1. Introduction

Lithium-ion batteries are rechargeable and widely recognized for their high energy density, long cycle life, and low self-discharge rates, which have revolutionized energy storage and usage, becoming a fundamental technology in modern society [1–3]. Conventional charging methods, such as constant current and constant voltage (CC/CV) techniques, often fail to address the complexities of lithium-ion battery dynamics, resulting in suboptimal charging performance and potential battery degradation over time [4]. To address these challenges, researchers have explored the application of advanced control algorithms and techniques for lithium-ion battery charging, aiming to improve efficiency, charging speed, and battery lifespan [5].

In this context, this paper presents a comparative analysis of three prominent intelligent control methods for lithium-ion battery charging: reinforcement learning (RL), fuzzy logic (FL), and classical proportional–integral–derivative (PID) control.

The RL-based controller was studied due to its ability to learn optimal control strategies through interactions with the battery model, and the power electronics. This last component was simplified using the small signal model to enhance the training process. The RL controller trains a neural network based on a reward function that penalizes current and voltage spikes to achieve greater stability in the charging process. This RL agent aims to control the input voltage to the battery; the neural network adjusts the output to activate the power electronics, evaluating the obtained response and maximizing the reward value through multiple interactions. In addition to the voltage response, the system also monitors the battery state, rewarding or penalizing based on current values and aggressive control actions. RL algorithms, such as Q-learning or deep Q-networks, can discover charging policies that minimize charging time, energy consumption, and battery degradation while ensuring safe operation within the battery's limitations [6,7].

On the other hand, FL controllers offer a flexible and intuitive way to incorporate expert knowledge and heuristic rules into the charging process. By defining linguistic variables such as state of charge, temperature, and charging rate, and establishing a set of inference rules, two FL controllers are designed: the first one regulates voltage, allowing it to remain stable despite different charging profiles; the second one regulates current, avoiding excessive spikes and maintaining a stable value over time. Additionally, fuzzy logic-based methods handle the nonlinearity and inherent uncertainty in lithium-ion battery dynamics better, leading to improved charging performance and extended battery life [8]. Finally, PID controllers can be tuned to optimize charging profiles, balancing factors such as charging time, energy efficiency, and battery health preservation.

This study will compare the three implemented controllers, considering that the choice of the most suitable intelligent control method for lithium-ion battery charging will depend on factors such as the specific application requirements, available computational resources, the desired level of complexity, and trade-offs between charging speed, energy efficiency, and battery lifespan. The three intelligent control methods will be evaluated using MATLAB-Simulink 2024-B , where the most relevant factors for achieving an efficient lithium battery charging method will be analyzed. It is important to note that MATLAB provides a detailed battery model for research purposes. The simulation algorithms will be available to the community at: https://github.com/PrediJos/Intelligent-control-strategies-for-lithium-battery-chargers (accessed on 26 September 2024). The remaining sections of this paper are organized as follows: Methodology, Experiments and Results, and Conclusions.

2. Methodology

Simulating lithium battery chargers in MATLAB-Simulink offers a robust platform for analyzing and optimizing battery charging systems [9]. Lithium batteries provide numerous advantages that make them the preferred choice across various applications. Their high energy density ensures a superior energy-to-weight ratio, making them ideal for portable electronics and electric vehicles, where space and weight are crucial factors [10]. Moreover, their low self-discharge rate allows them to retain charge over extended periods, making them suitable for long-term energy storage. Additionally, recent advancements in recycling technologies and reduced environmental impact contribute to lithium batteries being a more sustainable and eco-friendly energy storage solution. The following sections will outline control architectures designed to efficiently manage the lithium battery charging process.

2.1. *Averaged Small-Signal Converter Modelling*

Lithium-ion batteries require an effective design of the energy transmission system to avoid damage and guarantee optimal performance in its discharge process. Inefficient charging methods and components may affect the parameters of the battery, such as life cycle, capacity and efficiency, state of charge (SOC) and Health Status (SOH) [11,12]. This article relies on an isolated DC/DC converter power circuit for the energy transference from a source to the lithium-ion battery. The forward converter features source voltage step-down characteristics much like a DC/DC buck converter topology. in addition, the forward converter includes an extra stage between the input and the output with a transformer, which provides galvanic isolation, which offers greater safety and protection to the battery against possible system failures, such as short circuits or overloads [13]. The low signal model criterion will examine the behavior of the forward converter in the face of small disturbances near a stable operating point. Considering that the internal impedance of the battery will be the resistive component, the model design will consider the capacitive, inductive, and resistive components of the converter. According to Equation (1), the behavior of this circuit can be characterized as a second-order transfer function [14].

$$G(s) = \frac{1}{\left(1+\frac{s}{\omega_1}\right)\cdot\left(1+\frac{s}{\omega_2}\right)}, \tag{1}$$

where ω_1 and ω_2 are the circuit cut-off frequencies. The complete analysis of the circuit can be simplified by using the low-signal (low-Q) approximation, which considers the quality factor Q of a low value, this factor measures the ability of the circuit to filter frequencies close to the resonance frequency ω_0 [14], where the interactions between the inductive and capacitive components of the circuit are less evident. Equation (2) results from applying this criterion to Equation (1), making the transfer functions more manageable and controllable.

$$G(s) = \frac{1}{1+\frac{1}{Q\omega_0}s+\frac{1}{\omega_0^2}s^2}, \tag{2}$$

The converter model includes its working and transformation ratios yielding,

$$G(s) = N \cdot G_{g0}\frac{1}{1+\frac{1}{Q\omega_0}s+\frac{1}{\omega_0^2}s^2},\ G_{g0} = D,\ \omega_0 = \frac{1}{\sqrt{L\cdot C}},\ Q = R\cdot\sqrt{\frac{C}{L}} \tag{3}$$

where N is the transformer's transformation ratio, D is the duty cycle, L and C are the inductive and capacitive components of the output filter of the forward converter, and R is the battery impedance. Based on these parameters, the transfer function becomes,

$$G_s = \frac{1}{1+133x10^{-6}s+13.3x10^{-9}s^2} \tag{4}$$

The converter simulation uses idealized components, such as power supply, transformer, ideal diodes and switches without delays or internal resistance, with the aim of improving the simulation efficiency [15].

In Figure 1, two scenarios were analyzed: without load and another with a lithium-ion battery as load. For the no-load case, Figure 1a shows that the voltage and current curves obtained from the transfer function are smooth, although the converter output presents oscillations in its final value due to the charge and discharge cycles of the inductor. and capacitor. Despite this, both the transfer function and the converter present similar voltage and current trends over time. In the second scenario, with the battery as load, Figure 1c highlights that the main difference lies in the stabilization time. The idealization of the transfer function leads to a faster system response compared to the real converter, although the final output values are consistent.

One of the most notable features of using the transfer function is the simulation time. Although the converter model provides a more accurate representation of the real world, simulations when designing the controllers will be time consuming and will often provide similar results.

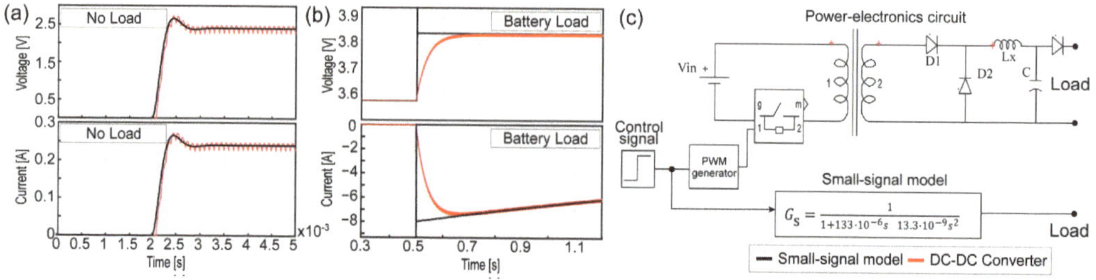

Figure 1. DC–DC converter averaged small-signal modeling: (**a**) no load model validation, (**b**) load model validation, and (**c**) converter scheme.

2.2. Control Architectures Description

The control strategies use the lithium battery model in MATLAB-Simulink as a load, which provides detailed technical parameters for accurate simulation and analysis of battery performance. These parameters include SOC, open circuit voltage (OCV), internal resistance, capacity, charge/discharge rates, thermal dynamics, voltage limits, self-discharge rate, and equivalent circuit components. The controllers will handle the supplied voltage and current during the charging process.

The control architectures are modeled with the same structure for a fair comparison. The CC/CV charging process begins with a current control phase, where the current is set at a safe level, usually a fraction of the battery's nominal capacity, in this control the battery voltage gradually increases as it accumulates charge, until the battery voltage reaches a threshold of 3.855 volts per cell, slightly below the maximum value. Then, the system switches to the voltage control phase, keeping the voltage stable at this level [16,17], in this voltage control, the current gradually decreases as the battery approaches full charge. Current reduction occurs because the voltage is held constant, resulting in a drop in charging current as the battery becomes almost fully charged. This CC/CV approach prevents overcharging, reduces battery stress, minimizes overheating risks, and extends battery life [16,17]. The control strategies that were assessed are summarized in Table 1, and will be detailed in the following sections.

Table 1. Proposed Control Architectures.

Controller.	Scenario 1	Scenario 2	Scenario 3
Voltage Current	Reinforcement Learning (RL) PI	Sugeno Fuzzy PD Sugeno Fuzzy PD + I	PID PID

2.2.1. Reinforcement Learning Architecture

Reinforcement learning (RL) enables the controller to learn optimal control strategies through interaction with the system [18], adapting to varying conditions and improving performance over time. This approach allows the controller to handle complex, nonlinear dynamics that traditional control methods might struggle with. The RL architecture is summarized in Figure 2b, where the trained agent is executed using the MATLAB's reinforcement learning toolbox. The current control was addressed with a classic PI. This choice was made due to the high computational load experienced during the training while using two agents. In addition, the battery model responds slowly to any action yielding an excessively long training time.

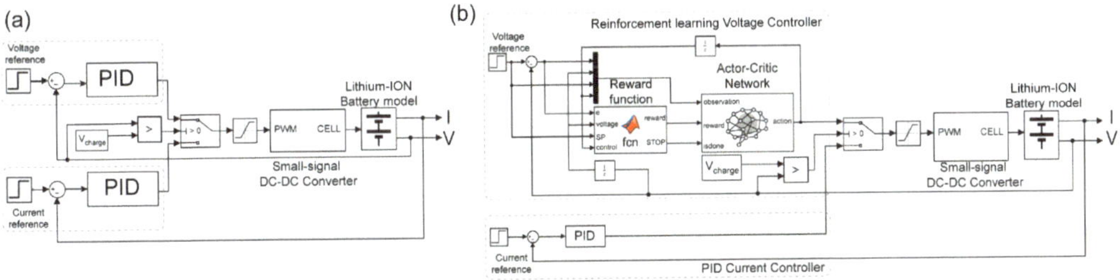

Figure 2. Control architectures: (**a**) classic PID and (**b**) reinforcement learning.

This should incentivize maintaining the voltage within a desired range while penalizing deviations from this range. Additionally, it should consider other objectives such as minimizing energy consumption, reducing oscillations, and ensuring stability. In this sense, the reward function is computed as,

$$R_k = r_1 \cdot (|e_k| \leq 0.5) + r_2 \cdot \left(|u_{k-1}| = V_{ref}\right) + p_1 \cdot (|e_k| > 0.5) + p_2 \cdot (V_{k-1} > V_{th}) \quad (5)$$

where r_x and p_x denote a reward and a penalization constant, respectively. Note that the penalizations are negative values that produce a decrease in the total reward. e_k is the voltage error at time k (current error), u_{k-1} and V_{k-1} are the control action and the voltage observation at time $k-1$ (previous samples). V_{ref} and V_{th} are voltage constants that denote the voltage reference and the maximum voltage overshooting threshold. The last constant is usually considered a 10% above the cell's voltage level. The proposed reward function promotes lower errors and actions closer to the voltage setpoint and penalizes large voltage observations and larger errors.

The agent´s policy learning architecture is composed by an actor–critic scheme as shown in Figure 3a. The actor network, responsible for selecting actions, can handle continuous action spaces effectively, which is crucial for tasks requiring fine-grained control, such as voltage regulation. The critic network, on the other hand, evaluates the actions by estimating the value function, providing feedback to the actor. This setup allows for stable and efficient learning by reducing variance in the policy gradient estimates. In this work, we use a Continuous Gaussian Actor Network (CGAN), which handles continuous action spaces by outputting the parameters of a Gaussian distribution—mean and standard deviation. This stochastic approach facilitates exploration, allowing the agent to try various actions and learn optimal policies. The actor–critic architecture enhances learning stability by using the critic to evaluate actions and reduce variance in policy updates. This method is particularly effective for tasks requiring fine-grained control, like voltage regulation in battery charging, where precise adjustments are crucial. The smooth and differentiable nature of the Gaussian distribution supports efficient gradient-based optimization, improving learning efficiency and performance in complex control environments.

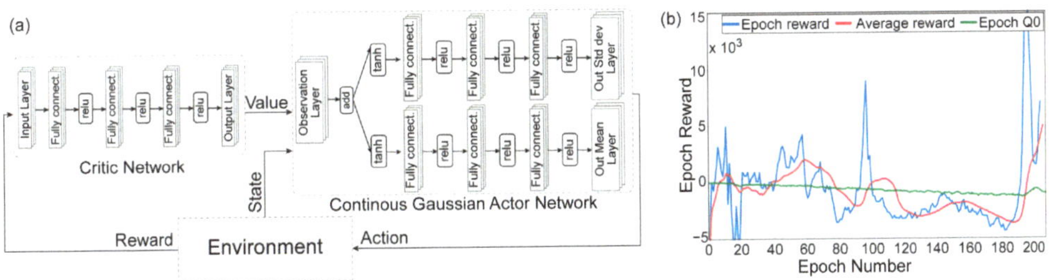

Figure 3. Reinforcement learning scheme: (**a**) actor–critic networks and (**b**) training results.

The actor–critic agent was setup with several fully connected layers (the universal function approximators) to handle the complex voltage relationships [19,20]. Most of the activation functions were setup as Rectified Linear Unit functions (RELU) for a computationally efficient training process, mitigate the vanishing gradient problem, and promote sparse activation of the layers, as shown in Figure 3a. The proposed agent was trained using the MATLAB's reinforcement learning toolbox. The training results displayed in Figure 3b showed that after 200 epochs the maximum average reward is achieved. It is worth noticing that each epoch has a duration of approximately 5 min in a standard PC, yielding a long training time. The constants in the reward function were set empirically to ensure a maximum learning. The most critical objective is the error reduction, hence, p_1 and p_2 were set to 200, and p_1 and p_2 were set to -10. Other values may enhance the learning process; however, the slow response of the battery difficulties the tuning process.

2.2.2. Fuzzy Architecture

The fuzzy Proportional-Derivative (PD) controller can handle nonlinearities and uncertainties in the system more effectively than traditional PD controllers, providing robust performance under varying operating conditions. It adapts to changes in battery characteristics, such as state-of-charge and temperature variations, ensuring stable and accurate voltage control. The fuzzy logic approach allows for smooth and gradual control actions, reducing the risk of overshooting and oscillations. The fuzzy architecture detailed in Figure 4a uses a fuzzy-PD to control the voltage and a fuzzy-PD + I that includes an integral action to control the current. The Fuzzy Inference Systems (FIS) were built with a Sugeno scheme, which is robust to system variations and uncertainties. It requires a simplified rule base, reducing design complexity, and ensures smooth output transitions, minimizing oscillations and overshooting for a more reliable and efficient charging process. In this work, we trust in normalized triangular membership functions to handle the error and its derivative, and the output uses three Sugeno normalized functions to handle the states minimum, zero and maximum as shown in Figure 4b. The inputs range in Figure 4b were constrained from -100 to 100 to avoid saturations due to the input variable´s variation. The input range is modified by the constants K_p and K_d, and the output range by K. In our notation, the subindices v and i denote the relation for the voltage and current controllers, respectively.

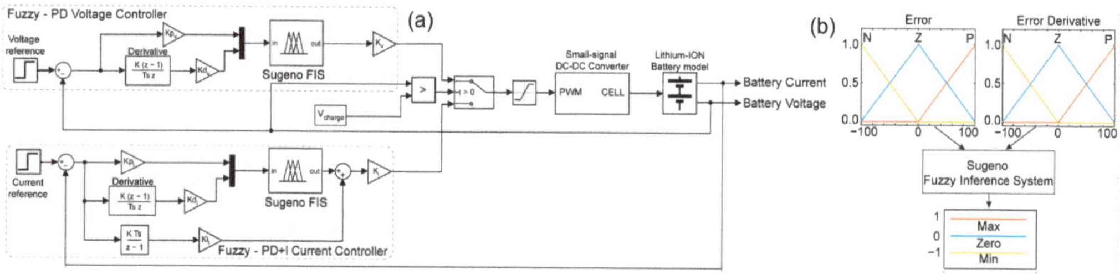

Figure 4. Fuzzy control architecture: (**a**) fuzzy PD and PD + I controllers and (**b**) membership functions.

2.2.3. Classic PID Architecture

Classic proportional–integral–derivative (PID) controllers are widely used in battery charging applications due to their simplicity, effectiveness, and proven performance. They are easy to implement and tune, making real-time adjustments to optimize charging conditions. PID controllers provide stable and accurate control by combining proportional, integral, and derivative actions, which ensures precise regulation of voltage and current. Their versatility allows them to be adapted to various battery types and charging scenarios. Additionally, PID controllers are cost-effective, requiring minimal computational resources, making them suitable for embedded systems and low-cost hardware implementations. The PID architecture only uses two classic controllers for voltage and current as shown in

Figure 2a. The internal structure of a classic PID was assumed to be well known and it was not detailed due to space limitations.

3. Experiments and Results

3.1. Performace Assessment

In the case of implemented drivers. The reinforcement learning reward function was determined using heuristic methods, varying each constant of a given condition in small steps until obtaining the function that provided the best training by trial and error, obtaining $r_1, r_2, p_1, p_2, = 200, -25, -10, 180$. For the fuzzy controller (PD + I), manual tuning was carried out by trial and error, starting with the adjustment of the parameter P until an adequate response was achieved, followed by the adjustment of the constants D and I, and finally the control signal K in that order, achieving for current control P, D, I, K_D are 20; 0.000001; 2; 0.297, and for voltage control P, D, K_D are 15; 0.0001;0.315. The PID controller was tuned using the MATLAB PID Tuner tool, which uses neural networks to find the optimal parameters, obtaining that for Current control $P, I_{PID} = 15; 5$, and for voltage control Current $P, I, D_{PID} = 22.5; 4.9; 0.03$ [21].

In the analysis of the controllers, several aspects are evaluated such as precision in voltage and current regulation, response time, stability and resistance to disturbances. For the controller based on neural networks trained with reinforcement learning, the system reached a stable nominal voltage without overshoots when the current was zero, see Figure 5b. However, when applying CC/CV control, the agent's lack of knowledge of the existing current caused fluctuations in the voltage, allowing the current to variably decrease while the battery continued to charge. The fuzzy controller, designed to maintain a stable charge, showed longer charge and stabilization times, but no spikes in the transition between current and voltage control, as shown in Figure 5c. The PID controller, although it responded quickly to errors, presented greater disturbances, and during the transition from voltage to current, an excess current was observed in Figure 5b, that could be dangerous for the battery. In terms of loading speed, the reinforcement learning controller was the fastest, although for practical use, it would need to consider the circulating current to improve its performance. Also, a chattering phenomenon occurred, which could damage the electronic components and reduce battery life, indicating the need to improve the RL and fuzzy architectures. The fuzzy controller, although safer, was slower and could be optimized by adjusting the inference rules. The PID showed intermediate performance, without depending on the operator's experience.

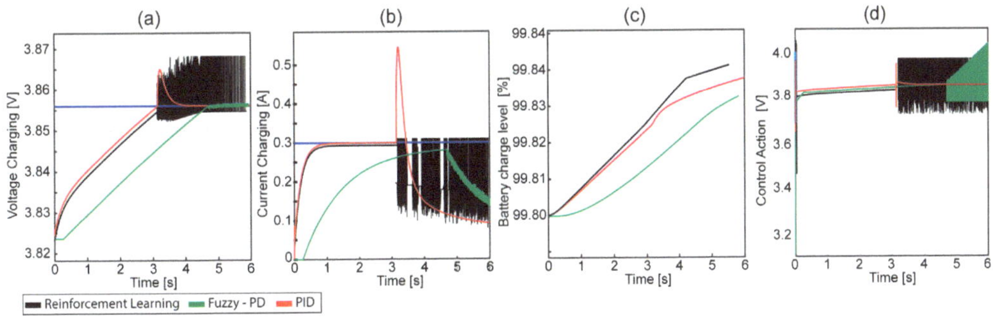

Figure 5. Performance comparison results: (**a**) voltage charging, (**b**) current charging, (**c**) battery charge level, and (**d**) control action.

3.2. Performace Metrics

To measure the performance of the different controllers, an analysis of the voltage control action entering the DC/DC converter was carried out, resulting in Table 2. For the RMS value, it is intended to verify the variation in the voltage control action, showing that the

RL controller has a slightly higher value, indicating greater variations in its control action. This suggests that it has a very fine-tuned response, reacting to even the smallest changes.

Table 2. Performance metrics in control action.

Controller	RMS Voltage [V]	RMS Current [A]	Charging Time	Simulation Time 6 s	Step Simulation Time
RL	3.9347	0.3	10,017.57 s	21,046.5 s	3.5 ms
Fuzzy PI + D	3.8601	0.3	18,401.40 s	1961.4 s	0.33 ms
PID	3.8601	0.3	12,933.57 s	87.90 s	0.015 ms

In addition to the analysis of the voltage delivered as a control action to the converter, the battery current has been evaluated with an RMS value for all controllers. In which a constant value of 0.3 A was obtained. This result suggests that, despite differences in control strategies and variations in control action, all controllers manage to maintain the output current within a desired range. This is indicative that although the RL controller shows greater variability in voltage control action, it does not compromise stability and consistency in terms of output current, which is crucial for the safe and efficient operation of the system.

4. Discussion

The results in Table 2 were obtained in the simulation of each controller which is configured for 6 s and for the three controllers, the one that took the longest to execute was the RL, which took 21,046 s, additionally, when compared with the other controllers applied to the DC/DC converter to regulate the voltage and current of a lithium-ion battery, it will be observed that the RL controller presented greater variations in its output, even when faced with small disturbances, causing the chattering phenomenon, which refers to high frequency oscillations in the control actions, which when applied as an activation signal in real power electronics can cause: overheating of the battery and converter components, noise in the sensor and a reduction in battery life. This variation in the RL controller is 0.02 V. To improve the controller and mitigate vibrations, options include adding low-pass filters at the controller output, modifying the neural network architecture, or using a hybrid control approach where a PID or fuzzy controller regulates the output of the neural network.

The RL controller has a high adaptability to different situations, due to the fact that it reacts to disturbances with greater sensitivity, but these variations suggest the need to include more parameters in the agent training to improve its ability to adapt to a greater number of scenarios, increase charging efficiency, and avoid possible instabilities. On the other hand, fuzzy controllers due to the inference rules placed seek to avoid high current peaks, so it performs a slower charge, while the PID reacts to the error by modifying the output signal and can adapt to dynamic changes.

Regarding the performance of the three controllers, their output was evaluated using the root mean square (RMS) value. The results show that all controllers maintain an average voltage similar to the nominal voltage of the battery, which is crucial to prevent overcharging and avoid overheating of the battery. Regarding current control, all controllers manage to reach the reference without presenting overshoots, in very short response times, also demonstrating good tolerance to external disturbances. Regarding voltage control, it was observed that the controllers manage to reach the reference consistently, which causes the current to begin to gradually decrease until it reaches zero, thus completing the charging cycle. However, it is important to note that, in the case of the reinforcement learning (RL)-based controller, a slight oscillation in the current was observed, which fluctuates between the current value and zero, which could require additional adjustments to improve its stability in the final phase of the charge.

5. Conclusions

Using PID, fuzzy, and agent-based controllers trained with reinforcement learning to regulate the charge of a battery offers different advantages and challenges. PID controllers are known for their simplicity and effectiveness in regulating voltage and current, providing a fast and stable response in many applications. Fuzzy controllers transfer certain experience to programmers by making the system adaptable to certain situations, depending on the number of inference rules and controllers applied. On the other hand, controllers based on neural networks with reinforcement learning can dynamically adapt to changing conditions and optimize their performance over time. Although the latter can offer significant improvements in terms of precision and efficiency, they also require a greater computational load and can result in longer simulation times; and for greater precision, they will require more complex neural network architectures that will increase simulation times. When charging lithium-ion batteries, each type of controller offers specific advantages and disadvantages. The controller based on reinforcement learning stands out for its high adaptability and ability to adjust to various loading conditions, thanks to its ability to handle small disturbances and optimize the loading process dynamically. However, it requires more extensive training and the inclusion of additional parameters to maximize its efficiency and avoid instabilities. On the other hand, fuzzy and PID controllers offer greater stability and lower variability, which can be beneficial in applications where more predictable and reliable performance is required. Although these controllers show less ability to adapt to changing conditions, their stability and simplicity may be suitable for less dynamic systems or where predictability is crucial.

Author Contributions: Conceptualization, P.R. and W.C.; methodology, P.R., W.C., and J.M.; software, P.R., W.C., J.M., and J.R.; validation, P.R., W.C., J.M., J.R, D.O., and L.S.; formal analysis, P.R.; investigation, P.R., W.C., and J.M.; resources, P.R.; data curation, W.C.; writing—original draft preparation, P.R.; writing—review and editing, J.M.; visualization, W.C.; supervision, J.R., D.O., and L.S. All authors have read and agreed to the published version of the manuscript.

Funding: This research received no external funding.

Institutional Review Board Statement: Not applicable.

Informed Consent Statement: Not applicable.

Data Availability Statement: Data are contained within the article.

Conflicts of Interest: The authors declare no conflicts of interest.

References

1. Nishi, Y. The Dawn of Lithium-Ion Batteries. *Electrochem. Soc. Interface* **2016**, *25*, 71–74. [CrossRef]
2. Zubi, G.; Dufo-López, R.; Carvalho, M.; Pasaoglu, G. The lithium-ion battery: State of the art and future perspectives. *Renew. Sustain. Energy Rev.* **2018**, *89*, 292–308. [CrossRef]
3. Takeuchi, E.S.; Leising, R.A. Lithium Batteries for Biomedical Applications. *MRS Bull.* **2002**, *27*, 624–627. [CrossRef]
4. Chen, K.; Zhang, K.; Lin, X.; Zheng, Y.; Yin, X.; Hu, X.; Song, Z.; Li, Z. Data-Enabled Predictive Control for Fast Charging of Lithium-Ion Batteries with Constraint Handling. *arXiv* **2022**, arXiv:2209.12862. [CrossRef]
5. Liu, K.; Zou, C.; Li, K.; Wik, T. Charging Pattern Optimization for Lithium-Ion Batteries with an Electrothermal-Aging Model. *IEEE Trans. Ind. Inform.* **2018**, *14*, 5463–5474. [CrossRef]
6. Park, S.; Pozzi, A.; Whitmeyer, M.; Perez, H.; Joe, W.T.; Raimondo, D.M.; Moura, S. Reinforcement Learning-based Fast Charging Control Strategy for Li-ion Batteries. In Proceedings of the 2020 IEEE Conference on Control Technology and Applications (CCTA), Montreal, QC, Canada, 24–26 August 2020; pp. 100–107. [CrossRef]
7. Chang, F.; Chen, T.; Su, W.; Alsafasfeh, Q. Control of battery charging based on reinforcement learning and long short-term memory networks. *Comput. Electr. Eng.* **2020**, *85*, 106670. [CrossRef]
8. Tian, N.; Fang, H.; Wang, Y. Real-Time Optimal Lithium-Ion Battery Charging Based on Explicit Model Predictive Control. *IEEE Trans. Ind. Inform.* **2021**, *17*, 1318–1330. [CrossRef]
9. MathWorks. Battery, MathWorks. Available online: https://la.mathworks.com/help/sps/powersys/ref/battery.html (accessed on 7 August 2024).
10. Takkalaki, N.; Desai, S.G.; Mishra, R.; Dubey, K. Design and Simulation of Lithium-Ion Battery for Electric Vehicle. In Proceedings of the 2021 12th International Conference on Computing Communication and Networking Technologies (ICCCNT), Kharagpur, India, 6–8 July 2021; pp. 1–6. [CrossRef]
11. Qi, Y.; Zhi, P.; Ye, H.; Zhu, W. Research on Lifetime Optimization of Unmanned Ship Lithium Battery Pack Power Supply System Based on Artificial Fish Swarm Algorithm. In Proceedings of the 2020 39th Chinese Control Conference (CCC), Shenyang, China, 27–29 July 2020; pp. 5379–5384. [CrossRef]

12. Chowdhury, S.; Bin Shaheed, M.N.; Sozer, Y. An Integrated State of Health (SOH) Balancing Method for Lithium-Ion Battery Cells. In Proceedings of the 2019 IEEE Energy Conversion Congress and Exposition (ECCE), Baltimore, MD, USA, 29 September–3 October 2019; pp. 5759–5763. [CrossRef]
13. Mohan, N.; Undeland, T.M.; Robbins, W.P. *Power Electronics: Converters, Applications, and Design*, 3rd ed.; Wiley: Hoboken, NJ, USA, 2002.
14. Erickson, R.W.; Maksimovic, D. *Fundamentals of Power Electronics*, 2nd ed.; Kluwer: Norwell, MA, USA, 2001.
15. MATLAB. Power Electronics Simulation, MathWorks. Available online: https://la.mathworks.com/discovery/power-electronics-simulation.html (accessed on 7 August 2024).
16. Aizpuru, I.; Iraola, U.; Canales, J.M.; Echeverria, M.; Gil, I. Passive balancing design for Li-ion battery packs based on single cell experimental tests for a CCCV charging mode. In Proceedings of the 2013 International Conference on Clean Electrical Power: Renewable Energy Resources Impact, (ICCEP), Alghero, Italy, 11–13 June 2013; pp. 93–98. [CrossRef]
17. Reddy, B.S.T.; Reddy, K.S.; Deepa, K.; Sireesha, K. A FLC based Automated CC-CV Charging through SEPIC for EV using Fuel Cell. In Proceedings of the 2020 International Conference on Recent Trends on Electronics, Information, Communication & Technology (RTEICT), Bangalore, India, 12–13 November 2020; pp. 177–183. [CrossRef]
18. Marahatta, A.; Rajbhandari, Y.; Shrestha, A.; Phuyal, S.; Thapa, A.; Korba, P. Model predictive control of DC/DC boost converter with reinforcement learning. *Heliyon* **2022**, *8*, e11416. [CrossRef] [PubMed]
19. MATLAB. rlACAgent. MathWorks, Inc. Available online: https://www.mathworks.com/help/reinforcement-learning/ref/rl.agent.rlacagent.html (accessed on 7 August 2024).
20. Ye, J.; Guo, H.; Mei, S.; Hu, Y.; Zhang, X. A TD3 Algorithm Based Reinforcement Learning Controller for DC-DC Switching Converters. In Proceedings of the 2023 International Conference on Power Energy Systems and Applications (ICoPESA), Nanjing, China, 24–26 February 2023; pp. 358–363. [CrossRef]
21. MATLAB. PID Tuner. MathWorks, Inc. Available online: https://www.mathworks.com/help/control/ref/pidtuner-app.html (accessed on 7 August 2024).

engineering proceedings

MDPI

Proceeding Paper

Fault Detection in Distribution Networks with Distributed Generation: A Practical Guide to the Morphological Median Filter for the Feature Extraction of Faults †

Verónica Rosero-Morillo [1,*], Le Nam Hai Pham [2], Sebastián Salazar-Pérez [3], Francisco Gonzalez-Longatt [4] and Eduardo Orduña [1]

1 Institute of Electrical Energy (IEE), National University of San Juan, San Juan 5400, Argentina; eorduna@iee-unsjconicet.org
2 Department of Electrical Engineering and ICT, University of South-Eastern Norway, 3918 Porsgrunn, Norway; le.pham@usn.no
3 Faculty of Electrical and Electronic Engineering (FIEE), National Polytechnic School, Quito 170143, Ecuador; gabriel.salazar01@epn.edu.ec
4 Digital Energy Systems Laboratory, Loughborough University, Loughborough LE11 3TU, UK; fglongatt@fglongatt.org
* Correspondence: vrosero@iee.unsj.edu.ar

† Presented at the XXXII Conference on Electrical and Electronic Engineering, Quito, Ecuador, 12–15 November 2024.

Abstract: In this paper, a signal processing method based on Mathematical Morphology (MM) is developed, designed to extract representative characteristics of signals that allow the identification and detection of various types of faults in distribution network feeders that incorporate distributed generation with inverter interfaces (IIDG). The goal is to improve the performance of the fault protection system, ensuring rapid, sensitive, and reliable detection. The fault detection method presented in this article employs a well-known signal processing filter, called the morphological median filter (MMF), applied to the current measured at the current transformer (CT) associated with the relay located at the head of a feeder in a medium-voltage distribution network with IIDG. The extracted characteristics will be used in future research to detect and classify events, such as short-circuit faults or operational manoeuvres, thus facilitating the implementation of protection strategies.

Keywords: distributed generation based on inverter interface; distribution networks; fault detection; mathematical morphology; structuring element

Citation: Rosero-Morillo, V.; Pham, L.N.H.; Salazar-Pérez, S.; Gonzalez-Longatt, F.; Orduña, E. Fault Detection in Distribution Networks with Distributed Generation: A Practical Guide to the Morphological Median Filter for the Feature Extraction of Faults. *Eng. Proc.* **2024**, *77*, 5. https://doi.org/10.3390/engproc2024077005

Academic Editor: Pablo Proaño

Published: 18 November 2024

1. Introduction

In recent years, growing concerns about sustainability and environmental issues, coupled with the need to exploit unconventional renewable energy sources and increased demand, have driven more significant distributed generation (DG) integration, especially in distribution networks [1]. Traditionally, the protection systems of these networks were not designed to handle a high penetration of DG, particularly renewable ones. Instead, they were intended for a passive unidirectional current flow, typical of synchronous machines, with which traditional overcurrent relays function effectively [2,3].

The algorithms of feeder overcurrent relays primarily rely on integral transforms, such as the Fourier transform, to determine the characteristic feature of a fault, namely, the current magnitude. However, the accuracy and speed of these algorithms can be compromised by the DC component and harmonic distortions introduced by renewable sources [3,4].

Although alternative signal processing techniques such as the Wavelet transform, Hilbert Huang transform, Stockwell transform, and principal component analysis have been widely explored in the literature for fault characteristic extraction, these techniques face significant limitations. They fail to accurately extract fault characteristics in the presence of noise, harmonics, and decaying DC components, which affects their effectiveness and precision. Additionally, most of these techniques' high computational load and complexity can cause further delays in the initial fault detection stage: feature extraction [3].

To overcome the inherent limitations of conventional signal processing methods, a feature extraction tool is needed that is effective without being affected by harmonic components, DC components, or noise, and that also avoids the issues of temporal delay and computational complexity associated with traditional methods. In this context, MM has established itself as a promising technique in signal processing due to its operational simplicity and robustness.

This tool modifies the shape of signals through morphological operations, the goal of which is to reveal waveform details that allow the detection of hard-to-detect faults; moreover, these operations also serve to eliminate noise and enhance the signal for detecting any anomalies. The application of MM for extracting transient features, using basic mathematical operations such as addition, subtraction, and calculating maximum and minimum values, facilitates faster processing of the signal. This contributes to reducing waiting times in feature extraction and decreases the computational load [5–7].

This paper describes the development of a morphological tool designed to precisely extract characteristics that represent faults. These characteristics will be used in future research to detect and classify events such as short circuits and operational manoeuvrers, thereby facilitating the implementation of protection strategies. The proposed method employs an MMF, a well-established technique in signal processing, to identify a characteristic index of faults that allows for effective detection. This method is applied to the current signal measured at the CT associated with the relay at the head of a feeder in a medium-voltage distribution network with IIDG. In this regard, the contribution of this paper focuses on two important aspects:

- It provides a detailed methodological guide on applying the application of MM to electrical signals, specifically to fault currents.
- It demonstrates the impact of the interaction of the structuring element with the signal and provides a method for extracting fault characteristics based on the MMF, which researchers in the field of electrical protections can easily replicate.

The structure of the rest of the article is as follows: Section 2 offers a description of the fundamentals of MM, including the structuring element and fundamental morphological operations; Section 3 details the method of feature extraction based on the MMF; Section 4 presents the findings along with their respective analysis; and Section 5 concludes the study.

2. Fundamentals of Mathematical Morphology

Mathematical Morphology is a signal processing method that effectively extracts fault-representative characteristics by applying a structuring element (SE) to the test signal through various morphological operations. These characteristics are crucial for detecting the presence or absence of a fault. During processing, MM extracts fault characteristics by moving and interacting the SE with each sample of the signal and its neighbouring samples. The structuring element is fundamental in signal processing, and its selection depends specifically on the application and the data being analysed. It is vital to appropriately select the length and shape of the SE for applications in electrical systems, as the extracted characteristics will reflect its shape [8]. There are a variety of shapes used in protection schemes, including flat, triangular, or combinations of these [9], as illustrated in Figure 1. Research has shown that flat-shaped SEs are less effective at extracting characteristics from signals affected by interferences such as noise or harmonics. In these cases, more suitable shapes would be triangular, or those that better fit the signal being processed [10,11]. Signal processing using this method relies on the fundamental operators of MM: erosion and dilation. The combination of these operations allows the development of advanced methods for extracting fault characteristics. Each of these operations is described below. Weighted dilation, represented by the mathematical operator "$\oplus$", is an operator that induces an expansion in each sample of the signal, depending on the selected structuring element. The weighted dilation of a signal $f(n)$ using a structuring element $g(m)$ is defined by (1).

$$f \oplus g(n) = \max_{m} \left\{ \frac{f(n-m)}{g(m)} \right\},\ 0 \leq (n-m) \leq n \tag{1}$$

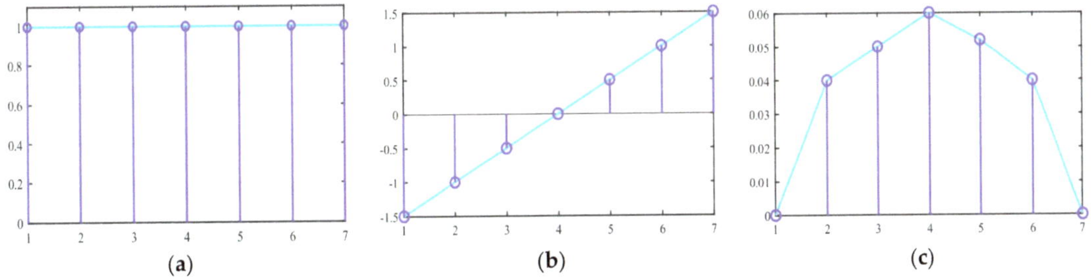

Figure 1. Types of structuring elements applied to signal processing: (**a**) flat; (**b**) straight lines; (**c**) semicircular.

To comprehend the operation of weighted dilation and how the SE moves through each sample of the signal, refer to (1) and Figure 1. This figure displays a signal $f(n) = [2.90, 3.83, 4.71, 5.56, 6.34, 7.07]$ of length $N = 6$, which is dilated by an SE $g(m) = [0.99, 1, 0.99]$ of length $M = 3$. The domains of g and f are $\mathfrak{D}_g = \left\{-\left(\frac{M+1}{2} - 1\right), \ldots, 0, \ldots, \left(\frac{M+1}{2} - 1\right)\right\}$ and $\mathfrak{D}_f = \{0, 1, 2, \ldots, N-1\}$, respectively, where N is the length of the signal under analysis, and M is the length of the structuring element, with the condition $M < N$.

By performing the respective operations on each sample of the signal with the corresponding samples of the structuring element, it is possible to observe that the origin of the SE, designated as the element $g(0)$, acts as a pivot. This pivot moves along each sample of the signal, and at each position, the maximum value is calculated by dividing each sample of the signal by the corresponding sample of the SE. Thus, the dilation operation is performed with adjacent samples at each position occupied by the pivot of the structuring element. During the dilation operation, the SE moves to the right of the signal $f(n)$, starting with $g(-1)$, followed by $g(0)$, and finally $g(1)$. It is important to note that the dilation of the initial sample, $f(0)$, cannot be calculated, as shown in Figure 2, because the SE does not reach its neighbouring samples. A similar case occurs with the final sample of the signal, $f(5)$.

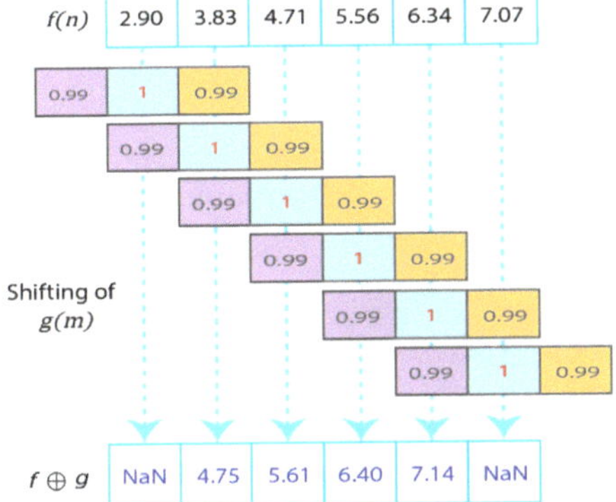

Figure 2. Weighted dilation of the signal.

The result of the weighted dilation of a sample $f(n)$, on a portion of the signal with a positive slope, is the next sample $f(n+1)$, as seen in Figure 2. Conversely, when dilating a portion of a sinusoidal signal with a negative slope, the result is the previous sample $f(n-1)$. These characteristics will be used in the following section to implement the morphological filter in fault detection.

On the other hand, the erosion operation, represented by $\ominus$, is a technique that contracts each sample of the signal depending on the chosen SE. Thus, the weighted erosion of a signal $f(n)$ by an SE $g(m)$ is defined through (2).

$$f \ominus g(n) = \min_{m} \left\{ \frac{f(n+m)}{g(m)} \right\}, \ 0 \leq (n-m) \leq n \tag{2}$$

The interaction of the SE with the signal during the weighted erosion operation is illustrated in Figure 3. Unlike weighted dilation, here the SE moves through each sample of the signal in the order of $g(-1)$, $g(0)$, and $g(1)$, moving to the right of the signal $f(n)$. Each movement of the SE shows that for a portion of the signal with a positive edge, the erosion of $f(n)$ results in the previous sample, namely $f(n-1)$. Conversely, for a signal with a negative edge, the erosion of $f(n)$ results in the next sample, $f(n+1)$.

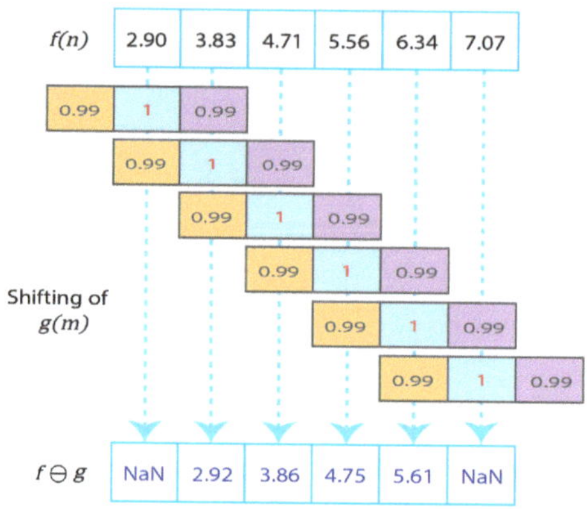

Figure 3. Weighted erosion of the signal.

3. Application of the Morphological Median Filter in Fault Characteristic Extraction

The combination of basic morphological operations, such as dilation and weighted erosion, leads to the development of morphological filters that facilitate the extraction of fault characteristics. Various morphological filters applied to fault characteristic extraction are mentioned in the literature as part of the detection process, including top-hat, morphological gradient algorithms, and the multiresolution morphological gradient, among others. In this context, the fault detection method presented in this article utilises the morphological median filter, a well-established signal processing approach based on calculating the average between dilation and weighted erosion. The logic implemented for fault detection is explained in detail below.

3.1. *Stage 1: Selection of the Structuring Element*

The SE was selected based on prior knowledge of the signal waveform. Given the sinusoidal characteristics of current signals, the SE is defined as specified in (3).

$$g_l = [\cos l\varphi, \ldots, \cos 2\varphi, \cos \varphi, 1, \cos \varphi, \cos 2\varphi, \ldots, \cos l\varphi] \tag{3}$$

In this equation, $\varphi = \omega \times \Delta T$, where ω is the fundamental angular frequency and ΔT represents the sampling interval. The term l denotes the l-th SE used in the signal processing, and the length of the SE is expressed as $2l+1$. For the design of the fault detector, two levels of SEs are selected: one with $l = 1$ and another with $l = 2$, corresponding to lengths of $m = 3$ and $m = 5$, respectively, as detailed in (4) and (5).

$$g_1 = [\cos\varphi, 1, \cos\varphi] \tag{4}$$

$$g_2 = [\cos 2\varphi, \cos\varphi, 1, \cos\varphi, \cos 2\varphi] \tag{5}$$

3.2. Stage 2: Calculation of Basic Dilation and Weighted Erosion Operations

Fault characteristics are derived by applying dilation and weighted erosion operators, as previously detailed in (1) and (2). These operations are performed on the same signal using two different levels, corresponding to the SEs $g1$ and $g2$. Figure 4 illustrates the process of moving these SEs during the erosion operation of the signal.

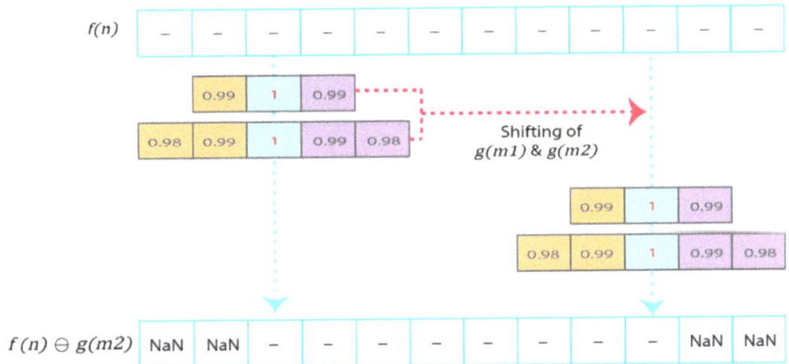

Figure 4. Calculation of basic weighted erosion operation.

3.3. Stage 3: Calculation of the Morphological Median Filter

The MMF, detailed in (6), calculates the average between dilation and weighted erosion. This calculation is performed using two SEs: the first is indexed as 1, with a length of three elements, and the second is indexed as 2, with five elements.

$$MMF_l(n) = \frac{f \oplus g_l(n) + f \ominus g_l(n)}{2} \tag{6}$$

The MMF becomes critically important because, as mentioned in the previous section, the operations of dilation and weighted erosion of the signal $f(n)$ with SE $g(n)$, defined according to Stage 1, result in either the previous sample $f(n-1)$ or the next sample $f(n+1)$ relative to $f(n)$. Therefore, under the normal operating conditions of the system, the sum of $f(n-1)$ and $f(n+1)$ should be equal to twice $f(n)$, a fact that is mathematically demonstrated in reference [12]. This property is expressed morphologically as the arithmetic mean of the dilation and erosion of $f(n)$, which under normal operating conditions would result in $f(n)$ itself. However, in the presence of a fault, this premise does not hold, serving as a primary indicator of a potential anomaly.

3.4. Stage 4: Calculation of the Average of the MMFs for the Two Levels

Depending on the levels considered for the development of the fault detector—in this case there are two—the average of the outputs from the MMFs corresponding to each level, identified as Level 1 and Level 2, is calculated. This procedure is described in (7).

$$mean_{MMF}(n) = \frac{MMF_l + MMF_2}{2} \tag{7}$$

3.5. Stage 5: Calculation of the Difference between the Average MMF(n) and the Signal f(n)

The difference calculated in this stage results from subtracting the current signal $f(n)$ from the average of the MMF outputs calculated in Stage 4. According to (8), it is anticipated that the difference $\Delta f(n)$ will be approximately zero under normal operating conditions, while in a fault situation, $\Delta f(n)$ will deviate from zero, thus serving as an initial indicator of an anomalous event in the system.

$$\Delta f(n) = |f(n) - mean_{MMF}(n)| \tag{8}$$

3.6. Stage 6: Obtaining the Fault Detection Characteristic

The fault detection characteristic of the MMF-based filter is determined by calculating the absolute difference between Δf of the current sample and Δf of the previous sample, as specified in (9). Under normal operating conditions, this characteristic value tends to be close to zero. However, this characteristic will increase significantly in the presence of a fault, providing a clear indication of an anomaly.

$$I_{fault}(n) = |\Delta f(n) - \Delta f(n-1)| \tag{9}$$

4. Simulation and Results

The distribution network selected for testing is a typical American network, as shown in Figure 5, characterized by two three-phase feeders in a radial configuration. The total length of the feeder is 20 km, so a short-line model was used for each line segment, with positive and zero sequence impedances of $z(1) = 0.1153 + j0.53\,\Omega/\text{km}$ and $z(0) = 0.413 + j1.043\,\Omega/\text{km}$, respectively. The nominal voltage of the network is 25 kV RMS line to line, operating at a frequency of 60 Hz. The feeder's total apparent power is 17 MVA, and it is connected to a 110 kV system through a step-down transformer from 110 kV to 25 kV. The IIDG is modelled using a three-level, three-phase Voltage Source Converter (VSC) with an LC output filter [13,14], which includes a fault response model with both negative and positive sequence injection for balanced and unbalanced faults [15,16]. The connection between the IIDG and the AC network is made through a delta-star step-up transformer, supplying the system with 8 MW, which represents a penetration level of 47.24%. The IIDG model implemented in MATLAB/Simulink is illustrated in Figure 6, while Table 1 provides details on its specific parameters.

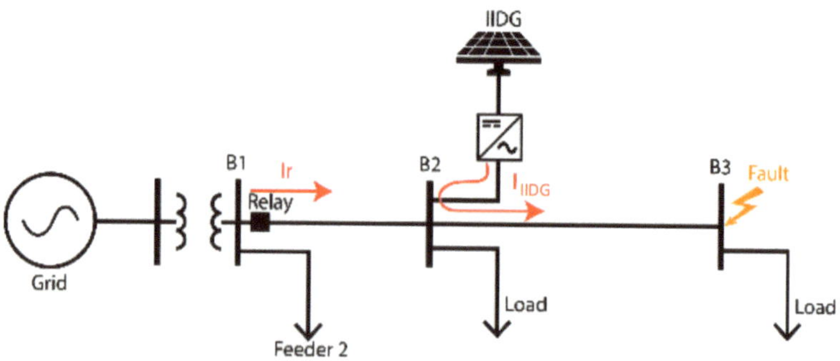

Figure 5. Network diagram under consideration.

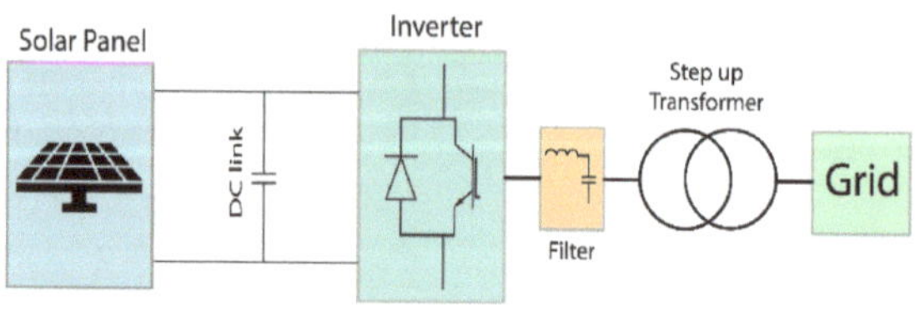

Figure 6. IIDG model.

Table 1. Main data of the test system.

Parameter	Value	Unit
Rated active power	8.0	MW
Rated DC voltage	6000	V
Rated AC grid voltage	25	kV
Rated AC grid frequency	60	Hz
Filter inductance, L	0.485	mH
Filter resistance, R	1.8	mΩ
Filter Capacitance, C	95	uF

To verify the functioning of the morphological extraction tool, different types of faults—three-phase, two-phase, and single-phase—were simulated at 0.2 s on bus 3, as shown in Figure 5. The signals taken at the CT of the feeder's head relay were sampled at a frequency of 3840 Hz, which corresponds to 64 samples per cycle of the fundamental frequency.

Figure 7 shows the phase A of a current during a simulated three-phase ABC fault at 0.2 s. The currents of phases A, B, and C were processed using the designed morphological tool, resulting in the Δf characteristic illustrated in Figure 8. The index appears at the onset of the fault with a maximum magnitude of approximately 50, compared to an average value of 2.3 before the event, which represents an increase of 21.74 times compared to the pre-fault current. This indicates that a high magnitude is a clear indicator of a fault. The tool has been successfully evaluated for detecting single-phase to ground, two-phase to ground, and three-phase faults.

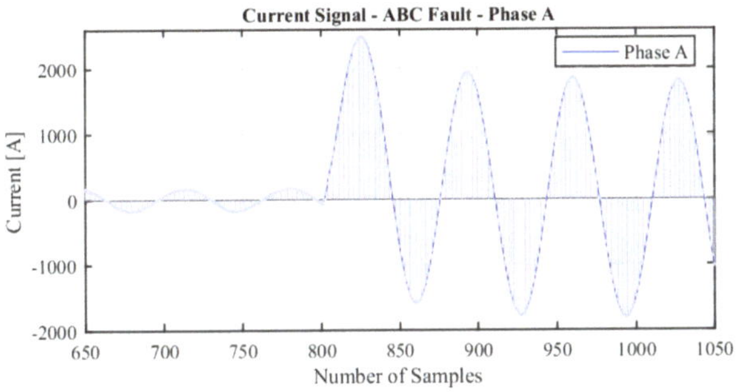

Figure 7. Current signal.

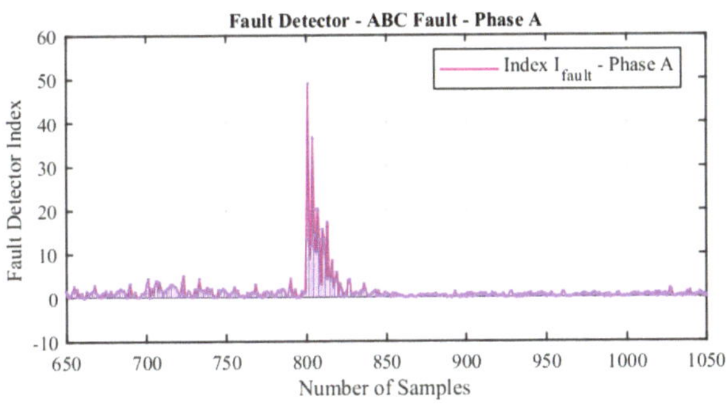

Figure 8. Fault detector index.

Regarding future work and based on reference [12], the magnitude of the index in each sample can be compared with a pre-established threshold based on the characteristic magnitude under normal conditions. If, under normal operation, the characteristic is of a specific value, it will be considered a fault when the index reaches 1.5 times that value. This method is independent of variations in the IIDG as it relies on comparisons with samples prior to the fault and will adjust based on the IIDG penetration.

Additionally, changes in the current signal due to variations in IIDG penetration can lead to false alarms, causing the algorithm to mistake these changes for faults and incorrectly activate the relay. To mitigate this problem, a counter could be implemented to record how often the fault characteristic exceeds the established threshold, helping to distinguish between a true fault and normal system variations.

5. Conclusions

The purpose of this scientific paper is to provide a detailed methodological guide on the effects of MM applied to electrical signals, specifically to fault currents. It also aims to demonstrate the impact of the interaction of the SE with the signal and to offer a replicable method for extracting fault characteristics based on the morphological median filter, intended for new researchers in the field of electrical protections.

It has been observed that there is no specific guide for selecting the geometry and appropriate size of the SE for specific applications; instead, this selection typically relies on an empirical trial-and-error approach using various SEs of different shapes and sizes. This methodology can alter important signal characteristics, such as edge displacement and signal distortion, leading to losses of crucial information and, as a result, a decrease in the accuracy of the feature extraction algorithm and, consequently, in fault detection. Therefore, it is recommended to specifically research a structuring element to optimize its application.

The approach described in this article employs the MMF to extract a representative fault index, enabling effective fault detection. The results have demonstrated that the method is effective in extracting relevant characteristics from fault signals, which can be used to identify faults in real time in distribution networks with inverter-based distributed generation. The reliability and exact timing of the detection will depend on the comparison method used, whether it is through predefined thresholds or advanced artificial intelligence techniques.

Author Contributions: Conceptualization, V.R.-M., F.G.-L. and E.O.; methodology, V.R.-M., F.G.-L. and E.O.; software, V.R.-M.; validation, V.R.-M., F.G.-L. and E.O.; writing—original draft preparation, V.R.-M., L.N.H.P. and S.S.-P.; writing—review and editing, V.R.-M., L.N.H.P. and S.S.-P.; visualization, V.R.-M.; supervision, F.G.-L. and E.O. All authors have read and agreed to the published version of the manuscript.

Funding: This research was supported by the German Academic Exchange Service with a scholarship for a Ph.D. scholarship of Funding programme/-ID: Third Country Programme Latin America, 2020 (57519668).

Institutional Review Board Statement: Not applicable.

Informed Consent Statement: Not applicable.

Data Availability Statement: The data presented in this study are available on request from thecorresponding author.

Conflicts of Interest: The authors declare no conflicts of interest.

References

1. Salazar, E.J.; Rosero, V.; Gabrielski, J.; Samper, M.E. Demand response model: A cooperative-competitive multi-agent reinforcement learning approach. *Eng. Appl. Artif. Intell.* **2024**, *133*, 108273. [CrossRef]
2. Norshahrani, M.; Mokhlis, H.; Bakar, A.H.A.; Jamian, J.J.; Sukumar, S. Progress on protection strategies to mitigate the impact of renewable distributed generation on distribution systems. *Energies* **2017**, *10*, 1864. [CrossRef]
3. Kavi, M.; Mishra, Y.; Vilathgamuwa, M. Morphological Fault Detector for Adaptive Overcurrent Protection in Distribution Networks With Increasing Photovoltaic Penetration. *IEEE Trans. Sustain. Energy* **2018**, *9*, 1021–1029. [CrossRef]

4. Sarangi, S.; Sahu, B.K.; Rout, P.K. Review of distributed generator integrated AC microgrid protection: Issues, strategies, and future trends. *Int. J. Energy Res.* **2021**, *45*, 14117–14144. [CrossRef]
5. Zhang, Y.; Ji, T.Y.; Li, M.S.; Wu, Q.H. Identification of Power Disturbances Using Generalized Morphological Open-Closing and Close-Opening Undecimated Wavelet. *IEEE Trans. Ind. Electron.* **2016**, *63*, 2330–2339. [CrossRef]
6. Salehi, M.; Namdari, F. Fault classification and faulted phase selection for transmission line using morphological edge detection filter. *IET Gener. Transm. Distrib.* **2018**, *12*, 1595–1605. [CrossRef]
7. Gush, T.; Bukhari, S.B.A.; Haider, R.; Admasie, S.; Oh, Y.S.; Cho, G.J.; Kim, C.H. Fault detection and location in a microgrid using mathematical morphology and recursive least square methods. *Int. J. Electr. Power Energy Syst.* **2018**, *102*, 324–331. [CrossRef]
8. Gautam, S.; Brahma, S.M. Guidelines for selection of an optimal structuring element for Mathematical Morphology based tools to detect power system disturbances. In Proceedings of the IEEE Power and Energy Society General Meeting, San Diego, CA, USA, 22–26 July 2012; pp. 1–6. [CrossRef]
9. Aminirad, M.; Shahrtash, S.M. A novel protection and fault location combined algorithm for mixed/hybrid lines. *IET Gener. Transm. Distrib.* **2022**, *16*, 2016–2031. [CrossRef]
10. Zhou, C.; Zou, G.; Zhang, S.; Zheng, M.; Tian, J.; Du, T. Mathematical Morphology Based Fault Data Self Synchronization Method for Differential Protection in Distribution Networks. *IEEE Trans. Smart Grid* **2022**, *14*, 2607–2620. [CrossRef]
11. Lopes, G.N.; Lacerda, V.A.; Vieira, J.C.M.; Coury, D.V. Analysis of Signal Processing Techniques for High Impedance Fault Detection in Distribution Systems. *IEEE Trans. Power Deliv.* **2021**, *36*, 3438–3447. [CrossRef]
12. Wu, Q.H.; Lu, Z.; Ji, T. *Protective Relaying of Power Systems Using Mathematical Morphology*; Springer: New York, NY, USA, 2009; ISBN 9781119130536.
13. Riquelme-Dominguez, J.M.; Martinez, S. Systematic Evaluation of Photovoltaic MPPT Algorithms Using State-Space Models Under Different Dynamic Test Procedures. *IEEE Access* **2022**, *10*, 45772–45783. [CrossRef]
14. Riquelme-Dominguez, J.M.; Martinez, S. Comparison of Different Photovoltaic Perturb and Observe Algorithms for Drift Avoidance in Fluctuating Irradiance Conditions. In Proceedings of the 2020 IEEE International Conference on Environment and Electrical Engineering and 2020 IEEE Industrial and Commercial Power Systems Europe (EEEIC/I&CPS Europe), Madrid, Spain, 9–12 June 2020; pp. 2–6. [CrossRef]
15. Rosero-Morillo, V.A.; Gonzalez-Longatt, F.; Riquelme-Dominguez, J.M.; Orduña, E. Impact of Inverter-Based Distributed Generation Integration on Current Seen by Overcurrent Relay: CIGGRE European MV Network Case. In Proceedings of the 2023 IEEE 41st Central America and Panama Convention (CONCAPAN XLI), Tegucigalpa, Honduras, 8–10 November 2023; pp. 1–6.
16. Rosero-morillo, V.A.; Orduña, E.; Riquelme-dominguez, J.M. Evaluation of Grid-Following Inverter Control Models for Fault Response and their Impact on Protection Devices. In Proceedings of the 2024 IEEE 22nd Mediterranean Electrotechnical Conference (MELECON), Porto, Portugal, 25–27 June 2024; pp. 467–472. [CrossRef]

engineering proceedings

MDPI

Proceeding Paper

Time-Series Modelling and Granger Causality Analysis of GDP and Energy Consumption: The Case of Ecuador 1965–2022 †

Fausto Valencia

Departamento de Energía Eléctrica, Facultad de Ingeniería Eléctrica y Electrónica, Escuela Politécnica Nacional, Quito 170525, Ecuador; fausto.valencia@epn.edu.ec

† Presented at the XXXII Conference on Electrical and Electronic Engineering, Quito, Ecuador, 12–15 November 2024.

Abstract: Knowledge of the causality between Energy Consumption and GDP is important because it leads to the future actions of policymakers, such as developing infrastructure in case GDP growth depends on Energy Consumption. Hence, the Granger Causality between the Gross Domestic Product and Energy Consumption of Ecuador is analysed in this research. For this purpose, a VAR model was developed with data from 1965 to 2022. Before including the time series inside the VAR model, ARIMA models were evaluated so that the need for differentiation and the use of dummy variables was detected. To ensure that the models include all possible information from the available data, the residuals were diagnosed until they did not have any autocorrelation between each other, there was no evidence of heteroskedasticity, and the residuals had a normal distribution. The Akaike Information Criteria and the Schwarz criteria indexes were compared to detect causality. The Granger *p*-value was also used to detect the probability of having null coefficients in the added time series. In the end, it was shown that Energy Consumption Granger causes Gross Domestic Product growth, but the same does not happen in the reverse direction. As a consequence, the government could support the development of energy infrastructure to incentivise economic growth.

Keywords: gross domestic product; energy consumption; time series; Granger causality; VAR model; ARIMA model

Citation: Valencia, F. Time-Series Modelling and Granger Causality Analysis of GDP and Energy Consumption: The Case of Ecuador 1965–2022. *Eng. Proc.* **2024**, *77*, 6. https://doi.org/10.3390/engproc2024077006

Academic Editor: Walter Vargas

Published: 18 November 2024

1. Introduction

People who are responsible for the economic and energy policies of a country make plans according to Gross Domestic Product (GDP) growth and the evolution of Energy Consumption. For example, they could target GDP growth and, in that line of planning, forecast the needed energy. Or, on the other hand, they could predict an amount of Energy Consumption (EC) in the coming years, which would cause the GDP to grow.

However, are we sure that GDP growth causes the increase in Energy Consumption? Or is it the other way around, i.e., Energy Consumption causes GDP growth? We do not know which variable is the cause and which variable is the consequence.

One approximation towards determining causality is the Granger Causality test. Although it is not a definitive test, it has been used over the years to test the relation between GDP and Energy Consumption [1–10].

There are four types of causality that have been found [11]:

- Energy Consumption causes GDP growth.
- GDP growth causes Energy Consumption.
- Growth is produced in both directions.
- There is no effect from either variable.

Ecuador has usually been included in groups of studies [12–16], or investigations have been conducted on other effects, like carbon emissions or exclusively oil [17]. Until the development of this research, the last study dedicated to Ecuador was performed by Pinzon with data from 1970 to 2015 [18]. This research constitutes an update to that investigation, with the advantage of performing the analysis in a way that is closer to an energetic rather than an economic perspective.

In this regard, the main objective of this research is to find out what kind of relationship between economic growth and Energy Consumption exists in Ecuador. This knowledge might constitute a step towards an improvement in the strategies related to economic and energetic development. The secondary objectives are to develop ARIMA models for GDP and Energy Consumption, where a strict procedure of residual analysis is performed.

2. Granger Causality

Say that there are two time series X and Y, modelled as a function of their past values, as shown in (1) and (2), respectively, where X_{t-1} is the value of X in one past period. Likewise, Y_{t-1} means the same for the series Y. ε is the residual that always exists since the noise cannot be predicted:

$$X_t = \beta_0 + \beta_1 X_{t-1} + \beta_2 X_{t-2} + \cdots + \varepsilon_x \tag{1}$$

$$Y_t = \alpha_0 + \alpha_1 Y_{t-1} + \alpha_2 X_{t-2} + \cdots + \varepsilon_y \tag{2}$$

Then, we say that Y_t Granger causes X_t if the time series of X_t with Y_t includes (3) (the VAR model), which has a better description of X_t than the series with X_t alone [19,20]:

$$X_t = \beta_{xy} + \beta_1 X_{t-1} + \beta_2 X_{t-2} + \cdots + \alpha_1 Y_{t-1} + \alpha_2 X_{t-2} + \cdots + \varepsilon_{xy} \tag{3}$$

Two time-series models can be compared by using the Akaike (AIC) [21] or the Schwarz (SC) [22] criteria, where a lower value is better. The AIC criterion is shown in (4), where k is the number of regressors (e.g., β_0, β_1, etc.) and n is the number of observations. The SC criterion is shown in (5):

$$AIC = e^{2k/n} \frac{\sum \varepsilon_i^2}{n} \tag{4}$$

$$AIC = n^{k/n} \frac{\sum \varepsilon_i^2}{n} \tag{5}$$

Each series must be well modelled individually before performing the Granger Causality analysis. First of all, both series must be stationary, i.e., the mean value must be constant. Additionally, the residuals must be spherical, i.e., they must be normally distributed, with zero mean value; their variance must be constant (homoscedasticity), and they must not be correlated [23,24].

The model for each time series may be in the form of an ARIMA(p,d,q) model, where p is the order for the $Xt - q$, i.e., regarding the lags used in the model, q is the order of the model if the residuals are used (or better known as the Moving Average, MA) and d is the order of differentiation. The orders of the ARIMA model must be kept inside the VAR model.

3. Methods

3.1. Data Collection

The GDP data were taken from the World Bank database [25]. It has records from the year 1965 to 2022, including every year. The Energy Consumption data were taken from the BP Energy outlook [26], which also keeps energy records every year. Coincidentally, the time period is the same as that of the GDP data. Both the GDP and Energy Consumption of Ecuador are shown in Figure 1.

There are outliers inside the data. Particularly, these data exist because there were sudden changes over time. For example, the 1999 bank crisis caused a deep decrease in the GDP and Energy Consumption. Something similar happened due to the COVID-19 pandemic. These data were treated with the inclusion of dummy variables, which is a technique used in econometrics to avoid influencing the structure of the data by paying exclusive attention to an outlier. Note that the outlier is not eliminated but is treated apart from the rest of the data.

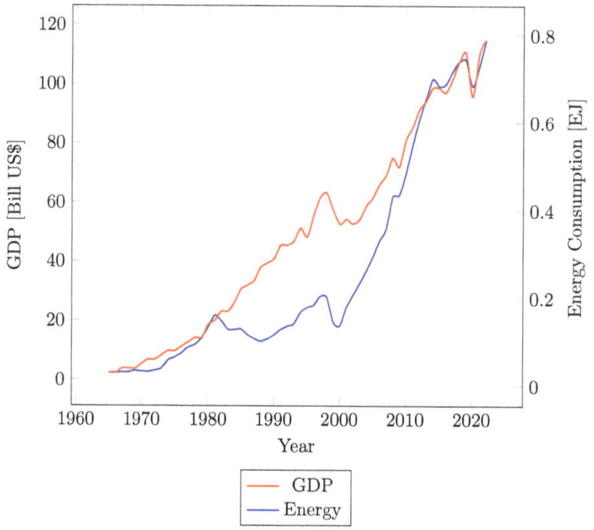

Figure 1. GDP and Energy Consumption for Ecuador.

Generally, the logarithm of the data was used. There is no obstacle in doing this because all the data are positive. This transformation has the advantage of reducing the range of the dependent variables and softening the exponential components of the time series.

3.2. Econometrics Model

In the end, the Granger test is the evaluation of a VAR model where both the GDP and Energy Consumption series are included and related. However, to have a dependable VAR model, both series must be well individually modelled.

The first thing that we must be sure of is that the series is stationary. The Unit Root test checks for this characteristic. In case the presence of a Unit Root is detected, we must determine if the trend is stochastic or deterministic. If it is the first case, the series is differentiated, and if it is the second case, the trend must be included in the model.

Then, in the correlogram of the stationary series (original or differentiated), we determine the possible model, AR or MA, and the lags that might be included. This is only a reference as the analysis of the model could change this.

The model is developed, e.g., ARIMA(1,1,1), and after that, the residuals are evaluated according to three concepts: an autocorrelation analysis, the autocorrelation of the squares, and normality. With these three characteristics, we are sure that the residuals are white noise and that no more information is kept inside them. After that, the impulse response is checked to see if the model is stable.

When the models for each time series (GDP and Economic Consumption) are well established, we include the variables into each other. This is the VAR model. We also analyse the residuals to see if they are not correlated if there is no heteroskedasticity and if they have a normal distribution. If the residuals are well behaved (spherical residuals) and if there is no Unit Root, the VAR model is accepted.

3.3. *Causality Test*

To determine if GDP growth causes Energy Consumption or vice versa, first of all, we compare the AIC and the SC values of the ARIMA and the VAR model. The better model has the lowest value. In other words, if the VAR model for GDP has a lower value of AIC than that of the ARIMA GDP model, we can conclude that Energy Consumption causes GDP growth.

There is also the Granger test that is directly performed. In that case, the null hypothesis tests if the coefficients of the included variable are zero. If not, there exists Granger causality.

4. Results

4.1. *Modelling the GDP*

A simpler time-series model resulted when using the logarithm of the GDP instead of the original data, which is shown in Figure 2.

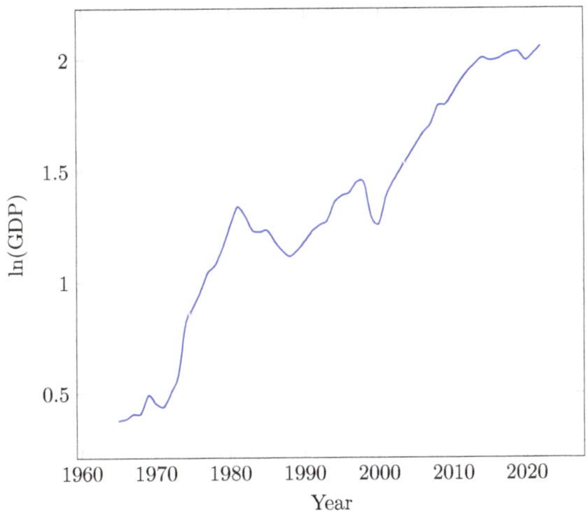

Figure 2. Logarithm of the GDP of Ecuador.

The Unit Root test for this series, considering only an intercept, shows a probability of 59.35% for $H0$. This means that the series is non-stationary. If a trend is added, the probability for $H0$ is 32.49%. The series is non-stationary with a stochastic trend.

The Unit Root test for the first difference, only with an intercept, shows that the probability for the null hypothesis $H0$ is 1.56%. Under a 5% level of confidence, the series could be considered stationary in its first difference, i.e., it is an $I(1)$ series.

The correlograms for the autocorrelation and the partial autocorrelation of the $I(1)$ series are shown in Figure 3a and Figure 3b, respectively. Note that in both correlograms, lag 1 is the most relevant. That characteristic suggests that the model might be an ARIMA(1,1,1). However, after some trial and error, the most suitable model was ARIMA(1,1,0).

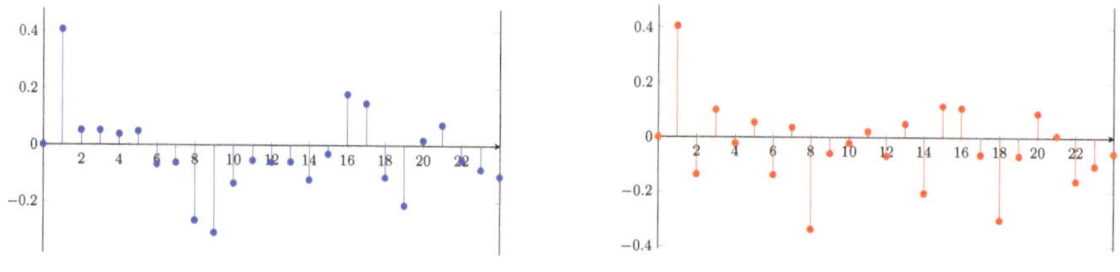

(**a**) (**b**)

Figure 3. (**a**) Autocorrelation of the $I(1)$ series—GDP. (**b**) Partial autocorrelation of the $I(1)$ series—GDP.

The ARIMA(1,1,0) model cannot explain the behaviour of the GDP on its own. It fails in the normality of the residuals. Two outliers exist in the residual (Figure 4) and correspond to the years 1974 and 1999. The first outlier indicates the year when Ecuador began to export oil [27]. The second outlier shows the 1990s bank crisis.

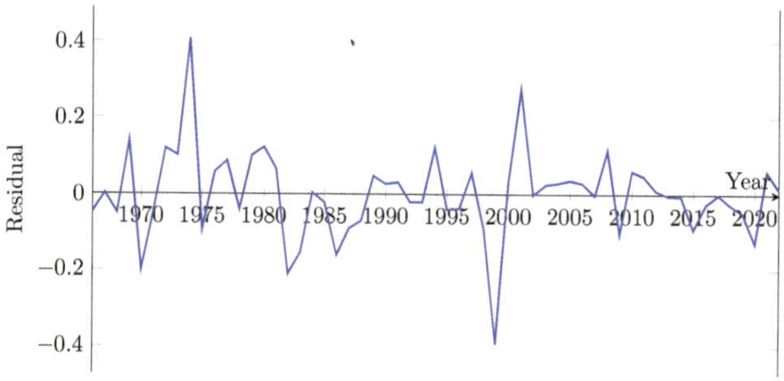

Figure 4. Residuals of the regression when no dummy variables are included.

To normalise the residuals, a dummy variable is included, specifically to aim at the outliers. The results are shown in Table 1. The coefficients of the variable AR(1) and the intercept C are significant. Only the dummy(-1) is not significant; however, this variable is kept so that the behaviour of the residuals is maintained.

Table 1. Regression results for the ARIMA(1,1,0) model of the GDP.

Variable	Coefficient	Std. Error	t-Statistic	Prob.
C	0.0744	0.0208	3.5710	0.0008
Dummy	−0.3175	0.0380	−8.3393	0
dummy(-1)	−0.0834	0.0806	−1.0348	0.3055
AR(1)	0.3926	0.1343	2.9234	0.0051
SIGMASQ	0.0076	0.0016	4.5908	0
R-squared	0.5114	Mean dependent var		0.0679
Adj. R-squared	0.4739	S.D. dependent var		0.1263
S.E. of regression	0.0916	Akaike info criterion		−1.855
Sum squared resid	0.4366	Schwarz criter.		−1.6762
Log likelihood	57.8808	Hannan–Quinn criter.		−1.7858
F-statistic	13.6110	Durbin–Watson stat.		1.9054
Prob (F-statistic)	0	Inverted AR Roots		0.39

The p-values for the autocorrelation of the residuals are shown in Figure 5. All of them are greater than 0.05, i.e., under 95% confidence, and the null hypothesis (H0: there is no correlation) is accepted. For the heteroskedasticity, the p-values for the squared residuals are also greater than 0.05 for all the cases, so the model does not present heteroskedasticity.

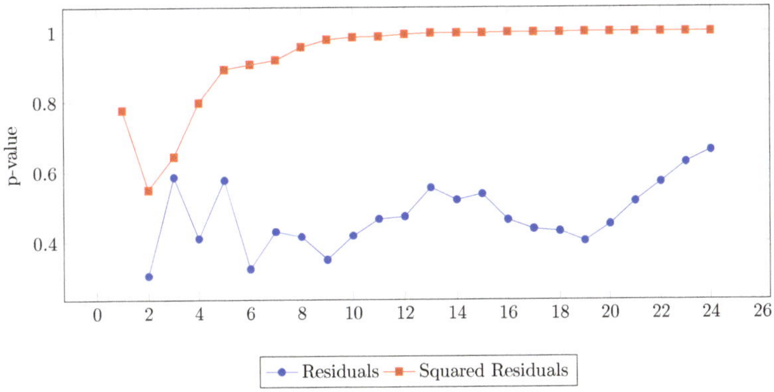

Figure 5. *p*-values for the residuals of the ARIMA(1,1,0) model for the GDP.

Now, we check the normality of the residuals. Once the dummy variables are introduced in the model, the distribution of the residuals is Gaussian, as the Jarque–Bera test shows under a probability of 79.65% (Figure 6a). The impulse response is shown in Figure 6b. The response is stabilised towards zero after a few periods, which is a sign that the phenomenon was modelled appropriately.

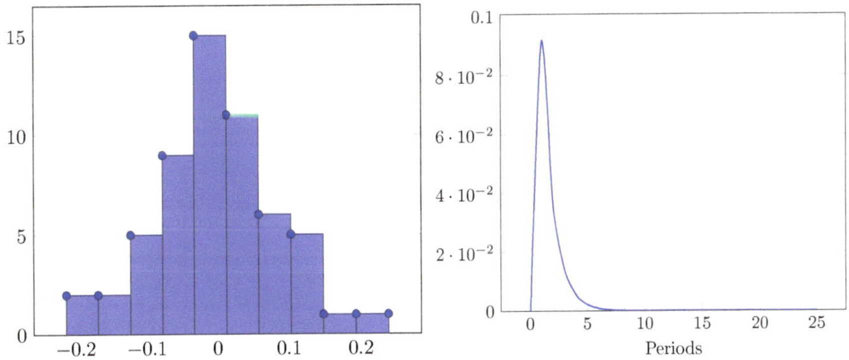

(**a**) (**b**)

Figure 6. (**a**) Histogram of the residuals of the ARIMA(1,1,0) model for Ecuador. (**b**) Impulse response of the residuals of the ARIMA(1,1,0) model for Ecuador.

4.2. Modelling Energy Consumption

For Energy Consumption, the logarithmic transformation of the series was used. The Unit Root test, with interception only, gave a probability of 0.94% of H0. The time series is stationary.

The correlograms for the autocorrelation and partial autocorrelation of the time series are shown in Figure 7a,b. It is seen that the first term of the partial autocorrelation is the most relevant. We can assume a model ARIMA(1,0,0).

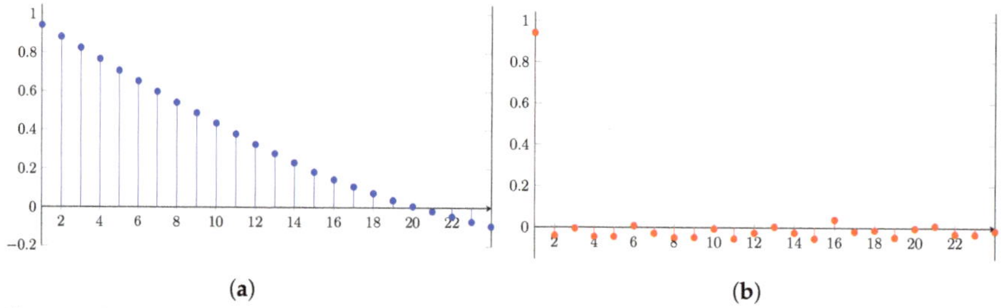

Figure 7. (**a**) Autocorrelation of the $I(1)$ series—Energy Consumption. (**b**) Partial autocorrelation of the $I(1)$ series—Energy Consumption.

In the regression results for ARIMA(1,0,0), a Unit Root was shown. This usually means that a difference is needed, although that was not evident in the initial Unit Root test. The regression results for the ARIMA(1,1,0) model are presented in Table 2. As happened with the GDP, for Energy Consumption, the dummy variables were needed too to normalise the residuals. The logarithmic model of Energy Consumption was stationary without any differentiation. As a consequence, the final model is almost a constant with noise, i.e., it is similar to white noise with media different from zero. This result is confirmed by the low R-squared and the low significance level of the AR(1) coefficient.

Table 2. Regression results for the ARIMA(1,0,1) model of Energy Consumption.

Variable	Coefficient	Std. Error	t-Statistic	Prob.
C	0.0591	0.0095	6.2418	0.0000
dummy	−0.1495	0.0455	−3.2872	0.0018
dummy(-1)	0.0082	0.0253	0.3229	0.7481
AR(1)	0.0205	0.1768	0.1161	0.9080
SIGMASQ	0.0035	0.0007	5.2052	0.0000
R-squared	0.2554	Mean dependent var		0.056656
Adj. R-squared	0.1981	S.D. dependent var		0.068727
S.E. of regression	0.0615	Akaike info criterion		−2.654476
Sum squared resid	0.1970	Schwarz criter.		−2.475261
Log likelihood	80.6526	Hannan–Quinn criter.		−2.584827
F-statistic	4.4582	Durbin–Watson stat.		1.994629
Prob (F-statistic)	0.0036	Inverted AR Roots		0.02

The p-values of the autocorrelation for the residuals and the squared residuals are shown in Figure 8. All of them show a probability greater than 5% for the null hypothesis, i.e., that there is no autocorrelation between the lags.

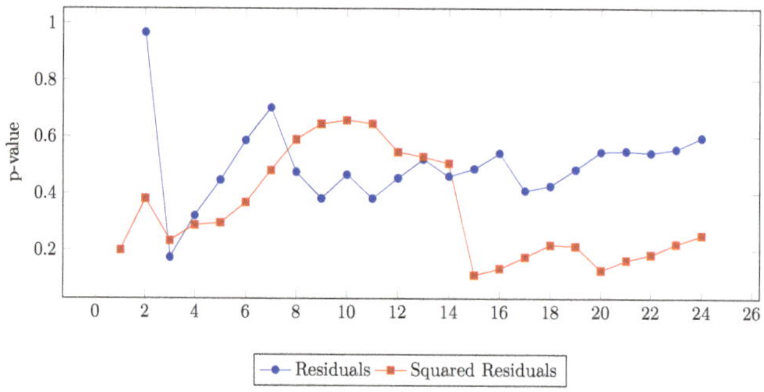

Figure 8. p-values for the residuals of the ARIMA(1,1,0) model for Energy Consumption.

The histogram of the residuals is shown in Figure 9a. The Jarque–Bera test gives a value of 4.28, i.e., a probability of 11.74% of H0: normal distribution, which, therefore, is accepted. The impulse test is shown in Figure 9b. The model is stable since it is damped in a few periods.

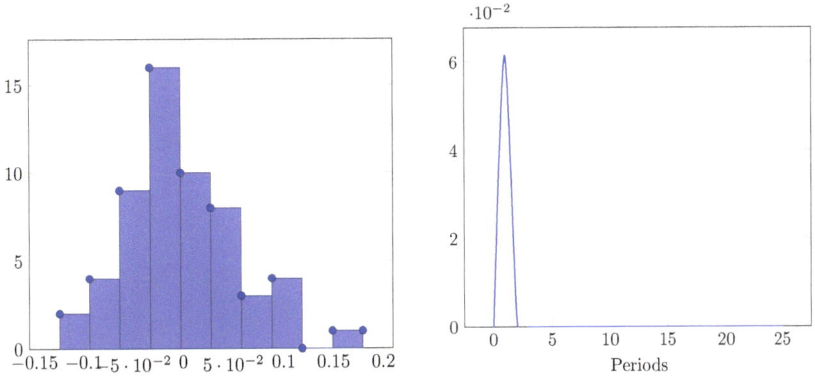

(**a**) (**b**)

Figure 9. (**a**) Histogram of the residuals of the ARIMA(1,1,0) model for Energy Consumption. (**b**) Impulse response for the ARIMA(1,0,1) model of Energy Consumption.

4.3. VAR Model and Granger Causality Analysis

An analysis with two lags is performed. The results are shown in Table 3. If we compare the Akaike and the Schwarz criteria with those of the ARIMA models, we can verify that, when the energy is inserted in the GDP equation, the model improves. The case of EC is special since it improves with the inclusion of GDP; however, in the regression results, the coefficients were non-significant, so the analysis was not valid.

Table 3. Akaike and Schwarz criteria with and without the inclusion of the independent variable.

	GDP	GDP (EC)	EC	EC (GDP)
Akaike info criterion	−1.855	−1.8313	−2.6541	−2.5837
Schwarz criterion	−1.6762	−1.6123	−2.4752	−2.3647

Finally, the Granger test corroborates the fact that only Energy Consumption affects the GDP and not the other way around, see Table 4. H0 in this case means that the independent variable does not Granger cause the dependent variable. For the GDP, the probability of H0 is 1.75%; hence, H0 is rejected, and we can say that Energy Consumption Granger causes GDP. On the other hand, for Energy Consumption, the probability of H0 is 97.45%, so we accept the null hypothesis and consider that the GDP does not cause Energy Consumption.

Table 4. Granger test for the VAR model of Energy Consumption and GDP.

Dependent Variable: D (log(GDP))				**Dependent Variable: D (log(energy))**			
Excluded	Chi-sq	df	Prob.	Excluded	Chi-sq	df	Prob.
D (log(energy))	8.0869	2	0.0175	D (log(GDP))	0.0516	2	0.9745

5. Conclusions

It was demonstrated that Energy Consumption Granger causes GDP growth and not the other way around. This result was confirmed by the Akaike Information Criterion and the Schwarz criterion, as well as the Grainger test itself. This result might help energy–economics planners in future activities towards the development of the economy of Ecuador.

In this research, data from Ecuador were exclusively used. This is a limitation since only 56 years were included. In a future study, panel data, which include relations with other countries of the region, might be employed. This could be an improvement due to the higher number of available data.

Through this research, the dependency of GDP growth on Energy Consumption was confirmed. This is an important result because it shows that policymakers must invest in energy infrastructure if they want to encourage an improvement in the Ecuadorian economy. The development of the electrical sector and better conditions for oil production may be seen as future implications of the results of this research.

One drawback of this study is that it is limited to the relationship between Energy Consumption and GDP. A more complete study should add other aspects of the macroeconomy, such as fiscal policy, productivity levels, technology development, etc.

Funding: This research received no external funding.

Institutional Review Board Statement: Not applicable.

Informed Consent Statement: Not applicable.

Data Availability Statement: The original data presented in this study are openly available in the World Bank National Accounts Data via https://data.worldbank.org and in the BP, Energy Outlook database via https://www.bp.com.

Conflicts of Interest: The author declares no conflicts of interest.

References

1. Beaudreau, B.C. On the methodology of energy-GDP Granger causality tests. *Energy* **2010**, *35*, 3535–3539. [CrossRef]
2. Tiwari, A.K. The asymmetric Granger-causality analysis between energy consumption and income in the United States. *Renew. Sustain. Energy Rev.* **2014**, *36*, 362–369. [CrossRef]
3. Mutascu, M. A bootstrap panel Granger causality analysis of energy consumption and economic growth in the G7 countries. *Renew. Sustain. Energy Rev.* **2016**, *63*, 166–171. [CrossRef]
4. Chiou-Wei, S.Z.; Chen, C.F.; Zhu, Z. Economic growth and energy consumption revisited—Evidence from linear and nonlinear Granger causality. *Energy Econ.* **2008**, *30*, 3063–3076. [CrossRef]
5. Sunde, T. Energy consumption and economic growth modelling in SADC countries: An application of the VAR Granger causality analysis. *Int. J. Energy Technol. Policy* **2020**, *16*, 41–56. [CrossRef]
6. Tran, B.L.; Chen, C.C.; Tseng, W.C. Causality between energy consumption and economic growth in the presence of GDP threshold effect: Evidence from OECD countries. *Energy* **2022**, *251*, 123902. [CrossRef]
7. Odhiambo, N.M. Trade openness and energy consumption in sub-Saharan African countries: A multivariate panel Granger causality test. *Energy Rep.* **2021**, *7*, 7082–7089. [CrossRef]
8. Rahman, M.H.; Ruma, A.; Hossain, M.N.; Nahrin, R.; Majumder, S.C. Examine the empirical relationship between energy consumption and industrialization in Bangladesh: Granger causality analysis. *Int. J. Energy Econ. Policy* **2021**, *11*, 121–129. [CrossRef]
9. Krkošková, R. Causality between energy consumption and economic growth in the V4 countries. *Technol. Econ. Dev. Econ.* **2021**, *27*, 900–920. [CrossRef]
10. Bayar, Y.; Sasmaz, M.U.; Ozkaya, M.H. Impact of trade and financial globalization on renewable energy in EU transition economies: A bootstrap panel granger causality test. *Energies* **2020**, *14*, 19. [CrossRef]
11. AlKhars, M.; Miah, F.; Qudrat-Ullah, H.; Kayal, A. A systematic review of the relationship between energy consumption and economic growth in GCC countries. *Sustainability* **2020**, *12*, 3845. [CrossRef]
12. Sanchez-Loor, D.A.; Zambrano-Monserrate, M.A. Causality analysis between electricity consumption, real GDP, foreign direct investment, human development and remittances in Colombia, Ecuador and Mexico. *Int. J. Energy Econ. Policy* **2015**, *5*, 746–753.
13. Cetin, M.; Ecevit, E. The dynamic causal links between energy consumption, trade openness and economic growth: Time series evidence from upper middle income countries. *Eur. J. Econ. Stud.* **2018**, *7*, 58–68.
14. Yoo, S.H.; Kwak, S.Y. Electricity consumption and economic growth in seven South American countries. *Energy Policy* **2010**, *38*, 181–188. [CrossRef]
15. Rodríguez-Caballero, C.V.; Ventosa-Santaulària, D. Energy-growth long-term relationship under structural breaks. Evidence from Canada, 17 Latin American economies and the USA. *Energy Econ.* **2017**, *61*, 121–134. [CrossRef]
16. Apergis, N.; Payne, J.E. Energy consumption and growth in South America: Evidence from a panel error correction model. *Energy Econ.* **2010**, *32*, 1421–1426. [CrossRef]
17. Jin, S.J.; Lim, S.Y.; Yoo, S.H. Causal relationship between oil consumption and economic growth in Ecuador. *Energy Sources Part B Econ. Plan. Policy* **2016**, *11*, 782–787. [CrossRef]
18. Pinzón, K. Dynamics between energy consumption and economic growth in Ecuador: A granger causality analysis. *Econ. Anal. Policy* **2018**, *57*, 88–101. [CrossRef]

19. Granger, C.W. Investigating causal relations by econometric models and cross-spectral methods. *Econom. J. Econom. Soc.* **1969**, 37, 424–438. [CrossRef]
20. Shojaie, A.; Fox, E.B. Granger causality: A review and recent advances. *Annu. Rev. Stat. Its Appl.* **2022**, 9, 289–319. [CrossRef]
21. Akaike, H. Akaike's Information Criterion. In *International Encyclopedia of Statistical Science*; Lovric, M., Ed.; Springer: Berlin/Heidelberg, Germany, 2011; p. 25.
22. Schwarz, G. Estimating the dimension of a model. *Ann. Stat.* **1978**, 6, 461–464. [CrossRef]
23. Gujarati, D.N.; Porter, D.C. *Basic Econometrics*, 5th ed.; McGraw-Hill: New York, NY, USA, 2009.
24. Franses, P.H.; Dijk, D.V.; Opschoor, A. *Time Series Models for Business and Economic Forecasting*, 2nd ed.; Cambridge University Press: Cambridge, UK, 2014.
25. Bank, W. World Bank National Accounts Data, and OECD National Accounts Data Files. 2024. Available online: https://data.worldbank.org (accessed on 2 May 2024).
26. BP. Energy Outlook. 2024. Available online: https://www.bp.com (accessed on 5 May 2024).
27. Quevedo, C. Ecuador: Petróleo y crisis económlca. In *El Sector Energético Ecuatoriano y la Caída de los Precios Internacionales del Petróleo*; FLACSO: Ciudad de Guatemala, Guatemala, 1986; pp. 91–150.

engineering proceedings

MDPI

Proceeding Paper

Sliding Mode Control for Trajectory Tracking of a TurtleBot3 Mobile Robot in Obstacle Environments †

Jacob Riera [1], Sebastián Cachiguango [1], Michael Pedraza [1], Gabriela M. Andaluz [1,2] and Paulo Leica [1,*]

[1] Departamento de Automatización y Control Industrial, Facultad de Ingeniería Eléctrica y Electrónica, Escuela Politécnica Nacional, Quito 170525, Ecuador; gabriela.andaluz@epn.edu.ec (G.M.A.)
[2] Department of Electronic Engineering and Communications, Universidad de Zaragoza, 44003 Zaragoza, Spain
* Correspondence: paulo.leica@epn.edu.ec
† Presented at XXXII Conference on Electrical and Electronic Engineering, Quito, Ecuador, 12–15 November 2024.

Abstract: The proposed work presents a Sliding Mode Controller (SMC) for trajectory tracking of a TurtleBot3 Burger mobile robot based on sliding mode. Trajectory tracking is performed in congested environments; therefore, an obstacle avoidance strategy is presented to prevent the robot from colliding with obstacles. A clear and detailed methodology is developed for the design of the SMC controller with a PID sliding surface for trajectory tracking that can be extrapolated to position control and posture control. The discontinuous part of the SMC is used to ensure that the robot tends to the desired reference whenever an obstacle appears. The stability analysis of the proposed controller is presented, and the experimental results are shown, demonstrating the good performance of the controller.

Keywords: TurtleBot3 mobile robot; sliding mode control; obstacle avoidance; trajectory tracking; robust control

Citation: Riera, J.; Cachiguango, S.; Pedraza, M.; Andaluz, G.M.; Leica, P. Sliding Mode Control for Trajectory Tracking of a TurtleBot3 Mobile Robot in Obstacle Environments. *Eng. Proc.* **2024**, 77, 7. https://doi.org/10.3390/engproc2024077007

Academic Editor: Pablo Proaño

Published: 18 November 2024

1. Introduction

Nowadays, robots have received remarkable consideration and application in various sectors, such as industrial, service, agriculture, and medicine, among others. The intelligent navigation of wheeled mobile robots is one of the important characteristics that have been of interest to researchers, together with their mobility capacity, their simple structure, their low cost, and their use in different environments that are difficult to access or that represent a risk to human beings [1,2]. This has enabled robots to play a crucial role in these fields, such as agriculture, industry, and medicine [3]. In recent years, research has increased into control strategies to address the challenges of controlling mobile robots, including trajectory tracking and path following, as well as comparative studies of controllers for tracking tasks.

In this context, several control strategies have been published concerning trajectory tracking of differential drive mobile robots. Classical techniques such as PID have been widely used for trajectory tracking control of mobile robots [4]. Advanced versions, such as PID fractional optimization employing particle swarm algorithms, have been developed [5,6]. Reference [7] proposes a retroactive PID to control both virtual velocity and angular velocity for trajectory tracking. Other approaches employ both the kinematic model combined with Lyapunov [8] for designing controllers that account for input saturation constraints. Additionally, ref. [9] employs a Lyapunov-based control for four-wheeled omnidirectional robots employing the dynamic model, and [10] a similar SMC is used for the four-wheeled robot at the simulation level. A proposed SMC aimed at suppressing the influence of electromechanical systems by feedback of the vortex and sliding mode algorithm is introduced in [11]. Meanwhile, ref. [12] considers a hybrid control approach that contemplates a virtual controller for the kinematic model to address stabilization and tracking, along with a fuzzy logic controller to handle uncertainties or uncertain parameters, combining the techniques to improve the robustness of the SMC and the flexibility of fuzzy logic to contribute to the control objective. Other control techniques have been widely implemented with an SMC due to their excellent results, such as the SMC controller presented in [13], which is designed for trajectory tracking for a finite time period, demonstrating its robustness in handling disturbances. Ref. [14] presents an SMC for trajectory tracking combined with an LQR algorithm to decrease the chattering. To achieve improved performance and effectiveness, the SMC has been combined with other control techniques. In [15], the combination of an SMC with neural networks to compensate for external disturbances in velocity tracking, along with a kinematic controller for pose tracking, allows for a significant improvement in handling variable conditions. Similarly, considering that obstacle avoidance plays an important role in mobile robot applications, several strategies have been developed for unstructured environments. Notable approaches include [16], which combines an online tuning of PID parameters without requiring a robot model for obstacle avoidance. Fuzzy logic has also been used for static obstacle avoidance [17], providing a flexible method to manage such events. Additionally, ref. [18] implements a detection algorithm based on YOLO-v4 combined with reinforcement learning for tuning PID controller parameters. Given its potential, the SMC has been considered alongside other algorithms for obstacle avoidance through bilateral teleoperation [19], denoting that the SMC can be effectively integrated with other control techniques to improve the results. It is worth noting that all the cited contributions have presented simulation-level results; thus, implementing a robust control algorithm, like the SMC, in real environments would allow for understanding the real behavior of these controllers, even more when environments with obstacles become a challenge, especially for the calibration and switching of the controller.

Therefore, this work presents the implementation of a Sliding Mode Controller for trajectory tracking of a TurtleBot3 Burger mobile robot in obstacle environments. Unlike previous studies, this work develops a clear and detailed methodology for the design of the SMC controller with a PID sliding surface for trajectory tracking, which can also be extrapolated to position control and posture control. Another contribution of this work is the use of the discontinuous part of the SMC to ensure that the robot tends to the desired reference every time an obstacle appears. A key challenge addressed in this study is the calibration of the controller parameters. The experimental implementation evidenced the criticality of the calibration of this controller; finally, the experimental results demonstrate the controller's strong performance.

This document is structured as follows. Section 2 describes the kinematic model of the mobile robot; Section 3 presents the design and stability analysis of the SMC controller for trajectory tracking and the implemented obstacle avoidance strategy; and Section 4 presents the experimental results in an environment with obstacles. Finally, the conclusions of this work are given.

2. Modeling

The mobile robot used in this work is illustrated in Figure 1, corresponding to a TurtleBot3 Burger that has several features [20]. As shown in Table 1, its use is for educational and research purposes. This mobile platform corresponds to a differential traction-type robot; its locomotion system is based on the use of two wheels.

Figure 1. TurtleBot3 Burger mobile robot.

Table 1. TurtleBot3 Burger mobile robot technical specifications.

Technical Feature	Description
Size (length × width × height)	138 mm × 178 mm × 192 mm
Weight (+SBC + battery + sensors)	1 kg
Maximum velocities	0.22 m/s, 2.84 rad/s
Single board computers	Raspberry Pi
Motors (2 units)	Dynamixel (XL430-W250-T)
Laser distance sensor (LDS)	360° LDS-1
Wheels (2 units)	Sprocket wheels for tire, diameter 66 mm
Battery	Lithium polymer 11.1 V, 1800 mAh
Operating time (battery)	2 h:30 min

The improved kinematic model of the mobile robot with a non-holonomic constraint is given by its compact form [21] as follows:

$$\dot{h} = J\,U \tag{1}$$

where $\dot{h} = [\dot{x}\ \dot{y}]^T$ is the time variation of position in x and y and $U = [u\ \omega]^T$ are the linear and angular velocities, respectively. J is the Jacobian rotation matrix and is given by the following:

$$J = \begin{bmatrix} cos(\psi) & -asin(\psi) \\ sin(\psi) & acos(\psi) \end{bmatrix} \tag{2}$$

ψ is the orientation of the mobile robot and a is the distance between the point of interest and the center of the axis joining the wheels.

3. Controller

3.1. Sliding Mode Controller (SMC)

The schematic of the Sliding Mode Control strategy is presented in Figure 2, where $h_d = [x_d\ y_d]^T$ is the desired position; $h = [x\ y]^T$ is the position of the robot; $\tilde{p} = (h_d - h) = [\tilde{x}\ \ \tilde{y}]$ is the tracking error of x and y, respectively; $\dot{\tilde{p}}$ corresponds to the first derivative of the position error; and $\ddot{\tilde{p}}$ represents the second derivative of the position error.

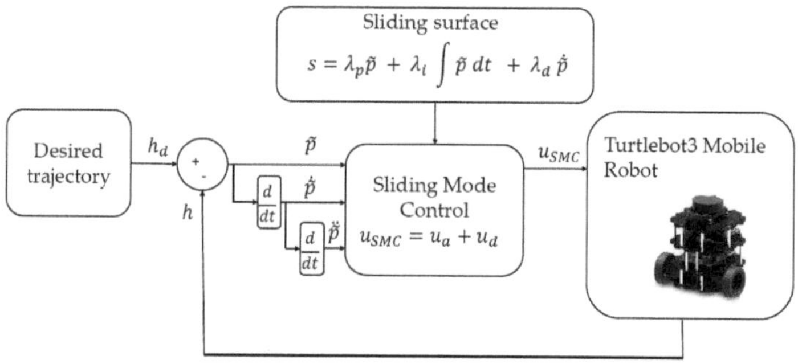

Figure 2. Sliding Mode Control scheme (SMC).

The Sliding Mode Controller is given by the following:

$$u_{SMC} = u_a + u_d \tag{3}$$

where u_a, u_d correspond to the continuous and discontinuous part, respectively.

For the design of the controller, the sliding surface s is set, which is given by the following:

$$s = \lambda_p \tilde{p} + \lambda_i \int \tilde{p}\, dt + \lambda_d \dot{\tilde{p}} \tag{4}$$

where λ_p, λ_i, and λ_d, are the proportional, integral, and derivative constants, respectively. Deriving (4), we have the following:

$$\dot{s} = \lambda_p \dot{\tilde{p}} + \lambda_i \tilde{p} + \lambda_d \ddot{\tilde{p}} \tag{5}$$

The velocity error is given by $\dot{\tilde{p}} = \dot{h}_d - \dot{h}$, replacing (5) as follows:

$$\dot{s} = \lambda_p \left(\dot{h}_d - \dot{h}\right) + \lambda_i \tilde{p} + \lambda_d \ddot{\tilde{p}} \tag{6}$$

Replace (1) in (6).

$$\dot{s} = \lambda_p \left(\dot{h}_d - JU\right) + \lambda_i \tilde{p} + \lambda_d \ddot{\tilde{p}} \tag{7}$$

For the system to remain inside the sliding surface, we define $\dot{s} = 0$, while to obtain u_a, we consider that $u_d = 0$. In a closed loop, we have $u_{SMC} = u_a$. Replacing (7), one has the following:

$$0 = \lambda_p \left(\dot{h}_d - Ju_a\right) + \lambda_i \tilde{p} + \lambda_d \ddot{\tilde{p}} \tag{8}$$

By clearing u_a, we obtain the following:

$$u_a = J^{-1}\left[\dot{h}_d + \frac{\lambda_i}{\lambda_p}\tilde{p} + \frac{\lambda_d}{\lambda_p}\ddot{\tilde{p}}\right] \tag{9}$$

To calculate u_d, we consider the Lyapunov candidate function $V = \frac{1}{2}s^T s$, where its derivative is $\dot{V} = s^T \dot{s}$; replacing (7) in V, we have the following:

$$\dot{V} = s^T \left[\lambda_p \left(\dot{h}_d - JU\right) + \lambda_i \tilde{p} + \lambda_d \ddot{\tilde{p}}\right] \tag{10}$$

Replace (3) and (9) in (10) and develop the following:

$$\dot{V} = -s^T\, J\, u_d \tag{11}$$

Therefore, u_d can be defined as follows:

$$u_d = J^{-1} k\, sig(s);\; k > 0 \tag{12}$$

Replace (12) in (11).

$$\dot{V} = -s^T\, k\, sig(s) \tag{13}$$

where $\dot{V} < 0$; hence, $s \to 0$ with $t \to \infty$. Analyzing (4) for $s = 0$ and since we have a polynomial of two degrees, we have $\tilde{p} \to 0$. When the roots are $r = \frac{-\lambda_p - \sqrt{\lambda_p^2 - 4\lambda_d\lambda_i}}{2\lambda_d}$, the following must be fulfilled: $\lambda_p^2 > 4\lambda_d\lambda_i$.

3.2. Obstacle Avoidance Strategy

Among the characteristics of the TurtleBot3 Burger is that it has a LiDar-type distance sensor capable of scanning 360° of its environment to perform a sweep for SLAM or navigation applications; it reaches a detection distance from 12 to 350 cm with a sampling rate of 1.8 kHz. For the implementation of the obstacle avoidance strategy, a 180° sweep seen from the front of the robot has been considered in order to avoid obstacles that may appear when the robot is executing the trajectory tracking.

Obstacle detection is performed when the LiDar sensor emits a laser light pulse and determines the distance γ_i through the delay between emission and bounce off an object; in this case, the obstacle towards the sensor is shown in Figure 3.

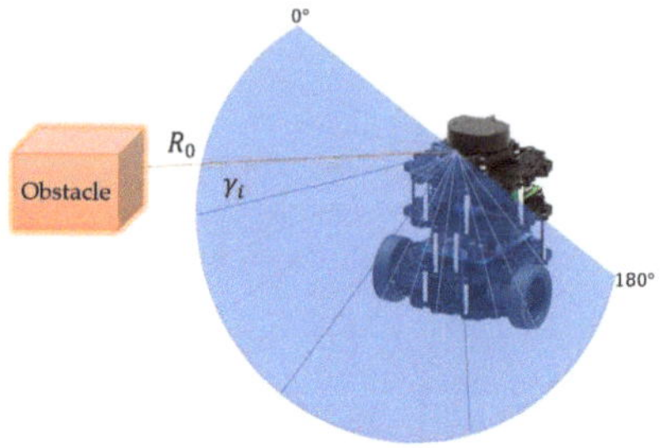

Figure 3. Obstacle avoidance scheme.

The following consideration is established to determine the detection of the obstacle:

$$u_{SMC} = \begin{cases} u_a + u_d, \gamma_i < R_o \\ u = 0.05\left[\frac{\text{m}}{\text{s}}\right],\ \omega = 0.5\left[\frac{\text{rad}}{\text{s}}\right], \gamma_i >\geq R_o \end{cases}$$

where R_o represents the distance from the robot to the obstacle. In the case of obstacle detection, only the LiDar beams that are smaller than the radius R_o are taken into consideration, which depends on the value that is set for the detection.

4. Test

In order to evaluate the trajectory tracking of the TurtleBot3 Burger mobile robot using the SMC controller, experimental tests were performed for a circular trajectory described by $h_d = [x_d\, y_d]$; $y_d = 0.5sin\left(\frac{t}{10}\right) + \frac{1}{2}$; and $x_d = 0.5\, cos\left(\frac{t}{10}\right) + \frac{1}{2}$. Based on the expertise in the management of the SMC controller, the initial parameters that were considered were $\lambda_{po} = 1$, $\lambda_{i0} = 1$ and $\lambda_{d0} = 0.1$. After the implementation of the tests, the respective calibration was carried out based on the index of the integral square error (ISE) of said parameters, and the values that were used in the tests were obtained as follows: $\lambda_p = 1$, $\lambda_i = 0.5$ and $\lambda_d = 0.000001$. For obstacle avoidance, $\gamma_i = 15$ cm. The experiment has a duration of 90 s and considers the incorporation of two obstacles within the desired trajectory. The physical obstacle used is a wooden box measuring 10 × 10 × 30 cm (width, length, and height) with a hollow structure that weighs 0.500 kg.

4.1. *SMC Controller Test*

This section presents the results of experimental tests for the trajectory tracking of the TurtleBot3 Burger mobile robot using an SMC controller considering obstacle avoidance.

4.1.1. Described Trajectory

Figure 4 shows the desired trajectory and the actual trajectory of the mobile robot during the experiment. With the results obtained, it is observed that at the beginning of the trajectory, the mobile robot is positioned to follow the circular trajectory. Once the robot is aligned with the trajectory, there is a perfect tracking of the trajectory. Likewise, in the presence of an obstacle, the robot executes the obstacle avoidance strategy, deviating its trajectory to avoid a collision; once it avoids the obstacle, the SMC control positions the robot again on the desired trajectory.

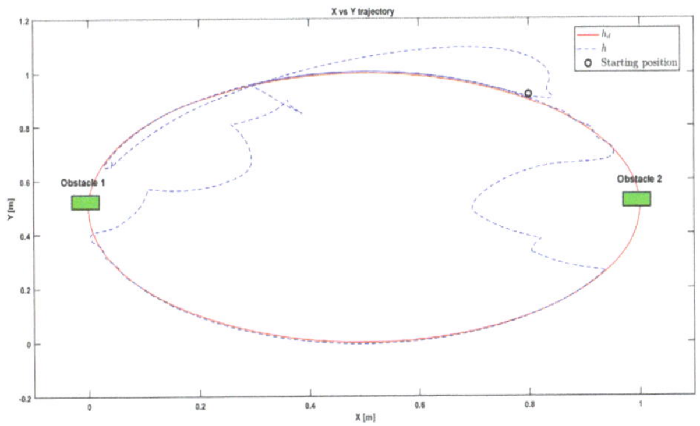

Figure 4. Circular trajectory described by the mobile robot.

4.1.2. Mobile Robot Velocities

Figure 5 shows linear velocity u. At the beginning, it exhibits high values of 0.2 m/s as the robot tries to align itself with the desired trajectory from the initial position. Once the robot reaches the desired trajectory, the velocity remains constant at approximately 0.025 m/s. Between 19 and 31 s, obstacle 1 appears, during which the linear velocity drops to 0 m/s due to the detection of obstacle 1; this causes the robot to stop and start and turn to avoid a collision. Once the robot turns and the obstacle disappears from its path, the robot attempts to return to the trajectory, resulting in spikes in linear velocity as it strives to reach the reference. However, as the robot turns, it encounters obstacle 1 again, causing fluctuations in linear velocity until it completely avoids the obstacle. A similar pattern occurs between 57 and 66 s, where obstacle 2 is present. In other intervals, the robot continues along the trajectory at a velocity of approximately 0.025 m/s.

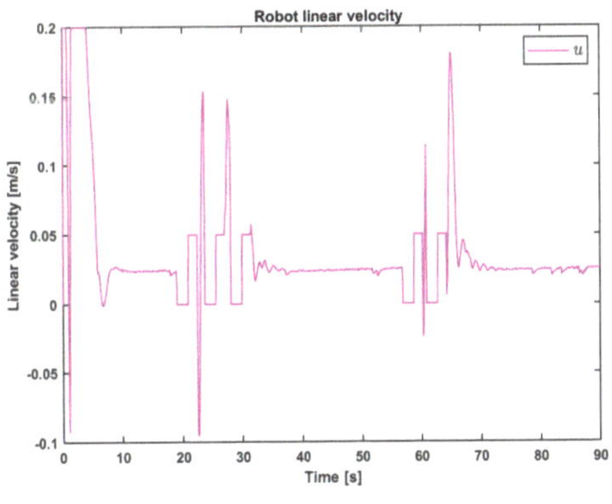

Figure 5. Robot linear velocity.

Figure 6 shows the angular velocity of the mobile robot. In the interval from 0 to 9 s, there are peaks in angular velocity as the TurtleBot3 Burger reaches the circular trajectory. During the interval from 9 to 19 s, the angular velocity fluctuates between 0.1 and 0.15 rad/s, indicating that the robot is following the circular trajectory. In the interval from 19 to 31 s, when obstacle 1 is present, the robot's angular velocity increases to 0.5 rad/s to avoid the obstacle, causing the robot to temporarily deviate from the trajectory. Once the robot turns and the obstacle is no longer in its path, it attempts to return to the trajectory, resulting in spikes in angular velocity as it seeks to reacquire the trajectory. As the robot turns, it encounters obstacle 1 again, leading to further changes in angular velocity until it fully avoids the obstacle. A similar pattern is observed between 57 and 66 s, where obstacle 2 is present. In other intervals, the robot continues following the trajectory with an angular velocity of approximately 0.05 rad/s.

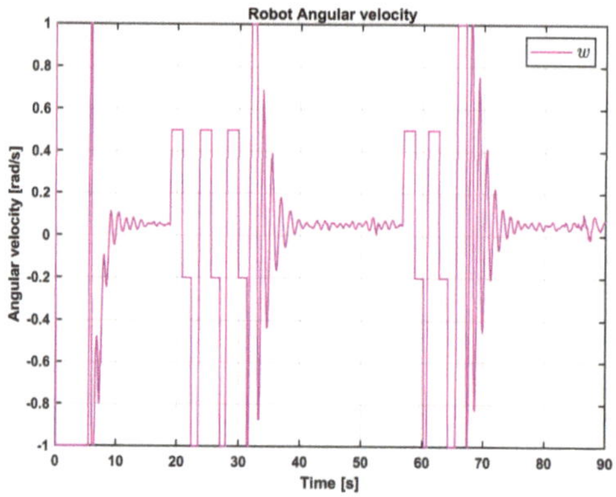

Figure 6. Robot angular velocity.

Figure 7 presents the position error, $\tilde{x}$. In the interval from 0 to 9 s, the mobile robot starts from the initial position and tries to align to the desired trajectory, generating an approximate maximum error of 0.9 m. In the interval from 9 to 19 s, the robot reaches the desired trajectory, which reduces the position error $\tilde{x}$ to approximately 0.05 m. However, at the instant from 19 s to 31 s, the robot detects obstacle 1, causing it to temporarily deviate from the circular trajectory, and the position error $\tilde{x}$ fluctuates until it returns to the desired trajectory. Similarly, this happens between 57 and 66 s, and the position error $\tilde{x}$ increases and shows oscillating behavior due to the avoidance of obstacle 2.

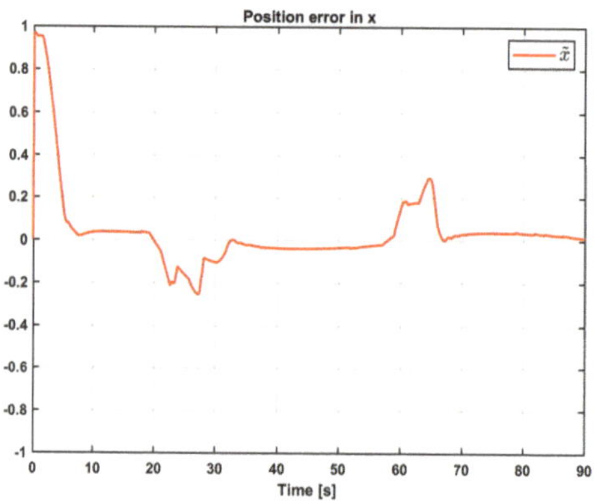

Figure 7. Position errors, $\tilde{x}$.

Figure 8 presents the position error, $\tilde{y}$. It is observed that in the interval from 0 to 9 s, the mobile robot is in search of the desired trajectory, which causes oscillations in the position error in y. During 9 to 19 s, the error gradually increases until it reaches a value of approximately 0.01 m. In the interval from 19 to 31 s, the y-position error presents oscillations due to the fact that the robot temporarily deviates from the trajectory while avoiding obstacle 1; the same happens in the interval from 57 to 66 s, where it detects obstacle 2. For the rest of the time, the error value is a maximum of approximately 0.02 m.

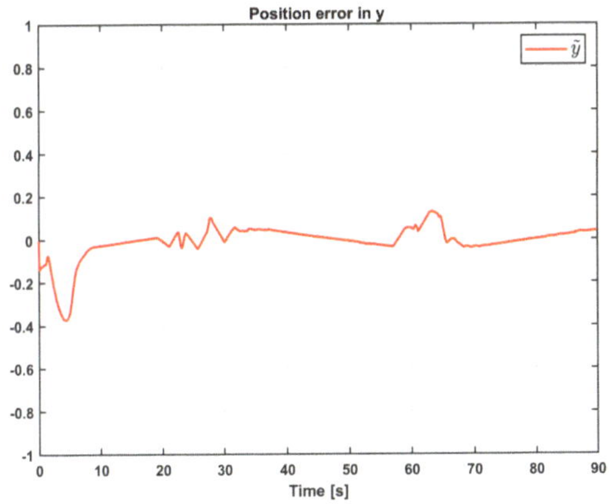

Figure 8. Position errors, $\tilde{y}$.

5. Discussion

The SMC controller proposed for the trajectory tracking of the mobile robot TurtleBot3 Burger presents good trajectory tracking without obstacles, and the linear and angular velocities do not present oscillations and tend to have values of 0.025 m/s and 0.15 rad/s, respectively. This allows the errors in x and y to have maximum values of 0.05 m, which are values expected by the odometry of the robot. When the obstacles are presented in the trajectory in the intervals from 19 to 31 s and 57 to 66 s, position errors in x and y are shown in Figures 7 and 8, causing the robot to temporarily deviate from the trajectory to avoid colliding with the obstacle, generating an error 0.3 m and 0.15 m in x and y, respectively, which is expected for the robot to avoid the obstacle. Likewise, the robot speeds in these intervals are oscillatory to avoid colliding with the obstacle.

6. Conclusions

This work presented the implementation of the Sliding Mode Controller for trajectory tracking of a TurtleBot3 Burger mobile robot in a congested environment. A clear and detailed methodology of the design of the SMC controller with a PID sliding surface for trajectory tracking is provided; this controller can be used for position control, and the methodology can be extrapolated to develop posture control. An obstacle avoidance strategy was implemented, which caused the robot to temporarily deviate from the trajectory. The discontinuous part of the SMC was utilized to ensure that the robot returns to the desired reference whenever it deviates from the intended trajectory to avoid obstacles. The calibration of the controller parameters was carried out based on experimental tests to achieve the smallest possible error. The experimental results demonstrate the controller's strong performance. The main disadvantage of the implemented controller is its challenging calibration, which could be addressed in future work through optimization techniques for controller calibration.

Author Contributions: This work presents the implementation of the Sliding Mode Controller for trajectory tracking of a TurtleBot3 Burger mobile robot in a congested environment. Conceptualization: G.M.A. and P.L.; investigation and methodology: S.C. and M.P.; design of control laws: G.M.A. and P.L.; software and validation: J.R. and S.C.; writing—original draft preparation: J.R., S.C. and M.P.; writing—review and editing: G.M.A. and P.L.; Supervision: P.L. All authors have read and agreed to the published version of the manuscript.

Funding: This research received no external funding.

Institutional Review Board Statement: Not applicable.

Informed Consent Statement: Not applicable.

Data Availability Statement: Data is contained within the article.

Acknowledgments: The authors would like to thank the GIECAR group and ARCI for their technical support for the work carried out and the ESPE-LA and Escuela Politécnica Nacional that through the PIEX-DACI-ESPE-24 project, autonomous control of aerial manipulator robots, have provided research hours for the development of the project.

Conflicts of Interest: The authors declare no conflicts of interest.

References

1. Thai, T.T.N.; Le, Q.T.T.; Nguyen, C.V.; Nguyen, M.T.; Nguyen, X.T.; Phan, G.T.; Nguyen, T.S.; Pham, V.C. Autonomous Mobile Robot: Navigating and Monitoring Fire Safety at Power Substations. In Proceedings of the 2022 7th National Scientific Conference on Applying New Technology in Green Buildings (ATiGB), Da Nang, Vietnam, 11–12 November 2022; pp. 117–182. [CrossRef]
2. Mohamed, J.; Sheik, A.; Venusamy, K.; Ramanathan, K. Optimization of an Intelligent and Survey Robot for Inspection and Fault Diagnosis Mechanism in the Manufacturing Sector. In Proceedings of the 2022 3rd International Conference on Electronics and Sustainable Communication Systems (ICESC), Coimbatore, India, 17–19 August 2022. [CrossRef]
3. Baskoro, C.H.A.H.B.; Saputra, H.M.; Mirdanies, M.; Susanti, V.; Radzi, M.F.; Aziz, R.I.A. An Autonomous Mobile Robot Platform for Medical Purpose. In Proceedings of the 2020 International Conference on Sustainable Energy Engineering and Application (ICSEEA), Tangerang, Indonesia, 18–20 November 2020; pp. 41–44. [CrossRef]
4. Ramasamy, S.; Senthilkumar, N.; Karpagam, S.; Ramani, U.; Swetha, B. Embedded PID Controller Design Based Self Adjusting Robot. In Proceedings of the 2022 Second International Conference on Artificial Intelligence and Smart Energy (ICAIS), Coimbatore, India, 23–25 February 2022; pp. 1531–1536. [CrossRef]
5. Du, J.; Song, B.; Xu, L. Design of Fractional-order PID Controller for Path Tracking of Wheeled Mobile Robot. In Proceedings of the 2021 China Automation Congress (CAC), Beijing, China, 22–24 October 2021; pp. 8019–8023. [CrossRef]
6. Salinas, L.R.; Santiago, D.; Slawiñski, E.; Mut, V.A.; Chavez, D.; Leica, P.; Camacho, O.; Salinas, L.R.; Santiago, D.; Slawiñski, E.; et al. P+d Plus Sliding Mode Control for Bilateral Teleoperation of a Mobile Robot. *Int. J. Control Autom. Syst.* **2018**, *16*, 1927–1937. [CrossRef]
7. Chen, B.; Cao, Y.; Feng, Y. Research on Trajectory Tracking Control of Non-holonomic Wheeled Robot Using Backstepping Adaptive PI Controller. In Proceedings of the 2022 7th Asia-Pacific Conference on Intelligent Robot Systems (ACIRS), Tianjin, China, 1–3 July 2022; pp. 7–12. [CrossRef]
8. Bai, J.; Sun, Z.; Chen, Y. Trajectory tracking control for wheeled mobile robots with input saturation. In Proceedings of the 2020 7th International Conference on Information, Cybernetics, and Computational Social Systems (ICCSS), Guangzhou, China, 13–15 November 2020; pp. 537–540. [CrossRef]
9. Andreev, A.; Peregudova, O. On the Trajectory Tracking Control of a Wheeled Mobile Robot Based on a Dynamic Model with Slip. In Proceedings of the 2020 15th International Conference on Stability and Oscillations of Nonlinear Control Systems (Pyatnitskiy's Conference) (STAB), Moscow, Russia, 3–5 June 2020; pp. 1–4. [CrossRef]
10. Moudoud, B.; Aissaoui, H.; Diany, M. Robust trajectory tracking control based on sliding mode of Differential Driving Four-Wheeled Mobile Robot. In Proceedings of the 2020 IEEE 6th International Conference on Optimization and Applications (ICOA), Beni Mellal, Morocco, 20–21 April 2020; pp. 1–5. [CrossRef]
11. Rassadin, Y.; Shinkaryuk, A. Modified Sliding Mode Control for Tracking Problem of Wheeled Mobile Robot. In Proceedings of the 2020 13th International Conference "Management of Large-Scale System Development" (MLSD), Moscow, Russia, 28–30 September 2020; pp. 1–3. [CrossRef]
12. Shi, W.; Xu, L.; Chen, S. Adaptive Dynamic Surface Control for Simultaneous Stabilization and Tracking of Wheeled Mobile Robot. In Proceedings of the 2020 39th Chinese Control Conference (CCC), Shenyang, China, 27–30 July 2020; pp. 381–386. [CrossRef]
13. Yu, H.; Ai, Z.; Sheng, N.; Liu, B. Finite-Time Trajectory Tracking Control of Mobile Robots Based on Dynamic Terminal Sliding Mode. In Proceedings of the 2022 34th Chinese Control and Decision Conference (CCDC), Hefei, China, 21–23 May 2022; pp. 4421–4426. [CrossRef]
14. Chatterjee, M.; Hanif, O.; Deshpande, N.G.; Stancu, A. Trajectory Tracking of a Nonholonomic Mobile Robot using Optimal Cascade Sliding Mode Controller. In Proceedings of the 2020 3rd International Conference on Intelligent Robotic and Control Engineering (IRCE), Oxford, UK, 10–12 August 2020; pp. 81–86. [CrossRef]
15. Gao, H.; Wang, X.; Hu, J. Adaptive Tracking Control of Mobile Robots based on Neural Network and Sliding Mode Methods. In Proceedings of the 2023 38th Youth Academic Annual Conference of Chinese Association of Automation (YAC), Hefei, China, 27–29 August 2023; pp. 962–967. [CrossRef]
16. Pastrana, M.A.; Oliveira, L.H.; Mendes, D.A.; Silva, D.L.; Mendoza-Peñaloza, J.; Muñoz, D.M. Implementation of a PID Controller using Online Tuning Applied to a Mobile Robot Obstacle following/Avoidance. In Proceedings of the 2024 Argentine Conference on Electronics (CAE), Bahia Blanca, Argentina, 7–8 March 2024. [CrossRef]
17. Adam, Y.M.; Sariff, N.B.; Algeelani, N.A. E-puck Mobile Robot Obstacles Avoidance Controller Using the Fuzzy Logic Approach. In Proceedings of the 2021 2nd International Conference on Smart Computing and Electronic Enterprise (ICSCEE), Cameron Highlands, Malaysia, 15–16 June 2021; pp. 107–112. [CrossRef]

18. Mashhouri, S.; Rahmati, M.; Borhani, Y.; Najafi, E. Reinforcement Learning based Sequential Controller for Mobile Robots with Obstacle Avoidance. In Proceedings of the 2022 8th International Conference on Control, Instrumentation and Automation (ICCIA), Tehran, Iran, 2–3 March 2022; pp. 1–5. [CrossRef]
19. Leica, P.; Balseca, J.; Cbascango, D. Controller Based on Null Space and Sliding Mode (NSB-SMC) for Bidirectional Teleoperation of Mobile Robots Formation in an Environment with Obstacles. In Proceedings of the 2019 IEEE Fourth Ecuador Technical Chapters Meeting (ETCM), Guayaquil, Ecuador, 13–15 November 2019. [CrossRef]
20. Robots e-Manual TurtleBot3. Available online: https://emanual.robotis.com/docs/en/platform/turtlebot3/features/ (accessed on 2 September 2024).
21. Leica, P.; Sagnay, B.; Poveda, F.; Camacho, O. Null-Space-Based Controller for Heterogeneous Robot Formation in Congested Environments. In Proceedings of the 2019 International Conference on Information Systems and Computer Science (INCISCOS), Quito, Ecuador, 20–22 November 2019; pp. 238–243. [CrossRef]

engineering proceedings

MDPI

Proceeding Paper

Sliding Mode Control Proposed Using a Clegg Integrator for Speed Control of a Three-Phase Induction Motor †

Pablo Proaño *, Renato Díaz, César Chillán, Jorge Medina, William Chamorro and José Zuñiga

Department of Industrial Automation and Control, National Polytechnic School, Quito 170525, Ecuador; victor.diaz@epn.edu.ec (R.D.); cesar.chillan@epn.edu.ec (C.C.); jorge.medinaj@epn.edu.ec (J.M.); william.chamorro@epn.edu.ec (W.C.); jose.zuniga@epn.edu.ec (J.Z.)

* Correspondence: pablo.proano@epn.edu.ec

† Presented at the XXXII Conference on Electrical and Electronic Engineering, Quito, Ecuador, 12–15 November 2024.

Abstract: This work presents a new sliding mode control strategy that integrates a Clegg integrator into a Proportional Integral Derivative (PID) sliding surface to address controller saturation (windup) and chattering. The proposed controller was implemented on a Programmable Logic Controller (PLC) to regulate the speed of a three-phase induction motor using a frequency inverter as the final control element. Conventional PID and sliding mode controllers (SMCs) were also implemented for performance comparison. This study included tests for sudden set-point changes and disturbances to evaluate the control strategies. The proposed controller demonstrated a faster response compared to the PID control and a smoother control action compared to the SMC. The results and conclusions discuss the advantages and limitations of the proposed control strategy in comparison to the conventional methods.

Keywords: sliding mode control; clegg integrator; hybrid control; PID controller; induction motor

Citation: Proaño, P.; Díaz, R.; Chillán, C.; Medina, J.; Chamorro, W.; Zuñiga, J. Sliding Mode Control Proposed Using a Clegg Integrator for Speed Control of a Three-Phase Induction Motor. *Eng. Proc.* **2024**, *77*, 8. https://doi.org/10.3390/engproc2024077008

Academic Editor: Jaime Cepeda

Published: 18 November 2024

1. Introduction

In control systems, Proportional-Integral-Derivative (PID) controllers have been widely used for decades due to their simplicity and effectiveness in a variety of industrial applications. Their ability to combine proportional, integral, and derivative actions allows for a balanced response that theoretically provides stability and precision. However, PID controllers have significant limitations, particularly in systems with nonlinear dynamics or varying operating conditions. Their performance can degrade due to issues like instability, overshoot, and sustained oscillations [1,2]. Despite their advantages in linear systems, PIDs struggle with nonlinearities such as saturation, and their performance can become unstable in fast systems [3].

To address these limitations, various control structures have been proposed, including adaptive gains [4], online tuning [5], and controllers with anti-windup actions [6,7]. Among the robust alternatives, the Sliding Mode Controller (SMC) is notable for its ability to handle disturbances and complex dynamics [8,9]. However, SMCs are prone to "chattering", a phenomenon of high-frequency oscillations that can reduce actuator lifespan [10]. Techniques like replacing the sign function with a sigmoid function can mitigate chattering, but this approach can introduce issues like wind-up and reduced controller speed [11].

Wind-up occurs when the integral action accumulates excess control effort due to the controller's output needing to reduce its magnitude, which leads to overshoot and prolonged settling times [12]. Various methods, such as neural networks [4] and fractional-order PID controllers [13], address wind-up but add complexity to controller design.

This work introduces a new sliding control approach using a Clegg integrator on a PID-type sliding surface to mitigate integral windup and chattering. The Clegg integrator reduces saturation effects by resetting the integral action when the error crosses zero, improving control precision and response speed [14]. The proposed approach will be evaluated through a closed-loop velocity control system implemented on a half-horsepower, three-phase motor. The controller was implemented in a Modicon M580 PLC that acts over a variable frequency drive. For comparison purposes, the proposed SMC control is assessed alongside a traditional PID controller, aiming to offer smoother control actions and enhanced actuator longevity.

2. Methodology

The Sliding Mode Controller (SMC) with a Clegg Integrator introduces a robust control strategy capable of handling systems with nonlinearities and uncertainties. The SMC provides robustness by driving the system's state to a predefined sliding surface and maintaining it there despite disturbances. Integrating the Clegg integrator, known for its reset mechanism that reduces overshoot and improves transient response, enhances the performance of the SMC. This combination effectively mitigates chattering, a common issue in conventional SMCs, and optimizes system response, making it a valuable approach in control systems where precision and robustness are critical. The following sections present details about the system identification and controller design for a velocity control application.

2.1. System Description and Identification

The control system consists of a three-phase induction motor driven by a frequency inverter with feedback from an encoder. A Schneider Electric Modicon M580 PLC is responsible for managing the process control. The test bench, depicted in Figure 1, includes the M580 PLC connected to expansion modules for analog inputs/outputs and fast counting driver, all integrated via a backplane to evaluate the controllers' performance.

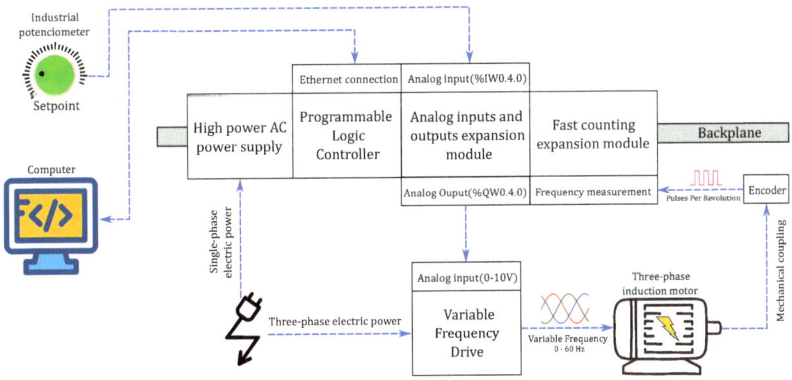

Figure 1. Schematic of implemented system.

The speed setpoint is given by an analog input on the expansion module, and an analog output generates a 0 to 10 V signal to control the variable frequency drive to adjust the motor speed. An encoder, attached to the motor shaft, provides pulse counts per revolution to the fast counting module in frequency mode, which is then used to calculate the motor's speed in revolutions-per-minute (RPM).

The control setup used for testing and comparing the controllers is shown in Figure 2. The main components, including the frequency drive, motor, encoder, and PLC, are highlighted. The PLC is housed in a protective case with external output connectors.

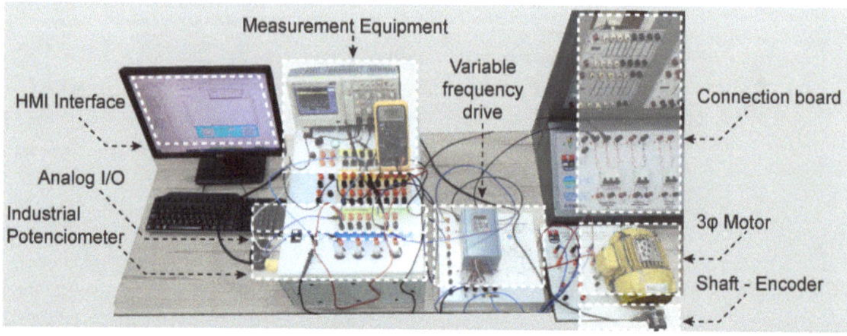

Figure 2. Plant implemented to test the designed controllers.

Model Approximation

Industrial processes often exhibit nonlinear behavior or are modeled by high-order equations, complicating controller design. Approximation techniques simplify by deriving a linear model that retains the key traits of the nonlinear process. One widely used approximation is the First Order Plus Dead Time (FOPDT) model, derived by applying a step input, measuring the output, and extracting 'Smith points' from the reaction curve [15]. This allows the nonlinear process to be effectively approximated by the FOPDT model, as shown in (1).

$$\frac{Y(s)}{U(s)} = \frac{K \cdot e^{-t_o \cdot s}}{\tau \cdot s + 1} \quad (1)$$

where $Y(s)$ is the controlled variable, $U(s)$ is the output variable, K is the steady-state gain, t_o is the system delay, and τ is the system time constant. The steady-state gain K is determined by $K = \frac{\Delta Y}{\Delta u}$, the system delay time t_o is found using $t_o = t_2 - \tau$, and the time constant τ is given by $\tau = 1.5 \cdot (t_2 - t_1)$. The "Smith two-point method" [15] is used to identify t_1 and t_2, the times at which the output reaches 28.3% and 63.2% of its final value, respectively.

The system's response, depicted in Figure 3, highlights nonlinearities that challenge controller development. For comparison, the response of the first-order model with delay to the same step input was simulated allowing a visual assessment alongside the measured data.

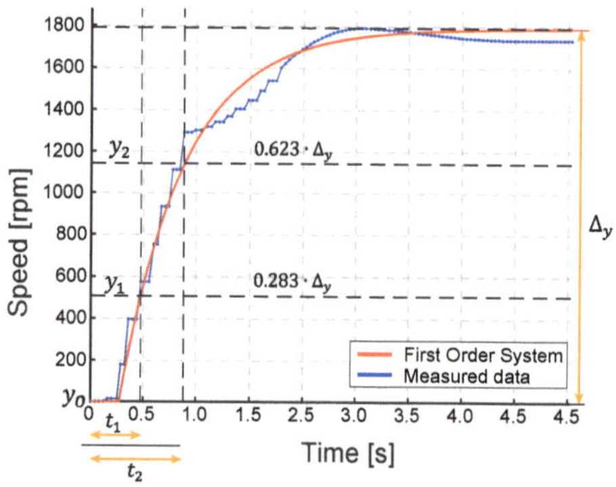

Figure 3. Response to a step input.

2.2. *Fundamental Aspects of the Controllers*

2.2.1. Proportional Integral Derivative Controller (PID)

PID controllers are the most commonly used feedback controllers, with over 90% of control loops employing them, primarily in PI form due to the rare use of derivative action [16]. Their widespread use stems from their simplicity and effectiveness. Designing a PID controller involves identifying the process model, designing the controller structure, and tuning the parameters [17]. In this case, the classical PID structure, as shown in Equation (2), will be utilized.

$$u(t)_{PID} = K_p \cdot \left(e(t) + \frac{1}{T_i} \cdot \int (e(t) \cdot dt) + T_d \cdot \frac{de(t)}{dt} \right), \quad (2)$$

where $u(t)_{PID}$ is the controller output, $e(t)$ is the system error, which is the difference between the setpoint and the motor's speed. T_i is the integral action time constant, and T_d is the derivative action time constant.

2.2.2. Sliding Mode Controller (SMC)

The Sliding Mode Controller (SMC) is a robust, nonlinear controller designed to handle modeling inaccuracies by driving the system state to and maintaining it on a predefined sliding surface [18].

The sliding surface use in this work $s(t)$ is defined as:

$$s(t) = \frac{de(t)}{dt} + \lambda_1 \cdot e(t) + \int \lambda_0 \cdot e(t)dt \quad (3)$$

where λ_1 and λ_0 are tuning constants.

SMC consists of two main components: a discontinuous part, u_d, which moves the system quickly towards the sliding surface, and a continuous part, u_c, which keeps the system on the surface and guides it to the desired state [18]. The SMC control equation is given by (4). The Sliding Mode Controller (SMC) described in [19] was used as a basis for the development of the proposed controller. The following key equations are presented:

$$u(t) = u_c(t) + u_d(t) \quad (4)$$

The continuous component $u_c(t)$ is derived from the sliding surface (4) and the system model approximation (1), as follows:

$$u_c(t) = \frac{\tau \cdot t_0}{K} \left(\frac{y(t)}{\tau \cdot t_0} + \lambda_0 e(t) \right) \quad (5)$$

The discontinuous component $u_d(t)$ is given by

$$u_d(t) = K_D \frac{s(t)}{|s(t)| + \rho} \quad (6)$$

where λ_1, λ_0, K_D, and ρ are constants used in the controller design. The constants λ_1 and λ_0 are associated with the sliding surface dynamics, K_D is a gain factor for the discontinuous part, and ρ is a small positive constant to avoid division by zero. The method for calculating these constants is proposed in [19].

2.3. *Control Proposal: SMC+CLEGG Integrator*

Chattering and high M_p are common issues in SMC due to its discontinuous control nature. Using a Clegg integrator (CI) can reduce these problems by providing smoother control and minimizing high-frequency oscillations.

The Clegg integrator, introduced by Clegg [20], is a nonlinear integrator with a reset action. It was further developed by Krishnan and Horowitz and generalized as FORE by Horowitz and Rosenbaum [21]. The CI resets its state to zero when the input crosses zero.

The PID+CI controller, shown in Figure 4, is a nonlinear/hybrid PID controller that maintains the characteristics of traditional PID controller. Its input is the error signal $e(s)$, the output is the control signal $u(s)$, k_p is the proportional gain, τ_d is the derivative constant, and τ_i is the integral time constant [14].

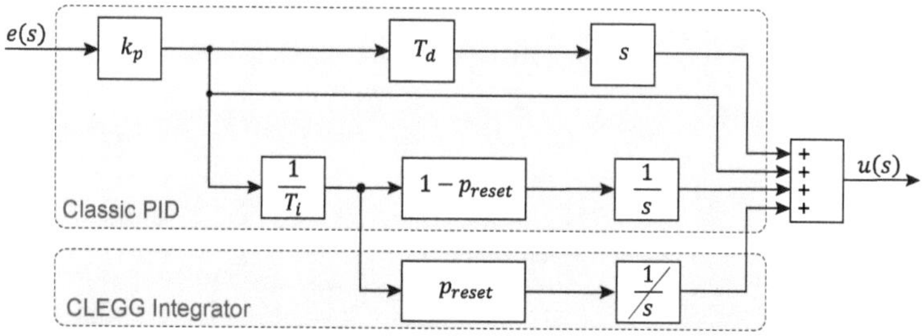

Figure 4. Structure of the PID+CI controller.

The parameter p_{reset}, or reset ratio, is a dimensionless constant with values $\epsilon \in [0, 1]$ that adjusts the weight of the CI term relative to the I term. A $PID + CI$ controller becomes a PID controller if the reset action is removed ($preset = 0$) and a $PD + CI$ controller if $preset = 1$. The reset should not apply to the entire integral term to preserve its fundamental asymptotic property [14].

The $PID + CI$ control scheme is described in (7), as follows:

$$(PID + CI)(s) = K_p\left(1 + T_d s + \frac{1 - p_{\text{reset}}}{T_i s} + \frac{p_{\text{reset}}}{T_i s}\right) \tag{7}$$

The reset action is indicated in (7) by setting $t_r = t_{current}$ when $e(t)$ crosses zero. Recall that the operator (/) denotes the CLEGG integrator.

Using the same structure as that used by [19], a new sliding surface proposal is presented in (8).

$$s_{PID+CI}(t) = \frac{de(t)}{dt} + \lambda_1 e(t) + \lambda_0\left((1 - p_{reset})\int_0^t e(t)dt + p_{reset}\int_{t_r}^t e(t)dt\right) \tag{8}$$

The reset action impacts only the discontinuous component of the controller, as the control law for the continuous component requires the derivative of the sliding surface, thereby excluding the Clegg integrator. Conversely, the discontinuous component uses the sliding surface in both the numerator and denominator of its equation.

The objective of the proposed approach is to reduce the significant overshoot and chattering, which are primarily caused by the discontinuous part of the controller.

The final control equation is presented in (9), as follows:

$$u(t) = \frac{\tau \cdot t_0}{K}\left(\frac{y(t)}{\tau \cdot t_0} + \lambda_0 e(t)\right) + K_D \frac{s_{PID+CI}(t)}{|s_{PID+CI}(t)| + \rho} \tag{9}$$

2.4. Evaluation Methods

To assess the efficacy of the controller and to juxtapose its performance with alternative control methodologies, the following indices will be employed [14]:

- Maximum overshoot (M_p): Evaluates the overshoot of the controlled variable, indicating the speed at which the control action can change or vary.
- Rise time (t_r): Measures how quickly the controller pushes the controlled variable towards the reference.
- Settling time (t_s): Determines the time required for the control action to stabilize the controlled variable within a 2% margin.
- Integral index of absolute error (IAE): Assesses the controller's performance by comparing the controlled variable with the reference; in this study, it is used to identify which controller more effectively ensures that the system follows the reference.
- Total variations in control efforts (TVu): Evaluates the strength of the control action, used here to compare the smoothness of different control strategies.

3. Implementation and Results

3.1. Implementation Details

The control assessment was carried out in the experimental setup displayed in Figure 2 where the controllers were implemented in the Modicon PLC. The control strategies evaluated were a PID controller, a sliding mode controller (SMC), and a combined SMC with a Clegg integrator. The controlled variable is the motor speed, which is handled by a frequency drive system.

To conduct a thorough and precise assessment of the system, two categories of tests were executed:

1. Sudden changes in the set point: Sudden modifications in the speed set point were applied to examine the system's capability to respond and adjust promptly to the newly established conditions. Also, sudden set point decreases, analogous to abrupt changes, and a swift reduction in the speed set point was initiated. This facilitated the evaluation of the system's stability and its proficiency in managing deceleration events.
2. Disturbances: A significant perturbation was introduced into the coupler connecting the encoder and the motor to assess the resilience of the system. This disturbance was simulated in the PLC by introducing a step change in the feedback from the encoder. This evaluation seeks to determine whether the system can sustain the desired set-point speed, even amidst external interferences or physical disturbances that may compromise its operational efficacy

The Software Matlab R2022b was used to obtain the system model, calculate the performance indices, and compare the results. For PLC programming and data acquisition, Control Expert version 15, with a Python add-on for MODBUS communication, was utilized.

3.2. Sudden Changes in the Set Point

Figure 5 illustrates the speed response to sudden positive and negative changes, as well as the controller's output in reaction to abrupt set-point modifications.

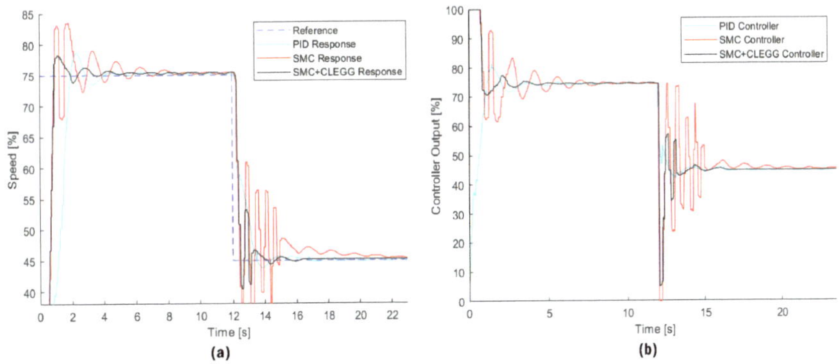

Figure 5. Results of a sudden set-point change test. (**a**) Speed of the three-phase motor, (**b**) controller output sent to the frequency variator.

Table 1 presents the performance indices observed during this test, along with the percentage improvement of the proposed approach compared to conventional PID and SMC controllers.

Table 1. Test results for different controllers: sudden changes in the set point.

Sudden Set-Point Change Test Results			
Metric	**PID**	**SMC**	**SMC+CLEGG**
Positive M_p	10.80%	4.80%	4.40%
Improvement M_p	145.45%	9.09%	-
t_s	3.7 s	5.8 s	1.599 s
Improvement t_s	131.39%	217.70%	-
t_r	1.934 s	0.926 s	0.926 s
Improvement t_r	108.86%	0.00%	-
Negative M_p	4.00%	36.00%	15.33
Improvement M_p	−73.91%	134.78%	-
t_s	2.105 s	7.81 s	2.69 s
Improvement t_s	−21.75%	190.33%	-
t_f	1.558 s	0.438 s	0.438 s
Improvement t_f	255.71%	0.00%	-
IAE	3553.53	3479.50	2409.30
Improvement IAE	-	-	-
TVu	1825.74	7396.00	3351.00
Improvement TVu	-	-	-

Figure 5a shows the results obtained for sudden set-point changes, both positive and negative, which show clear differences in the performance of the evaluated controllers. In positive changes, the SMC+Clegg controller proves to be the most effective, with overshoot reduced to 4.40% and the shortest settling time of 1.599 s, indicating fast convergence to the desired set point with greater stability compared to PID and SMC. While the PID shows a longer rise time (1.934 s) and a larger overshoot (10.80%), the SMC is characterized by steeper oscillations before stabilizing. This is reflected in the significant percentage improvements in response times for SMC+Clegg, positioning it as the fastest and most accurate option in scenarios with positive set-point changes.

For the negative changes, the SMC+Clegg once again demonstrates its superiority by offering shorter settling times and a better response to the chattering phenomenon. Although the PID has a shorter decay time (1.558 s), this benefit is overshadowed by its greater negative overshoot and lower overall stability. The SMC+Clegg, with a decay time reduced to 0.438 s, ensures fast adaptation without compromising stability, while the SMC struggles with oscillations and long settling times. These quantitative results reinforce that SMC+Clegg offers a better balance between response speed and robustness, making it suitable for applications that require high precision and stability against disturbances and sudden changes.

Figure 5b shows the control signal; the SMC+Clegg controller effectively minimizes oscillations and achieves the lowest IAE value (2409.30), indicating its superior ability to reduce accumulated error over time. However, while it demonstrates efficiency in maintaining precision, it is important to note that the PID controller achieves a lower TVu value (1825.74), suggesting that PID requires less overall control effort. Despite this, the SMC+Clegg controller strikes a more effective balance between precision and response speed, making it a robust choice for applications demanding high accuracy and stability in the presence of abrupt set-point changes.

3.3. Disturbance Testing

Figure 6 shows the speed response to disturbances and the corresponding output generated by the controller in reaction to these disturbances.

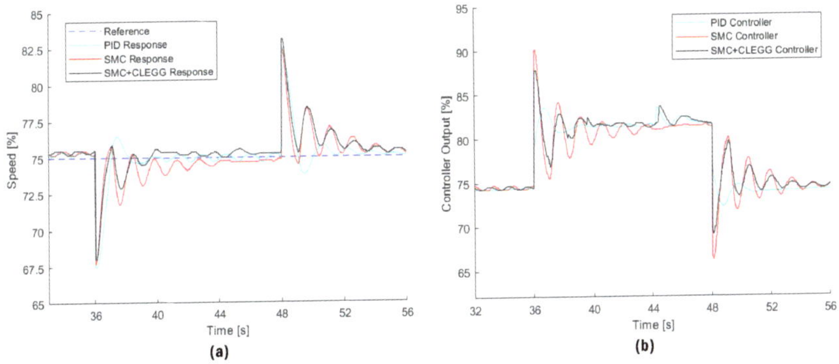

Figure 6. Results of disturbance test. (**a**) Speed of the three-phase motor, (**b**) controller output sent to the frequency variator.

Table 2 illustrates the various performance indexes and the comparisons with the proposed control that were observed during the disturbance tests.

Table 2. Disturbance test results.

Disturbance Test			
Metric	**PID**	**SMC**	**SMC+CLEGG**
Negative M_p	10.00%	9.60%	9.33%
Improvement M_p	7.14%	2.86%	-
t_s	1.512	3.334	1.97
Improvement t_s	−23.25%	69.24%	-
t_r	0.901	0.641	0.641
Improvement t_r	40.56%	0.00%	-
IAE	1423.04	1729.10	1262.49
Improvement IAE	-	-	-
TVu	1240.80	3089.00	2437.00
Improvement TVu	-	-	-

Figure 6a shows the response of the engine speed system in the presence of a disturbance. Three controllers are compared: PID, SMC (Sliding Mode Controller), and SMC+CLEGG. The PID controller shows a fast response, but with notable oscillations before stabilizing, with a negative Mp of 10% and an improvement of 7.14%. The SMC has a slower response, but with fewer oscillations at the beginning, obtaining a negative Mp of 9.60% and an improvement of 2.86%. Finally, the SMC+CLEGG offers an intermediate performance with a negative prominence of 9.33%, showing an improvement in the establishment time (ts) compared to the SMC, with a value of 1.97 s compared to the 3.334 s of the SMC.

Figure 6b analyzes the control signal generated by each controller in response to the disturbance. The PID controller requires quick and significant adjustments, which is reflected in a TVu of 1240.80 and an IAE of 1423.04. Although these values are high, they are lower than those obtained by the SMC, which presents a TVu of 3089.00 and an IAE of 1729.10, which indicates a greater effort in correcting the disturbance. The SMC+CLEGG, although it achieves a balance by reducing both the TVu (2437.00) and the IAE (1262.49), improving the traditional SMC, the PID still offers better performance in this particular test, with lower TVu and IAE values. It should be noted that in Figures 5 and 6, the speed does not stabilize at the set point but maintains a constant oscillation of approximately 0.5% in magnitude. This is due to electromagnetic noise from other elements connected in the laboratory and the vibrations of the mechanical coupling observed in Figure 2, effects that are unavoidable in the implementation of a real system.

4. Conclusions

In this work, a new sliding mode control proposal was designed and implemented on a PLC for speed control of an induction motor. Additionally, conventional PID and SMC were employed to compare the performance of the proposed control method using various performance indices.

The PID controller shows slower response times but is more stable initially, with less overshoot. The SMC provides faster response times but is characterized by higher overshoot and more chattering. The SMC+CLEGG controller effectively combines the speed of the SMC with improved stability, significantly reducing chattering due to the action of the Clegg integrator.

The SMC+CLEGG controller offers a smoother and more stable response compared to the SMC, minimizing chattering that could damage the actuators, thus prolonging their useful life. Although it has a higher TVu rating than the PID, it is still lower than the SMC. This is because sliding mode controllers allow the system to reach the set point more quickly, which requires more intense control actions. However, by reducing the windup effect of the conventional SMC, the SMC+CLEGG provides smoother control action. The SMC+CLEGG controller not only improves system accuracy and stability, but also contributes to greater energy efficiency. By minimizing oscillations and settling time, unnecessary motor work and therefore energy consumption is reduced. This is especially important in industrial applications where energy efficiency is a key factor in reducing operating costs and improving sustainability.

Author Contributions: Conceptualization, P.P.; methodology, P.P. and R.D.; software, C.C.; validation, R.D. and W.C.; formal analysis, R.D.; writing—original draft preparation, P.P.; writing—review and editing, J.Z., J.M. and W.C.; visualization, C.C.; supervision, P.P. All authors have read and agreed to the published version of the manuscript.

Funding: This research received no external funding.

Institutional Review Board Statement: Ethical review and approval were waived for this study due to the fact that no humans or animals were used.

Informed Consent Statement: Not applicable.

Data Availability Statement: Data are contained within the article.

Conflicts of Interest: The authors declare no conflicts of interest.

References

1. George, T.; Ganesan, V. Optimal tuning of PID controller in time delay system: A review on various optimization techniques. *Chem. Prod. Process. Model.* **2022**, *17*, 1–28. [CrossRef]
2. Alfaro, V.M.; Vilanova, R. PID control: Resilience with respect to controller implementation. *Front. Control. Eng.* **2022**, *3*, 1061830. [CrossRef]
3. Chehaidia, S.E.; Kherfane, H.; Cherif, H.; Boukhezzar, B.; Kadi, L.; Chojaa, H.; Abderrezak, A. Robust nonlinear terminal integral sliding mode torque control for wind turbines considering uncertainties. *IFAC-PapersOnLine* **2022**, *55*, 228–233. [CrossRef]
4. Hernandez-Barragan, J.; Rios, J.D.; Alanis, A.Y.; Lopez-Franco, C.; Gomez-Avila, J.; Arana-Daniel, N. Adaptive Single Neuron Anti-Windup PID Controller Based on the Extended Kalman Filter Algorithm. *Electronics* **2020**, *9*, 636. [CrossRef]
5. Memon, F.; Shao, C. An Optimal Approach to Online Tuning Method for PID Type Iterative Learning Control. *Int. J. Control Autom. Syst.* **2020**, *18*, 1926–1935. [CrossRef]
6. Lei, Z.; Zhou, Y. A kind of nonlinear PID controller for Refrigeration Systems based on Vapour Compression. *IFAC-PapersOnLine* **2018**, *51*, 716–721. [CrossRef]
7. Hasan, M.W.; Abbas, N.H. Disturbance Rejection for Underwater robotic vehicle based on adaptive fuzzy with nonlinear PID controller. *ISA Trans.* **2022**, *130*, 360–376. [CrossRef] [PubMed]
8. Liu, J.; Gao, Y.; Yin, Y.; Wang, J.; Luo, W.; Sun, G. Basic Theory of Sliding Mode Control. In *Sliding Mode Control Methodology in the Applications of Industrial Power Systems*; Springer International Publishing: Cham, Switzerland, 2020; pp. 11–25. [CrossRef]

9. Gambhire, S.J.; Ravi Kishore, D.; Londhe, P.S.; Pawar, S.N. Review of sliding mode based control techniques for control system applications. *Int. J. Dyn. Control* **2021**, *9*, 363–378. [CrossRef]
10. Li, S.; Yu, X.; Fridman, L.; Man, Z.; Wang, X. (Eds.) *Advances in Variable Structure Systems and Sliding Mode Control—Theory and Applications*; Springer International Publishing: Cham, Switzerland, 2018; Volume 115. [CrossRef]
11. Prado, A.J.; Herrera, M.; Dominguez, X.; Torres, J.; Camacho, O. Integral Windup Resetting Enhancement for Sliding Mode Control of Chemical Processes with Longtime Delay. *Electronics* **2022**, *11*, 4220. [CrossRef]
12. Abdedjebbar, T.; Zellouma, L.; Benchouia, M.T.; Krama, A. Adaptive linear neuron control of three-phase shunt active power filter with anti-windup PI controller optimized by particle swarm optimization. *Comput. Electr. Eng.* **2021**, *96*, 107471. [CrossRef]
13. Wang, X.; Gambier, A.; Vinagre, B.M. Fractional Order PID Control with Rate-limited Anti-windup for the Pitch System of Wind Turbines. In Proceedings of the 2020 IEEE Conference on Control Technology and Applications (CCTA), Montreal, QC, Canada, 24–26 August 2020; pp. 933–938. [CrossRef]
14. Díaz, R.; Camacho, O. A Comparison of Linear and Nonlinear PID Controllers Reset-Based for Nonlinear Chemical Processes with Variable Deadtime. In *Applied Technologies*; Botto-Tobar, M., Montes León, S., Camacho, O., Chávez, D., Torres-Carrión, P., Zambrano Vizuete, M., Eds.; Springer: Cham, Switzerland, 2021; Volume 1388, pp. 195–206. [CrossRef]
15. Gude, J.J.; García Bringas, P. Proposal of a General Identification Method for Fractional-Order Processes Based on the Process Reaction Curve. *Fractal Fract.* **2022**, *6*, 526. [CrossRef]
16. Borase, R.; Maghade, D.; Sondkar, S.; Pawar, S.N. A review of PID control, tuning methods and applications. *Int. J. Dyn. Control* **2021**, *9*, 818–827. [CrossRef]
17. Szczepanski, R.; Kaminski, M.; Tarczewski, T. Auto-Tuning Process of State Feedback Speed Controller Applied for Two-Mass System. *Energies* **2020**, *13*, 3067. [CrossRef]
18. Proaño, P.; Capito, L.; Rosales, A.; Camacho, O. A dynamical sliding mode control approach for long deadtime systems. In Proceedings of the 2017 4th International Conference on Control, Decision and Information Technologies (CoDIT), Barcelona, Spain, 5–7 April 2017; pp. 108–113. [CrossRef]
19. Camacho, O.; Smith, C.A. Sliding mode control: An approach to regulate nonlinear chemical processes. *ISA Trans.* **2000**, *39*, 205–218. [CrossRef] [PubMed]
20. Teel, A.R. Continuous-Time Implementation of Reset Control Systems. In *Trends in Nonlinear and Adaptive Control: A Tribute to Laurent Praly for his 65th Birthday*; Jiang, Z.P., Prieur, C., Astolfi, A., Eds.; Springer International Publishing: Cham, Switzerland, 2022; pp. 27–41. [CrossRef]
21. Moreno, J.C.; González, J.; Navarro, A.; Guzmán, J.L. New Tuning Rules of PI+CI Controllers for First-Order Systems. *Actuators* **2024**, *13*, 67. [CrossRef]

engineering proceedings

MDPI

Proceeding Paper

Impact of Distributed Generation Integration on Protection Devices: A Case Study in the CIGRE European Medium Voltage Network †

Verónica Rosero-Morillo [1,*], Sebastián Salazar-Pérez [2], F. Gonzalez-Longatt [3], Eduardo Salazar [1], Le Nam Hai Pham [4] and Eduardo Orduña [1]

[1] Institute of Electrical Energy (IEE), National University of San Jaun, San Juan 5400, Argentina; ejsalazar@irec.cat (E.S.); eorduna@iee-unsjconicet.org (E.O.)
[2] Faculty of Electrical and Electronic Engineering (FIEE), National Polytechnic School, Quito 170143, Ecuador; gabriel.salazar01@epn.edu.ec
[3] Digital Energy Systems Laboratory, Loughborough University, Loughborough LE11 3TU, UK; fglongatt@fglongatt.org
[4] Department of Electrical Engineering and ICT, University of South-Eastern Norway, 3918 Porsgrunn, Norway; le.pham@usn.no
* Correspondence: vrosero@iee.unsj.edu.ar
† Presented at the XXXII Conference on Electrical and Electronic Engineering, Quito, Ecuador, 12–15 November 2024.

Abstract: This study examines the impact of Inverter-Based Distributed Generators (IIDGs) on short-circuit currents detected by the main relay at the head of a radial feeder. It highlights how changes in fault currents induced by these inverter technologies can significantly affect the effectiveness and reliability of network protection systems. Key variables, such as the level of IIDGs penetration and the relative location of faults with respect to the relay, have been identified as influential factors. The significance of these findings lies in their contribution to a deeper understanding of how inverter fault responses and the integration of IIDGs alter fault currents. To mitigate the adverse effects associated with the insertion of IIDGs, a fault-detection and tripping strategy based on the inverse-time overcurrent curve is proposed. The suggested strategies not only improve fault detection accuracy but also ensure an appropriate response to variations in network conditions.

Keywords: European medium-voltage network; inverter fault response; inverter-based distributed generation; adaptive overcurrent relay

Citation: Rosero-Morillo, V.; Salazar-Pérez, S.; Gonzalez-Longatt, F.; Salazar, E.; Pham, L.N.H.; Orduña, E. Impact of Distributed Generation Integration on Protection Devices: A Case Study in the CIGRE European Medium Voltage Network. *Eng. Proc.* **2024**, 77, 9. https://doi.org/10.3390/engproc2024077009

Academic Editor: Jaime Cepeda

Published: 23 September 2024

1. Introduction

The increase in the penetration of IIDG in conventional distribution networks responds to the growing energy demand by leveraging their technical and economic benefits. IIDGs significantly contribute to improving energy quality, reliability, and the production of low-cost, carbon-free energy [1]. However, the inclusion of these generating units can alter the sensitivity of protection devices, especially overcurrent relays in feeders, which may suffer from partial to total blindness during faults [2].

To understand how distributed generators (DGs) modify fault characteristics and thereby develop or improve protection schemes, it is crucial to characterize and understand the response of DGs to such faults. This response varies depending on the type of DG and whether they are synchronous generators or inverter-based units. While the fault response of synchronous DGs has been widely studied, with uncontrolled fault currents that can reach between 5 and 10 times the generator's nominal current [3], inverter-based generators, being relatively new technologies, present a contribution to the fault current that is still under study and depends on the inverter control method. These IIDGs, during a fault, contribute a maximum of between 1.2 and 2 times the inverter's nominal current to prevent damage to electronic components, with the duration of this current injection varying depending on the severity of the voltage drop at the point of common coupling (PCC) [4].

In this context, a DG unit, regardless of its type, will contribute to the fault current and, no matter how insignificant this contribution may be, it will cause changes in the short-circuit current in other parts of the network, thus altering traditional short-circuit patterns [4]. The main negative effects impacting protections include reverse power flow, sympathetic tripping, and protection blindness, among others. Specifically, this article focuses on protection blindness, a phenomenon in which the fault current perceived by the main relay is reduced due to the connection of the IIDG, thus causing a delay in the operation of the protection or, in more severe cases, completely disabling the relay's ability to detect faults. The extent of this current reduction is influenced by certain factors, such as the location of the DG, its size, and the relative location of the fault. These factors are analyzed both analytically and through simulations by the authors of [5,6].

The literature has proposed solutions to mitigate the negative effects on protections, including protection blindness by immediately disconnecting DGs during a fault, modifying the protection system through the installation of new switches, changing distance relays, and using directional overcurrent or differential relays. Proposals also include limiting the capacity of the DG, installing fault-current limiters (FCLs), and adaptive protection schemes, among others. While these measures are partially effective, they present certain limitations. Disconnecting DGs can cause severe voltage drops and stability issues if the disconnection is massive. Also, disconnecting DGs during temporary faults is not economically beneficial [7]. Modifying the protection scheme increases costs and complicates coordination among devices. Limiting the capacity of IIDGs clearly does not represent a viable solution, as it contradicts incentive policies for transitioning to a more sustainable energy matrix and underutilizes an economical resource that should be increasingly exploited [8]. On the other hand, implementing FCLs to preserve the original relay settings requires a detailed analysis of the optimal impedance and the appropriate location of the FCLs. However, a major drawback is their high additional cost, which is undesirable for both utility companies and DG owners. These challenges highlight the advantages of another alternative, such as adaptive adjustments in overcurrent protection. The adaptive protection of the overcurrent relay has been extensively researched for its ability to adjust to variations induced by DGs, thus effectively addressing the adverse effects of their inclusion in protection systems, such as protection blindness and loss of coordination among different protection devices. This adaptability allows for managing fluctuations in fault currents without the need to replace existing protection systems, an option that would be economically unfeasible [9]. Based on a review of the existing literature, this document proposes a strategy to mitigate the blindness of relays in the face of the increasing penetration of IIDGs. This article contributes the following proposals:

- An adaptive adjustment of the pickup current in overcurrent relays is proposed in response to changes associated with the connection and disconnection of IIDGs. This adjustment, by increasing the current pickup multiplier above one, ensures that the relay maintains its sensitivity and is capable of detecting even the smallest faults in the system, thus mitigating delays in response time and protection blindness.
- A protection scheme specifically designed to combat protection blindness induced by the integration of IIDGs is introduced. This approach is based on modifying the current pickup multiplier used to calculate the tripping time of the protection. By adjusting this single variable, the variability of the penetration of IIDGs into the network, as well as their location and proximity to faults, is taken into account. This modification significantly improves the applicability of the scheme in digital overcurrent relays in emerging systems.

2. Fundamentals of the Inverse-Time Overcurrent Relay

The inverse-time overcurrent relay (OCR) serves as the primary protection in radial distribution feeders. It operates based on the principle that the relay's operation time is inversely proportional to the magnitude of the fault current flowing through it. Various families of inverse-time curves have been defined by international standards; their analytical expression and corresponding coefficients for the IEC 60255 and ANSI/IEEE standards are shown in Equation (1) and reference [10], [11] and [12], respectively. Here, t_{op} is the operation time of the relay, and TDS represents the time dial setting. I_{sc} is the fault current measured by the relay, while I_{pickup} is the relay's setting for fault detection. The constants A, B, and ρ are specified by international standards for each type of inverse-time curve.

$$t_{op} = \left[\frac{A}{\left(\frac{I_{sc}}{I_{pickup}} \right)^{\rho} - 1} + B \right] \times TDS \quad (1)$$

Inverse-time overcurrent relays (TOC) are configured with principal settings: (1) I_{pickup}, which defines the minimum fault current necessary to activate the relay, and (2) the TDS, which, along with the type of curve, establishes the relay's response time to a fault based on coordination with other protection devices. In traditional relays, TDS values can range from 0.5 to 10. The I_{pickup} setting is determined based on the maximum fault current and load current, thus ensuring that the protection's selectivity is maintained under any operational condition. Minimum fault current values are used solely to verify that the relay can detect the smallest fault within the protected zone. Equation (2) mathematically describes I_{pickup}, where I_{load} represents the maximum load current and k is a constant that prevents the relay from activating under normal conditions, such as temporary overloads, power transfers, and errors in current transformers (CTs). The value of k ranges between 1.5 and 2.

$$I_{pickup} = k \times I_{load} \quad (2)$$

The TOC serves as primary protection for its line section and as backup protection for the adjacent section, as illustrated in Figure 1. This implies that the relay's reach must extend from the minimum fault current at the remote end of the adjacent section to the maximum current at the point of installation of the head relay. However, establishing this range presents challenges because the I_{pickup}, calculated as a multiple of the load current, introduces uncertainty into the precise definition of the relay's reach, especially in modern networks where the renewable resource is variable.

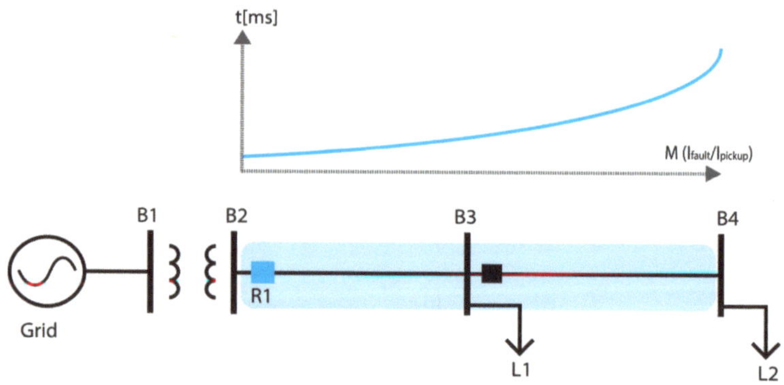

Figure 1. Reach of the overcurrent relay.

To mitigate the uncertainty in the relay's reach, the OC is analyzed using the pickup current multiplier (M). This approach ensures that the relay adequately responds to both significant and subtler faults, providing comprehensive and effective network protection.

From Equation (1), the operation time of the relay can also be expressed in terms of M, which is the ratio between I_{sc} and I_{pickup}, as specified in Equation (3).

$$t_{op} = \left[\frac{A}{(M)^{\rho} - 1} + B\right] \times TDS \quad (3)$$

In analyzing the multiplier M, it is considered that a fault is detected when Equation (2) is satisfied, thus defining the threshold that activates the relay, $M_{threshold}$, as shown in Equation (4). It is established that a fault condition exists when the threshold exceeds 1, while the system is considered to be operating normally if the threshold is less than 1. This allows for the graphical representation of the relay's curve as a function of time and M. Furthermore, a lower limit of 1.5 is set in Equation (4) based on the observation that, as illustrated in Figure 2, operation times for an M between 1 and 1.5 are excessively prolonged and, therefore, ineffective for protection purposes. Consequently, a fault is declared when M is greater than or equal to the threshold $M_{threshold}$ = 1.5 [13].

$$M_{threshold} = \frac{I_{sc}}{I_{pickup}} = \frac{(I_{sc} \mid I_{sc} = k \times I_{load})}{k \times I_{load}} = 1 \quad (4)$$

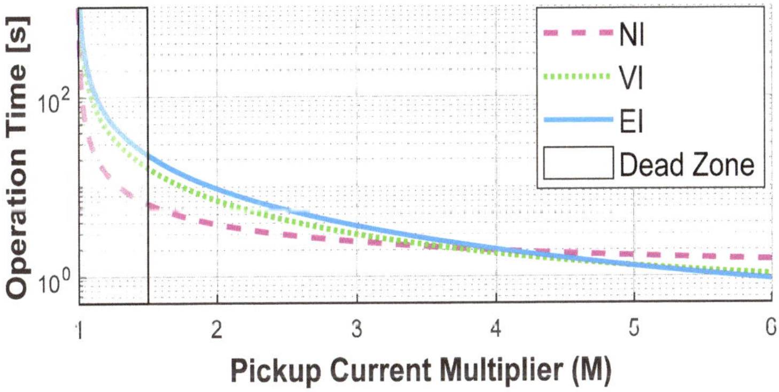

Figure 2. Operational time curves based on M vs. time.

3. Modeling

To assess the impact of IIDG on the protection at the feeder head, the well-documented CIGRE European MV distribution network was selected. This is a three-phase, symmetrical, and balanced network, whose single-line diagram is shown in Figure 3. The network includes main three-phase feeders with a radial configuration. This model was chosen for its adaptability and relevance to studies related to the integration of IIDGs [14]. In this analysis, IIDGs were incorporated at buses 5, 9, and 7 of the system, with capacities of 0.7, 0.5, and 1.5 MW, respectively, allowing for different levels of IIDG penetration at 16%, 33%, 44%, 50%, and 60% with the different combinations of the IIDGs.

The IIDG was modeled using a three-phase, two-level Voltage Source Converter (VSC) with an LC output filter, which includes a fault response model with injection of negative and positive sequence currents for unbalanced and balanced faults, respectively. The interconnection of the IIDG and the AC network is made through a delta-star step-up transformer, supplying the system with powers of 0.7, 0.5, and 1.5 MW. The IIDG model implemented in MATLAB–Simulink (https://www.mathworks.com/products/matlab.html, accessed on 6 September 2024) is illustrated in Figure 4, and details of its modeling are discussed by the authors in references [5,6].

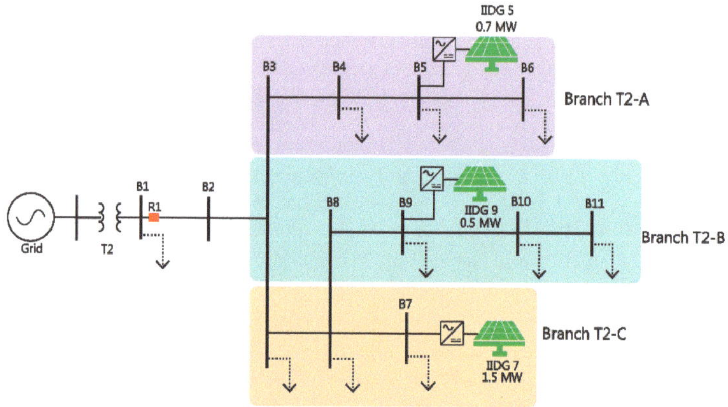

Figure 3. European CIGRE MV network model.

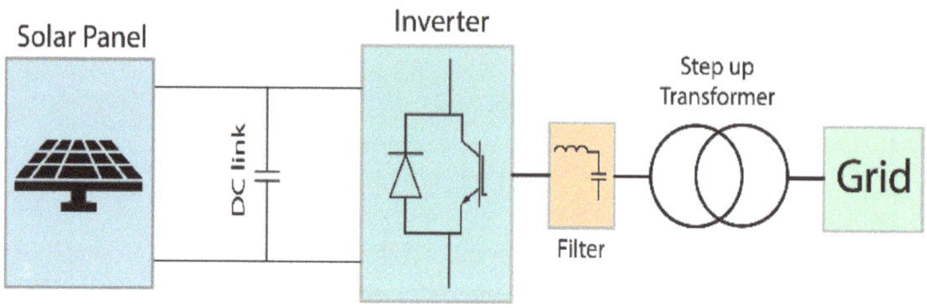

Figure 4. IIDG model.

4. Effects of Inclusion of IIDGs on the Overcurrent Relay

To evaluate the impact of various topological changes, such as the connections and disconnections of distributed generators, on the fault current perceived by the head relay, a sensitivity study was conducted on the CIGRE medium-voltage network with IIDG integration. This study considered three scenarios: pre-fault conditions for each level of IIDG penetration, fault conditions without IIDGs, and fault conditions with various levels of IIDG penetration. The results of this analysis provided the foundation for making adaptive adjustments to the head relay, which is essential for maintaining the effectiveness of the protection system amidst changes in network topology due to the integration of renewable generation sources.

4.1. Pre-Fault Condition

In this initial pre-fault scenario, a reduction in the RMS current magnitudes detected by the head relay is observed as the penetration of IIDGs increases. This trend is evident in Table 1, where the highest current is recorded without the connection of IIDGs and the lowest is recorded with 60% penetration. Although this pattern is consistent across all three phases, for illustrative purposes, the variation of phase A is graphically presented in Figure 5.

Table 1. Pre-fault current magnitudes for different levels of penetration.

Phases	Penetration Level					
	0%	**16%**	**33%**	**44%**	**50%**	**60%**
Ia [A]	132.865	114.472	94.9389	82.6763	78.7549	66.5616

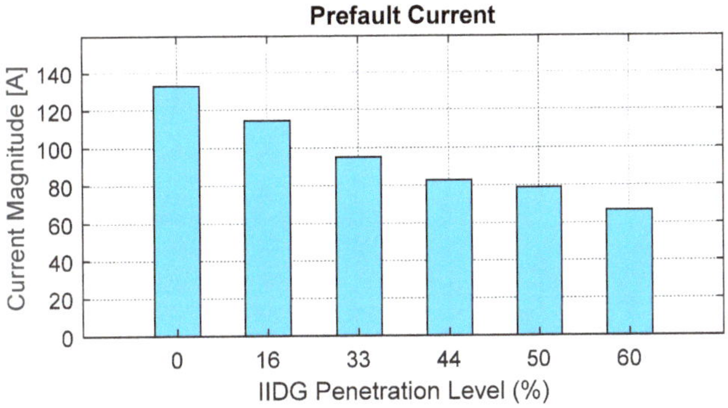

Figure 5. Variation of pre-fault current for different levels of penetration—phase A.

Given the fluctuation of currents due to the connection or disconnection of IIDGs, it is crucial to adaptively adjust the I_{pickup} based on the current state of the network instead of using a fixed threshold, as is customary in networks without IIDGs. As illustrated in Table 1, if the I_{pickup} is set at twice the pre-fault current—resulting in 265.73 A—a $M_{threshold}$ of 1.5 is obtained. However, maintaining a constant I_{pickup} results in significant deficiencies when faced with variations in IIDG penetration; for example, at a penetration level of 16%, the calculation of $M_{threshold}$ is $(1.5 \times 2 \times 114.472)/(2 \times 132.865) \approx 1.29$, which is clearly below the value of 1.5. Similarly, $M_{threshold}$ values for penetration levels of 33%, 44%, 50%, and 60% are 1.07, 0.93, 0.89, and 0.75, respectively. This indicates that a fixed I_{pickup} value for different network states causes relay underreach or blindness, which worsens with increased DG penetration, as shown in Figure 6, where the $M_{threshold}$ for all levels of penetration falls below 1.5.

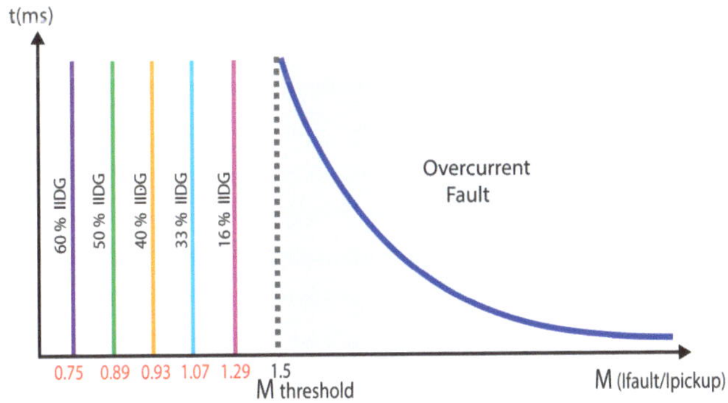

Figure 6. Effect of considering a fixed I_{pickup} for different levels of penetration.

An effective solution to this challenge involves adapting the I_{pickup} based on the load current corresponding to the current state of the system (connection or disconnection of IIDGs). This ensures that the $M_{threshold}$ consistently remains at 1.5, allowing the protection system to adequately respond to the smallest fault in all topological states of the network. In this way, the overcurrent curve will always operate with an $M_{threshold}$ of 1.5, regardless of the connections or disconnections of the IIDGs. This adaptive adjustment resolves the issue of underreach or blindness of protections, a common consequence of the inclusion of IIDGs, and enhances the reliability and effectiveness of the protection system in the evolving network.

4.2. Fault Current Behaivior Without IIDGs

For this analysis, simulations of various short-circuit fault conditions (1φ, 2φ-g, and 3φ) were conducted along feeder T2. The fault currents recorded by the head relay at each fault location are detailed in Table 2 for branches T2-A, T2-B, and T2-C. The results indicate that the fault current decreases as the fault location moves away from the feeder head, a trend consistently observed across all phases affected by the different types of simulated faults. To illustrate this behavior, Figure 7 shows the three-phase currents detected by the head relay for a three-phase fault with a fault resistance of 0.01 Ω at different points along branch T2-C. This phenomenon is attributed to the increased distance between the substation and the fault point, which increases the total impedance of the circuit and, consequently, reduces the magnitude of the fault current.

Table 2. Fault currents observed by the header relay [A]—0% DG.

Location			B1	B2	B3	B4	B5	B6	B8	B9	B10	B11	B7
Branch T2-A			X	X	X	X	X	X					
Branch T2-B			X	X	X				X	X	X	X	
Branch T2-C			X	X	X				X				X
Types of Faults	1φ	B	630.999	572.573	493.159	484.914	477.735	459.736	475.29	471.225	462.066	458.373	456.564
	2φ-g	B	5258.74	2568.59	1406.18	1322.36	1253.91	1098.66	1239.13	1203.97	1127.15	1097.25	1075.72
		C	4456.26	2067.22	1128.54	1064.93	1013.38	897.682	1001.65	975.229	917.937	895.774	880.627
	3φ	A	5635	2677.5	1439.27	1352.58	1282.11	1123.43	1266.55	1230.43	1151.91	1121.47	1100.04

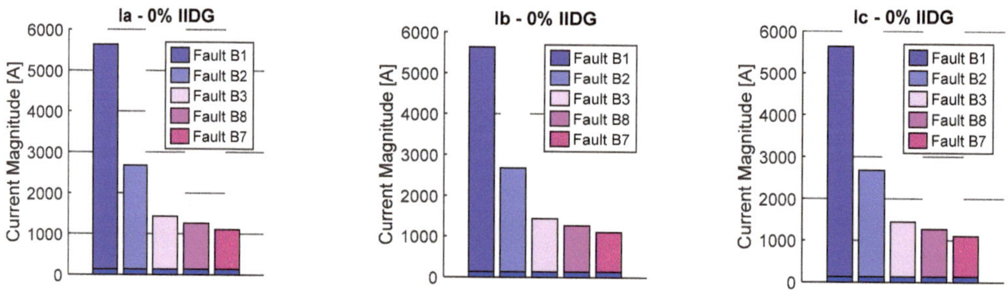

Figure 7. Three-phase fault—branch T2-C—0% IIDG penetration level.

4.3. Fault Current Behavior with IIDGs

This study examines the influence of the fault location on the current detected by the head relay while maintaining a constant level of IIDG penetration. Branches T2-A, T2-B, and T2-C of the feeder were analyzed, as shown in Figure 3. For each branch, the amplitudes of the fault currents observed by the head relay were measured, revealing that the magnitude of the fault current decreases as the fault point moves away from the feeder head. This phenomenon was consistently manifested across all branches, levels of DG penetration, and types of faults, which is similar to observations in systems without DGs. The current results for branch T2-C, with a 50% DG penetration, and the decreasing trend for three-phase faults are shown in Figure 8, while Table 3 presents the current results for feeder T2-C with a 50% IIDG penetration. Additionally, it was noted that short-circuit currents are lower in scenarios with DGs compared to those without DGs. The minimum current detected by relay R1 occurred with the maximum DG penetration, while the maximum fault current was recorded in the absence of DGs, especially near the substation during a three-phase fault.

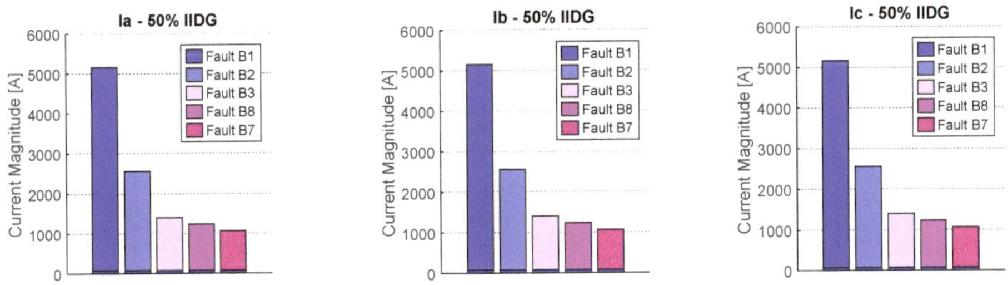

Figure 8. Three-phase fault—branch T2-C—50% IIDG penetration level.

Table 3. Fault currents observed by the header relay [A]—50% DG.

Location			B1	B2	B3	B4	B5	B6	B8	B9	B10	B11	B7
Branch T2-A			X	X	X	X	X	X					
Branch T2-B			X	X	X				X	X	X	X	
Branch T2-C			X	X	X				X				X
Types of Faults	1φ	B	571.457	521.264	460.37	451.874	444.521	423.41	444.201	439.455	428.853	424.559	428.05
	2φ-g	B	4849.18	2473.65	1381.71	1298.96	1231.32	1074.83	1219.27	1183.69	1105.8	1075.53	1059.65
		C	4047.37	1967.81	1096.48	1036.28	987.186	875.458	975.944	950.821	895.669	874.151	859.481
	3φ	A	5157.78	2561.66	1403.93	1318.43	1248.29	1087.3	1235.91	1199.19	1119.55	1088.43	1073.1

Considering the calculation of the I_{pickup} multipliers, using Equation (4) and considering the adaptive I_{pickup} depending on each level of DG penetration as detailed in study 1 of this section, results were obtained as shown in Table 4 for simulated faults in feeder T2-C at the 50% level of renewable generator penetration. From the results obtained, it can be concluded that regardless of the level of IIDG penetration in the network, the fault current will tend to decrease as it moves away from the feeder head, thereby maintaining the concept of the inverse-time overcurrent curve, as faults further from the substation will have a longer operation time of the relay-associated breaker, while for faults closer to the feeder head, this time decreases. On the other hand, as reflected in Table 5, the data indicate that the fault current perceived by the head relay significantly decreases as the level of IIDG penetration increases, reaching the lowest value with 60% penetration. This suggests that a higher level of IIDG penetration leads to reduced fault currents, thus extending the operation time on the overcurrent curve and allowing the network to tolerate these currents for an extended period. In contrast, 0% IIDG penetration results in the maximum fault currents, which require a quicker relay response to ensure adequate network protection, as shown in Figure 9.

Table 4. I_{pickup} Multiplier—50% DG.

Location			B1	B2	B3	B4	B5	B6	B8	B9	B10	B11	B7
Branch T2-A			X	X	X	X	X	X					
Branch T2-B			X	X	X				X	X	X	X	
Branch T2-C			X	X	X				X				X
Types of Faults	1φ	B	5.44	4.96	4.38	4.30	4.23	4.03	4.23	4.19	4.08	4.04	4.08
	2φ-g	B	46.18	23.56	13.16	12.37	11.73	10.24	11.61	11.27	10.53	10.24	10.09
		C	38.54	18.74	10.44	9.87	9.40	8.34	9.29	9.05	8.53	8.32	8.19
	3φ	A	49.12	24.40	13.37	12.56	11.89	10.35	11.77	11.42	10.66	10.37	10.22

Table 5. M_{max} calculated for each level of DG penetration.

	0% IIDG	16% IIDG	33% IIDG	44% IIDG	50% IIDG	60% IIDG
I pre-fault [A]	132.87	114.47	94.94	82.68	78.75	66.56
I pickup [A]	265.73	228.94	189.88	165.35	157.51	133.12
I fault max [A]	5635.00	5461.90	5309.78	5204.11	5157.78	5030.21
M_{max} [pu]	31.81	35.79	41.95	47.21	49.12	56.7

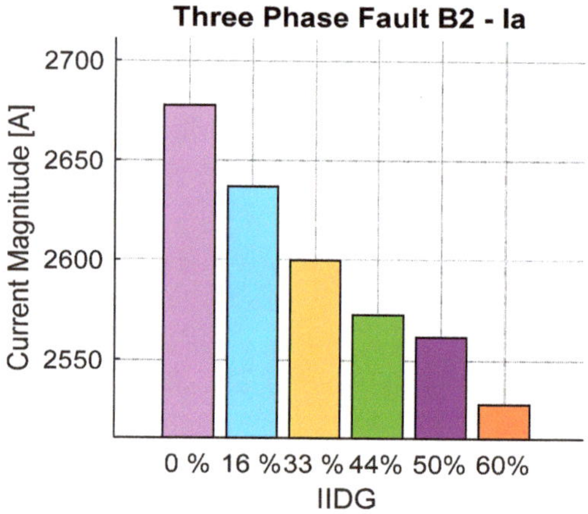

Figure 9. Current magnitude for different levels of penetration.

In this context, the multiplier M, detailed in Table 5, should tend to be lower for higher levels of IIDG penetration, while higher values of M are expected in scenarios without IIDG inclusion. Contrary to expectations, the simulation results show that M is lower when the fault current is at its maximum, implying a delay in the circuit breaker's operation in the presence of high and severe currents, which should be interrupted almost instantaneously to protect the network. Additionally, M increases as the level of IIDG penetration increases, resulting in faster activation for lower currents. This behavior is contrary to what is desirable according to the logic of a typical overcurrent curve, where faults with high currents are expected to trigger faster actions, while faults with lower currents allow for a longer time interval before the breaker intervenes, as seen in Table 5 and Figure 10.

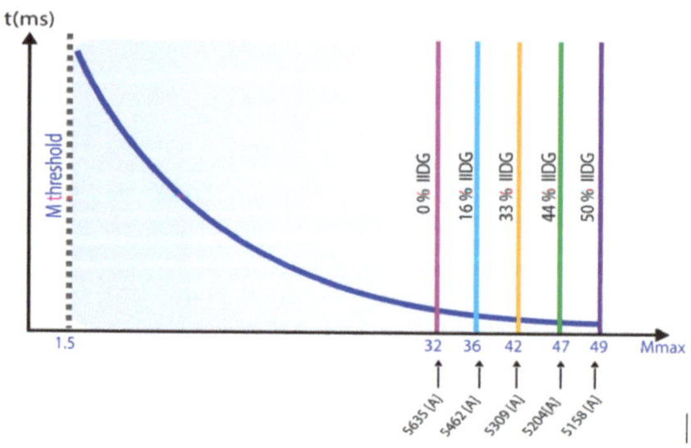

Figure 10. Overcurrent curve—M_{max}.

Therefore, it is evident that merely adjusting the pickup current for each system state—including the various states of connection or disconnection of the IIDGs—while effectively resolving the issues of protection blindness and delay in tripping also introduces a complication, as the M calculated for a level of penetration is greater than for a system without IIDGs, which contradicts the conventional overcurrent protection theory. To address this discrepancy, the M used for fault detection is adjusted. The mathematical expression for this adjustment is based on the premise that the maximum multiplier (M_{max}) is established from a simulated three-phase fault at the relay location for the scenario without connected DGs, as shown in Equation (5).

$$M_{max} = \frac{1.5 \times I_{fault\_3\varphi}^{0\%\ IIDG\_B1}}{2 \times I_{prefault}^{0\%\ IIDG}} \quad (5)$$

From M_{max}, it is inferred that the M calculated for variations in IIDG penetration will be lower than this value. Thus, with the increase in IIDG penetration, any fault along the feeder will result in a lower current magnitude compared to M_{max}. Consequently, all of the trip multipliers (M_{trip}) are calculated based on the maximum pickup current, as per Equation (6).

$$M_{trip} = \frac{I_{fault}}{I_{fault\_3\varphi}^{0\%\ IIDG\_B1}} \times M_{max} + M_{pickup} \quad (6)$$

Tables 6 and 7 present the results of the M_{trip} calculations obtained from Equations (5) and (6), respectively. These calculations were performed for two fault locations on the T2-C branch: the fault at B1, which is closest to the substation, and the fault at B7, which is the furthest away. The purpose of these calculations is to verify the proper functionality of the proposed current multiplier, thus ensuring that the calculated values exceed 1.5. This guarantees that regardless of the type and location of the fault, the relay is sufficiently sensitive to detect faults under variations in the DGs.

As observed in Table 6, the multipliers calculated from Equation (5), which only considers the adjustment of the adaptive pickup current without the proposed modification, were found to be greater than 1.5. However, an erroneous trend is identified in these values, as the multiplier increases in a way that is directly proportional to the increase in the DG penetration level. This is conceptually incorrect, as a higher multiplier value implies a reduced operation time theoretically for higher currents. However, as demonstrated throughout this article, the higher the level of DG penetration, the lower the current perceived by the relay, suggesting that the multiplier should decrease as the level of penetration increases, a trend not observed in Table 6.

Table 6. M_{trip}—fault at B1 and B7—Equation (5).

IIDG	Fault B1				Fault B7			
	1φ	2φ-g		3φ	1φ	2φ-g		3φ
	b	b	c	a	b	b	c	a
0%	3.56	29.68	25.15	15.11	2.58	6.07	4.97	6.21
16%	4.01	33.48	28.22	17.28	2.90	6.99	5.72	7.12
33%	4.66	39.34	33.00	20.54	3.47	8.44	6.85	8.58
44%	5.23	44.33	37.05	23.34	3.93	9.65	7.82	9.78
50%	5.44	46.18	38.54	24.40	4.08	10.09	8.19	10.22
60%	6.29	53.37	44.35	28.48	4.75	11.87	9.61	11.99

Table 7. M_{trip}—fault at B1 and B7—Equation (6).

IIDG	Fault B1				Fault B7			
	1φ	2φ-g		3φ	1φ	2φ-g		3φ
	b	b	c	a	b	b	c	a
0%	5.06	31.18	26.65	33.31	4.08	7.57	6.47	7.71
16%	4.95	30.35	25.82	32.33	4.00	7.52	6.43	7.64
33%	4.83	29.61	25.08	31.47	3.98	7.53	6.39	7.63
44%	4.76	29.08	24.56	30.88	3.94	7.50	6.37	7.59
50%	4.73	28.87	24.35	30.61	3.92	7.48	6.35	7.56
60%	4.65	28.23	23.72	29.89	3.88	7.45	6.32	7.51

On the other hand, incorporating the proposed variation in Equation (6), M_{trip} is determined for faults at bars B2 and B7 for different levels of penetration, as indicated in Table 7. The results clearly show that the pickup multiplier decreases as the level of penetration increases, corroborating the premise that faults occurring under an increasing level of penetration produce lower fault currents and, therefore, require a longer operation time. Analyzing the two fault locations—the one closest to the substation at bar B2 and the farthest at B7—it is observed that the M_{trip} calculated are lower for the fault farther from the head, while they increase for the fault closer to the substation. This confirms that the adjustment of the multiplier adequately considers the dynamic variation in the magnitude of the feeder current, which is influenced by various factors, such as different levels of IIDG penetration and the location of the fault.

5. Discussion and Conclusions

This study explores the impact of integrating IIDGs on the short-circuit currents detected by the main relay at the head of a radial feeder. An adaptive fault detection and tripping strategy based on the inverse-time overcurrent curve is proposed, which not only improves the accuracy of fault detection but also ensures an appropriate response to the connection and disconnection variations of the IIDGs.

The results indicate that regardless of the level of IIDG penetration, the fault current decreases as it moves away from the feeder head, maintaining the principle of the overcurrent curve, which stipulates longer operation times for distant faults and reduced times for closer faults. Additionally, it was observed that as the level of IIDG penetration increases, the fault current perceived by the relay significantly decreases, which was reflected in lower values of the pickup current multiplier when the fault current was at its maximum, causing a delay in the operation of the protection system. Conversely, an increase in IIDG penetration resulted in faster activation for lower currents.

In response to this behavior, the calculation of the pickup current multiplier was adjusted to ensure that regardless of the network state, the adjusted overcurrent curve maintains a threshold of 1.5, thus facilitating effective fault detection. Once a fault is detected, the M_{trip} parameter is calculated to determine the operating time of the breaker associated with the relay. This ensures that the protection system acts correctly, thus providing a rapid response in the presence of high currents and allowing for a longer interval before breaker intervention for lower currents.

Author Contributions: Conceptualization, V.R.-M., F.G.-L. and E.O.; methodology, V.R.-M., F.G.-L. and E.O.; software, V.R.-M.; validation, V.R.-M., F.G.-L. and E.O.; writing—original draft preparation, V.R.-M., L.N.H.P., S.S.-P. and E.S.; writing—review and editing, V.R.-M., L.N.H.P., S.S.-P. and E.S.; visualization, V.R.-M.; supervision, F.G.-L. and E.O. All authors have read and agreed to the published version of the manuscript.

Funding: This research was supported by the German Academic Exchange Service with a scholarship for a Ph.D. scholarship of Funding programme/-ID: Third Country Programme Latin America, 2020 (57519668).

Institutional Review Board Statement: Not applicable.

Informed Consent Statement: Not applicable.

Data Availability Statement: Not applicable.

Conflicts of Interest: The authors declare no conflicts of interest.

References

1. Salazar, E.J.; Rosero, V.; Gabrielski, J.; Samper, M.E. Demand response model: A cooperative-competitive multi-agent reinforcement learning approach. *Eng. Appl. Artif. Intell.* **2024**, *133*, 108273. [CrossRef]
2. Rizwan, M.; Gao, C.; Yan, X.; Ahmad, S.; Zaindin, M. An approach to disparage the blindness of backup protection in grid connected renewable energy sources system by inducing artificial fault current. *Int. J. Electr. Power Energy Syst.* **2023**, *153*, 109185. [CrossRef]
3. Meskin, M.; Domijan, A.; Grinberg, I. Impact of distributed generation on the protection systems of distribution networks: Analysis and remedies—Review paper. *IET Gener. Transm. Distrib.* **2020**, *14*, 5816–5822. [CrossRef]
4. Norshahrani, M.; Mokhlis, H.; Bakar, A.H.A.; Jamian, J.J.; Sukumar, S. Progress on protection strategies to mitigate the impact of renewable distributed generation on distribution systems. *Energies* **2017**, *10*, 1864. [CrossRef]
5. Rosero-morillo, V.A.; Orduña, E.; Riquelme-dominguez, J.M. Evaluation of Grid-Following Inverter Control Models for Fault Response and their Impact on Protection Devices. In Proceedings of the 2024 IEEE 22nd Mediterranean Electrotechnical Conference (MELECON), Porto, Portugal, 25–27 June 2024; pp. 467–472. [CrossRef]
6. Rosero-Morillo, V.A.; Gonzalez-Longatt, F.; Riquelme-Dominguez, J.M.; Orduña, E. Impact of Inverter-Based Distributed Generation Integration on Current Seen by Overcurrent Relay: CIGGRE European MV Network Case. In Proceedings of the 2023 IEEE 41st Central America and Panama Convention (CONCAPAN XLI), Tegucigalpa, Honduras, 8–10 November 2023; pp. 1–6.
7. Vijitha, K.; Selvan, M.P.; Raja, P. Short circuit analysis and adaptive zonal protection of distribution system with distributed generators. In Proceedings of the 2015 International Conference on Energy, Power and Environment: Towards Sustainable Growth (ICEPE), Shillong, India, 12–13 June 2015. [CrossRef]
8. Telukunta, V.; Pradhan, J.; Agrawal, A.; Singh, M.; Srivani, S.G. Protection challenges under bulk penetration of renewable energy resources in power systems: A review. *CSEE J. Power Energy Syst.* **2017**, *3*, 365–379. [CrossRef]
9. Uma, U.U.; Nmadu, D.; Ugwuanyi, N.; Ogah, O.E.; Eli Chukwu, N.; Eheduru, M.; Ekwue, A. Adaptive overcurrent protection scheme coordination in presence of distributed generation using radial basis neural network. *Prot. Control Mod. Power Syst.* **2023**, *8*, 63. [CrossRef]
10. Kavi, M.; Mishra, Y.; Vilathgamuwa, M. Morphological Fault Detector for Adaptive Overcurrent Protection in Distribution Networks with Increasing Photovoltaic Penetration. *IEEE Trans. Sustain. Energy* **2018**, *9*, 1021–1029. [CrossRef]
11. International Electrotechnical Commission. IEC Standard for Measuring Relays and Protection Equipment—Part 151: Functional Requirements for Over/Under Current Protection. IEC 60255-151. 2009. Available online: https://webstore.iec.ch/en/publication/1166 (accessed on 28 August 2009).
12. IEEE Standard for Inverse-Time Characteristics Equations for Overcurrent Relays. 2018. Available online: https://ieeexplore.ieee.org/document/8635630 (accessed on 5 February 2019).
13. Enríquez, A.C. *Overcurrent Relay Advances for Modern Electricity Networks*; Elsevier: Amsterdam, The Netherlands, 2022.
14. Rosero-morillo, V.A. Advanced Grid–Tied Inverter Modeling in Matlab-Simulink for Protection Studies in Modern Distribution Networks. In Proceedings of the IEEE PES General Meeting, Seattle, WA, USA, 21–25 July 2024.

engineering proceedings

MDPI

Proceeding Paper

Anxiety Detection Using Consumer Heart Rate Sensors †

Soraya Sinche *, Jefferson Acán and Pablo Hidalgo

Departamento de Electrónica, Telecomunicaciones y Redes de Información (DETRI), Escuela Politécnica Nacional, Quito 170525, Ecuador; jefferson.acan@epn.edu.ec (J.A.); pablo.hidalgo@epn.edu.ec (P.H.)

* Correspondence: soraya.sinche@epn.edu.ec

† Presented at the XXXII Conference on Electrical and Electronic Engineering, Quito, Ecuador, 12–15 November 2024.

Abstract: Increasingly, humans are exposed to different activities at work, at home, and in general in their daily lives that generate episodes of stress. In many cases, these episodes could produce disorders in their health and reduce their quality of life. For this reason, it is crucial to implement mechanisms that can detect stress in individuals and develop applications that provide feedback through various activities to help reduce stress levels. Physiological parameters, such as galvanic skin response (GSR) and heart rate (HR) are indicative of stress-related changes. There exist methodologies that use wearable sensors to measure these stress levels. In this study, a sensor of blood volume pulse (BVP) and an electrocardiography (ECG) sensor were utilized to obtain metrics like heart rate variability (HRV) and pulse arrival time (PAT). Their features were extracted, processed, and analyzed for anxiety detection. The classification performance was evaluated using decision trees, a support vector machine (SVM), and meta-classifiers to accurately distinguish between "stressed" and "non-stressed" states. We obtained the best results with the SVM classifier using all the features. Additionally, we found that the ECG AD8232 sensor provided more reliable data compared to the photoplethysmography (PPG) signal obtained from the MAX30100 sensor. Therefore, the ECG is a more accurate tool for assessing emotional states related to stress and anxiety.

Keywords: anxiety detection; stress recognition; ECG; PPG; heart rate variability

1. Introduction

Anxiety and stress are common in daily life and can impact performance. According to the World Health Organization (WHO), depression and anxiety cases have increased by 50% from 1990 to 2013 [1]. The WHO has also recommended investing in mental health treatment to enhance both individual well-being and economic growth [1]. Technology allows for the monitoring of physical and mental health through Quality-of-Life Technology (QoLT) [2,3], we can use devices like the MAX30100 oximeter (Maxim Integrated, San Jose, CA, USA) and AD8232 (Analog Devices, Wilmington, MA, USA) ECG sensors to assess anxiety. Physiological indicators such as heart rate (HR) [2–6], blood pressure (BP) [7,8], and heart rate variability (HRV) [6,8] can reflect stress levels.

This article analyzes students' behavior before a test or evaluation about their stress and anxiety levels, using ECG and BVP sensors, with data collected by an ESP32 (Espressif, Shanghai, China). The ESP32 was chosen for its ability to handle multiple sensors and ADC converters with a higher resolution, facilitating connection and accuracy in reading. Twenty volunteers were monitored for five minutes before and after an evaluation event, and features were extracted from ECG and PPG signals using Matlab 2018B.

Section 2 discusses heart rate variability and its link to anxiety and reviews the features extracted from the ECG and PPG signals. Section 3 describes the data collection, while Section 4 explains the methodology. Finally, Section 5 presents the results and discussion.

Citation: Sinche, S.; Acán, J.; Hidalgo, P. Anxiety Detection Using Consumer Heart Rate Sensors. *Eng. Proc.* **2024**, 77, 10. https://doi.org/10.3390/engproc2024077010

Academic Editor: Jackeline Abad

Published: 18 November 2024

2. Analysis of Anxiety with Custom Sensors

Affective states can be analyzed by monitoring physiological signals, reflecting the influence of the Autonomic Nervous System (ANS) on respiratory and cardiovascular functions [8].

The ANS regulates involuntary activities like heart contraction and blood pressure [3] and consists of the parasympathetic nervous system (PNS) and the sympathetic nervous system (SNS) [7]. The PNS fosters a "rest and repose" state, while the SNS triggers a "fight or flight" response [8]. Stress increases SNS activity and decreases PNS activity [3,7]. Both HR and HRV are affected by these systems [3].

This study examines heart activity and oximetry responses to stress using the MAX30100 oximeter (Maxim Integrated, San Jose, CA, USA) and AD8232 ECG sensor (Analog Devices, Wilmington, MA, USA).

2.1. Heart Activity

Heart rate variability (HRV) measures the variation between consecutive heartbeats that can be analyzed in both the frequency and time domains [9]. It serves as a non-invasive tool to assess the ANS and is a useful stress indicator [3]. Factors like anxiety, stress, physical exercise, and heart disease activate the SNS, leading to increased HR [9]. HRV is derived from ECG and BVP signals.

Measurements of BVP can be realized using a photoplethysmography (PPG) sensor. The MAX30100 sensor can be configured to record a PPG signal at a frequency of 100 Hz. The finger is placed on the sensor, and it sends infrared light into the skin and measures the amount of light reflected. There is a direct relation between the amount of light reflected and the amount of blood that is present [3].

The AD8232 sensor captures ECG signals by detecting the electrical activity of the heart through body electrodes (Figure 1). It is ideal for HRV studies, as it provides cardiac data that help identify patterns related to stress, anxiety, and other physiological states related to the autonomic nervous system. The ECG is a periodic signal that shows the electrical activity produced by an impulse of ions flowing through the heart muscles [3].

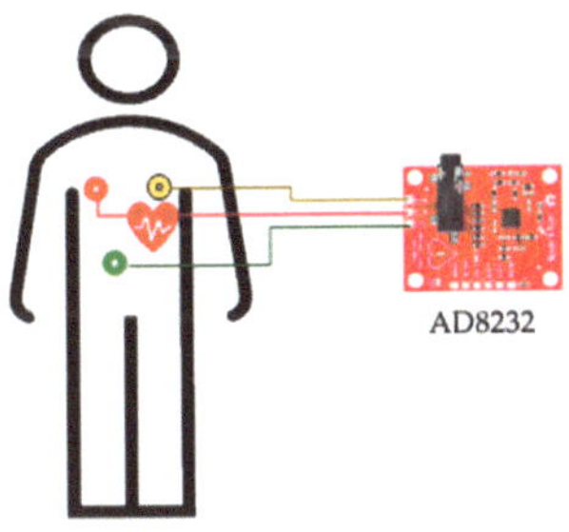

Figure 1. Measuring with a triode ECG sensor.

In the time domain, the features of the ECG signal can be obtained using statistical and geometric methods. The features extracted statistically are as follows:

- *RR interval:* Distance between two R-peaks consecutives (Figure 2) of the ECG signal.
- *Heart Rate:* Number of R-peaks measured inside a period (beats per minute bpm).

$$HR = \frac{number\ of\ R-peaks}{Time\ (minutes)} \quad (1)$$

- $\overline{RR}$: Mean value of normal RR intervals, where RR_i is the interval i-th between two R-peaks consecutives.

$$\overline{RR} = \frac{1}{N}\sum_{i=1}^{N} RR_i \quad (2)$$

- *SDNN:* Standard deviation of all normal RR intervals; where RR_i is the interval i-th between R-peaks consecutives, N is the total number of normal RR intervals inside a period and $\overline{RR}$ is the mean value of all normal RR intervals [10].

$$\overline{RR} = \frac{1}{N}\sum_{i=1}^{N} RR_i \tag{3}$$

- *RMSSD:* Root mean square successive difference of normal RR intervals.

$$RMSSD(ms) = \sqrt{\frac{1}{N-1}\sum_{i=1}^{N-1}(RR_{i+1} - RR_i)^2} \tag{4}$$

- *pNN50%:* Number of successive differences of normal RR intervals (ΔRR) which differ by more than 50 ms. It is expressed as a percentage of the total number of normal RR intervals [9], where ΔRR_j is the difference j-th between two RR intervals.

$$\Delta \mathrm{RR}_j = \mathrm{RR}_{\mathrm{i}+1} - \mathrm{RR}_{\mathrm{i}} \tag{5}$$

$$\mathrm{NN50\%} = \text{number of } \Delta RR_j > 50 \text{ ms} \tag{6}$$

$$pNN50\% = \frac{NN50\%}{total\ number\ of\ \Delta RR_j} \times 100 \tag{7}$$

- *SDSD:* Standard deviation of differences between the adjacent normal RR intervals, where $\overline{\Delta RR}$ is the mean of all values ΔRR_j [11].

$$SDSD\ (ms) = \sqrt{\frac{1}{N-1}\sum_{j=1}^{N-1}\left(\Delta RR_j - \overline{\Delta RR}\right)^2} \tag{8}$$

In the case of geometric methods, the features can be the HRV triangular index and the triangular interpolation of NN interval histogram (TINN) [6,9,11]. To calculate these features, the series of normal RR intervals should be converted into geometric patterns (Figure 2), where the duration of normal RR intervals serves as the x-axis of the plot and the number of normal RR intervals serves as the y-axis [11].

- HVR triangular index is measured as the relation between the total number of all normal *RR intervals* and the *Y* value, where *Y* is the maximum value of the sample density function [11].
- TINN is given by the difference between M and N values shown in Figure 2.

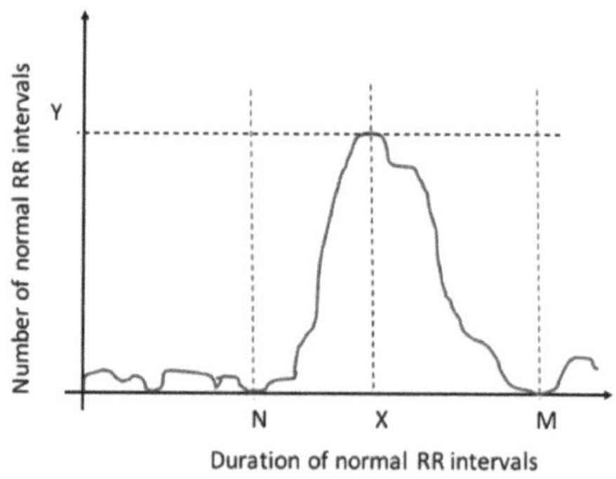

Figure 2. Sample density distribution function [11].

In the frequency domain, the short-term or long-term recordings can be executed. In short-term recordings (2 to 5 min), the power distribution and the central frequency of very low frequency (VLF, f < 0.04 Hz) [11], low frequency (LF, 0.04–0.15 Hz) and high frequency (HF, 0.15–0.4 Hz) [3,11,12] bands can be analyzed and quantified; other features can be obtained like the ratio of power in the LF to HF bands and total spectral power of all RR intervals up to 0.4 Hz [4–6,8,11,13]. For long-term recording (24 h), calculations can include total spectral power and the power in the ultra-low frequency (ULF, f $\leq$ 0.003 Hz), VLF (0.003–0.04 Hz), LF, and HF bands [11].

2.2. Pulse Arrival Time

Pulse arrival time (PAT) is the interval that a pulse wave takes to travel the distance from the heart to some distal place on the body [14]. It can be determined as the time delay between the R-peak of ECG and the peak value of an arterial pulse waveform Since PAT and HR are correlated with blood pressure, features related to PAT variability may be useful for analyzing their relationship with a person's stress level. PAT features used can be the mean of PAT, standard deviation (SD) of PAT, SD of PAT variability PAT-V, and RMS of PAT-V.

3. Collected Data

3.1. Participants

The collected dataset was realized in twenty student volunteers between 19 and 27 years old, including two women and eighteen men. The measurements were taken for five minutes in the following two stages: minutes before an evaluation event and after this event. Consent was obtained from all volunteers before the measurement protocol started.

3.2. Equipment

The MAX30100 sensor and the AD8232 module were used for data acquisition. To measure ECG, the AD8232 sensor was placed on the left side of the student's chest. The MAX30100 sensor was positioned on the student's index finger to obtain the PPG signal and heart rate data (Figure 3).

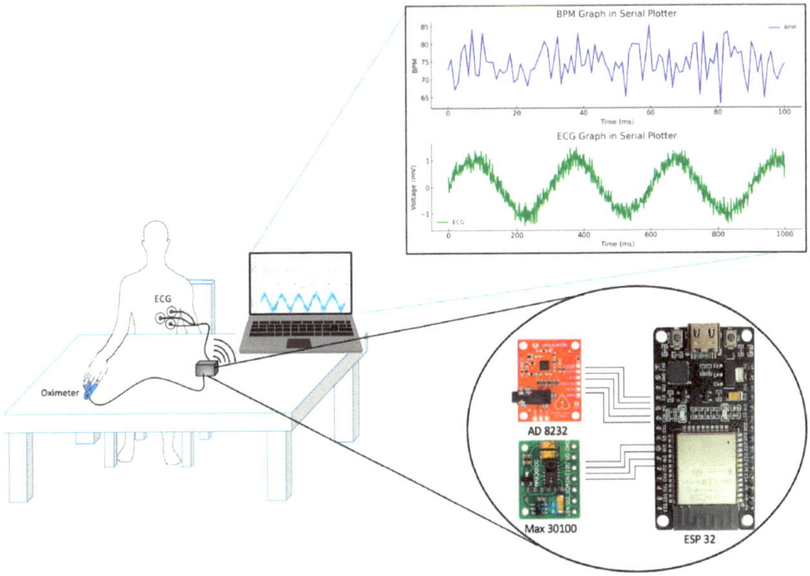

Figure 3. Connections with MAX-30100, AD8232, and ESP32.

3.3. Measure Protocol

The measures were divided into the following two stages over five minutes: one stage, a few minutes before the evaluation activity, and the second stage, after this activity. Each stage was executed in the same manner as follows:

- Each student filled out the "Cognitive Test Anxiety Scale" of 27 questions [15].
- Triode ECG and oximeter were placed.
- Each student sat still for five minutes, with their arm resting on a table and hand outstretched.

3.4. Extract Features

In this work, we extracted five features from the ECG and PPG data recording, and four were calculated with the PAT value. All the measurements before the evaluation activity were associated with a "stressed" state. At the same time, the measures after an evaluation activity were associated with a "non-stressed" state.

Table 1 shows the features calculated. Each feature set consisted of 20 values for the "stressed" state and 20 values for the "non-stressed state".

Table 1. Features analyzed.

Signals	Features
ECG	HR (bps), HRV (ms), SDNN (ms), RMSSD (ms), pNN0–50%
PPG	HR (bps), HRV (ms), SDNN (ms), RMSSD (ms), pNN0–50%
PAT	Mean of PAT, Standard Deviation of PAT, Standard Deviation of PATV, RMS of PATV

3.5. Anxiety Level Calculation

An anxiety test [15] was applied before the measurement process, which consisted of 27 questions. Students were asked to score for each question ranging from A to D, where A = "Not at all typical of me", B = "Only somewhat typical of me", C = "Quite typical of me", and D = "Very typical of me". The weights to calculate the score for A, B, C, and D were 1, 2, 3, and 4 points, respectively. The possible range of scores was 27 to 108 points. The score calculated was characterized by the following three anxiety levels: less than 33 as "Low", 33–66 as "Moderate", and greater than 66 as "High" (Table 2).

Table 2. The score obtained in the anxiety test.

No. Student	Score	Anxiety Level	No. Student	Score	Anxiety Level
1	58	"Moderate"	11	60	"Moderate"
2	50	"Moderate"	12	71	"High"
3	63	"Moderate"	13	63	"Moderate"
4	47	"Moderate"	14	55	"Moderate"
5	57	"Moderate"	15	56	"Moderate"
6	65	"Moderate"	16	64	"Moderate"
7	55	"Moderate"	17	58	"Moderate"
8	54	"Moderate"	18	67	"High"
9	49	"Moderate"	19	68	"High"
10	66	"Moderate"	20	66	"Moderate"

4. Methods

4.1. ECG and PPG Signals Obtained from ESP32

ECG data collected with an AD8232 sensor was sent to ESP32 using pin 34 as analogic input, while PPG data collected with a MAX30100 sensor was sent to ESP32 using I^2C. Schematic of the connection between the MAX30100, AD8232, and ESP32.

Figure 4 shows the connections between the ESP32, sensors, OLED display, and electrodes on the test subject. The program on the ESP32 initializes the sensors and display, providing instructions and real-time readings. The MAX30100 sensor detects heartbeats, with data continuously updated on the OLED screen. Simultaneously, the AD8232 module monitors heart electrical activity, transmitting the ECG signal to the ESP32. Signals indicating electrode contact loss (LO+ and LO−) are managed to ensure data accuracy. The reset button allows system reinitialization at any time, easily enabling new measurements.

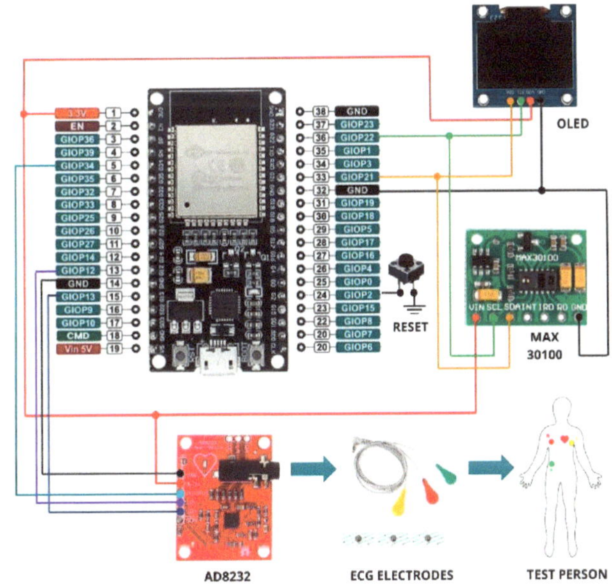

Figure 4. Wiring diagram implemented for data acquisition.

4.2. Heart Rate and Features Obtained from ECG and PPG Signals

Before feature extraction, signals were preprocessed in Matlab using a 4th order Butterworth low-pass filter to eliminate high-frequency noise and a five-sample median filter to remove peaks and unwanted noise.

The ECG signal's periodic nature allowed for the calculation of HRV features from the R–R intervals. Matlab was used to segment the ECG data and detect R-peaks, with RR intervals as shown in Figure 5a. The HR was derived from ECG R-peaks and calculated with Equation (1). ECG features like SDNN, RMSSD, SDSD, and pNN50% were computed with Matlab.

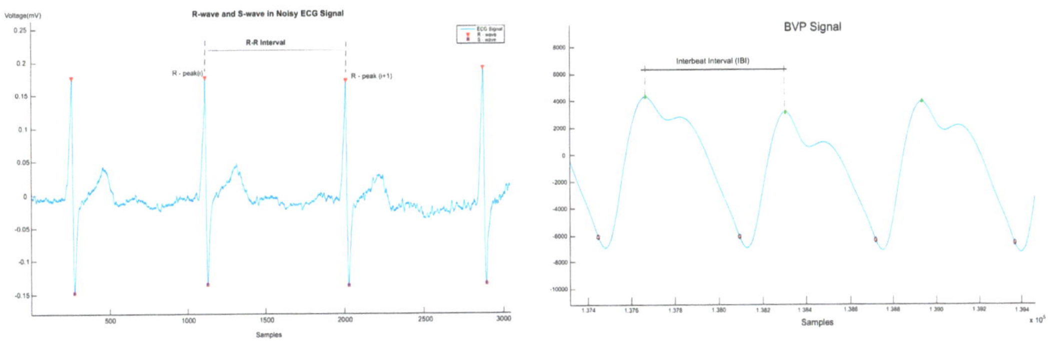

(**a**) (**b**)

Figure 5. (**a**) R–R interval in the ECG signal; (**b**) IBI in the BVP signal [16].

For PPG signals, beat peaks and intervals were calculated using Matlab and represented in Figure 5b. On the other hand, Figure 6 represents the PAT calculated by synchronizing ECG and PPG signals; the green and red markers represent the R-peak of the ECG signal and the maximum peak of the BVP signal, respectively.

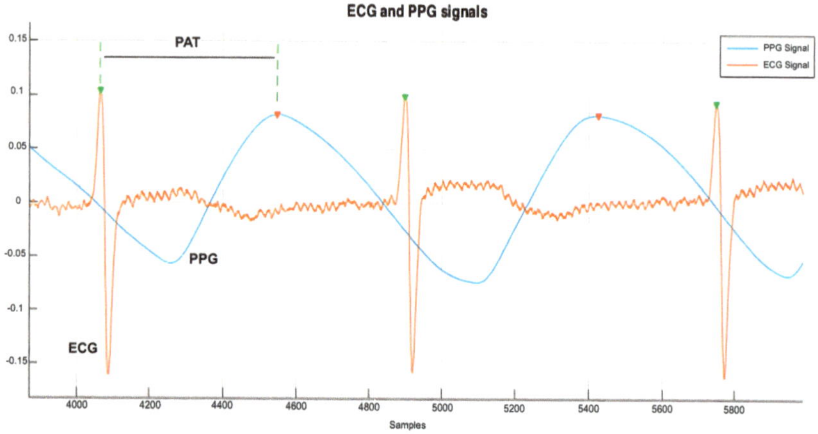

Figure 6. Representation of pulse arrival time [16].

4.3. Statistical Evaluation Methods

The Kolmogorov–Smirnov test (KST) assessed if each feature followed a normal distribution with a 5% significance level. Table 3 shows that ECG and PAT features met this requirement, while SDNN, SDSD, and RMSSD from PPG did not. Despite this, parametric methods were used as only 3 of 14 features rejected normality.

Table 3. Results of Kolmogorov–Smirnov test.

	Features	"Stressed"		"Non-Stressed"	
		h	p	h	p
ECG signal	HR	0	0.999	0	0.892
	SDNN	0	0.262	0	0.785
	RMSSD	0	0.591	0	0.712
	SDSD	0	0.591	0	0.715
	pNN50	0	0.997	0	0.704
PPG signal	HR	0	0.497	0	0.984
	SDNN	1	0.001	1	0.005
	RMSSD	1	0.004	1	0.013
	SDSD	1	0.012	1	0.004
	pNN50	0	0.164	0	0.629
PAT	Mean	0	0.824	0	0.824
	SD	0	0.692	0	0.519
	SD-PATV	0	0.890	0	0.727
	RMS_PATV	0	0.890	0	0.726

This paired study compared stress levels before and after a stressor, evaluating the same subjects two times. Paired Student t-test was used, assuming a normal distribution. The null hypothesis stated no significant difference in stress levels, while the alternative suggested changes due to evaluation.

Table 4 shows that for the ECG features, the null hypothesis was rejected for pNN50. For SDNN, RMSSD, and SDSD, the null hypothesis was rejected due to t-stat values outside the confidence interval. Significant differences were also found in the PAT features and SDNN and RMSSD from the PPG. The other features met the null hypothesis.

Table 4. Results of *t*-test.

	Features	T-test				
		h	p	Tstat	C.I.	SD
ECG signal	HR	0	0.186	1.371	−1.370; 6.570	8.482
	SDNN	0	0.145	1.520	−0.005; 0.034	0.042
	RMSSD	0	0.114	1.655	−0.006; 0.050	0.060
	SDSD	0	0.114	1.656	−0.006; 0.050	0.060
	pNN50	1	0.043	2.164	0.268; 16.113	16.923
PPG signal	HR	0	0.682	0.417	−3.422; 5.121	9.127
	SDNN	0	0.159	1.465	−0.062; 0.351	0.442
	RMSSD	0	0.085	1.817	−0.041; 0.577	0.660
	SDSD	0	0.085	1.812	−0.041; 0.578	0.578
	pNN50	0	0.469	0.740	−5.300; 11.094	17.514
PAT	Mean	0	0.089	−1.793	−0.036; 0.003	0.042
	SD	0	0.469	0.739	−0.016; 0.034	0.053
	SD-PATV	0	0.341	0.976	−0.015; 0.041	0.060
	RMS_PATV	0	0.341	0.976	−0.015; 0.041	0.060

5. Results and Discussion

5.1. Anxiety Score and Relation with ECG and PPG Features

According to the Cognitive Test Anxiety Scale, 3 subjects scored over 66, indicating "high" anxiety, while 17 had moderate anxiety. Figure 7a shows HR changes before and after the evaluation event. Ten students experienced a significant HR increase, four had a decrease, and six showed minimal variation.

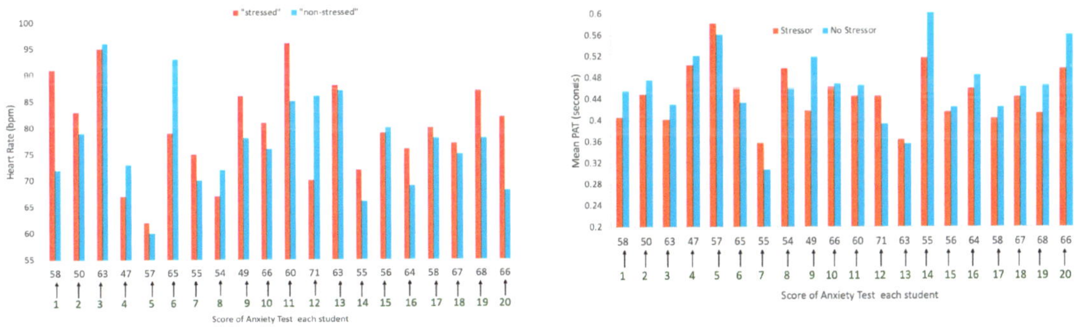

(a) (b)

Figure 7. (**a**) HR of each student before and after the activity evaluation related to the score of tests.; (**b**) Mean of PAT related to the score of the test.

Figure 7b shows that nine students had a decrease in the mean PAT before the evaluation event, five experienced an increase, and three showed minimal change.

5.2. Relation between HR Calculated with ECG and PPG Signal

If we make the correlation between the values of HR obtained with the ECG signal and the value calculated using the PPG signal, we can see that it has a linear tendency in Figure 8.

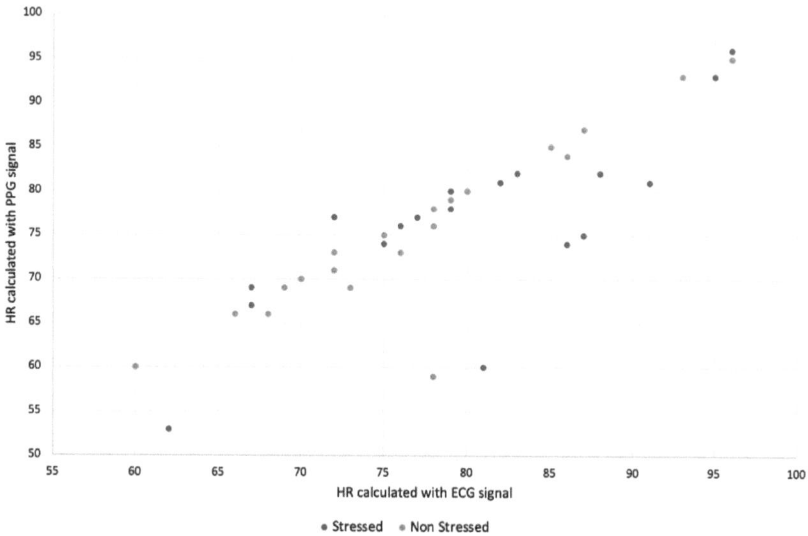

Figure 8. Relation between HR calculated with ECG and PPG signals.

5.3. Classification Performance

Each ECG, PPG, and PAT feature was calculated with a 180 s window, shifting every 6 s. The data were split into 70% training and 30% testing, with the training set containing 633 instances. Based on the t-test results, 11 out of 14 features were selected. Classification performance was evaluated using 10-fold cross-validation on the training set. The results from our experiments are displayed in Table 5. The best results are with the SMO-RBF classifier using all the features with a value of 99.84%. If we compare each feature, the best result is with the PAT features.

Table 5. Stress classification accuracies with different classifiers.

Features	Accuracy (%)		
	J48	SVM	Vote
ECG	87.34	84.51	84.51
PPG	72.35	64.14	63.99
PAT	95.90	93.53	95.58
All	96.99	99.84	99.05

6. Conclusions

This study analyzed features of ECG, PPG, and PAT signals, finding that the ECG AD8232 sensor provided more reliable data than the PPG signal obtained from the MAX30100 sensor. This indicates that ECG is a more accurate tool for assessing emotional states related to stress and anxiety.

To enhance the accuracy and reliability of the results, it is crucial to increase the sample size. A larger sample would enable a more robust and representative population analysis. No direct relationship was observed between the anxiety test values and the features of the signals, suggesting that current tests may not fully capture the variability in emotional responses.

Frequency domain analysis was not included in this study, which could be a promising direction for future research. Additionally, while the MAX30100 sensor is useful for measuring oxygen saturation and heart rate, it showed less precision than ECG for detecting subtle changes. The AD8232, on the other hand, provided detailed measurements of the cardiac activity, excelling in the heart rate variability (HRV) assessment.

In conclusion, while both sensors have their strengths, the combined use of the MAX30100 and AD8232 can offer a more comprehensive view of emotional and physical states. Future studies should consider including frequency domain analysis and expanding the sample size to improve the understanding of the relationship between physiological signals and stress.

Author Contributions: Conceptualization, S.S.; methodology, S.S., J.A., and P.H.; software, S.S.; validation, S.S., J.A., and P.H.; writing—original draft preparation, S.S.; writing—review and editing, S.S., J.A., and P.H. All authors have read and agreed to the published version of the manuscript.

Funding: This research received no external funding.

Institutional Review Board Statement: Not applicable.

Informed Consent Statement: Informed consent was obtained from all subjects involved in the study.

Data Availability Statement: The raw data supporting the conclusions of this article will be made available by the authors upon request.

Conflicts of Interest: The authors declare no conflict of interest.

References

1. World Health Organization Investing in Treatment for Depression and Anxiety Leads to Fourfold Return. Available online: https://www.who.int/news/item/13-04-2016-investing-in-treatment-for-depression-and-anxiety-leads-to-fourfold-return (accessed on 9 October 2024).
2. Sandulescu, V.; Andrews, S.; Ellis, D.; Bellotto, N.; Mozos, O.M. Stress Detection Using Wearable Physiological Sensors. In Proceedings of the International Work-Conference on the Interplay Between Natural and Artificial Computation (IWINAC 2015), Elche, Spain, 1–5 June 2015; pp. 526–532.
3. Sharma, N.; Gedeon, T. Objective Measures, Sensors and Computational Techniques for Stress Recognition and Classification: A Survey. *Comput. Methods Programs Biomed.* **2012**, *108*, 1287–1301. [CrossRef] [PubMed]
4. Liu, D.; Ulrich, M. Listen to Your Heart: Stress Prediction Using Consumer Heart Rate Sensors. 2013. Available online: https://cs229.stanford.edu/proj2013/LiuUlrich-ListenToYourHeart-StressPredictionUsingConsumerHeartRateSensors.pdf (accessed on 9 October 2024).
5. Jung, S.-J.; Chung, W.-Y. Non-Intrusive Healthcare System in Global Machine-to-Machine Networks. *IEEE Sens. J.* **2013**, *13*, 4824–4830. [CrossRef]
6. Zenonos, A.; Khan, A.; Kalogridis, G.; Vatsikas, S.; Lewis, T.; Sooriyabandara, M. HealthyOffice: Mood Recognition at Work Using Smartphones and Wearable Sensors. In Proceedings of the 2016 IEEE International Conference on Pervasive Computing and Communication Workshops (PerCom Workshops), Sydney, Australia, 14–18 March 2016; pp. 1–6.
7. Fernandes, A.; Helawar, R.; Lokesh, R.; Tari, T.; Shahapurkar, A.V. Determination of Stress Using Blood Pressure and Galvanic Skin Response. In Proceedings of the 2014 International Conference on Communication and Network Technologies, Sivakasi, India, 18–19 December 2014; pp. 165–168.
8. Barreto, A.; Zhai, J.; Adjouadi, M. Non-Intrusive Physiological Monitoring for Automated Stress Detection in Human-Computer Interaction. In Proceedings of the Human-Computer Interaction (HCI 2007), Rio de Janeiro, Brazil, 20 October 2007; pp. 29–38.
9. Rajendra Acharya, U.; Paul Joseph, K.; Kannathal, N.; Lim, C.M.; Suri, J.S. Heart Rate Variability: A Review. *Med. Biol. Eng. Comput.* **2006**, *44*, 1031–1051. [CrossRef] [PubMed]
10. Wang, H.M.; Huang, S.C. SDNN/RMSSD as a Surrogate for LF/HF: A Revised Investigation. *Model. Simul. Eng.* **2012**, *2012*. [CrossRef]
11. Malik, M. Heart Rate Variability: Standards of Measurement, Physiological Interpretation, and Clinical Use. *Circulation* **1996**, *93*, 1043–1065. [CrossRef]
12. Karim, N.; Hasan, J.; Ali, S.S. Heart Rate Variability—A Review. *J. Basic Appl. Sci.* **2011**, *7*, 71–77.
13. Zheng, Y.; Wong, T.; Leung, B.; Poon, C. Unobtrusive and Multimodal Wearable Sensors to Quantify Anxiety. *IEEE Sens. J.* **2016**, *16*, 1. [CrossRef]
14. Allen, J. Photoplethysmography and Its Application in Clinical Physiological Measurement. *Physiol Meas* **2007**, *28*, R1–R39. [CrossRef] [PubMed]

15. Cassady, J.C.; Johnson, R.E. Cognitive Test Anxiety and Academic Performance. *Contemp Educ Psychol* **2002**, *27*, 270–295. [CrossRef]
16. Maita, S.L.S.; Silva, J.M.S. New Models of Reliability in the New Generation of Internet of Things. Ph.D. Thesis, Universidade de Coimbra, Coimbra, Portugal, 2020.

engineering proceedings

Proceeding Paper

Direction-of-Arrival (DOA) Estimation Based on Real Field Measurements and Modified Linear Regression †

Luis Antonio Flores *, Ismael Lomas, Lenin Guachalá, Pablo Lupera-Morillo , Robin Álvarez and Ricardo Llugsi

Departamento de Electrónica, Telecomunicaciones y Redes de Información (DETRI), Escuela Politécnica Nacional, Quito 170525, Ecuador; dennis.lomas@epn.edu.ec (I.L.); lenin.guachala@epn.edu.ec (L.G.); pablo.lupera@epn.edu.ec (P.L.-M.); robin.alvarez@epn.edu.ec (R.Á.); ricardo.llugsi@epn.edu.ec (R.L.)

* Correspondence: luis.flores04@epn.edu.ec

† Presented at the XXXII Conference on Electrical and Electronic Engineering, Quito, Ecuador, 12–15 November 2024.

Abstract: This study applied modified linear regression in machine learning (ML) to predict the direction of arrival (DoA) in cellular networks using field measurements and radiofrequency parameters. Models were developed from base station data, with preprocessing for pattern identification and formula adjustments to improve the accuracy across angle ranges. Machine learning, tested here as an additional method to traditional techniques, achieved a root mean square error (RMSE) of 3.63 to 17.93, demonstrating enhanced adaptability. While requiring substantial data and computational resources, this approach highlights machine learning's potential as a valuable tool for DoA estimation in cellular networks.

Keywords: DoA; machine learning; cellular network; linear regression; real radio frequency measurements

Citation: Flores, L.A.; Lomas, I.; Guachalá, L.; Lupera-Morillo, P.; Álvarez, R.; Llugsi, R. Direction-of-Arrival (DOA) Estimation Based on Real Field Measurements and Modified Linear Regression. *Eng. Proc.* **2024**, *77*, 11. https://doi.org/10.3390/engproc2024077011

Academic Editor: Christian Tipantuña

Published: 18 November 2024

1. Introduction

Direction-of-arrival (DoA) estimation is a fundamental topic of considerable interest in wireless communications and radar-based applications. In this paper, we propose a new methodology based on real field measurements for acquiring radiofrequency parameters, analyzing the radio infrastructure characteristics of the base station (BS) and employing a machine learning (ML) technique to address the challenges in DoA estimation caused by imperfect arrays. Nowadays, industries are also using traditional methods to detect the DoA, such as those employed in fifth-generation fighter planes, achieving impressive results. Compared to existing methods, the proposed ML approach using linear regression achieves lower complexity. Numerous direction estimation algorithms have been proposed, each with distinct characteristics. Generally, existing DoA estimation methods can be roughly classified into super-resolution methods and Fourier transform (FT)-based methods. Recently, ML and deep learning techniques have been explored for DoA estimation, leveraging multiple signal classification (MUSIC), estimation of signal parameters via rotational invariance techniques (ESPRIT), and signal sparsity in the spatial domain of an antenna array model. Deep learning-based DoA super-resolution methods often use sampled received signals or the covariance matrices of received signals as inputs for feature extraction. In [1], a deep learning-based method is proposed where the input is the covariance matrix and the output is the spectrum from which the DoA is estimated. A sparse loss function is used to train the network. In [2], the study explores the use of deep neural networks (DNNs) for estimating the DoA of wireless signals. Techniques such as batch learning [3] and various optimization strategies are employed during the training phase. The correlation matrix serves as the input to the DNN, and the output represents the probability of an incident wave's presence in each direction. We use the correlation matrix because it is a statistical measure indicating the extent of a linear relationship between two variables, as shown in [4]. This technique is also chosen because it is a common method for describing simple relationships without inferring causality, as shown in [5]. Several key parameters are considered in configuring the system, including the antenna structure, signal frequency, signal-to-noise ratio (SNR), the number of signal sources, and the architecture of the deep neural network, which encompasses the number of intermediate layers and units per layer.

The methodology described in [6] utilizes an oversized lens-loaded cavity antenna with reconfigurable mode mechanisms. This includes a computational layer employing ML to optimize the antenna's state. In contrast, this research uses a mobile device to acquire data, which are then processed to develop a DoA estimation model. Fourier transform-based methods have also been proposed, such as in [7], where the short-time Fourier transform and spatial time-frequency distributions (STFD) matrix are used alongside the MUSIC algorithm to estimate source DoAs [8]. The novelty of this research lies in using mobile devices and ML to create reliable models for DoA behavior and prediction.

Instead, it uses real radio frequency measurements from the mobile station (MS), considers radio frequency infrastructure characteristics, and applies ML techniques. The paper is organized as follows. Section 2 reviews the DoA estimation method, Section 3 presents the results, and Section 4 concludes the paper. The key acronyms are listed in Table 1.

Table 1. List of acronyms.

Acronym	Definition
BS_{LAT}	Base Station Latitude
BS_{LON}	Base Station Longitude
M_{LAT}	Mobile Latitude
M_{LON}	Mobile Longitude
RSSI	Received Signal Strength Indicator
$RSSI_{strongest}$	Received Strongest Signal Intensity
RSRQ	Reference Signals Received Power
$RSSN_r$	Reference Signal Signal-to-Noise Ratio
D	Distance
DoA	Direction of Arrival

2. Methodology

The methodology for analyzing terrain in open fields with various obstacles involves a multifaceted approach. Techniques such as adaptive directional time–frequency distributions [9] and multiple screening k-means clustering for multiple sources [10] can be utilized. In this study, we propose using both contemporary measurement tools and advanced computational techniques. Data acquisition is conducted using NetMonitor Pro, Google Maps, Google Earth, and Android-based smartphones, ensuring comprehensive coverage and accurate geospatial referencing. Additionally, ML methodologies implemented in RStudio are employed to develop predictive models, allowing for the extraction of valuable insights from the collected data [11,12].

2.1. Algorithm for DoA Estimation

This algorithm includes the blocks for the DoA estimation process using a ML technique. The process begins with the selection of coverage areas, identifying regions of interest for analysis. Next is the analysis of the existing infrastructure, where the network infrastructure is evaluated and sectorization within the area is established. Following this, measurements are executed by collecting field data with mobile devices, resulting in a dataset of 19,419 measurements. After data collection, data preprocessing takes place to generate additional attributes and estimate the actual DoA from the collected data. The ML technique involves using modified linear regression to generate the DoA model. A critical step is the feedback in the flow diagram, which entails adjusting the model through parameter selection and formula modification until the error is minimized. Finally, the process concludes with the obtaining of the resulting model, leading to the creation of a final model for each range of angles, as shown in Figure 1.

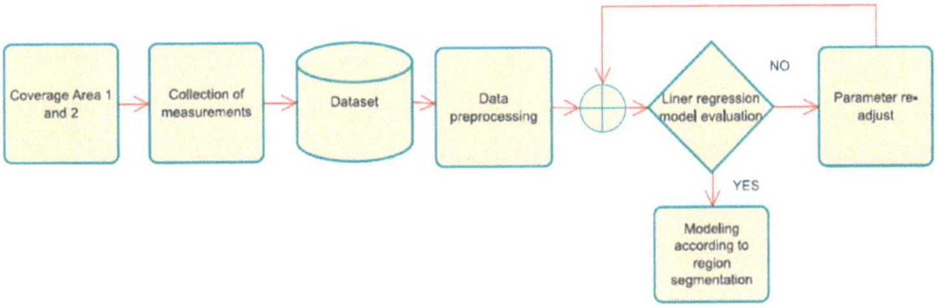

Figure 1. Process flow diagram of DoA Estimation.

The Friis equation (Equation (1)) [13] states that the far-field received signal power depends inversely on the square of the distance between the transmitter and receiver, and it is directly related to the gains of the antennas involved

$$P_{Rx} = \frac{P_{Tx} \cdot G_{Tx} \cdot G_{Rx} \cdot \lambda^2}{(4\pi D)^2} \quad (1)$$

where P_{Rx} is the received power, P_{Tx} is the transmitted power, G_{Tx} and G_{Rx} are the gains of the transmitting and receiving antennas, respectively, λ is the wavelength and D is the distance between the transmitting and receiving antennas. Equation (1) emphasizes the importance of the distance and relative DoA in signal reception, indicating that antenna gains depend on the DoA to optimize the signal strength and accuracy in different directions.

2.2. Data Collection Method

Data were collected in Quito, Ecuador, where a mobile device gathered data from mobile network operator towers. In Coverage Area 1, the measurements followed a circular and radial pattern around the BS, collecting approximately 72 samples at 5° intervals. Three circular measurements were taken in Zone 1, with radial measurements conducted outward from and returning to the BS, and circular measurements made in both the clockwise and counterclockwise directions, as shown in Figure 2a. This systematic approach ensures thorough coverage for accurate analysis of the signal propagation characteristics.

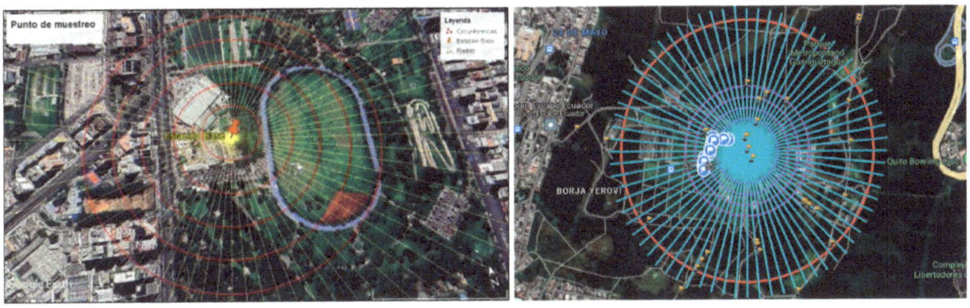

(a) (b)

Figure 2. (**a**) Map of the final radial and circular routes for measurements in Coverage Area 1. (**b**) Map showing the final radial and circular routes for measurements in Coverage Area 2 (Google Maps).

In Coverage Area 2, radial and circular measurements were also taken every 5 degrees, with two measurements per radial path: one from the BS to the handover point and one back [8].

Circular measurements were taken every 50 m from the BS, selected for its accessibility to ensure efficient data collection. A thorough inspection assessed the actual coverage area of the selected BS, allowing for segmentation based on the data collection method. Tools like Google Earth Pro version 7.3, Google Maps version 6.2, and Wikiloc version 3.40.10 were used for this purpose. Data were transferred from Wikiloc to Google Earth Pro and then to Google Maps via a KMZ file, guiding the execution of the radial and circular measurements in Coverage Area 2, as shown in Figure 2b.

For the radial measurements, a total of 72 radials resulted in 144 measurements (72 outward and 72 returning). The extent of the circular measurements varied based on the BS's coverage and path accessibility, constrained by factors such as steep slopes, inaccessible wooded areas, dense vegetation, irregular paths, and restricted zones like water treatment facilities.

2.3. Infrastructure Characteristics (Antenna Orientations)

To compare the estimated and actual DoA angles, we calculated the true DoA using Equation (2) with parameters from Table 2. This calculation employs the Haversine formula [14] to determine the distance between the BS and the mobile device (M) based on the latitude and longitude (Figure 3). The angular deviation was then computed using the Pythagorean theorem applied to latitude and longitude variations, with the dotted lines representing the reference axis relative to the BS.

$$\tan(\propto) = \frac{Dist(\Delta \mathrm{LAT})}{Dist(\Delta LON)} \tag{2}$$

where:

1. *Dist*: is the distance of the variation in longitude or latitude.
2. ΔLAT: latitude variation.
3. Δ*LON*: longitude variation.

Table 2. Considerations for analyzing the angle of the actual DoA.

Case	Considerations	θ Value
1	$BS_{LAT} < M_{LAT} \wedge BS_{LON} < M_{LON}$	$\theta = 90° - \alpha$
2	$BS_{LAT} > M_{LAT} \wedge BS_{LON} < M_{LON}$	$\theta = 90° + \alpha$
3	$BS_{LAT} > M_{LAT} \wedge BS_{LON} > M_{LON}$	$\theta = 270° - \alpha$
4	$BS_{LAT} < M_{LAT} \wedge BS_{LON} > M_{LON}$	$\theta = 270° + \alpha$

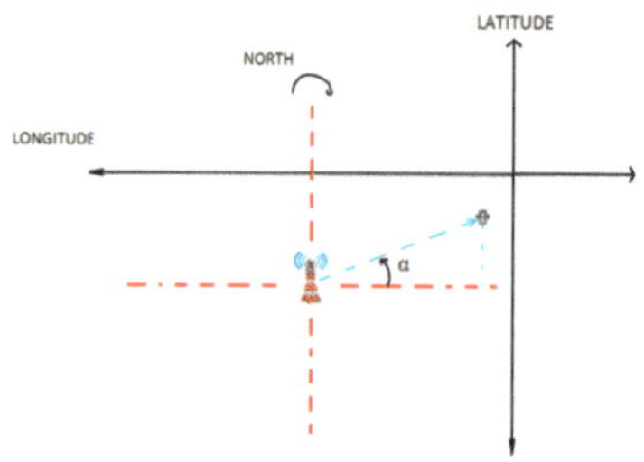

Figure 3. The reference system was positioned at the analyzed BS.

The BS in Coverage Area 1 is approximately 35 m tall and strategically positioned to optimize signal propagation. Its antennas are oriented and divided into five sectors to accommodate the dense urban landscape, ensuring adequate coverage for users in the area. Figure 4a illustrates this sectorization.

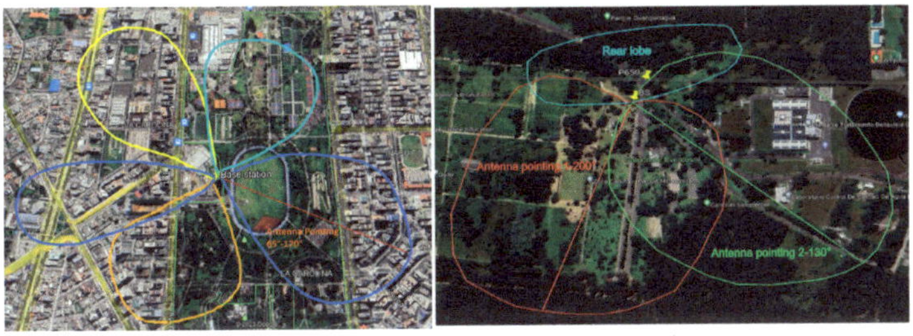

(a) (b)

Figure 4. (**a**) Coverage Area 1 and antenna orientation. (**b**) Coverage Area 2 and antenna orientation.

For Coverage Area 2, shown in Figure 4b, key characteristics such as the antenna height (approximately 9 m), number of antennas, and orientation are crucial for the network coverage and signal quality. With two antennas using a sectorized approach, Antenna 1 has an azimuth of 200°, and Antenna 2 has an azimuth of 130°, both vital for effective link connections. These attributes directly impact the DoA predictions, essential for optimizing the LTE network service quality. ML techniques, particularly supervised learning with linear regression, can enhance the DoA prediction accuracy by identifying the radio frequency (RF) parameter patterns.

3. Results

3.1. DoA Estimation in Coverage Area 1

3.1.1. Analysis of Radiofrequency Parameters and DoA Ranges

The BS's coverage area is divided into several geographical regions based on the cell's sector distribution. Figure 5a illustrates the RSSI values at different distances across various DoA ranges, with the DoA divided into 20° increments from 0° to 360°. The reference signal received quality (RSRQ) at different distances, shown in Figure 5b, also varies across three DoA ranges, suggesting potential relationships between these parameters. The figures indicate that both the RSSI and RSRQ are dependent on the distance within each DoA range.

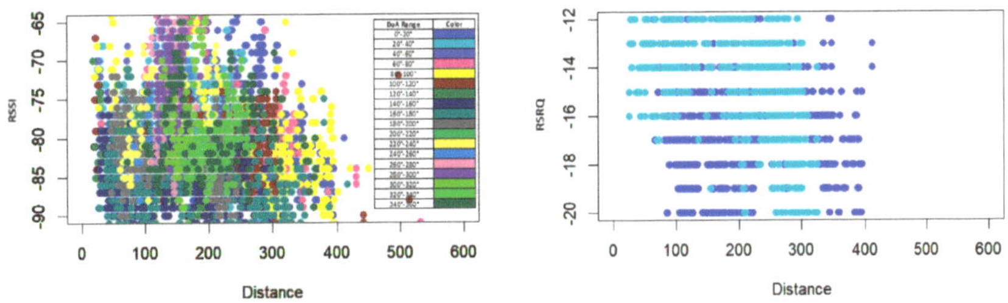

(a) (b)

Figure 5. (**a**) RSSI dispersion for different DoAs as a function of the distance. (**b**) RSRQ for three DoA ranges as a function of the distance in Coverage Area 1.

3.1.2. Correlation Matrix

Multiple ranges were selected to accommodate the sectors provided by the BS, delineated as follows: Range 1: 0°–65°, Range 2: 65°–80°, Range 3: 80°–170°, Range 4: 170°–250°, Range 5: 250°–305°, Range 6: 305°–360°.

Due to irregularities at the sector boundaries, additional models were developed. Figure 6a shows distinct correlation coefficients among the parameters, reflecting the spatial variability and signal propagation complexities. These variations require tailored models to accurately capture the signal behavior across sectors and ranges. Figure 6b presents the correlation matrix for the 340–360 degree range, where the correlation values between the DoA and other variables are higher than in Figure 6a.

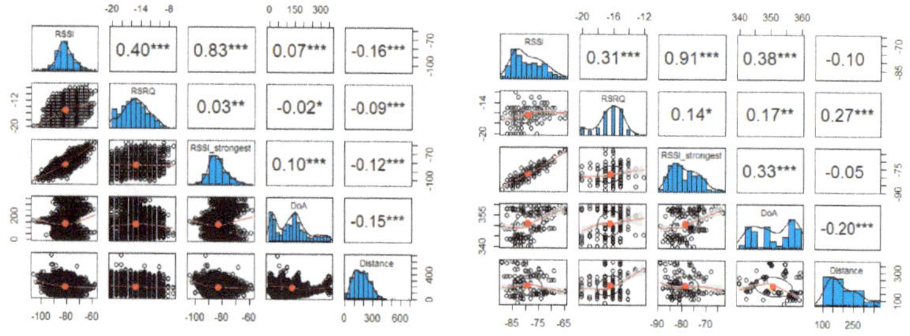

(a) (b)

Figure 6. (**a**) General correlation matrix in Coverage Area 1. (**b**) Correlation matrix in a 340–360 degree range within Coverage Area 1. The more asterisks, the more significant the result.

3.1.3. Model Summary Using DoA Ranges

The proposed analysis effectively mitigates the influence of certain parameters, facilitating the formulation of a DoA estimation model in a five-sector scenario. The models presented in Table 3 characterize the signal propagation within specified ranges, detailing the range covered, the corresponding derived equation, and the associated errors for each entry.

Table 3. Summary of the DoA estimation models in Coverage Area 1.

Range	Equation [1]	Evaluation	
		RMSE	R^2
0°–65°	$DoA = 129.6 - 3.74 \times 10^{-7}(D \times RSRQ \times RSSI \times RSSI_{strongest} - RSSI)$	15.78	0.37
65°–80°	$DoA = 48.770943 - 0.476991 \times RSSI + 0.103709 \times RSRQ$	3.63	0.15
80°–170°	$DoA = 244.19740 - 0.53929 \times RSSI - 0.80461 \times RSRQ$	16.68	0.54
170°–250°	$DoA = 326 - 1.065 \times 10^{-6}(D \times RSRQ \times RSSI \times RSSI_{strongest} - RSSI)$	13.15	0.71
250°–305°	$DoA = 114.6 + 7.454 \times 10^{-6}(D \times RSRQ_j \times RSSI \times RSSI_{strongest} - RSSI)$	10.71	0.47
305°–360°	$DoA = 321.7 + 3.416 \times 10^{-4}D^2 + 1.802 \times 10^{-38} \times RSSI^{20}{}_{strongest} + 1.54 \times 10^{-5} \times RSSI^3$	14.54	0.26

[1] Model for DoA estimation.

3.2. DoA Estimation in Coverage Area 2

3.2.1. Analysis of Radiofrequency Parameters and DoA Ranges

In the analysis of the RSSI as a function of the distance within various DoA ranges, the correlation between the RSSI values and the distance from the BS shows a non-linear trend [15]. The dataset, segmented into discrete DoA ranges, reveals that the RSSI values generally decrease with an increasing distance, consistent with the free-space path loss model. However, this attenuation is not uniform across all the DoA ranges [16].

Figure 7 depicts a scatter plot of the RSSI values, color-coded for different DoA ranges determined by the mobile device's geometric position relative to the BS. Each DoA range represents a specific angular sector around the BS. The colors in the scatter plot visually segment these sectors. Certain DoA ranges show more pronounced decreases in the RSSI with the distance, indicating a directional dependence of the signal attenuation. This variation could be due to factors such as the antenna radiation patterns, environmental obstacles, or multipath effects relative to the BS. Overall, the RSSI's dependence on the distance varies across the DoA ranges, reflecting the complex interaction between the propagation environment and the antenna characteristics. This study's approach, using segmented DoA ranges, offers a detailed understanding of this relationship and supports the development of predictive models tailored to the directional nature of wireless signal transmission.

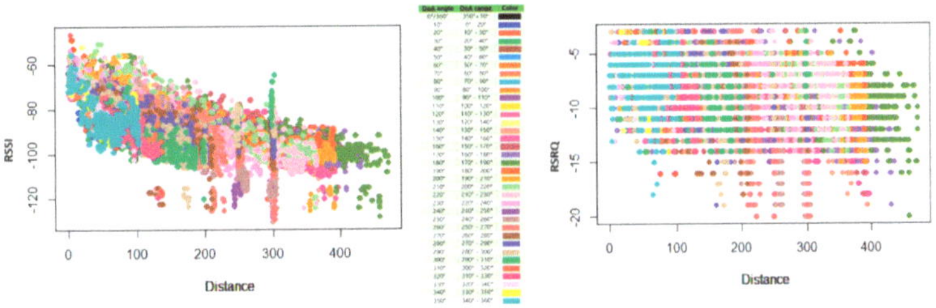

(a) (b)

Figure 7. (**a**) RSSI dispersion for different DoAs as a function of the distance. (**b**) Scatter plot of RSRQ for different DoAs as a function of the distance in Coverage Area 2.

3.2.2. Correlation Matrix

Based on Figure 8a, the decision to create multiple models for each specified range, rather than a single model, was driven by observing the varying trends in the RF parameters as a function of the DoA. It was concluded that three distinct models should be developed for each designated range. The analysis indicated that a single model is inadequate for accurately predicting the DoA across all the observed ranges due to differing trends in the RF parameters. Figure 8b shows the correlation matrix for a specific range of 340 to 360 degrees, where the correlation values between the DoA and other variables increase compared to Figure 8a. To address this, separate models are designed for the following specified ranges: Range 1: 0°–150°, Range 2: 150°–220°, Range 3: 220°–300°, Range 4: 300°–360°.

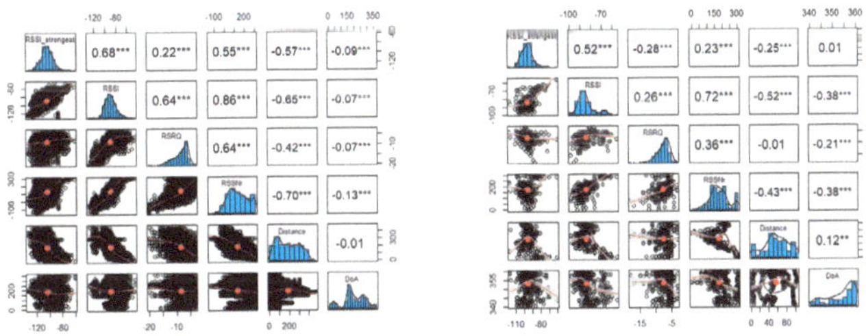

(a) (b)

Figure 8. (**a**) General correlation matrix in Coverage Area 2. (**b**) Correlation matrix in a 340–360 degree range within Coverage Area 2. The more asterisks, the more significant the result.

For each range, a new dataset is created to filter the DoA within the specified range. Correlation matrices are then obtained based on these filters [17]. Custom models are developed for each range, using all the attributes and fitted according to the correlation matrix for its respective range, as proposed in [18]. The accuracy of each fitted model is evaluated within its defined range, allowing assessment of the model performance across different DoA segments [19]. This approach, necessitated by the non-uniform behavior of the RF parameters at different angles [20], ensures that the predictive performance is optimized for each DoA range [21].

3.2.3. Model Summary Using DoA Ranges

The table summarizes the results of various models used to estimate the DoA in Coverage Area 2. The models are categorized into two types: "General" and "Ranges".

Each category includes different equations for estimating the DoA, with the effectiveness evaluated using two well-known metrics: RMSE (root mean square error) [22], and R^2 (coefficient of determination).

As shown in Table 4, the general model covers the full angular range, while specific equations are designed for segmented ranges. The model efficacy is assessed using the RMSE, which measures the prediction accuracy, and the R^2, indicating the variance explained by the model. The segmented approach shows that the tailored models yield more accurate estimations within specific angular intervals, enhancing the model performance understanding across different ranges.

Table 4. Summary of the DoA estimation models in Coverage Area 2.

Range	Equation [1]	Evaluation	
		RMSE	R^2
0°–150°	$(-4.969 \times 10^{-7} \times D + 9.921 \times 10^{-5} \times RSSI) \times RSSN_r \times RSRQ \times RSSI_{strongest} \times RSSI - 3.487 \times 10^3$	17.93	0.89
150°–220°	$2.693 \times 10^{-7} \times RSSN_r \times RSSI \times RSSI_{strongest} \times RSRQ \times D + 3.737 \times 10^3$	13.12	0.55
220°–300°	$(-1.528 \times 10^{-3} \times D - 8.819 \times 10^{-4} \times RSSN_r) \times RSSI \times RSSI_{strongest} + 1.791 \times 10^{-8} \times RSSN_r \times RSRQ \times D \times RSSI \times RSSI_{strongest} + 2.136 \times 10^3$	17.27	0.34
300°–360°	$(-8.540 \times 10^{-5} \times RSSI_{strongest} - 1.791 \times 10^{-5} \times D) \times RSSN_r \times RSRQ \times RSSI + 3.270 \times 10^3$	14.41	0.47

[1] Model for DoA estimation.

4. Conclusions

This paper addresses the DoA estimation problem in a real cellular network, proposing a novel method based on real radio frequency measurements and known radio infrastructure characteristics. Unlike traditional methods that use antenna array signals as input, this approach leverages changes in the radio frequency parameters within the spatial domain of cell coverage. The modified linear regression models for a five-sector coverage area outperform those for a three-sector coverage area. Future research should explore the analysis of radio frequency measurements, particularly because mobile devices are frequently connected to the considered BS. Additionally, iterative refinement of the estimation algorithm based on experimental feedback, considering communication obstructions, and working with an isolated BS offer promising directions. During data collection, events such as cell reselection and handover affected the DoA estimation accuracy. Future work will focus on theoretical analysis of the proposed method, improvements in the RMSE, and testing the method in dynamic environments to ensure its robustness and accuracy.

Author Contributions: Conceptualization, L.A.F.; methodology, I.L., L.G. and P.L.-M.; formal analysis, L.A.F.; investigation, L.A.F., P.L.-M. and R.Á.; resources, L.A.F., P.L.-M. and R.L.; data curation, P.L.-M. and R.L.; writing—original draft, L.A.F., P.L.-M. and R.Á.; writing—review and editing, R.L., P.L.-M. and L.A.F.; visualization, I.L. and L.G.; supervision, R.L. and P.L.-M.; project administration, L.A.F. and P.L.-M.; funding acquisition, L.A.F. All authors have read and agreed to the published version of the manuscript.

Funding: This research received no external funding.

Institutional Review Board Statement: Not applicable.

Informed Consent Statement: Not applicable.

Data Availability Statement: No new data were created or analyzed in this study. Data sharing is not applicable to this article.

Acknowledgments: The authors appreciate the sponsorship of the National Polytechnic School in carrying out this research work.

Conflicts of Interest: The authors declare no conflicts of interest.

References

1. Yuan, Y.; Wu, S.; Wu, M.; Yuan, N. Unsupervised learning strategy for direction-of-arrival estimation network. *IEEE Signal Process. Lett.* **2021**, *28*, 1450–1454. [CrossRef]
2. Kase, Y.; Nishimura, T.; Ohgane, T.; Ogawa, Y.; Kitayama, D.; Kishiyama, Y. Performance Analysis of DOA Estimation of Two Targets Using Deep Learning. In Proceedings of the 2019 22nd International Symposium on Wireless Personal Multimedia Communications (WPMC), Lisbon, Portugal, 24–27 November 2019; pp. 1–6. [CrossRef]
3. Llugsi, R.; El Yacoubi, S.; Fontaine, A.; Lupera, P. A novel approach for detecting error measurements in a network of automatic weather stations. *Int. J. Parallel Emergent Distrib. Syst.* **2022**, *37*, 425–442. [CrossRef]
4. Llugsi, R.; Yacoubi, S.E.; Fontaine, A.; Lupera, P. A novel Encoder-Decoder structure for Time Series analysis based on Bayesian Uncertainty reduction. In Proceedings of the 2021 IEEE Latin American Conference on Computational Intelligence (LA-CCI), Temuco, Chile, 2–4 November 2021; pp. 1–6. [CrossRef]
5. Llugsi, R.; Fontaine, A.; Lupera, P.; Yacoubi, S. Deep learning to implement a statistical weather forecast for the Andean city of Quito. In Proceedings of the IEEE ANDESCON 2020, Quito, Ecuador, 13–16 October 2020; pp. 1–6.
6. Abbasi, M.; Yurduseven, O.; Akinsolu, M.; Fusco, V.; Liu, B.; Khalily, M.; Cotton, S.; Imran, M. Machine learning-assisted direction-of-arrival accuracy enhancement technique using oversized lens-loaded cavity. *IET Microw. Antennas Propag.* **2022**, *16*, 305–315. [CrossRef]
7. Olmos, B.; Lupera-Morillo, P.; Llugsi, R.; Jaramillo, E. 3d Doa estimation and the clustering in a multipath environment using root music, esprit and k-means algorithms. In Proceedings of the 2019 International Conference on Information Systems and Software Technologies (ICI2ST), Quito, Ecuador, 13–15 November 2019; pp. 9–15.
8. Muirragui, C.; Lupera-Morillo, P.; Llugsi, R.; Villamar, V. Optimized performance evaluation of a q-learning hard handover algorithm for load balancing. In Proceedings of the 2021 IEEE Microwave Theory and Techniques in Wireless Communications (MTTW), Riga, Latvia, 7–8 October 2021; pp. 74–79.
9. Eranti, P.; Barkana, B. An overview of direction-of-arrival estimation methods using adaptive directional time-frequency distributions. *Electronics* **2022**, *11*, 1321. [CrossRef]
10. Zhang, Y.; Wang, D.; Cui, W.; You, J.; Li, H.; Liu, F. DOA-based localization method with multiple screening k-means clustering for multiple sources. *Wirel. Commun. Mob. Comput.* **2019**, *2019*, 5643752. [CrossRef]
11. Ge, S.; Li, S.; Rum, S. Deep learning approach in doa estimation: A systematic literature review. *Artif. Intell. Edge Comput. Mob. Inf. Syst.* **2021**, *2021*, 6392875. [CrossRef]
12. Jenkinson, G.; Abbasi, M.; Molaei, A.; Yurduseven, O.; Fusco, V. Deep learning-enabled improved direction-of-arrival estimation technique. *Electronics* **2023**, *12*, 3505. [CrossRef]
13. Huang, Y. *Antennas: From Theory to Practice*; John Wiley and Sons: Hoboken, NJ, USA, 2008.
14. Longley, P.; Goodchild, M.; Maguire, D.; Rhind, D. *Geographic Information Science and Systems*, 4th ed; John Wiley and Sons: Hoboken, NJ, USA, 2015.
15. Chen, B.; Ma, J.; Zhang, L.; Zhou, J.; Fan, J.; Lan, H. Research progress of wireless positioning methods based on RSSI. *Electronics* **2024**, *13*, 360. [CrossRef]
16. Li, B.; Liu, S.; Zhao, D.; Hu, B.-J. DOA Estimation of unknown emitter signal based on time reversal and coprime array. *Sensors* **2019**, *19*, 1398. [CrossRef]
17. Palomo-Alonso, A.; Casillas-Pérez, D.; Jiménez-Fernández, S.; Portilla-Figueras, A.; Salcedo-Sanz, S. CoSeNet: A novel approach for optimal segmentation of correlation matrices. *Digit. Signal Process.* **2023**, *144*, 104270. [CrossRef]
18. Li, J.; Wang, F.; Jiang, D. DOA estimation based on real-valued cross correlation matrix of coprime arrays. *Sensors* **2017**, *17*, 638. [CrossRef] [PubMed]
19. Hasib, M.; Kandeepan, S.; Rowe, W.; Al-Hourani, A. Directionof-arrival (doa) estimation performance for satellite applications in a multipath environment with rician fading and spatial correlation. *Sensors* **2023**, *23*, 5458. [CrossRef] [PubMed]
20. Sathish, K.; Chinthaginjala, R.; Kim, W.; Rajesh, A.; Corchado, J.M.; Abbas, M. Underwater wireless sensor networks with rssi-based advanced efficiency-driven localization and unprecedented accuracy. *Sensors* **2023**, *23*, 6973. [CrossRef] [PubMed]
21. Ahmed, T.; Tiang, J.; Mahmud, A.; Gwo, C.; Do, D. Deep reinforcement learning-based adaptive beam tracking and resource allocation in 6g vehicular networks with switched beam antennas. *Electronics* **2023**, *12*, 2294. [CrossRef]
22. Cui, L.; Zhao, P.; Li, X.; Li, B.; Wang, K.; Bu, X.; Fei, S.; Yang, Y. A new model for doa estimation and its solution by multi-target intermittent particle swarm optimization. *Syst. Sci. Control Eng.* **2021**, *9*, 87–95. [CrossRef]

engineering proceedings

MDPI

Proceeding Paper

Analysis of Energy Requirements for Massive Integration of Electric Buses in Ambato City, Ecuador †

Paola Quintana *, Angélica Vaca Yánez, Henry Acurio, Sebastián Villacrés and Verónica Guayanlema

Instituto de Investigación Geológico y Energético-IIGE, Av República E7-263, Quito 170518, Ecuador; angelicavacay@gmail.com (A.V.Y.); henry.acurio@geoenergia.gob.ec (H.A.); svillacres1892@gmail.com (S.V.); veronica.guayanlema@geoenergia.gob.ec (V.G.)

* Correspondence: paola.quintana@geoenergia.gob.ec; Tel.: +593-939-243-184

† Presented at the XXXII Conference on Electrical and Electronic Engineering, Quito, Ecuador, 12–15 November 2024.

Abstract: Ecuador's Energy Efficiency Law mandates that "as of 2030, all vehicles incorporated into urban public transport services must be electric". This legal framework sets the stage for the country's transition to electric mobility. This research examines the energy requirements for transitioning Ambato's public bus fleet to electric vehicles, considering various technical and operational factors. The analysis evaluates the current fleet size, the expected lifespan of buses, daily operational hours, average speed, and the specific characteristics of the city's public transport routes. Furthermore, this study delves into the technical aspects of energy consumption in electric public transport by calculating the driving force necessary to operate buses across different terrains and routes. Factors such as bus weight, passenger load, road gradient, and acceleration patterns are analyzed to assess their impact on energy consumption and vehicle range. Additionally, this study investigates the potential for regenerative braking and the necessary charging infrastructure, offering a comprehensive assessment of how these systems would function within Ambato. By forecasting future vehicle requirements and projecting growth in urban mobility, this study estimates the total energy demand for a fully electric public bus fleet. The potential for integrating renewable energy sources into the city's grid is also explored, ensuring that the transition to electric mobility not only decreases reliance on fossil fuels but also supports cleaner energy sources. This research serves as a crucial step in understanding the infrastructure and policy changes required for the successful implementation of electric public transport in Ambato and similar Ecuadorian cities.

Keywords: efficiency; energy; electric buses; public transport; Ecuador

Citation: Quintana, P.; Yánez, A.V.; Acurio, H.; Villacrés, S.; Guayanlema, V. Analysis of Energy Requirements for Massive Integration of Electric Buses in Ambato City, Ecuador. *Eng. Proc.* **2024**, *77*, 12. https://doi.org/10.3390/engproc2024077012

Academic Editor: Jaime Cepeda

Published: 18 November 2024

1. Introduction

Ecuador has emerged as a leading nation in Latin America in promoting sustainable transport, proactively advancing policies to accelerate the adoption of electric vehicles to support the decarbonization of its economy. A cornerstone of these efforts is the enactment of the Energy Competitiveness Law, which mandates that by 2030, all newly integrated vehicles in the urban public transportation system must be electric [1]. The National Transit Agency, in collaboration with transportation associations, has established a service lifespan of 22 years for public buses [2]. While the shift to electric vehicles presents substantial challenges, particularly in terms of investment and technological innovation, it also generates increased electricity demand for the transitioning fleet, as the Energy Efficiency Law will be fully applicable to public transport by 2030.

Ecuador generates approximately 36,683 GWh of electricity annually, with 74.4% sourced from renewable energy, predominantly hydropower [3]. This study offers a focused analysis of the energy requirements and infrastructure necessary to transition public transport to electric power in Ambato, a mid-sized city characterized by significant topographical and mobility challenges.

Existing research on electric vehicle integration has largely concentrated on larger metropolitan areas or regions with relatively flat terrain, leaving a gap in understanding the specific challenges faced by mid-sized cities such as Ambato. These cities, which are often characterized by more complex topographies and diverse energy demands, require specialized research [4]. Ambato, a key industrial hub in the Ecuadorian highlands, serves as a focal point for this analysis, offering insights into the complexities of electric public transport implementation in regions with varied terrain [5]. This study also seeks to address the lack of detailed projections regarding the impact of vehicle fleet transitions on local energy systems and infrastructure, particularly in Ecuador's Highlands [6].

Mobility in the city of Ambato presents high travel times due to the continuous vehicular congestion generated on the streets and avenues, which brings with it high rates of vehicular congestion, a particular that increases year after year. According to the Sustainable Urban Mobility Plan of the city of Ambato, trips based on home to work represent 48%, followed by home-shopping trips at 17% and recreation at 12%. These values do not include trips to external areas, that is, to rural areas and neighboring cantons. For this reason, special measures need to be defined to improve mobility for the citizens [7].

Successfully transitioning Ambato's public transportation system to electric vehicles will require careful planning and significant investment in energy infrastructure. Despite the topographical challenges, such a transition promises for both environmental and economic benefits. This research incorporates detailed projections of energy demand based on factors such as vehicle lifespan, charging infrastructure requirements, and the integration of renewable energy, thereby providing a comprehensive framework for electric vehicle deployment in Ecuador's highlands.

2. Materials and Methods

The methodological framework begins with the collection of data, considering initial parameters and collection methods, followed by a detailed description of the study scenario [8]. Mathematical calculations were employed to determine the force required to move the bus fleet, ultimately yielding estimates of the energy necessary for large-scale integration of electric buses in Ambato, as illustrated in Figure 1.

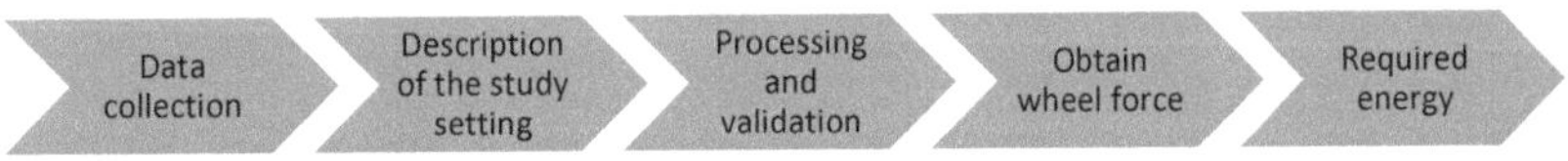

Figure 1. Methodological diagram.

2.1. Initial Considerations

This study evaluated the energy demand associated with the current conditions of Ambato's urban transportation system, projecting future energy requirements as buses near the end of their useful lives. A thorough understanding of vehicle dynamics is crucial for analyzing energy demand, as factors such as vehicle mass, route characteristics, and terrain must be considered. The technical specifications of a standard bus model, widely used in public transport, served as a baseline for these calculations.

The initial considerations, based on the requirements of vehicle dynamics equations, involved key point coefficients and ideal parameters used in the calculations [9]. These also incorporated the technical specifications of a standard bus model commonly used by public transportation companies. According to data provided by the Municipal Government of Ambato, the HINO model AK chassis is the most widely used unit in the public passenger transportation sector.

In order to establish a mathematical model for energy demand analysis, the following variables were considered: slope data, number of vehicles, urban transportation route analysis, and the useful life of the vehicle fleet. In addition, the variability of speed and slope along the route, influenced by elapsed time, was incorporated in conjunction with the initial parameters outlined in Table 1 below.

Table 1. Units for Magnetic Properties.

Characteristic	Symbol	Value	Unit
Mass bus bodywork	M	14,200	[kg]
Drag coefficient	Cd	0.73	[-]
Gravity	g	9.81	[m/s^2]
Front area	A	7.42	[m^2]
Air density (calculated)	pa	0.85898	[kg/m^3]
Dynamic radius	Rd	0.54	[rad]
Pressure	p	71	[Pa]
Temperature	T	15	[°C]
Constant air	Ra	0.287	[kl/kg·K]
Slope	θ	Variable	[rad]

2.2. Data Collection

Data were collected using GPS (GPS Garmin Montana 65) devices to track route performance at one-second intervals. This information was processed using georeferenced software to determine critical variables such as speed, altitude, and route profile. The study scenario is further described in terms of Ambato's topography, road infrastructure, and spatial dynamics, which play a fundamental role in shaping energy demand. The data were downloaded with DNR GPS software version 7.6 and processed with the following georeferenced information shown in Table 2.

Table 2. Parameter obtained from Garmin GPS Montana 650.

Parameter	Unit
Speed	m/s
Altitude	[m.a.s.l.]
Weather	[s]
Latitude	°
Length	m
Data frequency	f = 1 [s]

2.3. Description of the Scenario Under Study

Owing to its strategic location, the city of Ambato possesses distinctive characteristics that set it apart from other cities in Ecuador. Furthermore, it is regarded as the logistical hub for the country's commercial activities. It is located at an altitude of 2600 m, in a hollow formed by plateaus, settled on plains of volcanic deposits where the urban sprawl is growing rapidly, generating a dynamic of population displacement from rural to urban. Internal migration from the country to the city happened due to its commercial and manufacturing vocation that generates greater development of the territory [10]. Insights into energy consumption on slopes, including increased demand uphill and energy regeneration downhill, were applied to Ambato's topographical challenges for more accurate energy assessments [11].

The morphology of Ambato has irregular slopes, which prevent urbanization and the generation of road infrastructure easily, as well as the passage of the Ambato River that crosses the city transversely, sectoring it into different platforms [10]. Figure 2 shows the distribution of slopes and existing urban transport routes.

In relation to urban mobility, the sustainable urban mobility plan mentions that 48% of Ambato's trips are concentrated on public transport (buses, taxis, school transport, taxi routes, and informal vehicles), while 52% correspond to private transport (car, motorcycle, bicycle, walk, and other). Additionally, non-motorized or sustainable trips cover only 13% of participation (walking and cycling), while motorized or non-sustainable trips represent 87% of the total [7].

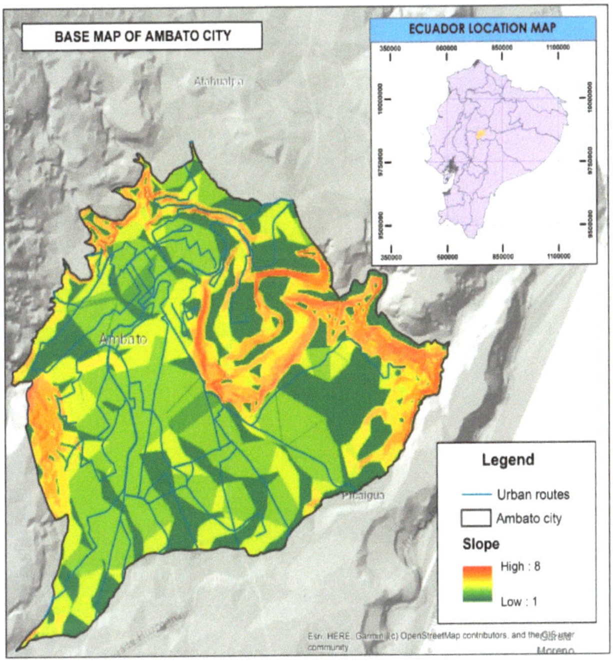

Figure 2. Base Map of Ambato City (author's own work).

2.4. Calculation of Force Acting on the Wheel

The energy consumption of an electric vehicle is directly dependent on the force required to move its mass over a given distance within a specific time frame. As a result, vehicle dynamics play a crucial role in estimating the energy requirements for such vehicle force exerted on the wheel, which is determined by the sum of four key opposing forces: drag resistance, inertia resistance, rolling resistance, and slope resistance. These forces are essential components in the overall calculation of the vehicle's energy and power requirements. Thus, the computation of both energy and power is intrinsically linked to the force exerted on the vehicle [12,13].

The calculation of the wheel force is the result of the sum of four forces that oppose movement: drag resistance, inertia resistance, rolling resistance, and slope resistance, which are part of the energy and power calculation of the vehicle. Consequently, energy and power calculation are part of the force of a vehicle [13,14].

It is possible to perform an energy demand analysis to find out what energy is required for a car to travel a given distance at a given speed at each time.

Table 3 details the equations involved in the wheel force analysis with all its considerations.

To calculate the positive energy required for the battery, the initial consideration is the torque demanded by the motor. Torque is a critical factor in determining the amount of energy necessary to propel the vehicle. Mathematically, torque can be expressed as a function of the longitudinal force acting on the vehicle, the rolling resistance force—accounting for the friction between the tires and the road surface—and the loaded radius of the wheels, which represents the effective radius under load conditions. These factors are combined in the following equation:

$$\tau_x = F_x \times R_d \tag{1}$$

where the torque provided by the motor is τ_x

- force on the wheel is $\rightarrow F_x$
- dynamic radius is $\rightarrow R_d$

Table 3. Vehicular dynamics equations for propulsive force calculation [12,13].

<table>
<tr><th colspan="2">F_x = Propulsive force [N]</th></tr>
<tr><td colspan="2">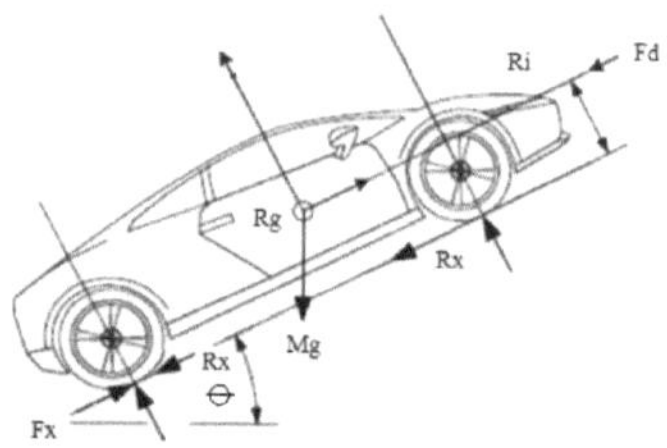
</td></tr>
<tr><td colspan="2">$F_x = F_d + R_x + R_i + R_g$</td></tr>
<tr><th colspan="2">F_d = Dragforce [N]</th></tr>
<tr><td>$F_d = \frac{1}{2} C_d \rho_a A V^2$</td><td>$C_d$ = Coefficient of drag [-]
ρ_a = Air density [kg/m^3]
A = Frontal area of the vehicle [m^2]
V = Velocity [m/s]</td></tr>
<tr><th colspan="2">R_x = Rolling resistance [N]</th></tr>
<tr><td>$R_x = f_r Mg Cos\theta$</td><td>f_r = Rolling resistance coefficient [-]
M = Mass [kg]
g = gravity [m/s^2]
θ = Road slope [rad]</td></tr>
<tr><th colspan="2">R_i = Resistance due to inertia</th></tr>
<tr><td>$R_i = Ma$</td><td>M = Mass [kg]
a = Acceleration [m/s^2]</td></tr>
<tr><th colspan="2">R_g = resistance to slope [N]</th></tr>
<tr><td>$R_g = Mg Sen\theta$</td><td>M = Mass [kg]
g = gravity [m/s^2]
θ = Road slope [rad]</td></tr>
</table>

Once the torque demanded by the motor is obtained, the power output of the electric motor can be obtained as follows [7]:

$$P_m = F_x \times V \tag{2}$$

where P_m is the delivered power of the motor, F is the force on the wheel, and V is the vehicle speed.

Finally, it is possible to determine the energy required from the battery for the motor to operate against the route requirements by means of Equation (3):

$$E_x = P_x \times \Delta t \tag{3}$$

where E_x is the energy required, P_m is the delivered power of the motor, and Δt is the instantaneous time.

3. Analysis of Established Routes

In the analysis of the 13 established routes, the number of buses assigned to each route, as well as their respective operating frequencies, were taken into account, as shown in Table 4. The analysis includes the number of daily trips each bus completes and the total number of cycles per day on the evaluated routes. This data allows for the precise determination of the energy required to complete the daily operational demands for each route.

Table 4. Evaluate routes with number of buses and circulation frequency.

Route	Buses [No.]	Laps Bus [No.]	Laps Total Route [No.]
1	11	8	88
2	22	2	44
3	11	9	99
4	11	9	99
5	10	5	50
6	21	4	84
7	18	7	126
8	29	6	174
9	26	7	182
10	25	5	125
11	16	7	112
12	16	7	112
13	15	5	75

The data regarding routes and operating frequencies were obtained from the transit and transportation administration of the city of Ambato, based on their contracts with private urban bus companies. This information is detailed in Table 4 and illustrated through the public transport route density map in Figure 3 [15].

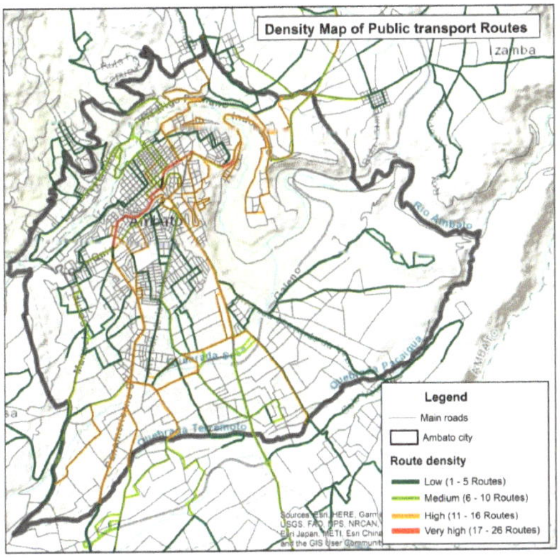

Figure 3. Density map of Public Transport Routes (author's own work).

All evaluated routes exhibit similar geographical and topographical characteristics, with both urban and rural transportation systems converging in the city's central area and extending toward its peripheral regions. The altitude profiles of each route, derived from GPS data, were considered to account for the varying elevations across the city's transport network.

Once the data were collected, the distances traveled along the evaluated routes were analyzed based on their operating frequency. Position, altitude, and instantaneous speed data were gathered to inform the mathematical model used to evaluate the energy demand for each of the analyzed routes.

The individual assessment of the routes provides insights into the maximum elevation points as well as the total distances the buses must travel to complete each circuit, as presented in Table 5.

Table 5. Characteristics of evaluated routes.

Route	Altitude Max.	Altitude Min.	Distance Lap [Km.]	Distance Bus [Km.]	Distance Total [Km.]
1	2665.90	2522.20	38.20	305.60	3361.60
2	2853.50	2518.70	25.90	51.80	1139.60
3	2892.40	2470.10	38.20	343.80	3781.80
4	2800.60	2468.50	38.20	343.80	3781.80
5	2746.00	2479.50	38.20	191.00	1910.00
6	2707.30	2521.40	40.20	160.80	3376.80
7	2920.00	2553.70	31.00	217.00	3906.00
8	2844.70	2470.10	62.00	372.00	10,788.00
9	3025.10	2479.70	62.00	434.00	11,284.00
10	2729.80	2522.70	38.20	191.00	4775.00
11	2716.40	2469.50	18.20	127.40	2038.40
12	2862.10	2522.30	28.00	196.00	3136.00
13	2745.20	2561.30	35.40	177.00	2655.00

4. Results

The results indicate the energy demand for the evaluated routes, along with their respective performance metrics expressed in kilometers traveled per kilowatt-hour consumed. This analysis takes into account the distance per lap, the distance per bus per day, and the total distance covered by the buses assigned to each route, as illustrated in Table 6. Specifically, the cumulative distance traveled across all routes totaled 55,934 km, necessitating an energy consumption of 72.14 GWh per day. While the average energy efficiency of 5.43 km/kWh was notable, it revealed significant variations across the routes, primarily attributable to the steep gradients and challenging topography encountered.

Table 6. Distance covered and energy demand.

Parameter	Value	Unit
Distance covered (Lap)	493.7	[km]
Distance covered (Bus)	3111.2	[km]
Distance covered (Route)	55,934	[km]
Energy demand (Lap)	0.67	[GWh]
Energy demand (Bus)	4.23	[GWh]
Energy demand (Daily)	72.14	[GWh]
Performance (Average)	5.43	[km/kWh]

Figure 4 presents the results derived from the analysis of the 13 evaluated routes, detailing the distance traveled on each route along with its corresponding performance metrics. The maximum distance recorded for a single route was 62.00 km, while the minimum distance was 18.20 km, resulting in an average distance of 37.98 km across all routes. The average energy efficiency of the buses was measured at 5.43 km per kilowatt-hour. However, this value exhibited considerable variation due to the differing slope characteristics inherent to each route.

After processing the data through the mathematical model of vehicle dynamics for the evaluated routes, the resulting consumption figures and their corresponding energy efficiencies were derived. The results obtained are presented in Table 7.

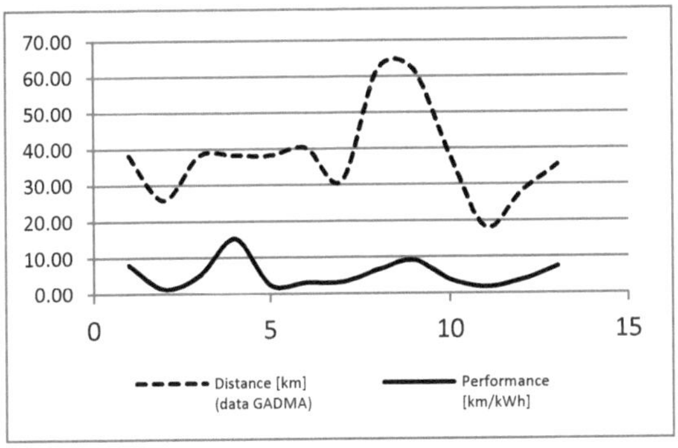

Figure 4. Density map of Public Transport Routes.

Table 7. Daily Energy Demand (routes).

Route	Lap	Bus	Total	Performance
	[kWh]	[kWh]	[kWh]	[km/kWh]
1	37.78	302.24	3324.64	8.09
2	37.4	74.8	1645.6	1.39
3	66.2	595.8	6553.8	5.19
4	22.56	203.04	2233.44	15.24
5	78.94	394.7	3947	2.42
6	52.66	210.64	4423.44	3.05
7	68.06	476.42	8575.56	3.19
8	56.94	341.64	9907.56	6.53
9	47.82	334.74	8703.24	9.08
10	50.72	253.6	6340	3.77
11	76.68	536.76	8588.16	1.66
12	54.32	380.24	6083.84	3.61
13	24.12	120.6	1809	7.34
Total	0.67	4.23	72.14	[GWh]

These findings have critical implications for the design and deployment of charging infrastructure in Ambato. For example, routes with lower performance, such as Route 2 and Route 11, will require more frequent charging points to maintain operational continuity, increasing the energy demand and the strain on the city's electric grid. Additionally, the projected daily energy requirement of 72.14 GWh highlights the need for robust energy planning and investment in renewable energy sources to sustainably support the transition to electric public transportation.

5. Conclusions

The variables analyzed in this research included instantaneous speed and slope, both of which were evaluated as functions of time and position, utilizing data obtained through GPS technology. These variables, in conjunction with the technical specifications of the transportation means and the ideal parameters for calculations, informed the vehicle dynamics equations used to ascertain the energy demand necessary for the operation of electric buses on the evaluated routes.

The data supplied by the city administration of Ambato, along with the comprehensive database generated during the study, facilitated an in-depth understanding of the public transport system's characteristics, encompassing route distribution, frequency, and the geomorphology of the road network within the study area. The findings confirmed that altitude significantly influences energy demand, which is particularly notable given that the highest elevation recorded was 3025 m above sea level and the longest route extended 62 km.

This research aimed to assess the energy requirements for transitioning to electric buses within the unique topographical context of Ambato. The results corroborate the initial hypothesis that the city's varied altitudes and slopes contribute to higher energy consumption compared to flatter terrains. Specifically, the findings indicate that uphill segments substantially increase energy demand, whereas downhill sections can facilitate energy regeneration, as evidenced by an average efficiency of 5.43 km/kWh. These results align with previous studies (e.g., Energy Efficiency and Emissions Reduction in Mountainous Regions, 2021), which demonstrated that the performance of electric vehicles in hilly regions is highly contingent upon elevation changes and road gradients.

In contrast to studies conducted in flatter regions, where energy consumption patterns are generally more uniform, the energy demands observed in Ambato are markedly higher on routes characterized by steeper inclines, such as Route 2, which recorded an efficiency of 1.39 km/kWh. This variation underscores the necessity for adapting energy and infrastructure planning to address the unique geographical challenges faced by cities like Ambato.

The findings of this study have substantial implications for policymakers and urban planners engaged in the advancement of electric mobility. The projected daily energy demand of 72.14 GWh for the city's urban transport routes highlights the urgent need for robust energy infrastructure, particularly in areas with challenging topographies. Policymakers should prioritize investments in electric grid enhancements and the strategic placement of charging stations, especially in high-elevation zones where energy demands are amplified.

Moreover, the spatial dynamics of Ambato—where all routes converge in the city center—must be integrated into transport planning to optimize routing and minimize energy consumption. The model developed in this research provides a valuable tool for predicting energy demand in other cities across Ecuador, contingent upon the availability of similar data regarding transport routes and terrain. Additionally, planners should consider the advantages of incorporating renewable energy sources to satisfy the increased electricity demands without exacerbating carbon emissions.

This study contributes to a broader understanding of the energy requirements for electric public transport systems in cities facing challenging geographical conditions. It illustrates the necessity for tailored infrastructure solutions and adaptive planning strategies to ensure a successful transition to electric mobility in urban environments such as Ambato.

Author Contributions: Conceptualization, P.Q. and H.A.; methodology, S.V.; software, A.V.Y.; investigation, all authors; resources, The Deutsche Gesellschaft für Internationale Zusammenarbeit (GIZ); writing—original draft preparation, V.G.; writing—review and editing, V.G., P.Q., H.A.; project administration, P.Q. All authors have read and agreed to the published version of the manuscript.

Funding: This research was funded by The Deutsche Gesellschaft für Internationale Zusammenarbeit (GIZ); grant number [81255293].

Institutional Review Board Statement: Not applicable.

Informed Consent Statement: Not applicable.

Data Availability Statement: Data are contained within the article.

Conflicts of Interest: The authors declare no conflict of interest.

References

1. Asamblea Nacional República del Ecuador. Proyecto de Ley de Competitividad, Segundo Suplemento N° 475—Registro Oficial. 2024; p. 17. Available online: https://www.registroficial.gob.ec/index.php/registro-oficial-web/publicaciones/suplementos/item/19990-segundo-suplemento-al-registro-oficial-no-475 (accessed on 1 July 2024).

2. Agencia Nacional de Tránsito. Resolución No. 026-DIR-2022-ANT, 2022, Amplíense los Años de Vida Útil en Vehículos del Transporte Terrestre Público y Comercial en Ecuador, Quinto Suplemento del Registro Oficial. No. 236, Quito, Ecuador. Available online: https://www.edicioneslegales-informacionadicional.com/webmaster/directorio/SIE-%20TRANSCOMERC-23-01.pdf (accessed on 21 February 2023).
3. Ministerio de Energía y Minas. *Balance Energético Nacional 2023*; Ministerio de Energía y Minas: Quito, Ecuador, 2024. Available online: https://www.recursosyenergia.gob.ec/5900-2/ (accessed on 2 September 2024).
4. Li, Z.; Zhang, Y. Public recharging infrastructure location strategy for promoting electric vehicles: A bi-level programming approach. *J. Clean. Prod.* **2018**, *172*, 2720–2734. [CrossRef]
5. Wyatt, D.W.; Li, H.; Tate, J.E. The impact of road grade on carbon dioxide (CO_2) emission of a passenger vehicle in real-world driving. *Transp. Res. Part D* **2014**, *32*, 160–170. [CrossRef]
6. CANFACC Camara Nacional de Fabricantes de Carrocerías. Carroceros de Tungurahua preocupados por la llegada de buses eléctricos. *Diario La Hora*. 15 March 2019. Available online: https://www.lahora.com.ec/noticias/carroceros-de-tungurahua-preocupados-por-la-llegada-de-buses-electricos/ (accessed on 16 May 2022).
7. Gobierno Autónomo Descentralizado Municipalidad de Ambato. Plan de Movilidad Urbana Sostenible. 2014, p. 511. Available online: https://www.mobiliseyourcity.net/sites/default/files/2022-06/Plan%20de%20Movilidad%20Urbana%20Sostenible%20de%20Ambato%20-%20Resumen%20ejecutivo.pdf (accessed on 2 February 2024).
8. Pamula, T.; Pamula, D. Prediction of Electric Buses Energy Consumption from Trip Parameters Using Deep Learning. *Energies* **2022**, *15*, 1747. [CrossRef]
9. Sennefelder, R.; Martín-Clemente, R.; González-Carvajal, R. Energy Consumption Prediction of Electric City Buses Using Multiple Linear Regression. *Energies* **2023**, *16*, 4365. [CrossRef]
10. Gobierno Autónomo Descentralizado Municipalidad de Ambato. Ordenanza de Aprobación del Plan de Desarrollo y Ordenamiento Territorial 2050 y del Plan de Uso y Gestión del Suelo 2033 del Cantón de Ambato. 2021; p. 527. Available online: https://gadmatic.ambato.gob.ec/lotaip/2024/junio/articulo%2021/RESOLUCIONES%20Y%20ORDENANZAS/ORDENANZA%20QUE%20ACTUALIZA%20EL%20PLAN%20DE%20DESARROLLO%20Y%20ORDENAMIENTO%20TERRITORIAL%202050%20Y%20EL%20PLAN%20DE%20USO%20Y%20GESTION%20DE%20SUELO%202033%20DEL%20CANTON%20AMBATO.pdf (accessed on 15 December 2023).
11. Puma-Benavides, D.S.; Cevallos-Carvajal, A.S.; Masaquiza-Yanzapanta, A.G.; Quinga-Morales, M.I.; Moreno-Pallares, R.R.; Usca-Gomez, H.G.; Murillo, F.A. Comparative Analysis of Energy Consumption between Electric Vehicles and Combustion Engine Vehicles in High-Altitude Urban Traffic. *World Electr. Veh. J.* **2024**, *15*, 355. [CrossRef]
12. Gillespie, T.D. *Fundamentals of Vehicle Dynamics*; Society of Automotive Engineers: Warrendale, PA, USA, 1992; Volume 400. [CrossRef]
13. Angelo, E.; Angelo, G.; Santos, P.; Andrade, D. Numerical Study of the Influence of Elements inside the Wheelhouse on the Passenger Vehicle Aerodynamic. *Open J. Fluid Dyn.* **2015**, *5*, 199–207. [CrossRef]
14. Villacrés, A.S.; Fernández, E. Analysis of Energy Efficiency Electric Vehicles Using a Driving Cycle in an Established Royte in the City of Ambato. In *Systems, Smart Technologies and Innovation for Society, Proceedings of CITIP'S, Volume 2*; Springer: Cham, Switzerland, 2016; pp. 63–79.
15. de Ambato, G.A.D.M. Contratos y Adendas con las Cooperativas de Pasajeros Para el Servicio de Transporte en la Ciudad de Ambato. 2018. Available online: https://gadmatic.ambato.gob.ec/lotaip/2023/junio/anexo%20literal%20s/ACT-002-E-2023-03-24.pdf (accessed on 25 December 2023).

engineering proceedings

Proceeding Paper

Cascade Control Based on Sliding Mode for Trajectory Tracking of Mobile Robot Formation †

Alejandro Camino [1], Andrés Villegas [1], Esteban Pérez [1], Richard López [1], Gabriela M. Andaluz [1,2,*] and Paulo Leica [1]

1 Departamento de Automatización y Control Industrial, Facultad de Ingeniería Eléctrica y Electrónica, Escuela Politécnica Nacional, Quito 170525, Ecuador; andres.villegas01@epn.edu.ec (A.V.); esteban.perez01@epn.edu.ec (E.P.); paulo.leica@epn.edu.ec (P.L.)

2 Department of Electronic Engineering and Communications, Universidad de Zaragoza, 44003 Zaragoza, Spain

* Correspondence: gabriela.andaluz@epn.edu.ec

† Presented at XXXII Conference on Electrical and Electronic Engineering, Quito, Ecuador, 12–15 November 2024.

Abstract: An innovative cascade control strategy is presented in this work, based on sliding mode control (SMC) for trajectory tracking of the formation of mobile robots. The proposed strategy was compared with five alternative control approaches: PID control, inverse dynamics, and other SMC-based structures. The objective was to evaluate the most effective control technique by analyzing the integral of squared error (ISE) index. Additionally, robustness tests were carried out by varying the parameters of the dynamic model of the mobile robot and analyzing the response of the controllers to perturbations in the modeling. The results show that the PD-SMCV controller provides the best performance in trajectory tracking and robustness against disturbances, demonstrating significant superiority over the evaluated methods for maintaining a stable mobile robot formation under dynamic conditions.

Keywords: Lyapunov; inverse dynamics; SMC; formation; mobile robot; robustness; cascade control

Citation: Camino, A.; Villegas, A.; Pérez, E.; López, R.; Andaluz, G.M.; Leica, P. Cascade Control Based on Sliding Mode for Trajectory Tracking of Mobile Robot Formation. *Eng. Proc.* **2024**, *77*, 13. https://doi.org/10.3390/engproc2024077013

Academic Editor: Pablo Proaño

Published: 18 November 2024

1. Introduction

Nowadays, with the development of technology and the increased use of robotic systems, the need to control multiple robots to perform a cooperative task in formation has increased, with the objective that all robots execute different tasks such as cargo transportation and navigation, among others [1–3]. The design and evaluation of controllers for robotic formation have been a subject of interest by researchers. Several studies have been conducted, as in [4], which presents a leader–follower robot formation system, employing a fully actuated system (FAS) method to improve stability and convergence by eliminating uncertainties and handling disturbances. In [5], formation control using a null space-based sliding mode control (NSB-SMC) is discussed, which improves robustness in environments with static and dynamic obstacles. On the other hand, for uncertainty and disturbance handling, an adaptive SMC is proposed in [6] to improve the accuracy, stability, and convergence of tracking errors under different conditions, and in [7], this type of controller is applied to sliding steering mobile robots (SSMRs) used in various mobile robotic systems with different morphologies that do not have steerable wheels.

Other techniques, such as the stage-by-stage evolutionary fuzzy obstacle boundary following (OBF) control of the three robots proposed in [8], were implemented through a cooperative behavioral supervisor to coordinate the learned OBF behavior, while [9] introduces fuzzy logic to replace switching control in traditional sliding mode control (SMC). Models such as Markov have also been employed within robot formation; for example, [10] proposes a static controller for multi-agent systems with Markovian topologies, ensuring stability in the face of imperfect information. Furthermore, [11] investigates tracking protocols with variable delays, while [12] presents a fault-tolerant approach for mobile manipulators. These works highlight the robustness of Markov models in complex environments.

In robot formation, the leader–follower system is employed, where the leader robot sets the trajectory and the followers adjust their positions accordingly [13,14]. This setup can facilitate coordination but also introduces vulnerabilities to possible leader failures. In [15], the author describes a multi-leader scheme, using an optimization method for the leaders and an SMC-like controller for the followers. The SMC can be combined with other types of control such as fuzzy control, which reduces the chattering present in the SMC, as demonstrated in [16]. When integrated with a PD control, it achieves a combination of the simplicity of the PD control and the robustness of the SMC, decreasing the chattering present in the SMC control studied in [17].

In the present work, in contrast to the literature reviewed on leader–follower robot-based formation control using traditional controllers, which focused on a single control strategy, a novel approach is proposed by implementing a cascade control strategy for mobile robot formation control. This strategy takes advantage of different controllers and is compared with five different control schemes, allowing the evaluation of the system behavior in the outer loop for the formation control and in the inner loop for the dynamic compensation of the robots. In addition, by working with the dynamic model of mobile robots, factors such as inertia, external forces, and interactions with the environment are taken into account to predict how the robot will behave in different conditions, allowing more precise control, improving stability, maneuverability, and response to system perturbations.

This work is organized as follows: Section 2 presents the dynamic model of the three mobile robots performing the formation and the formation system model; Section 3 details the design of the formation controllers: (i) PD without dynamic compensation (PD), (ii) PD with dynamic compensation (PD-DI), (iii) PD with dynamic sliding mode controller (PD-SMCV), (iv) SMC without dynamic compensation (SMC), and (v) SMC with dynamic sliding mode controller (SMC-SMCV); Section 4 describes the simulation tests and results. The stability analysis, discussion, and comparison of all the implemented controllers are presented by obtaining the integral of squared error (ISE) index, which helps to verify and evaluate the best controller for this application. Finally, Section 6 provides the conclusions of this work.

2. Modeling

2.1. Kinematic and Dynamic Model of the Mobile Robot

The real operation of a mobile robot can be represented by different mathematical models, starting with the kinematic model, which represents more simply the operation of the robot, since it does not consider friction forces or physical disturbances. The kinematic model is defined by the following:

$$\begin{bmatrix} \dot{x} \\ \dot{y} \\ \dot{\psi} \end{bmatrix} = \begin{bmatrix} cos\psi \;\; -a\ sin\psi \\ sin\psi\ a\ cos\psi \\ 0\,1 \end{bmatrix} \begin{bmatrix} u \\ \omega \end{bmatrix} \tag{1}$$

where ψ is the orientation of the robot, ω is the angular velocity of the robot, u is the linear velocity of the mobile robot, $\dot{x}$ is the component of the velocity in x, $\dot{y}$ is the component of the velocity in y, $\dot{\psi}$ is the angular velocity, and a is the distance between the point of interest $(x,\ y)$ from the axis joining the wheels.

On the other hand, the dynamic model of the robot can also be used, which considers factors such as friction forces, masses, dimensions, etc. This model is closer to the actual operation of the robot. The dynamic model is defined in [18].

2.2. Mobile Robot Formation Model

The objective of the proposed formation is based on the fact that three mobile robots $\mathcal{R}\{r_1, r_2, r_3\}$ will form a geometrical figure, in this case, a triangle, as shown in Figure 1. (x_c, y_c) follows a proposed trajectory, and it also controls the formation variables, i.e., the distances between the leader robot and the other two robots.

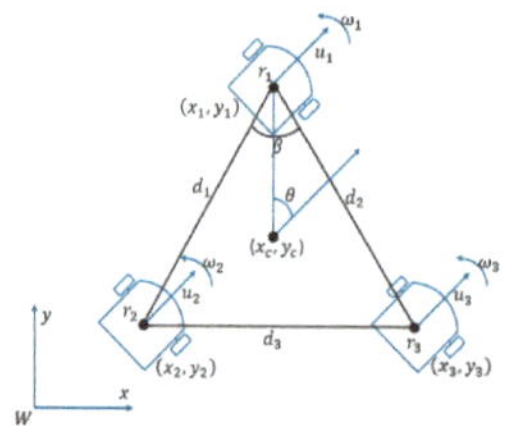

Figure 1. Formation diagram of three mobile robots r_1, r_2, and r_3.

In the diagram, (x_c, y_c) is the position of the centroid of the formation in the world coordinate frame {W}; d_1 is the distance between r_1 and r_2; d_2 is the distance between r_1 and r_3; d_3 is the distance between r_2 and r_3; β is the angle opposite to the segment d_3; and θ is the formation pose angle with respect to $\{W\}$. On the other hand, the control of the pose variables seeks to achieve trajectory tracking based on the calculation of the centroid of the geometric figure established in the algorithm.

The formation model of the mobile robot system is given in [19]:

$$\dot{h} = J_R U \tag{2}$$

where $\dot{h} = \begin{bmatrix}\dot{x_1}\ \dot{y_1}\ \dot{x_2}\ \dot{y_2}\ \dot{x_3}\ \dot{y_3}\end{bmatrix}^T$ are the variations of the robot positions in the formation; $U = [u_1\ \omega_1\ u_2\ \omega_2\ u_3\ \omega_3]^T$ are the velocities of the robots in the formation; and J_R is the Jacobian of the rotational arrays for the three-robot system. To determine the model in formation variables [19], the geometry or shape of the formation is described by $q_f = [d_1\ d_2\ \beta]^T$, and the posture geometry is given by $q_p = [x_c\ y_c\ \theta]^T$, mathematically defined in (3).

$$q_p = \begin{bmatrix} \frac{x_1+x_2+x_3}{3} \\ \frac{y_1+y_2+y_3}{3} \\ atan\left(\frac{\frac{2}{3}x_1-\frac{1}{3}(x_2+x_3)}{\frac{2}{3}y_1-\frac{1}{3}(y_2+y_3)}\right) \end{bmatrix} \quad q_f = \begin{bmatrix} \sqrt{(x_1-x_3)^2+(y_1-y_3)^2} \\ \sqrt{(x_1-x_2)^2+(y_1-y_2)^2} \\ acos\left(\frac{d_1{}^2+d_2{}^2-d_3{}^2}{2d_1d_2}\right) \end{bmatrix} \tag{3}$$

The relationship of the robot's velocity to the formation variables can be expressed as follows:

$$\dot{q} = J J_R U \tag{4}$$

where J is the formation Jacobian matrix and $\dot{q} = \begin{bmatrix} q_f\ q_p \end{bmatrix}^T$ represents the derivative of the shape and pose variables of the given formation.

3. Controllers

This section details the five different cascade controllers that were developed. The outer controller is responsible for managing the formation, while the inner controller handles the robot's dynamic compensation. The developed controllers are (i) PD formation controller without dynamic compensation (PD); (ii) SMC formation controller without dynamic compensation; (iii) PD formation controller with dynamic compensation (PD-DI); (iv) PD formation controller with dynamic sliding mode control (PD-SMCV); and (v) sliding mode formation control with dynamic sliding mode dynamic compensation (SMC-SMCV). Figure 2 shows the control diagram without dynamic compensation, which includes only the outer controller scheme. This scheme is used for the PD and SMC formation controllers. Additionally, Figure 3 shows the control diagram for both formation and dynamic compensation control, using a cascade controller configuration. This scheme is employed by the PD-DI, PD-SMCV, and SMC-SMCV controllers.

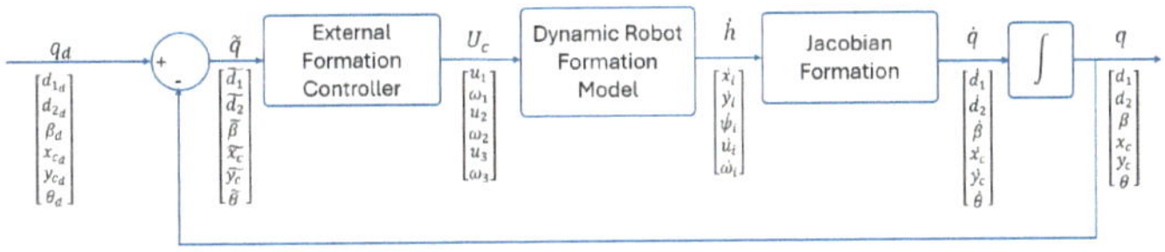

Figure 2. Formation control scheme without dynamic compensation.

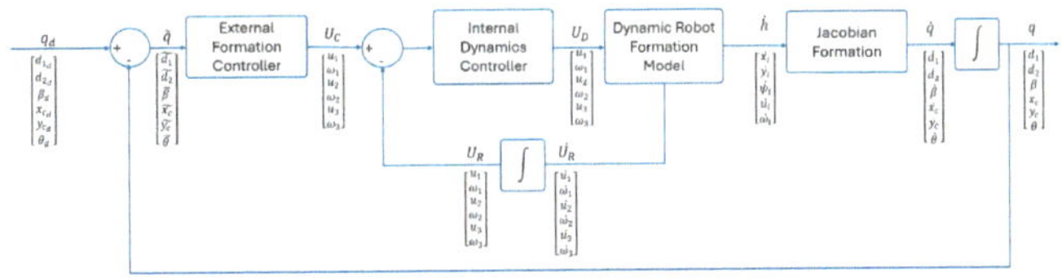

Figure 3. Formation cascade control scheme with dynamic compensation.

3.1. PD Formation Controller without Dynamic Compensation (PD)

Based on the scheme shown in Figure 2, the PD formation controller without dynamic compensation employs the following control law [20]:

$$U_c = (JJ_R)^{-1}\left(\dot{q}_d + k_1\tilde{q}\right) \tag{5}$$

where $\tilde{q} = q_d - q$ is the formation error variable; k_1 is a positive constant; and $\dot{q}_d$ is the derivative of the desired formation reference.

3.2. SMC Formation Controller without Dynamic Compensation

The SMC control law is defined by the following:

$$u_{smc} = u_{ac} + u_{dc} \tag{6}$$

where u_{ac} corresponds to the continuous part of the control law; u_{dc} is the discontinuous part of the control law, where each part is defined as follows:

$$u_{ac} = (JJ_R)^{-1}\left(\dot{q}_d + \frac{\lambda_{iF}}{\lambda_{pF}}\tilde{q}\right) ;\ u_{dc} = (JJ_R)^{-1}\frac{\delta_c}{\lambda_{pF}}\frac{s}{|s|+0.5} \tag{7}$$

To obtain the continuous part, a sliding mode surface s is defined as follows:

$$s = \lambda_{pF}\tilde{q} + \lambda_{iF}\int\tilde{q} \tag{8}$$

where λ_{iF}, λ_{pF} are the constants that correspond to the integral and proportional parts of the surface, respectively, and δ_c is a positive constant. For this controller, the control diagram shown in Figure 2 is employed, where the formation control is defined by the SMC controller [5].

3.3. PD Formation Controller with Dynamic Compensation (PD-DI)

This controller uses the cascade controller scheme shown in Figure 3, where the outer formation control is given by (5), while the inner control for dynamic compensation is defined as [21]:

$$u_{DI} = M\sigma + \eta \tag{9}$$

where M is a diagonal matrix formed by the values ϕ_1 y ϕ_2, while η is given by the following:

$$\eta = \begin{bmatrix} 0 & 0 & \omega_{ref}^2 & u_{ref} & 0 & 0 \\ 0 & 0 & 0 & 0 & u_{ref}\omega_{ref} & \omega_{ref} \end{bmatrix} \quad (10)$$

where $\phi = \begin{bmatrix} \phi_1 & \phi_2 & \phi_3 & \phi_4 & \phi_5 & \phi_6 \end{bmatrix}^T$ is a vector formed by parameters that describe the dynamic behavior of the robot. Their values are as follows [18]: $\phi_1 = 0.4072$, $\phi_2 = 0.2937$, $\phi_3 = -0.0287$, $\phi_4 = 0.9979$, $\phi_5 = -0.0004$, and $\phi_6 = 0.9865$. Lastly, σ:

$$\sigma = \dot{U}_c + Q_D tanh\left(\tilde{v}\right) \quad (11)$$

Here, $\tilde{v} = [U_c - U_r]$ represents the angular and linear speed error, with U_r being the robot's angular and linear velocity and U_c being the output of the outer control loop.

3.4. PD Formation Controller with Dynamic Sliding Mode Control (PD-SMCV)

The inner loop controller for PD-SMCV control is given by the following:

$$u_{smcV} = u_{aV} + u_{dV} \quad (12)$$

The sliding mode surface s is defined by the following:

$$s = \lambda_{pV}\,\tilde{p} + \lambda_{iV}\int \tilde{p} \quad (13)$$

By differentiating (13), we obtain the following:

$$\dot{s} = \lambda_{pV}\,\dot{\tilde{p}} + \lambda_{iV}\,\tilde{p} \quad (14)$$

where $\dot{\tilde{p}} = \dot{U}_c - \dot{U}_r$, $\dot{U}_r$ is the derivative of the angular and linear velocity of the robot, and U_c is the derivative of the output of the outer control loop. For the system to remain in the sliding mode surface s, it is required that $\dot{s} = 0$. To obtain u_{av}, it is considered that $u_{dV} = 0$; hence, in a closed loop, $u_{smcV} = u_{aV}$. By solving (16), the following expression is obtained:

$$u_{aV} = M\left(\frac{\lambda_{iV}}{\lambda_{pV}}\tilde{p} + \dot{U}_c\right) + \eta \quad (15)$$

To obtain u_{dV}, the Lyapunov candidate $V = \frac{1}{2}s^T s$ is defined. Thus, the derivative of V is given by $\dot{V} = s^T\dot{s}$. In a closed loop, $u_{smcV} = U_r$. By substituting (12), (14), and (15) into $\dot{V}$, the following expression is obtained:

$$\dot{V} = s^T\left(-\lambda_{pV}M^{-1}u_{dV}\right) \quad (16)$$

Thus, u_{dV} is defined by the following:

$$u_{dV} = \left(\lambda_{pV}M^{-1}\right)k_V\; sig(s); k_V > 0 \quad (17)$$

Substituting (17) into (16), it is verified that $\dot{V} < 0$, ensuring that $s \to 0$ as $t \to \infty$. Analyzing (13) at $s = 0$ and considering that the surface is defined by a first-degree polynomial function, it is determined that $\tilde{p} \to 0$ when $\frac{\lambda_{iV}}{\lambda_{pV}} > 0$.

3.5. Sliding Mode Formation Control with Dynamic Sliding Mode Dynamic Compensation (SMC-SMCV)

For this controller, the scheme shown in Figure 3 is used. The outer formation control is defined by the SMC controller described in Section 3.2, while the inner control loop is managed by a controller defined as SMCV, which is given by the SMS controller in Section 3.4.

3.6. Stability Analysis of the PD Formation Controller with Dynamic Compensation (PD-DI)

For the stability analysis of the cascade system, the following Lyapunov is used:

$$V_C = \frac{1}{2}\tilde{v}^T\tilde{v} + \frac{1}{2}\tilde{q}^T\tilde{q} \tag{18}$$

From this, the following is derived:

$$\dot{V}_C = \tilde{v}^T\dot{\tilde{v}} + \tilde{q}^T\dot{\tilde{q}} \tag{19}$$

The definition of errors is replaced as follows:

$$\dot{V}_C = \tilde{v}^T\left(\dot{U}_c - \dot{U}_r\right) + \tilde{q}^T\left(\dot{q}_d - \dot{q}\right) \tag{20}$$

In a closed loop, replacing (5) and (9) in (20) and developing the expression, the following is obtained:

$$\dot{V}_C = -\tilde{v}^T Q_D tanh\left(\tilde{v}\right) - \tilde{q}^T k_1 \tilde{q} \tag{21}$$

It is shown that $\dot{V}_C < 0$ whereby $\tilde{v} \to 0$ *and* $\tilde{q} \to 0$ with $t \to \infty$ are verified.

3.7. Stability Analysis of the PD Formation Controller with Dynamic Sliding Mode Control (PD-SMCV)

For the stability analysis of the cascaded system, we employed the Lyapunov (18), (19), and (20) in a closed loop and replaced (5), (15), and (17) in (20); it was shown that $s \to 0$ with $t \to \infty$, and after developing the expression, the following is obtained:

$$\dot{V}_C = -\tilde{v}^T Q_D tanh\left(\tilde{v}\right) - \tilde{q}^T \frac{\lambda_{iF}}{\lambda_{pF}}\tilde{q} \tag{22}$$

It is shown that $\dot{V}_C < 0$ whereby $\tilde{v} \to 0$ and $\tilde{q} \to 0$ with $t \to \infty$ are verified.

3.8. Stability Analysis of the Sliding Mode Formation Control with Dynamic Sliding Mode Dynamic Compensation (SMC-SMCV)

For the stability analysis of the cascade system, the following Lyapunov is used:

$$V_C = \frac{1}{2}\tilde{p}^T\tilde{p} + \frac{1}{2}\tilde{q}^T\tilde{q} \tag{23}$$

By differentiating it in a closed loop and replacing (7), (15), and (17) in (23), it is also shown that $s \to 0$ with $t \to \infty$, and after developing the expression, the following is obtained:

$$\dot{V}_C = -\tilde{p}^T \frac{\lambda_{iV}}{\lambda_{pV}}\tilde{p} - \tilde{q}^T \frac{\lambda_{iF}}{\lambda_{pF}}\tilde{q} \tag{24}$$

It is shown that $\dot{V}_C < 0$ whereby $\tilde{p} \to 0$ and $\tilde{q} \to 0$ with $t \to \infty$ are verified.

4. Results

To evaluate the trajectory tracking of the formation of the three mobile robots using the five proposed control techniques, two experiments were conducted: Experiment 1:

trajectory tracking of the formation of the robots with the five proposed controllers without perturbations; and Experiment 2: trajectory tracking of the formation of robots with the five proposed controllers with perturbations, to evaluate the robustness. The methodology implemented to define the parameters of the controllers was to select small initial values for the controllers to avoid abrupt control actions. The constants were assumed to have the following values $\lambda_{pF} = \lambda_{iF} = \lambda_{pV} = \lambda_{iV} = \delta_C = k_V = 0.1$, $Q_D = [10;\ 01]$, and $k_1 = [11111]$. For each controller, the parameters were modified according to the error curves obtained, and the integral of squared error (ISE) value was calculated. For the smallest ISE values, the following parameters were obtained: PD formation controller: diagonal 6×6 matrix k_1 with values [11251]; SMC formation controller: $\lambda_{pF} = 1$, $\lambda_{iF} = 1$, and $\delta_C = 0.6$; dynamic compensation controller: diagonal 2×2 matrix $Q_D = [70; 07]_{2x2}$; and SMCV controller: $\lambda_{pV} = 1$, $\lambda_{iV} = 1$, and $k_V = 1.5$. The parameters of the trajectory used in the tests were $x_d(t) = 1 + 10\,cos(0.01t)$ and $y_d(t) = 1 + 10\,sin(0.01t)$.

4.1. Experiment 1: Path Following without Disturbance

This experiment had a duration of 50 s. Figure 4 shows the pose and shape errors. Analyzing the results, the controller with the best performance (faster stabilization) was the PD-SMCV controller, which implemented PD formation control and SMC dynamic compensation. It can also be observed that controllers with SMC formation control exhibited oscillatory and abrupt variations in error, resulting in longer settling times.

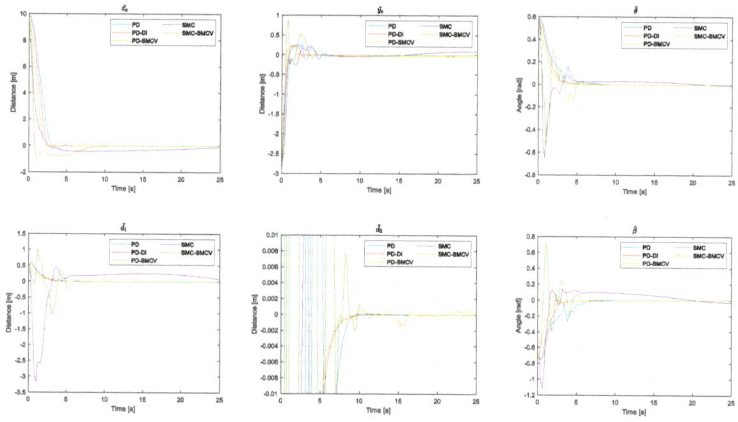

Figure 4. Formation errors.

Figure 5 shows the trajectory followed by the centroid of the triangle formation of the robots. For the trajectory, a circle with a radius of 10 m was set. In this graph, it can be observed that the PD, PD-DI, and PD-SMC controllers describe a smoother path with less oscillation and similar settling times. The SMC and SMC-SMCV controllers resulted in a rougher response, with a more pronounced overshoot at the start of the path. Analyzing this figure, it can be observed that, consistent with the error analysis, the SMC-SMCV controller provided a smoother response with a shorter settling time.

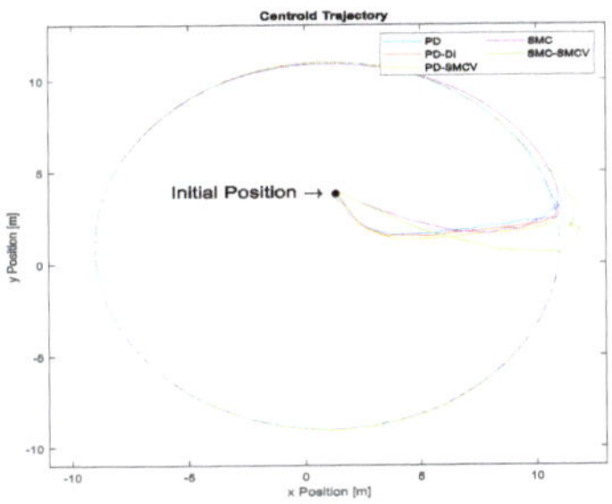

Figure 5. Centroid formation trajectory (Without Disturbance).

4.2. Experiment 2: Path Following Applying Dynamic Disturbance

From $t = 50$ s onwards, a disturbance was introduced to evaluate the robustness of the controllers. This disturbance was meant to simulate an external force that alters the dynamic behavior of the robots, affecting the path-following capabilities of the controllers. To modify the dynamic properties of the robot model, the parameters were multiplied by a constant $k = 15$.

Figure 6 shows the formation errors. In this experiment, it can be observed that the PD-SMCV controller resulted in lower errors and faster settling times, indicating that this controller better corrected the dynamic perturbations of the system compared to the other controllers, which exhibited higher errors. The PD-SMCV controller presented errors of less than 0.5 m for the centroid positions, while the PD controller, which lacked dynamic compensation, resulted in errors greater than 6 [m] for the centroid position. In contrast, the PD-DI, SMC, and SMC-SMCV controllers presented errors of less than 2 m. Additionally, it can be observed that controllers implementing the SMC formation control exhibited significantly higher errors and oscillations in the formation variables θ, d_1, and β. Figure 7 shows the centroid formation trajectory. It is observed that most of the controllers exhibited a significant deviation between the path followed and the reference, except for the PD-SMCV controller, which effectively corrected the disturbance and aligned closely with the reference.

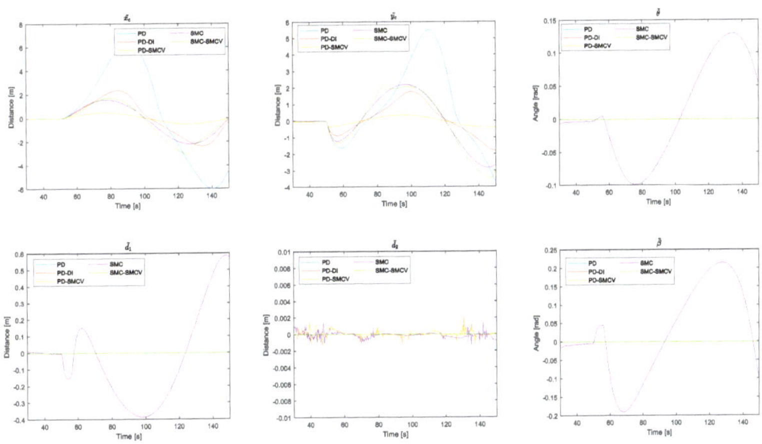

Figure 6. Formation errors.

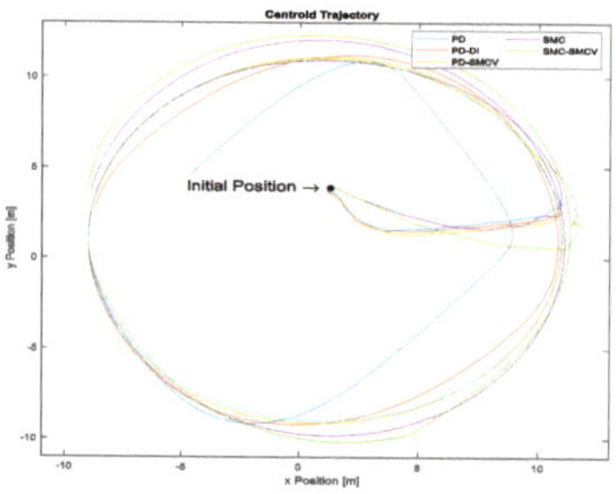

Figure 7. Centroid formation trajectory (With Disturbance).

5. Discussion

Based on the results in Tables 1 and 2, it is determined that the PD-DI and PD-SMCV controllers resulted in a more robust performance in following the desired trajectory and maintaining the specified formation. These controllers have lower ISE values, indicating reduced error and higher accuracy in formation control. In addition, the SMC-SMCV controller showed poor performance in all experiments analyzed, worsening especially when larger perturbations were applied to the dynamic parameters. This indicates that, although SMC controllers are useful due to their robustness, combining SMC controllers for formation control (outer loop) and dynamic compensation (inner loop) may be disadvantageous. Finally, based on the obtained plots and ISE values, the controller with the best results was the PD-SMCV controller.

Table 1. ISE for each controller in the path following of the formation without disturbance.

PD	0.536	2.034	0.719	207.505	10.258	0.453
PD-DI	0.361	1.770	0.617	187.605	7.691	0.319
PD-SMCV	0.481	1.358	1.104	156.313	6.212	0.480
SMC	0.476	1.960	0.479	38.100	4.479	0.727
SMC–SMCV	6.195	1.377	1.494	44.890	4.940	0.599

Table 2. ISE for each controller in the path following of the formation with disturbance.

Controller	$\tilde{d}_1$	$\tilde{d}_2$	$\tilde{\beta}$	$\tilde{x}_c$	$\tilde{y}_c$	$\tilde{\theta}$
PD	0.3726	2.0811	0.7767	1.80E+03	708.0329	0.3766
PD-DI	0.3307	1.8614	0.5798	328.021	99.8266	0.2986
PD-SMCV	0.211	1.93	0.3269	79.8945	8.5865	0.2118
SMC	20.4928	1.4243	2.8142	221.6967	269.3052	0.9635
SMC–SMCV	1.3112	1.5882	0.557	275.4384	335.2991	0.271

6. Conclusions

Based on the results of the experiments performed, the PD-SMCV controller demonstrated the best performance compared to the other controllers without disturbance. In the robustness test with the introduction of a disturbance, the controllers without dynamic compensation did not perform well. Although the SMC controller lacked dynamic compensation control, it showed remarkable robustness, outperforming the PD controller. According to ISE indicators, the PD-SMCV controller, which combined PD forming control with SMCV dynamic compensation for the inner loop, not only quickly reduced the error in forming variables, but also maintained stable and accurate robot forming in the presence of disturbances. In summary, the PD-SMCV controller excels in its ability to deliver superior performance in robotic formation trajectory tracking and disturbance management, reaffirming its effectiveness by combining advanced control techniques with dynamic compensation.

Author Contributions: Conceptualization, G.M.A. and P.L.; investigation and methodology, A.C., A.V., E.P. and R.L.; design of control laws, G.M.A. and P.L.; software and validation, A.C., A.V., E.P. and R.L.; writing—original draft preparation, A.C., A.V., E.P. and R.L.; writing—review and editing, G.M.A. and P.L.; supervision, P.L. All authors have read and agreed to the published version of the manuscript.

Funding: This research received no external funding.

Institutional Review Board Statement: Not applicable.

Informed Consent Statement: Not applicable.

Data Availability Statement: Data are contained within the article.

Acknowledgments: The authors would like to thank the GIECAR group and ARCI for the technical support for the work carried out, as well as the ESPE-LA and Escuela Politécnica Nacional under the PIEX-DACI-ESPE-24 project and PIS-23-09 project.

Conflicts of Interest: The authors declare no conflicts of interest.

References

1. An, X.; Wu, C.; Lin, Y.; Lin, M.; Yoshinaga, T.; Ji, Y. Multi-Robot Systems and Cooperative Object Transport: Communications, Platforms, and Challenges. *IEEE Open J. Comput. Soc.* **2023**, *4*, 23–36. [CrossRef]
2. Huang, T.; Li, B.; Shah, A.; Qin, N.; Huang, D. Fuzzy Sliding Mode Control for a Quadrotor UAV. *DDCLS* **2019**, *8*, 672–677. [CrossRef]
3. Liu, Q.; Nie, Z.; Gong, Z.; Liu, X.-J. An Omnidirectional Transportation System With High Terrain Adaptability and Flexible Configurations Using Multiple Nonholonomic Mobile Robots. *IEEE Robot. Autom. Lett.* **2023**, *8*, 6060–6067. [CrossRef]
4. Liu, Z.; Li, P. Robust multi-mobile robot formation control: A fully actuated system control approach. In Proceedings of the 2024 3rd Conference on Fully Actuated System Theory and Applications (FASTA), Shenzhen, China, 10–12 May 2024. [CrossRef]
5. Leica, P.; Balseca, J.; Cabascango, D.; Chávez, D.; Andaluz, G.; Andaluz, V.H. Controller Based on Null Space and Sliding Mode (NSB-SMC) for Bidirectional Teleoperation of Mobile Robots Formation in an Environment with Obstacles. In Proceedings of the 2019 IEEE Fourth Ecuador Technical Chapters Meeting (ETCM), Guayaquil, Ecuador, 11–15 November 2019. [CrossRef]
6. Liu, H.; Nie, J.; Sun, J.; Chen, G.; Zou, L. Adaptive sliding Mode Control for Nonholonomic Mobile Robots based on Neural Networks. In Proceedings of the 2019 Chinese Control and Decision Conference (CCDC), Nanchang, China, 3–5 June 2019. [CrossRef]
7. Mevo, B.B.; Saad, M.R.; Fareh, R. Adaptive Sliding Mode Control of Wheeled Mobile Robot with Nonlinear Model and Uncertainties. In Proceedings of the 2018 IEEE Canadian Conference on Electrical & Computer Engineering (CCECE), Quebec, QC, Canada, 13–16 May 2018. [CrossRef]
8. Juang, C.-F.; Lu, C.-H.; Huang, C.-A. Navigation of Three Cooperative Object-Transportation Robots Using a Multistage Evolutionary Fuzzy Control Approach. *IEEE Trans. Cybern.* **2022**, *52*, 3606–3619. [CrossRef] [PubMed]
9. Yu, D.; Chen, C.L.P.; Xu, H. Fuzzy Swarm Control Based on Sliding-Mode Strategy with Self-Organized Omnidirectional Mobile Robots System. *IEEE Trans. Syst. Man Cybern. Syst.* **2022**, *52*, 2262–2274. [CrossRef]
10. Rodriguez-Canale, E.S.; Costa, O.L.V. Formation Static Output Control of Linear Multi-Agent Systems with Hidden Markov Switching Network Topologies. *IEEE Access* **2021**, *9*, 132278–132289. [CrossRef]
11. Liu, X.; Deng, F.; Wei, W.; Wan, F. Formation Tracking Control of Networked Systems with Time-Varying Delays and Sampling Under Fixed and Markovian Switching Topology. *IEEE Trans. Control Netw. Syst.* **2022**, *9*, 601–612. [CrossRef]
12. Kang, Y.; Li, Z.; Dong, Y.; Xi, H. Markovian-Based Fault-Tolerant Control for Wheeled Mobile Manipulators. *IEEE Trans. Control Syst. Technol.* **2012**, *20*, 266–276. [CrossRef]
13. Qin, Z.; Xiang, S.; Sun, S.; Cai, Y.; Wang, S. Formation Control and Obstacle Avoidance for Multi-Robot Systems. In Proceedings of the 2022 First International Conference on Cyber-Energy Systems and Intelligent Energy (ICCSIE), Shenyang, China, 14–15 January 2023. [CrossRef]
14. Alfaro, A.; Morán, A. Leader-Follower Formation Control of Nonholonomic Mobile Robots. In Proceedings of the 2020 IEEE ANDESCON, Quito, Ecuador, 13–16 October 2020. [CrossRef]
15. Adderson, R.; Akbari, B.; Pan, Y.; Zhu, H. Multileader and Role-Based Time-Varying Formation Using GP Inference and Sliding-Mode Control. *IEEE/ASME Trans. Mechatronics* **2024**. [CrossRef]
16. Boo, J.; Chwa, D. Fuzzy Integral Sliding Mode Observer-Based Formation Control of Mobile Robots with Kinematic Disturbance and Unknown Leader and Follower Velocities. *IEEE Access* **2022**, *10*, 76926–76938. [CrossRef]
17. Lu, S.; Zhao, J. Research on Tracking Control of Circular Trajectory of Robot Based on the Variable Integral Sliding Mode PD Control Algorithm. *IEEE Access* **2020**, *8*, 204194–204202. [CrossRef]
18. Andaluz, G.M.; Andaluz, V.H.; Terán, H.C.; Arteaga, O.; Chicaiza, F.A.; Varela, J.; Ortiz, J.S.; Pérez, F.; Rivas, D.; Sánchez, J.S.; et al. Modeling Dynamic of the Human-Wheelchair System Applied to NMPC. In *Intelligent Robotics and Applications. ICIRA 2016*; Springer: Cham, Switzerland, 2016. [CrossRef]
19. Andaluz, G.M.; Leica, P.; Herrera, M.; Morales, L.; Camacho, O. Hybrid Controller based on Null Space and Consensus Algorithms for Mobile Robot Formation. *Emerg. Sci. J.* **2022**, *6*, 429–447. [CrossRef]
20. Moya, V.; Espinosa, V.; Chávez, D.; Leica, P.; Camacho, O. Trajectory tracking for quadcopter's formation with two control strategies. 2016 IEEE Ecuador Technical Chapters Meeting, Guayaquil, Ecuador, 12—14 October 2016; pp. 1–6. [CrossRef]
21. Morales, B.; Carelli, R. Robot control with inverse dynamics and non-linear gains. *Lat. Am. Appl. Res.* **2003**, *33*, 393–397.

engineering proceedings

MDPI

Proceeding Paper

Design and Construction of a Controlled Solid-State Relay with Variable Duty Ratio for DOMOTIC Applications †

Jorge Medina *, Kevin Barros, William Chamorro and Juan Ramírez

Departamento de Automatización y Control Industrial, Facultad de Ingeniería Eléctrica y Electrónica, Escuela Politécnica Nacional, Quito 170525, Ecuador; kevin.barros02@epn.edu.ec (K.B.); william.chamorro@epn.edu.ec (W.C.); juan.ramirezd@epn.edu.ec (J.R.)

* Correspondence: jorge.medinaj@epn.edu.ec

† Presented at the XXXII Conference on Electrical and Electronic Engineering, Quito, Ecuador, 12–15 November 2024.

Abstract: This paper proposes the design and construction of the prototype of a solid-state relay (SSR) that is controlled remotely through an interface developed in an Android application using a WIFI connection. Likewise, the prototype has a system for measuring electrical variables such as voltage, current, and power factor, whose values are also visualized in the application for monitoring the system's load. Experimental results demonstrate the effective control of various load profiles, including resistive and resistive–inductive loads. The SSR successfully regulates the firing angle of an electronic device called TRIAC, allowing precise control over the load. Key features include a network snubber and heatsink, enhancing the durability and reliability of the system. The main contribution of this work is the integration of IoT-based remote control and monitoring with a robust SSR design, offering enhanced functionality and reliability for domotic applications. This integration facilitates improved productivity, resource management, and equipment monitoring in smart home environments, addressing the current gap in the availability of intelligent SSR solutions.

Keywords: solid-state relay; optoisolator; TRIAC; network snubber; heat sink; power measurement; WIFI; remote drive; AC power; Android app

Citation: Medina, J.; Barros, K.; Chamorro, W.; Ramírez, J. Design and Construction of a Controlled Solid-State Relay with Variable Duty Ratio for DOMOTIC Applications. *Eng. Proc.* **2024**, *77*, 14. https://doi.org/10.3390/engproc2024077014

Academic Editor: Jaime Cepeda

Published: 18 November 2024

1. Introduction

Solid-state relays (SSR) are essential devices used to control AC or DC electrical circuits by switching them on or off. These devices can handle a wide range of loads, from low-power light bulbs to industrial electric motors, allowing the activation of a power load using a simple control signal [1,2]. The advent of the Internet of Things (IoT) has revolutionized both domestic tasks and industrial operations, particularly through domotic applications. These applications enable wireless connectivity between devices, facilitating real-time data exchange via technologies such as WIFI. Integrating IoT into daily activities enhances productivity, resource management, equipment monitoring, etc. [3–5]. Using wireless technologies, it is possible to convert ordinary homes into smart places such that there is the opportunity to reduce energy use while maintaining optimal comfort and convenience [6].

The proliferation of smart homes and the Internet of Things has fueled a demand for intelligent and interconnected devices that enhance comfort, energy efficiency, and automation [7]. However, traditional SSRs provide reliable switching but lack the flexibility and intelligence needed for seamless integration into modern smart home ecosystems. In the context of domotic applications, which emphasize automation, energy efficiency, and remote control, conventional SSRs often fall short due to their limited ability to adjust load characteristics and fully integrate them into smart home systems [8]. Researchers have explored various approaches to enhance the capabilities of SSRs for domotic applications [9]. For instance, Setiyo Budiyanto et al. [10] proposed a system for controlling home appliances using Arduino with a web server for remote control and monitoring. The need for a variable duty ratio SSR arises from the requirement to finely tune the power supplied to different types of loads, improving both energy efficiency and system reliability. Conventional SSRs with fixed duty ratios lack the necessary adaptability. Additionally, the absence of remote monitoring and control features in these traditional systems limits their usability in modern IoT-enabled smart homes. In this context, this work presents the design and construction of a remotely controlled solid-state relay prototype with a variable duty ratio for domotic applications, proposing an efficient and improved alternative to electromechanical relays [11]. The prototype includes a measurement system for voltage, current, and power factor, all of which can be monitored via an Android application using WIFI. The system is designed to handle a domestic load of up to 120 V at 10 A, utilizing a power semiconductor to regulate the RMS voltage. Additional features include a network snubber and a heatsink to extend the system's lifespan. The primary contribution of this work lies in integrating IoT-based remote control and monitoring with a robust SSR design, providing enhanced functionality and reliability for domotic applications. This combination improves productivity, resource management, and equipment monitoring in smart home environments, effectively addressing the current lack of intelligent SSR solutions. The following sections detail the prototype's design, experimental setup, results, and conclusions.

2. Materials and Methods

The approach of this work is quantitative, as it involves measuring electrical variables such as voltage, current, and power factor. Additionally, mathematical methods are used to size active and passive elements that are part of the circuitry of this controlled solid-state relay. This section deals with the characteristics of solid-state relays using compact, reliable, high-speed semiconductors with an increased service life, which allows them to feed the load despite a sinusoidal voltage curve [12].

2.1. Solid-State Relays

A solid-state relay is a switching device that has no moving parts. When a control signal is applied, these devices provide a path for the AC load current in the same manner as the moving contacts of a mechanical switch [13]. The switching carried out by synchronous SSRs is established at the zero crossing of the alternating current cycle. For this reason, the probability of generating some type of electronic noise is reduced such that this type of relay is significantly used in environments where devices that are susceptible to radio frequency interference (RFI) must be controlled [14,15]. A solid-state relay uses an optocoupler for its trigger circuit. This device, utilizing photodiodes and LEDs, provides a reliable switching mechanism without moving parts [16]. According to the level of voltage, current, and switching frequency that must be handled, a photosensitive thyristor (SCR) or a triode for alternating current (TRIAC) can be used to replace the phototransistor located on the output side [17]. A complete structure of an SSR (see Figure 1) has a larger quantity of additional elements than the LED and the photosensitive semiconductor since it requires a power semiconductor, which activates the load and other circuits such as snubbers and zero crossing detectors.

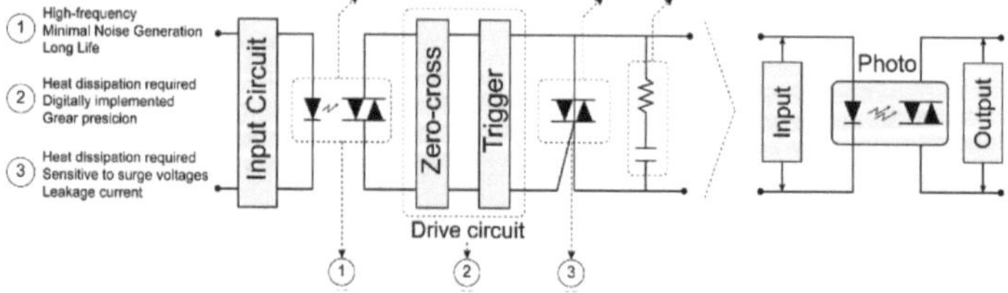

Figure 1. SSR internal structure [17].

When compared to conventional mechanical relays, a solid-state relay (SSR) has a longer lifespan due to the presence of optical isolation and the absence of mechanical components. However, an SSR may be more prone to failure from overloads, making the use of snubber networks highly important in its implementation [14].

2.2. *Power Semiconductor*

The incorporation of power semiconductor-driven SSRs with IoT software, enhanced power management strategies, and state-of-the-art reliability testing procedures highlight their benefits and possibilities in various applications. This study allows for an analysis of the crucial role played by the power semiconductor in controlling the load voltage in the prototype [18]. To select the power semiconductor for the prototype of the controlled solid-state relay, it is essential to consider the voltage to be handled as well as the type of current and its value. In this case, it is proposed to supply a load no greater than 120 V @ 10 A through an alternating current (AC) control.

The solid-state relay circuit connects and disconnects system loads based on a control signal. In this context, the TRIAC, a device from the thyristor family, is selected as the controlled power semiconductor. This element can be used in the prototype since it is used to switch alternating current as it is a bidirectional semiconductor, as proposed in [19]. TRIAC meets the design requirements in terms of voltage and current levels, as well as the value of the switching frequency, which is 120 Hz.

2.3. *Design of the Solid-State Relay*

2.3.1. Zero-Crossing Detector Circuit

The sinusoidal signal of the AC supply allows the TRIAC to deactivate at each zero-crossing point as the current value drops below the holding current. Therefore, to ensure proper firing angle control, it is essential to determine the points where the zero crossings of the voltage signal corresponding to the AC supply occur. Figure 2a shows the schematic of the circuit implemented for the zero-crossing detector.

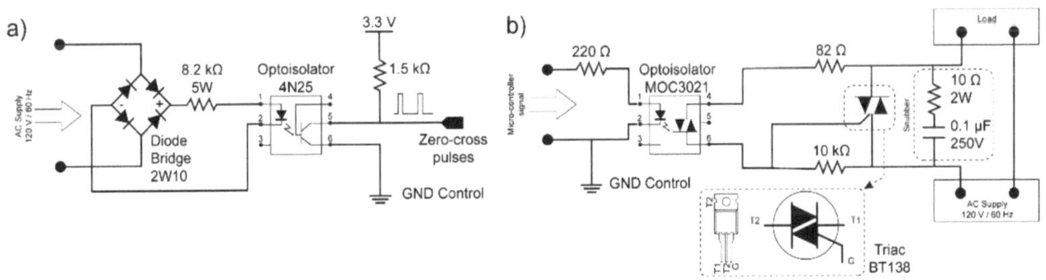

Figure 2. Implemented circuits: (**a**) zero-crossing detector, (**b**) controlled triggering.

2.3.2. TRIAC Selection

In order to select a suitable model for the TRIAC, it is fundamental to consider the voltage and current values that the semiconductor must withstand in the on state (fully ON) and in the cut-off region (fully OFF). On state current that the TRIAC must handle corresponds to the maximum RMS value of 10 A, which is variable according to the load requirements. Also, the voltage that the device must withstand is the phase-neutral voltage of the AC supply, which has a maximum RMS value of 120 V; that is, the repetitive peak voltage that the TRIAC will handle results in a value of 169.71 V. According to the previously mentioned values, TRIAC to be used in the solid-state relay prototype is BT138 model. The main characteristics of the selected TRIAC can be shown in Table 1.

Table 1. Main parameters of TRIAC BT138 [20].

Parameter	Symbol	Value
Repetitive peak off-state voltage	V_{DRM}	600 V
RMS on-state current	$I_{T(RMS)}$	12 A
Gate trigger voltage	V_{GT}	1.5 V
Operating junction temperature	T_J	125 °C
Thermal resistance junction to mounting base	$R_{\theta j-c}$	1.5 K/W

2.3.3. TRIAC Trigger Circuit

The circuit detailed in Figure 2b allows the TRIAC to be safely triggered using an optoisolator, which is activated through one of the microcontroller's pins. Figure 2b shows the schematic of the circuit implemented for triggering the BT138 TRIAC. It is fundamental to mention that the TRIAC gate is excited with a phase control based on the activation of the system through the firing angle of the semiconductors. Control of the supply voltage is carried out by varying the firing angle of the thyristor gate. Alpha firing angle (α) is in a range between 0° and 180°. The implementation of the circuit for phase control turns out to be simple and reliable [21].

2.3.4. Snubber Network and Heat-Sink

The need for implementing a snubber network in the prototype arises from issues related to the unwanted triggering characteristics of the TRIAC, which occur when the load is significantly reactive, potentially exceeding the critical rate of change of the commutating voltage (dV_{com}/dt) [22]. To solve this problem, the implementation of an RC snubber network between terminals MT1 and MT2 is required since this configuration limits abrupt voltage variations. The prototype requirements in terms of voltage and current levels are $V_O = 169.71$ V, $I = 10$ A, and $f_s = 120$ Hz. In addition, we used the Quick Snubber Design proposed by Cornell Dubilier in [23]. The complete circuit of the controlled solid-state relay is shown in Figure 2b.

To enhance the protection of the controlled solid-state relay prototype, a heat sink is proposed to maintain the necessary thermal flow and prevent the semiconductor from reaching its melting point. For this purpose, it is essential to know the value of the TRIAC power dissipation, which is influenced by the load current and can be calculated using the equations from [24,25]. In this case, TRIAC power dissipation is $P = 13.74$ W. This is based on that value and other parameters, which are ambient temperature, junction temperature, and the thermal resistances of the system. In addition, we used a direct contact and the analysis of [24]; the junction-to-ambient thermal resistance is $R_{thj-a} = 5.53$ °C/W. Accordingly, the model of the heat sink is TEA40 [26], whose thermal resistance is 5.6 °C/W.

2.4. *Power Measurement System*

For selecting a power measurement system that focuses on acquiring voltage, current, and power factor data of the load, the requirements of the maximum RMS values 120 V @ 10 A were considered. So, a PZEM-004T V3.0 AC communication module (see Figure 3) is used, which measures AC voltage, current, active power, frequency, power factor, and the energy of the application to which it is connected. Data obtained are sent to a microcontroller through WIFI wireless communication, which presents the information in an Android application for system monitoring.

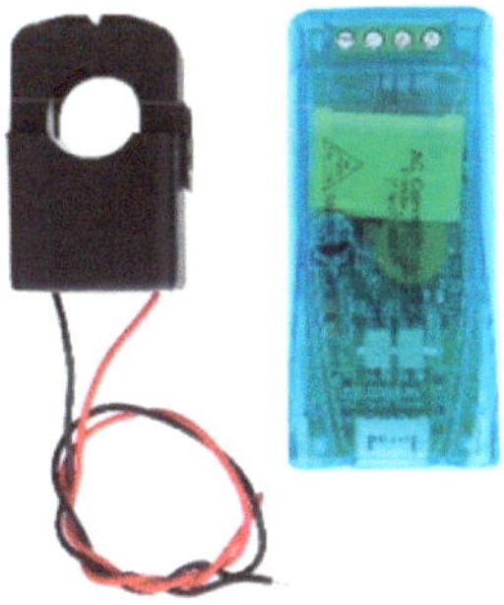

Figure 3. PZEM-004T V3.0 AC module [27].

2.5. Android Application for Remote Driving and Monitoring of the Solid-State Relay Using WIFIi

To design the Android application, the Arduino Cloud interface was selected, as it supports developing projects with embedded systems through Wi-Fi connections. Among the compatible devices are ESP32 microcontrollers since these devices have an integrated WIFI module [28]. The configuration of ESP32 boards in Arduino Cloud is carried out using a device ID and a secret key. These tools, along with WIFI credentials, are ideal for connecting to the Arduino database. In this way, it is possible to generate the operation as well as remote monitoring of the prototype through a wireless network with an Internet connection. The Android application design is shown in Figure 4.

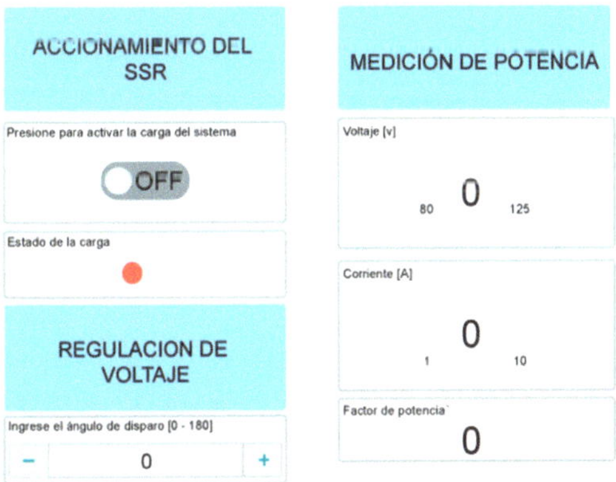

Figure 4. Arduino Cloud interface.

3. Results and Discussion

This section shows the operation of the prototype with different configurations of the system load, such as a resistive load (R) and a resistive–inductive load (RL). The results demonstrate the system's ability to regulate the firing angle of the TRIAC and manage the load effectively.

For the prototype to function properly, it is important to consider that losses are generated during operation. These losses are caused by the elements within the SSR's internal structure. For this case study, the main losses are found in the power semiconductor, which is a TRIAC BT138 with a value of 13.74 W, as well as in the resistance that is located after the diode bridge, that is, at the AC supply with a value of 0.82 W. These losses have a total estimated value of 14.56 W, which represents 1.21% of the maximum power that can be supplied to the load. This is considering that the maximum power is 1.2 kW when there is no phase lag between the voltage and the current.

3.1. Resistive Load (R) Tests

To validate the prototype's functionality, a 100 W incandescent light bulb was used as a resistive load. The firing angle (α) was regulated through the Android application, allowing control over the bulb's illumination. It was possible to analyze and select the appropriate components to carry out the design and implementation of the power, control, measurement, and remote communication circuit of the system. By connecting the bulb as a load, the firing angle (α) can be regulated throughout its range, that is, from 0° to 180°, where 0 electrical degrees determines that the load is fully on, with the bulb receiving the complete mains voltage and displaying a full sine waveform. The trigger angle is regulated with the Android application, as shown in Figure 5a.

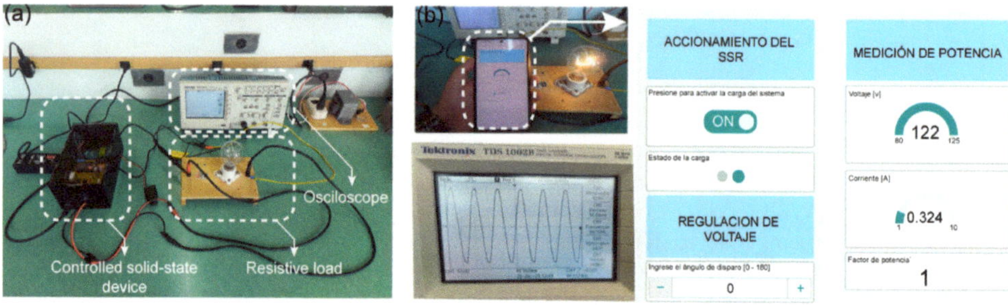

Figure 5. Experiment setup: (**a**) setup for resistive load, (**b**) measurements at 0 degrees.

The results of the voltage, current, and power factor measurements using the PZEM-004T V3.0 module are shown in Figure 5b. On the other hand, at a firing angle of 180 electrical degrees, the load is off, as shown in Figure 6a, as the bulb does not receive any voltage and remains unlit. This demonstrates the system's ability to completely control the load state through the firing angle adjustment.

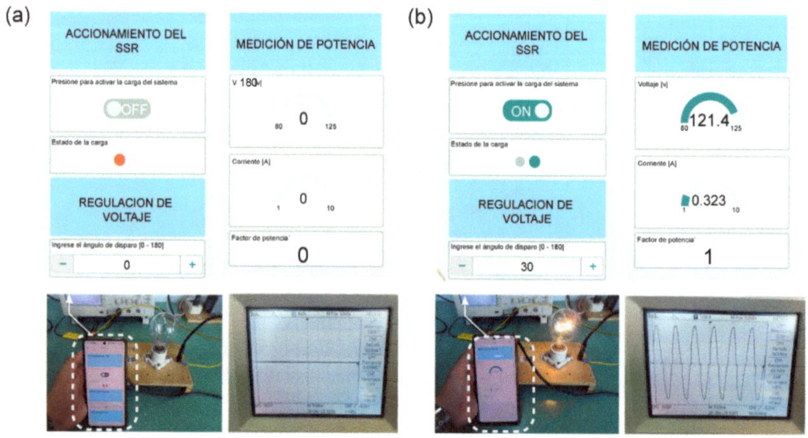

Figure 6. Experiments at different shooting angles: (**a**) 180 degrees, (**b**) 30 degrees.

For intermediate angles, such as 30°, the bulb's illumination varies accordingly. Figure 6b shows the results at 30°, indicating precise control over the load's voltage and power factor, monitored through the Android application.

3.2. Resistive–Inductive Load (RL) Test

In this test, a 250 mH inductor was added in series with the 100W incandescent bulb, creating a resistive–inductive load, as Figure 7a shows. Various firing angles were tested to observe the system's response to the inductive component.

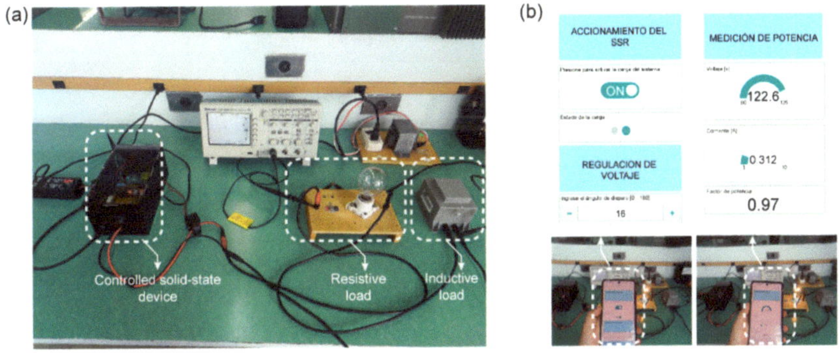

Figure 7. Experiment setup: (**a**) setup for rL load, (**b**) operation at 16 degrees.

After connecting the inductor, several tests of the firing angle were conducted. For values less than 16 electrical degrees, the voltage regulation was affected. This is due to the inductive load, which generates a phase lag between the voltage and current. Consequently, the current conduction lasts for more than 180 electrical degrees, and the TRIAC deactivates every time the current drops to zero. In this way, the electrical current is extinguished at the β angle. Because an extinction angle exists, the critical firing angle (α_c), can be defined as the limit value to go from continuous conduction mode to discontinuous conduction mode and vice versa. This parameter is very important to the operation of the prototype because it corresponds to the angle from which the AC control can be made. One may note from Figure 7b that the power factor no longer corresponds to 1, which is due to the presence of the inductor.

3.3. Prototype Tests and Theoretical Results Comparison

In order to verify the operation of the controlled solid-state relay, the following tables are shown below as a summary where the voltage, current, and power factor results obtained through the prototype and mathematical calculations are compared. Table 2 shows the measurements when there is a resistive load is used as the output of the system. The results are significantly similar for each of the firing angles. The main differences are because in the tests with the prototype, there are variations in the resistive value of the bulb, which changes as its temperature increases. The theoretical results were carried out using Equations (1)–(3), which are proposed in [29].

$$V_{rms} = V_s\left[\frac{1}{\pi}\left(\pi - \alpha + \frac{sin2\alpha}{2}\right)\right]^{1/2} \tag{1}$$

$$I_{rms} = \frac{V_{rms}}{R} = \frac{V_s}{R}\left[\frac{1}{\pi}\left(\pi - \alpha + \frac{sin2\alpha}{2}\right)\right]^{1/2} \tag{2}$$

$$pf = \sqrt{\frac{1}{\pi}\left(\pi - \alpha + \frac{sin2\alpha}{2}\right)} \tag{3}$$

where V_{rms} is the RMS voltage across the load, V_s is the mains voltage, α is the firing angle, R is the value of the bulb resistance, I_{rms} is the load RMS current, and pf is the load power factor. It is very important to consider that the approximate value of the bulb resistance is 370 Ω.

Table 2. Results with a resistive load.

Firing Angle (α)	Prototype Results			Theoretical Results			Percentage Errors		
	Voltage [V]	Current [A]	Power Factor	Voltage [V]	Current [A]	Power Factor	Voltage	Current	Power Factor
0°	122	0.324	1	120	0.324	1	1.67	0	0
30°	121.4	0.323	1	118.26	0.32	0.99	2.66	0.94	1.01
50°	115.9	0.156	1	115.5	0.152	0.94	0.35	2.63	6.38
70°	103.8	0.142	0.98	101.36	0.137	0.85	2.41	3.65	15.29
180°	0	0	0	0	0	0	-	-	-

Similarly to the previous case, Table 3 shows the measurements obtained when the system has an RL load as output. Based on this information, continuous conduction is achieved for firing angles less than 16°. Likewise, the firing angle for testing the prototype cannot be less than the value mentioned above since the load voltage and current will be sinusoidal. The theoretical results were calculated using Equations (4)–(8), which are also proposed in [29].

$$V_{rms} = V_s\left[\frac{1}{\pi}\left(\beta - \alpha + \frac{sin2\alpha}{2} - \frac{sin2\beta}{2}\right)\right]^{1/2} \tag{4}$$

$$I_{rms} = \frac{V_s cos\phi}{R}\left[\frac{\theta}{\pi} - \frac{sin\theta cos(2\alpha + \theta + \phi)}{\pi cos\phi}\right]^{1/2} \tag{5}$$

$$\theta = \beta - \alpha \tag{6}$$

$$\phi = tan^{-1}\left(\frac{\omega L}{R}\right) \tag{7}$$

$$pf = cos\phi \tag{8}$$

where V_{rms} is the RMS voltage across the load, V_s is the mains voltage, β is the extinction angle, α is the firing angle, ϕ is the impedance angle, θ is the driving angle, ω is the angular frequency, R is the value of the bulb resistance, L is the value of the inductor, I_{rms} is the load RMS current, and pf is the load power factor. It is very important to consider that the approximate value of the extinction angle is 196°.

Table 3. Results with a resistive–inductive load.

Firing Angle (α)	Prototype Results			Theoretical Results			Percentage Errors		
	Voltage [V]	Current [A]	Power Factor	Voltage [V]	Current [A]	Power Factor	Voltage	Current	Power Factor
16°	122.6	0.312	0.97	120	0.316	0.97	2.17	1.27	0
25°	119.4	0.309	0.97	117	0.306	0.97	2.05	0.98	0
70°	105.9	0.275	0.95	100.42	0.263	0.97	5.46	4.56	2.06
80°	98.5	0.266	0.93	96.29	0.252	0.97	2.3	5.56	4.12

As can be seen from the Table 4, while commercial SSRs are robust and reliable for specific applications, the developed SSR stands out by integrating IoT features (WIFI) for remote monitoring and control, which is particularly beneficial for smart home environments in domotic applications.

Table 4. Comparison with other SSR options.

Device	Max Current	Max Voltage	Control Type	Snubber	Remote Control	Datasheet Reference
Developed SSR	10 A	120 V	TRIAC	Yes	WIFIi	---
Omron G3ZA Series	10 A	240 V	TRIAC	Yes	No	[30]
Infineon FET SSR	5 A	600 V	FET	No	No	[31]
Seeed Studio S108T02 Series	8 A	600 V	TRIAC	Yes	No	[32]

4. Conclusions

The design and construction of a controlled solid-state relay (SSR) with a variable duty ratio for domotic applications were successfully completed. This project demonstrated the feasibility of integrating IoT-based remote control and monitoring with a robust SSR design, providing enhanced functionality and reliability for smart home environments. Experimental results, as shown in Tables 2 and 3, validate the system's capability to regulate the firing angle for both resistive and resistive–inductive loads. The close alignment between experimental and theoretical results highlights the precision and reliability of the proposed method in controlling different load profiles.

The proposed system can be employed for automated control of household devices, efficient energy management, and real-time monitoring of electrical parameters, making it particularly useful for smart home setups that prioritize energy efficiency and automation. Additionally, the system's ability to handle different load profiles, combined with features such as a network snubber and heatsink for enhanced durability, makes it a robust and adaptable solution for diverse domotic applications.

Furthermore, the comprehensive evaluation provided in this study shows that the SSR system not only supports seamless integration into smart home environments but also enhances the management of various electrical loads with high accuracy. Future work could explore its integration with renewable energy sources and the further optimization of control algorithms to improve system performance, extending its potential applications in both residential and industrial smart grid environments

Author Contributions: Conceptualization, K.B. and J.M.; methodology, K.B., J.M. and W.C.; software, K.B., J.M. and J.R.; validation, K.B., W.C., J.M. and J.R.; formal analysis, J.M. and K.B.; investigation, K.B., J.M., W.C. and J.R.; resources, K.B.; data curation, W.C.; writing—original draft preparation, K.B.; writing—review and editing, J.M.; visualization, W.C.; supervision, J.M. and J.R. All authors have read and agreed to the published version of the manuscript.

Funding: This research received no external funding.

Institutional Review Board Statement: Not applicable.

Informed Consent Statement: Not applicable.

Data Availability Statement: Data are contained within the article.

Conflicts of Interest: The authors declare no conflicts of interest.

References

1. Revista ElectroIndustria. ¿Qué es un Relé de Estado Sólido? 2020. Available online: https://www.emb.cl/electroindustria/articulo.mvc?xid=3691&ni=que-es-un-rele-de-estado-solido# (accessed on 18 July 2023).
2. GEYA. Aplicaciones y Usos de Relés de Estado Sólido. 2022. Available online: https://www.geya.net/es/8-solid-state-relay-applications-and-uses/#:~:text=Los%20rel%C3%A9s%20de%20estado%20s%C3%B3lido%20se%20utilizan%20ampliamente%20en%20los,se%20pueden%20utilizar%20otros%20dispositivos (accessed on 18 July 2023).
3. Cerem Comunicación. Por qué el Internet de las Cosas (iot) Promete Cambiar la Vida del ser Humano. 2022. Available online: https://www.cerem.es/blog/porque-el-internet-de-las-cosas-iot-promete-cambiar-la-vida-del-ser-humano#:~:text=Con%20IoT%20se%20puede%20programar,las%20personas%20a%20nivel%20mundial (accessed on 18 July 2023).

4. Shaw, K. ¿Qué es el Wi-Fi y Por Qué es tan Importante? 2020. Available online: https://concepto.de/wifi/#:~:text=El%20wifi%20da%20acceso%20a%20Internet%20a%20lo,o%20en%20cada%20piso%20de%20un%20edificio%20peque%C3%B1o. (accessed on 18 July 2023).
5. International Journal of Engineering and Advanced Technology (IJEAT). IoT Based Relay Operation. 2019. Available online: https://www.ijeat.org/wp-content/uploads/papers/v9i1/A1415109119.pdf (accessed on 29 July 2024). [CrossRef]
6. Miletiev, R.; Yordanov, R. Compact smart solid-state relay with WiFi connectivity. In Proceedings of the 2022 International Conference on Emerging Trends in Engineering, Technology, and Science (ETETSS), Sozopol, Bulgaria, 13–15 September 2022; pp. 67–72. [CrossRef]
7. Risteska Stojkoska, B.L.; Trivodaliev, K.V. A review of Internet of Things for smart home: Challenges and solutions. *J. Clean. Prod.* **2017**, *140*, 1454–1464. [CrossRef]
8. Mocrii, D.; Chen, Y.; Musilek, P. IoT-based smart homes: A review of system architecture, software, communications, privacy, and security. *Internet Things* **2018**, *1–2*, 81–98. [CrossRef]
9. Alaa, M.; Zaidan, A.A.; Zaidan, B.B.; Talal, M.; Kiah, M.L.M. A review of smart home applications based on Internet of Things. *J. Netw. Comput. Appl.* **2017**, *97*, 48–65. [CrossRef]
10. Budiyanto, S.; Silalahi, L.M.; Silaban, F.A.; Salamah, K.S.; Rahayu, F.; Wahyuddin, M.I.; Andryana, S. Design of control and monitoring tools for electricity use loads, and home security systems with internet of things system based on Arduino Mega 2560. *IOP Conf. Ser. Mater. Sci. Eng.* **2020**, *909*, 012020. [CrossRef]
11. Fernández, M.; Perpiñà, X.; Rebollo, J.; Vellvehi, M.; Sánchez, D.; Cabeza, T.; Llorente, S.; Jordà, X. Solid-State Relay Solutions for Induction Cooking Applications Based on Advanced Power Semiconductor Devices. *IEEE Trans. Ind. Electron.* **2019**, *66*, 1832–1841. [CrossRef]
12. Abduraimov, E.; Khalma, D. Development of contactless solid state voltage relay. In Proceedings of the 2020 RSES International Conference, Kazan, Russia, 21–26 September 2020. [CrossRef]
13. Mu, J.; Wang, L. Research on main circuit topology for a novel DC solid-state circuit breaker. In Proceedings of the IEEE Conference on Industrial Electronics and Applications, Taichung, Taiwan; 2010; pp. 926–930. [CrossRef]
14. OMEGA. Relés de Estado Sólido. 2021. Available online: https://es.omega.com/temperature/pdf/SSRL240_660.pdf (accessed on 19 September 2023).
15. Crydom SSR Ltd. Output Switching Functions of Solid-State Relays. 2011. Available online: https://pim.galco.com/Manufacturer/Crydom/TechDocument/Application%20Note/ssr_switching_functions_app3.pdf (accessed on 19 September 2023).
16. Barkana, A.; Cook, G.; McVey, E.S. A Solid-State Relay. *IEEE Trans. Ind. Gen. Appl.* **1973**, *IGA-9*, 589–594. [CrossRef]
17. Schweber, B. Cómo Regular la Corriente o el Voltaje de Forma Segura y Eficiente con los SSR. DigiKEY, 26 February 2019. Available online: https://www.digikey.com/es/articles/how-to-safely-and-efficiently-switch-current-or-voltage-using-ssrs (accessed on 19 September 2023).
18. Venugopal, K.B.; Srivani, S.G. Design and Analysis of MOSFET's as Solid-State Relays for Precise AC Load Control. In Proceedings of the 2023 International Conference on Smart Technologies in Computing, Electrical and Electronics (CSITSS), Bangalore, India, 2–4 November 2023. [CrossRef]
19. Fernández, M.; Perpiñá, X.; Vellvehi, M.; Jordà, X.; Cabeza, T.; Llorente, S. Analysis of solid state relay solutions based on different semiconductor technologies. In Proceedings of the 2017 19th European Conference on Power Electronics and Applications (EPE'17 ECCE Europe), Warsaw, Poland, 11–14 September 2017. [CrossRef]
20. Philips Semiconductors. Triacs BT138 Series. 1997. Available online: https://pdf1.alldatasheet.com/datasheet-pdf/view/16781/PHILIPS/BT138.html (accessed on 21 November 2023).
21. Valle, C. Diseño, Simulación y Control de Temperatura de un Horno Eléctrico Empleando Resistencias—Control Electrónico al Lado AC. Bachelor's Thesis, Escuela Politécnica Nacional, Quito, Ecuador, 2022.
22. González, M. Dispositivos Semiconductores de Potencia. 2015. Available online: http://orga.blog.unq.edu.ar/wp-content/uploads/sites/71/2015/06/Dispositivos-de-potencia.pdf (accessed on 30 September 2023).
23. Cornell Dubilier. Application Guide Snubber Capacitors. Available online: https://www.cde.com/resources/technical-papers/igbtAPPguide.pdf (accessed on 13 November 2023).
24. Philips Semiconductors. AN10384 Triacs: How to Calculate Power and Predict Tjmax. 2005. Available online: https://www.farnell.com/datasheets/1760767.pdf (accessed on 22 November 2023).
25. Dillon, E.; Palma, P. Diseño de Un Sistema Automático de Regulación de Voltaje y Frecuencia Generados por una Micro-Central Hidro-Eléctrica. Bachelor's Thesis, Escuela Politécnica Nacional, Quito, Ecuador, 2014.
26. DISIPA. Catálogo General 3.0. 2019. Available online: https://disipa.net/catalogos-1 (accessed on 14 November 2023).
27. Peacefair Electronic Technology. PZEM-004T V3.0 User Manual. Available online: https://innovatorsguru.com/wp-content/uploads/2019/06/PZEM-004T-V3.0-Datasheet-User-Manual.pdf (accessed on 24 November 2023).
28. Espressif Systems. ESP32 ESP-IDF Programming Guide. 2023. Available online: https://docs.espressif.com/projects/esp-idf/en/latest/esp32/esp-idf-en-master-esp32.pdf (accessed on 26 November 2023).
29. Rashid, M.H. *Electrónica de Potencia: Circuitos, Dispositivos y Aplicaciones*, 2nd ed.; Prentice Hall Hispanoamericana: Naucalpan de Juárez, Mexico, 1995.

30. OMRON. Technical Explanation for Solid-state Relays. Available online: https://www.ia.omron.com/data_pdf/guide/18/ssr_tg_e_9_2.pdf (accessed on 14 October 2024).
31. INFINEON. From Electromechanical Relays to Robust Semiconductor Solutions: Solid-State Relays with Optimized Superjunction FET Technology. Available online: https://www.infineon.com/dgdl/Infineon-Solid-state_relays_with_optimized_SJ_FET_technology-Whitepaper-v01_00-EN.pdf?fileId=8ac78c8c7bb971ed017bbaf30d3c02af&da=t (accessed on 14 October 2024).
32. SHARP. "S108T02 Series". Available online: https://files.seeedstudio.com/wiki/Grove-Solid_State_Relay/res/S208t02_datasheet.pdf (accessed on 14 October 2024).

engineering proceedings

MDPI

Proceeding Paper

AI-Driven Vishing Attacks: A Practical Approach †

Fabricio Toapanta *,‡, Belén Rivadeneira ‡, Christian Tipantuña ‡ and Danny Guamán ‡

Facultad de Ingeniería Eléctrica y Electrónica, Escuela Politécnica Nacional, Quito 170525, Ecuador; belen.rivadeneira@epn.edu.ec (B.R.); christian.tipantuna@epn.edu.ec (C.T.); danny.guaman@epn.edu.ec (D.G.)
* Correspondence: fabricio.toapanta@epn.edu.ec; Tel.: +593-987084772
† Presented at the XXXII Conference on Electrical and Electronic Engineering, Quito, Ecuador, 12–15 November 2024.
‡ These authors contributed equally to this work.

Abstract: Today, there are many security problems at the technological level, especially in telecommunications. Cybercriminals invade and steal data from any system using vector attacks such as phishing through scam mail, fake websites and phone calls. This latter form of phishing is called vishing (phishing using voice). Through vishing and using social engineering techniques, attackers can impersonate family members or friends of potential victims and obtain information or money or a specific target objective. Traditionally, to carry out vishing attacks, attackers imitated the vocabulary, voice and tone of a person known to the victim. However, with current artificial intelligence (AI) tools, obtaining synthetic voices similar or identical to the person to be impersonated is more straightforward and precise. In this regard, this paper, using ChatGPT and three AI-enabled applications for voice synthesis presents a practical approach for deploying vishing attacks in an academic environment to identify the limitations, implications and possible countermeasures to mitigate the effects on Internet users. Results demonstrate the effectiveness of vishing attacks, and the maturity level of the employed AI tools.

Keywords: cybersecurity; ChatGPT; social engineering; artificial intelligence; LLM; vishing

1. Introduction

Phishing attacks attempt to obtain personal information such as usernames, passwords and credit card details by pretending to be a trustworthy entity in an electronic communication. Communication purporting to be from popular social websites, auction sites, online payment processes or IT administrators is commonly used to lure the unsuspecting public. Furthermore, phishing emails may contain links to websites that are infected with malware [1]. Vishing attacks, a particular case of phishing, are also used to obtain private information using phone calls [2]. Even today, artificial intelligence (AI) tools are used to clone voices, so the victim hears a familiar voice, increasing the probability that the launched attack succeeds. Moreover, with suitable feedback (e.g., by obtaining personal data), AI solutions can impersonate a person and imitate their vocabulary, jargon or expressions. In this regard, this paper uses AI tools to present a practical approach for deploying vishing attacks. In this paper, ChatGPT was used to generate the text that will later be used to create audio dialogs (scripts) to send to victims. Texts from the victims, which could be obtained through social engineering or espionage techniques in real-world environments, have previously fed ChatGPT. Three AI-enabled tools have been used to generate the synthetic voices. The tests were conducted in an academic environment, and the targets were relatives or friends of university students who were aware of the nature of the experiments and their implications. Results demonstrated the effectiveness of the vishing attacks, and all collected information could be used not only to know the maturity level of the AI tools to perform this vector attack but also the possible countermeasures and actions to avoid or mitigate the effects on the users. At the end of the experiments, all individuals involved were informed about the study, the tools employed and the results obtained.

Citation: Toapanta, F.; Rivadeneira, B.; Tipantuña, C.; Guamán, D. AI-Driven Vishing Attacks: A Practical Approach. *Eng. Proc.* **2024**, *77*, 15. https://doi.org/10.3390/engproc2024077015

Published: 18 November 2024

2. Literature Review

In this work, it was possible to identify how phishing attacks, especially those that escape detection by security tools, can be generated using ChatGPT. Phishing websites, after being hosted on a web domain, are widely shared through several online communication resources, such as emails, SMS, social media, etc., and indexed by search engines. Since these attacks can cause a lot of harm within a short duration [3], threat intelligence actors, such as anti-phishing bots, utilize various automated, rule-based [4] and Machine Learning ML-based approaches [5] to identify phishing websites.

2.1. Threat Model

An attacker provides multiple prompts to ChatGPT with the intention of generating the source code for a phishing website that: (a) closely imitates the design of a popular organization's website, (b) employs various regular and evasive tactics to deceive users into sharing their sensitive information and (c) incorporates mechanisms for transmitting the obtained credentials back to the attacker [5]. These prompts are specifically designed to bypass ChatGPT's content filtering measures, making it difficult to detect the malicious intent. Once the attacker receives the generated source code, they proceed to host it on a domain, creating a live phishing website that poses a significant risk to unsuspecting users. Generating phishing attacks using ChatGPT provides the attacker with the following advantages:

Rapid Deployment: Attackers take advantage of the low cost and ease of use of ChatGPT to quickly iterate on their phishing attacks, making it more difficult for security vendors to identify and counter them at scale [6].

Technical Expertise Variability: The ease of use of ChatGPT allows attackers with varying levels of technical expertise to generate phishing attacks which employ various evasive techniques that can avoid anti-phishing detection, such as text encoding, browser fingerprinting or clickjacking [6].

Hosting and Accessibility: Attackers can utilize free hosting platforms to deploy their phishing websites, further lowering the barriers to entry and making large-scale attacks more feasible [6].

2.2. Ethical Considerations

This paper acts as a proof-of-concept to demonstrate the feasibility of generating evasive vishing attacks using ChatGPT. Since the authors are in the process of releasing their set of prompts used to design these attacks to OpenAI, they also describe their methodology for generating these attacks to encourage the development of more generalized approaches to detect and mitigate the possibility of such attacks in ChatGPT and other large language models (LLMs).

3. Background Technologies

This section will review some technologies used to carry out this work.

3.1. Large Linguistic Model (LLM)

Large Language Models (LLMs) represent an evolution from language models. Initially, language models were statistical in nature and laid the groundwork for computational linguistics. The advent of transformers has significantly increased their scale. This expansion, along with the use of extensive training corpora and advanced pre-training techniques is pivotal in areas such as AI for science, logical reasoning and embodied AI. These models undergo extensive training on vast datasets to comprehend and produce text that closely mimics human language. Typically, LLMs are endowed with hundreds of billions, or even more, parameters, honed through the processing of massive textual data. They have spearheaded substantial advancements in the realm of Natural Language Processing (NLP) [7] and find applications in a multitude of fields (e.g., risk assessment [8], programming [8], vulnerability detection [8], medical text analysis [9] and search engine optimization [7]).

Large language models (LLMs) can be used to analyze cyber threat intelligence (CTI) data from cybercrime forums, which contain extensive information and key discussions about emerging cyber threats. However, to date, the level of accuracy and efficiency of LLMs for such critical tasks has yet to be thoroughly evaluated. One study, hence, assessed the accuracy of an LLM system built on the OpenAI GPT-3.5-turbo model [9] to extract information. Various ways to enhance the model were uncovered, such as the need to help the LLM distinguish between stories and past events, as well as the need to be careful with verb tenses in prompts. Nevertheless, the results of this study highlight the efficiency and relevance of using LLMs for cyber threat intelligence [9].

3.2. *How LLMs Work?*

At a basic level, LLMs are based on ML, a subset of AI that refers to feeding a program a large amount of data to train it to identify functions from that data without human intervention [8]. LLMs use a type of ML called deep learning. Deep learning models can be trained to recognize distinctions without human intervention, although some human adjustment is usually necessary [8]. Deep learning uses probability to "learn". For example, in the sentence "The quick brown fox jumped over the lazy dog", the letters "e" and "o" are the most common, appearing several times each. From this, a deep learning model might conclude (correctly) that these characters are among those most likely to occur in Spanish text. A deep learning model cannot conclude anything based on a single sentence. However, after analyzing billions of sentences, it could learn enough to predict how to logically finish an incomplete sentence or even generate its sentences [8].

3.3. *Social Engineering (SE)*

This term has a long history, but it can be summarized by saying that social engineering is the name given to the various manipulation techniques cybercriminals use to obtain sensitive information from users. First, it was taken from computer science, specifically hackers, and second, it was taken from personal. For hackers or people engaged in computer security in general, this method is the process of using cognitive strategies and social skills to engage a specific target [9].

SE jeopardizes the security of networks, regardless of the type of perimeter security in place, from firewalls, cryptographic methods, virus detection system, etc., as it exploits the possibility that humans trust other humans more than computers or technology, thus making humans the weakest link in the security chain. Malicious activities carried out through human interactions psychologically influence a person to divulge confidential information or violate security procedures. Because of these human interactions, SE attacks are the most powerful, as they threaten all systems and networks and cannot be prevented by software or hardware solutions if people are not trained to prevent them. Cybercriminals choose these attacks when there is no way to hack a system without technical vulnerabilities [8].

Social Engineering Attacks

Given that SE is a technique that uses emotions, feelings, vulnerabilities and other mental processes and human behavior, it becomes a very dangerous tool to affect the potential victim; for example, by impersonating family members, technical support people, co-workers or trusted individuals in order to appropriate personal data and passwords or to impersonate the identity of the deceived person using memories, moments or personal information [7].

Although SE attacks differ from each other, they have a common pattern with similar phases. The pattern involves four stages [9,10]:

- Gather information about the target;
- Develop a relationship with the target;
- Exploit the available information and execute the attack; and
- Exit without leaving a trace.

Figure 1 illustrates the different stages of an SE attack.

Figure 1. Stages of SE attack, sources [11].

4. Vishing

Voice phishing is a form of phone criminal attack carried out using social engineering with the use of telephone system to look at private personal and financial information; for the use of financial work it is also referred as "vishing" [12].

Vishing is short for "voice phishing", which involves defrauding people over the phone and enticing them to divulge sensitive information. In this definition of vishing, the attacker attempts to obtain the victim's data and use it for their own benefit—typically, for their own financial advantage [12].

Voice Cloning

Voice cloning is the process in which a computer is used to generate the speech of a real person, generating a replica of their specific and unique voice through AI. This technique has become popular thanks to advances in AI and ML, which allow the replication of a person's voice with surprising accuracy; this has allowed the creation of increasingly realistic synthetic voices, and it can have both positive and negative implications [13]. For this process in general, the programs are based on artificial neural networks, which are ML systems that can identify complex patterns in training data. To clone a voice, the program first needs to be trained with a large amount of voice data from the person to be cloned, the more data provided, the more accurate the cloning will be [14].

Once the program has been trained, it can generate new audio clips that sound like the person in question. This is achieved by manipulating voice parameters such as fundamental frequency, syllable duration and intonation [15].

The expressive speech cloning framework is a multi-speaker Text-to-Speech (TTS) model conditioned by speaker encodings and style aspects of speech. Style conditioning in expressive TTS models is usually done by learning a dictionary of latent style vectors called Global Style Tokens (GSTs). Although GSTs can learn meaningful latent codes when trained on a dataset with a large variation of utterances, they have been empirically found to provide limited style control when trained on a large multi-speaker dataset with mostly neutral prosody [16].

A high-level schematic of the expressive speech cloning framework with the main components being trained separately is shown in Figure 2 [17].

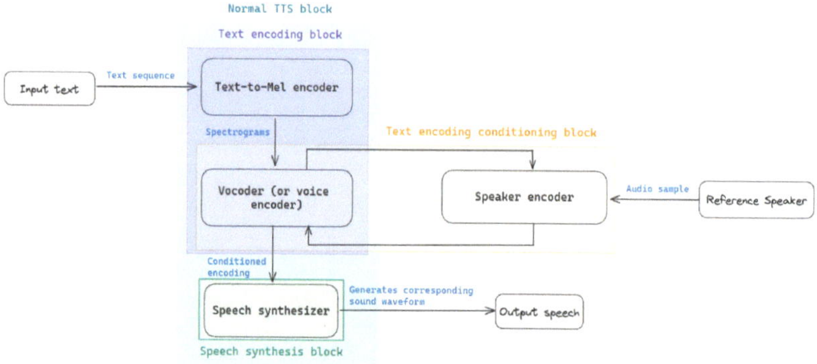

Figure 2. Expressive voice cloning model: Generic TTS model, sources [17].

5. Attack Methodology

To carry out the present project, first, the appropriate methodology is established to execute a vishing attack in a controlled environment. Also, the research participants are identified, where the target population of this study is composed of university students and other individuals.

Since this is a research project, and, to ensure the safety of the participants, informed consent was signed. In this document, participants give their consent for the use of their voice, as well as the information requested, including telephone numbers and the personal data of the persons on whom the analysis will be carried out.

The research is classified as qualitative, as it seeks to explore people's behavior through the simulation of three distinct scenarios. The first scenario involves pretending to be an agent of a financial or state entity to request financial and/or personal information. The second scenario involves simulating being a family member or friend in need of financial assistance. Finally, the third scenario consists of posing as a sales agent offering promotions, prizes or the buying and selling of goods or services.

Similarly, for this project, a conceptual model was used that includes the main aspects of SE, such as psychological factors, technical factors and sensitive information like bank data, personal information, names of family members, workplace, etc., all of which meet the established objectives as shown in Figure 3.

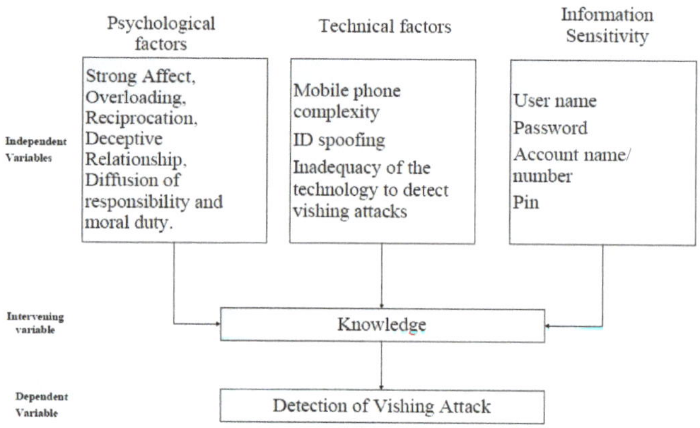

Figure 3. Conceptual model of social engineering factors for a vishing attack, based on [16].

5.1. Research Desing

For this study, a participatory action research (PAR) design was used, as its purpose is practical; to enable participants to respond to a problem using their resources (knowledge and reflection, intervention, action and resolution) [18]. A study of this nature is proposed with at least fifty participants for qualitative research, as it helps the researcher build and maintain a close relationship and thus improve the "open" relationship. A frank exchange of information [18] can help to mitigate some bias and validity.

For this purpose, an objective table is established in which six phases are established by which the type of attack and the possible results or actions to be obtained are classified and prioritized. These are the same ones that are described in Table 1, and they set the limits of the investigation, bearing in mind that the information to be received is of absolute confidentiality.

Table 1. Vishing Attack Classification and Action Phase.

Convention Phase		**Action Phase**
Specification / **Data**		**Operation**
Buy—Sell/Services	• Banks • Personals • Reference Person	• Bank transfer
Courier calls	• Family • Relationship	• Transfer (PayPal packages PayPal customs)
Banks operations	• Credit card (Photography Text)	
Home calls	• Access credentials • Application for purchase • Preparing special events	
Laboral calls	• Access credential (Physic Servers PC)	• Users
Personals calls	• Warranty • References • Contacts • Photo blackmail Extortion tactics	• Discover personal data • Conversations • Intimate photos • Threats, intimidation to obtain money or other benefit

5.2. Test Environment

For the testing environment, several regulatory guidelines were established to carry out the attacks; these are:

- The attacks will be carried out within a specified timeframe and with a limited scope regarding the dialogues used for the proposed scenarios.
- The use of phrases, jargon and words that do not discriminate against or insult the victim.
- The conversation should not exceed topics that are intimate, sexual, or of any other nature that might incite the victim to actions that could harm their physical appearance or psychological state.
- The attack will only focus on the scenarios presented, with the prior express authorization of the participants.
- The attacks will only take place on weekdays and at times that do not disrupt the victims' activities or at night.
- For the attacks, only the information provided by the participants will be used; no additional information may be used without prior authorization.

- The attacks will only be carried out on the phone numbers provided by the participants.

Additionally, it is important to highlight that a participant cannot be targeted in other scenarios, as a measure to prevent the victim from realizing that they are the object of a vishing attack.

5.3. Screenplay Composition

At this stage and as mentioned above, ChatGPT will be used; this AI tool will create the dialogues to be used in the calls and/or audio messages, using the cloned and synthesized voices of the participants so that at the time of the attack they are as realistic as possible.

Three main scenarios will be developed for this project based on three scenarios:

- Request for confidential information: generate content so that the victim accedes to giving sensitive information: mainly bank records, credit cards, etc.
- Request for financial assistance: generate content using SE to encourage the victim to agree to deliver financial assistance, using a known voice and with information obtained through social networks and from participants so that the victims offer no resistance.
- Request for information of a buy–sell service: generate content to obtain information about some good or service that the victim may possess, with the aim of trying to get more accurate information.

5.4. Procedure for Attack

Figure 4 shows the flowchart of the vishing attack procedure, which consists of three stages. In the first stage, content is generated using words that the victim commonly uses, entering them into ChatGPT to create the dialog. In the second stage, selected AI applications are used to clone and synthesize the voices, generating the voice message.

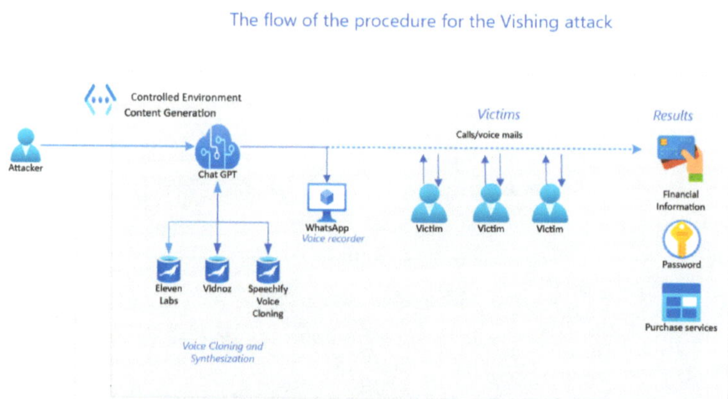

Figure 4. Flowchart of the procedure to be followed for vishing attacks.

This message is sent via WhatsApp or played during a call to the victim. Finally, in the third stage, the results are expected according to the scenarios proposed, i.e., if the victim agrees to provide the requested information or help.

6. Results

In the analysis of the attacks carried out, various types of responses were collected. The most common issue is the lack of response from the victims to audio messages and calls. Another observed reaction is, upon not recognizing the numbers from which the attacks were made, the victims chose to block those contacts.

Finally, to a lesser extent, some people responded to the audio messages and calls and even agreed to fulfill the requests made.

As can be seen in Table 2, the effectiveness percentage is calculated, and the total number of calls or audio messages sent using voices that were cloned from the different applications is counted, as are the number of blocked numbers, unanswered messages or successful attacks.

Table 2. Result of the vishing attack on the three applications.

Action/Application	Eleven Labs	Speechify Voice Cloning	Vidnoz
Attacks Performed	50	50	50
Successful	35	26	29
Locked	2	14	7
No Answer	13	10	14
% Success	70%	52%	58%

Based on the collected data, Figure 5 graphically shows the results of the attacks and the applications with which they were performed.

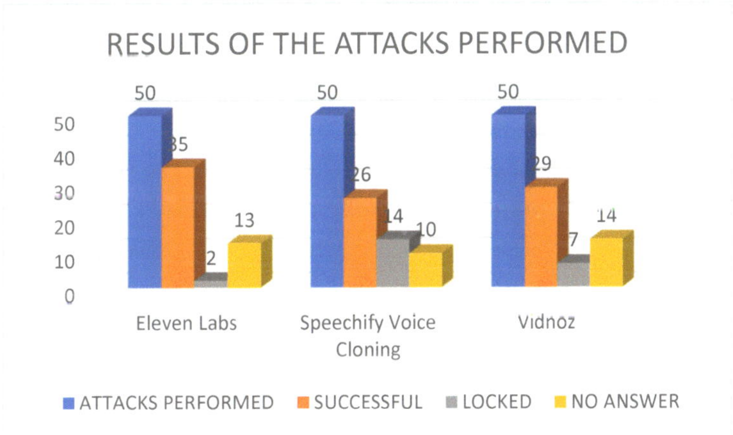

Figure 5. Overall results of vishing attacks for each application.

7. Conclusions and Lessons Learned

In conclusion, this document highlights that vishing represents a significant current global issue. The study shows that ChatGPT can generate both common vishing attacks and more specialized ones. Although it focuses on fifty individuals, this type of attack has proven effective and this methodology can easily be extrapolated to other categories of evasive attacks. Currently, the supervision of malicious content generated by large-scale language models, such as ChatGPT, is in its early stages, allowing attackers to exploit this resource to quickly implement new attacks.

Additionally, the process of voice generation and cloning can be significantly accelerated using the ChatGPT API. As for the AI applications for voice cloning and synthesis, all three applications had phonetic and word pronunciation errors. This resulted in improper phonetics, producing sounds in a spelled-out way or in a robotic manner. This problem occurred in some attacks when using unfamiliar terms for the speech synthesis applications.

This study makes three contributions:

First, an analysis of the phenomenon of social engineering is offered from theoretical and practical approaches. Through artificial intelligence, it is possible to generate content and replicate the voices of family members who are in emergency situations, thereby moving and appealing to people's solidarity.

Second, it delves into the use of artificial intelligence applications that not only replicate and synthesize voices but also create dialogues that incorporate common words and jargon, enhancing the authenticity of the interaction.

Third, it is noteworthy that artificial intelligence has various capabilities, among which is the ability to transform texts into audible dialogues. This technology can replicate any tone of voice in such a way that it becomes difficult to distinguish between the generated voice and a real voice.

The first lesson learned is that there is a limit to the number of words that can be cloned; when exceeding 1500 words, applications tend to translate inaccurately. The second lesson relates to the number of times voices can be cloned and synthesized. Since this is a paid tool, the number of available clones is limited; in the case of the applications used, the limit is 500 clones per month. Additionally, it has been observed that some uncommon words are pronounced incorrectly by the application when converted to audio, even sounding robotic at times, as they sometimes just spell them out. Although this is an emerging field and there is a lack of information about it, voice cloning applications are still under development. However, their future looks promising, with encouraging growth prospects.

Author Contributions: Conceptualization, F.T. and C.T.; validation, D.G. and B.R.; writing—original draft preparation, F.T.; writing—review and editing, C.T., B.R. and D.G. All authors have read and agreed to the published version of the manuscript.

Funding: This research received no external funding.

Institutional Review Board Statement: Not applicable.

Informed Consent Statement: Informed consent was obtained from all subjects involved in the study.

Data Availability Statement: Data is contained within the article.

Conflicts of Interest: The authors declare no conflicts of interest.

References

1. Rainie, L.; Anderson, J.; Connolly, J. Cyber-attacks likely to increase. *Digit. Life* **2019**, *2025*, 33–35.
2. Fischer, E.A. *Cybersecurity Issues and Challenges: In Brief*; Congressional Research Service: Washington, DC, USA, 2019; pp. 3–6.
3. Ali, B.; Awad, A. Cyber and physical security vulnerability assessment for IoT-based smart homes. *Sens. J.* **2018**, *18*, 817. [CrossRef] [PubMed]
4. Jang-Jaccard, J.; Nepal, A. A survey of emerging threats in cybersecurity. *J. Comput. Syst. Sci.* **2014**, *80*, 973–993. [CrossRef]
5. Ahlam, F.; Aymen, A. An Effective Blockchain-Based Defense Model for Organizations against Vishing Attacks. *Appl. Sci.* **2022**, *12*, 13020. [CrossRef]
6. Chen, Z.; Liu, B. Synthesis Lectures on Artificial Intelligence and Machine. In *Lifelong Machine Learning*, 2nd ed.; Springer: Berlin/Heidelberg, Germany, 2021; pp. 1–145.
7. Batta, M. Machine Learning Algorithms—A Review. *Int. J. Sci. Res. (IJSR)* **2020**, *9*, 381–390.
8. Aroyo, A.M.; Rea, F.; Sandini, G.; Sciutti, A. Trust and social engineering in human robot interaction: Will a robot make you disclose sensitive information, conform to its recommendations or gamble? *IEEE Robot. Autom Lett.* **2018**, *3*, 3701–3708. [CrossRef]
9. Friedman, A.A.; West, D.M. *Privacy and Security in Cloud Computing*; Issues in Technology Innovation; The Center for Technology Innovation: Washington, DC, USA, 2018; pp. 1–13.
10. Goodall, J.; Lutters, W.; Komlodi, A. Developing Expertise for Network Intrusion Detection. *Inf. Technol. People* **2009**, *22*, 92–108. [CrossRef]
11. Mukkamala, S.; Sung, A.; Abraham, A. Cyber security challenges: Designing efficient intrusion detection systems and antivirus tools. *Enhancing Comput. Secur. Smart Technol.* **2020**, 125–163.
12. Sun, N.; Zhang, J.; Rimba, P.; Gao, S.; Zhang, L.Y.; Xiang, Y. Data-driven cybersecurity incident prediction: A survey. *IEEE Commun. Surv. Tutor.* **2018**, *21*, 1744–1772. [CrossRef]
13. Sigler, K. Crypto-jacking: How cyber-criminals are exploiting the crypto-currency boom. *Comput. Fraud Secur.* **2018**, *2018*, 12–14. [CrossRef]
14. Kalniņš, R.; Puriņš, J.; Alksnis, G. Security evaluation of wireless network access points. *Appl. Comput. Syst.* **2017**, *21*, 38–45. [CrossRef]
15. Pokrovskaia, N. Social engineering and digital technologies for the security of the social capital' development. In Proceedings of the International Conference of Quality Management, Transport and Information Security, St. Petersburg, Russia, 24–30 September 2017; pp. 16–19.

16. Janczewski, L.J.; Lingyan, R.F. Social Engineering-Based Attacks: Model and New Zealand Perspective. In Proceedings of the International Multiconference on Computer Science and Information Technology, Wisla, Poland, 18–20 October 2010; pp. 847–853.
17. Salahdine, F.; Kaabouch, N. Social Engineering Attacks: A Survey Future. *Future Internet* **2019**, *11*, 89. [CrossRef]
18. Rodríguez, M.E. La Investigación Acción Participativa Compleja como Transmétodo Rizomático Transcomplejo en la Transmodernidad. *Rev. Int. Form. Profesores* **2020**, *5*, e020026.

engineering proceedings

MDPI

Proceedings

Communication System Comparison of IoT Applications Using Custom-Designed Antennas: A Basic Experimental Study †

Marco Vinueza Bustamante *, Jordan Guillén Arteaga, Carlos Yépez Vera, Aldrin Reyes Narváez and Hernan Barba Molina

Departamento de Electrónica, Telecomunicaciones y Redes de Información, Escuela Politécnica Nacional, Quito 170143, Ecuador; jordan.guillen@epn.edu.ec (J.G.A.); carlos.yepez01@epn.edu.ec (C.Y.V.); aldrin.reyesn@epn.edu.ec (A.R.N.); hernan.barba@epn.edu.ec (H.B.M.)

* Correspondence: marco.vinueza01@epn.edu.ec

† Presented at the XXXII Conference on Electrical and Electronic Engineering, Quito, Ecuador, 12–15 November 2024.

Abstract: A comparative study of the performance of a communication system for IoT applications is presented. The experiment is based on the bit error rate, which is obtained by varying the distance between two transceiver modules, each attached to a microcontroller Arduino Uno. Four scenarios are considered for our experimentation. Each scenario is mainly characterized by interchanging radiator elements which are attached to the transceiver modules. For this, two antennas are designed and implemented: a modified shape-optimized Landstorfer Yagi-Uda antenna and a printed turnstile antenna. The measurements show good agreement, with simulations having gain values of about 9 dBi and 3 dBi for the quasi Yagi-Uda structure and the turnstile antenna, respectively. System performance tests are conducted to compare the performance of the commercial solution at various distances to custom-designed antennas. These tests aim to evaluate the improvement achieved using a new set of antennas. The key to this solution is the use of a high-directivity antenna for data transmission and a circular polarized omnidirectional antenna for reception, which shows an improvement of around 60% in terms of the bit error rate during data transmission compared to the pair of commercial antennas included in the RF module.

Keywords: IoT; Landstorfer antenna; turnstile antenna; BER; performance; wireless

Citation: Vinueza Bustamante, M.; Guillen Arteaga, J.; Yépez Vera, C.; Reyes Narváez, A.; Barba Molina, H. Communication System Comparison of IoT Applications Using Custom-Designed Antennas: A Basic Experimental Study. *Eng. Proc.* **2024**, 77, 16. https://doi.org/10.3390/engproc2024077016

Academic Editor: Christian Tipantuña

Published: 18 November 2024

1. Introduction

Wireless communication allows a link between different nodes through electromagnetic waves, microwaves, or light signals [1]. Furthermore, this technique is highly adaptable due to the use of unguided channels, which enable mobility and facilitate communication across challenging geographic areas. However, these channels are subject to various perturbations, including distance, interference, and atmospheric conditions, which could degrade signal quality.

Telecommunication companies design and implement equipment for general-purpose applications, utilizing standardized antennas that enable communication between devices. While these generic solutions provide basic connectivity, they are significantly affected by channel perturbations, as they are not optimized for specific use cases. Consequently, these generic systems struggle to maintain long-distance communication with a low bit error rate (BER) with the same power as custom-designed equipment. Specially, in IoT communication systems, wireless communications have been utilized for location tracking, data logging, and efficient transportation, which have significantly benefited from the use of custom-designed equipment [2].

Many approaches to the evaluation of the performance of wireless communication systems in real-world scenarios have been carried out as part of field tests. Firstly, an IoT network has been tested in terms of its Received Signal Strength Indicator (RSSI) and Packet

Error Rate (PER) inside a university campus [3]. This study led the authors to propose an IoT network deployment for this area. Next, an experimental evaluation of an IoT network using LoRa for campus deployment is reported in [4], which focuses on the link quality, quantified through various parameters in different scenarios. Furthermore, over-the-air (OTA) performance evaluations of IoT wireless devices are also made, providing solutions to ensure efficient approaches [5].

In any IoT network evaluation, the hardware involved plays a crucial role. Commercial RF modules used as part of sensor nodes are commonly provided with conventional dipole antennas designed to radiate into free space at their operation frequency. This fact is related mainly because of its ease of use and the bandwidth used (up to 160 MHz), which is sufficient for most wireless applications [6]. In this context, many studies have been carried out utilizing merely the provided commercial dipole. Transmission methods have been analyzed for the IoT in both short- and long-range networks, encompassing technologies such as ESP-NOW, nRF24L01 modules, and local Wi-Fi access points [7]. In [8], the development of a wireless communication system using the nRF24L01 transceiver module and an Arduino Uno is described. The system operates at a frequency of 2.4 GHz, supports data transmission rates between 250 kbps and 2 Mbps, and has a communication range of approximately 80–100 m. The design allows for both transmitting and receiving data, using a compact setup that integrates the NRF24L01 with an Arduino Uno microcontroller.

Regarding the antennas attached to the transceiver modules, a microstrip patch antenna has been designed for IoT devices operating within the 868 MHz frequency band [9]. In order to obtain ultra-wide bandwidth, a U-shaped UWB antenna has been designed and analyzed for IoT applications [10].

In this paper, a performance comparison of IoT applications based on a basic experimental study at 2.4 GHz is proposed, which changes between commercial dipoles and two custom-designed antennas. Section 2 provides a description of the system used for the tests, the antenna structures, and the experimental approach. The results of the antenna design and the performance of the proposed system are presented in Section 3. Conclusions are depicted in Section 4.

2. Methodology

This section describes the methodology used to achieve this communication system comparison.

2.1. Equipment System Description

Figure 1 shows the proposed experimental setup of the IoT system used to evaluate the performance of a long-distance IoT communication system operating at 2.4 GHz. The system employs two RF modules, namely an NRF24L01 model; two computers; two microcrontrollers (Arduino Uno); and the custom-designed antennas under test. The system comprises two primary units—the transmission unit (TxU) and the reception unit (RxU)—each configured with specific components to optimize overall performance.

2.2. Antennas Under Test

Figure 2 shows the structures of the antennas, which are attached to the RF modules in different scenarios to carry out the IoT performance evaluation. A modified shape-optimized Landstorfer Yagi-Uda antenna (LaYUA) and a turnstile antenna composed of two printed dipole radiators (TuSA) are presented as custom-designed radiators for the experiment. Additionally, the commercial dipole antenna included in the RF modules is utilized for in this experimentation.

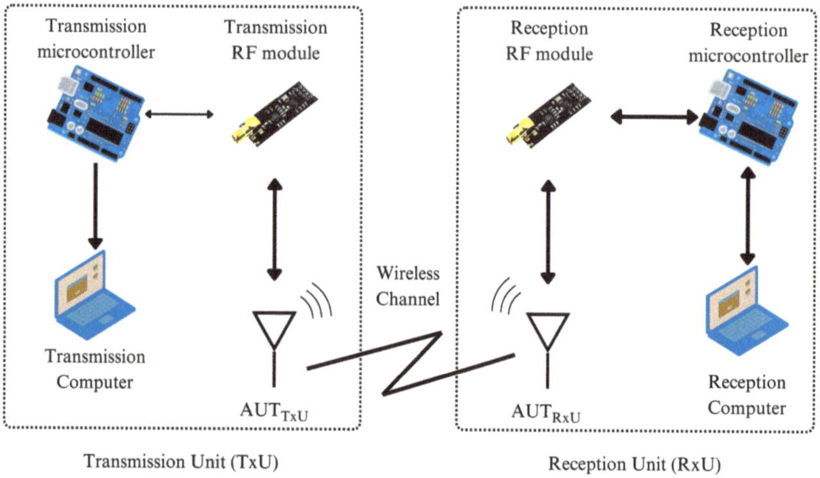

Figure 1. Proposed experimental setup used to evaluate the IoT communication system's performance by using custom-designed antennas.

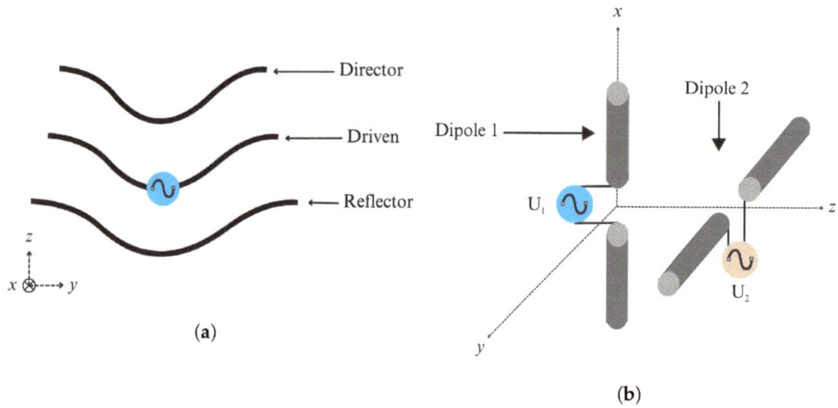

Figure 2. Antennas under test. (**a**) Modified shape-optimized Landstorfer Yagi-Uda antenna (LaYUA). (**b**) Printed turnstile antenna (TuSA) with 90°-phase-shift feeding between U_1 and U_2 sources, realized with a coaxial line.

The LaYUA radiator is a scaled version, at 2.4 GHz, of the antenna reported in [11] and serves as the radiator for transmission. The analysis of a turnstile antenna is detailed in [12,13], with the key difference being that the 90°-phased dipole is achieved through a coaxial line. Due to its low gain, this antenna operates at the reception side. Lastly, the commercial dipole corresponds to a 2.4 GHz antenna with an RP-SMA connector that is to be attached to the RF module.

2.3. Experimental Approach

According to Figure 1, the units TxU and RxU are linked through a wireless channel. On both sides, a computer is connected to a microcontroller (Arduino Uno), which in turn interfaces with an RF module (NRF24L01). These RF modules operate at 2.4 GHz and are equipped with both a low-noise amplifier and a power amplifier to enhance their transmission and reception capabilities. This RF module employs the SPI (Serial Peripheral Interface) for communication between the module and the microcontroller. Thus, the NRF24L01 implements its proprietary wireless communication protocol, operating in the 2.4 GHz ISM band. The latter includes features such as addressing, error control, and the capability to handle multiple channels and addresses, enabling, in this way, robust wireless data transmission and reception [14].

In order to ensure the relevance of the measurements and minimize interference, the TxU and RxU were strategically placed at distances varying from 10 m to 100 m in an open-field environment, specifically chosen to reduce obstructions and minimize external interference from buildings, vegetation, or other sources of signal reflection and attenuation. The tests were conducted under clear weather conditions, with environmental factors of a 18.7 °C temperature and 65.8% humidity. Additionally, a key feature taken into account in this experiment is the tilt angle of the linear polarized antennas (LaYUA and commercial dipole) to achieve polarization matching. This adjustment is essential for accurately evaluating the system's performance metrics. The system is tested under a transmit power level of 0 dBm and a data rate of 2 Mbps to evaluate the bit error rate (BER) characteristic across different scenarios, including urban fading profiles, which will be discussed in the following sections. A low BER implies high-quality communication, which is essential for reliability in IoT applications. Furthermore, it is critical for assessing system efficiency, as a high error rate requires retransmissions, thereby increasing latency and energy consumption in battery-dependent IoT devices.

A key component of this setup is the use of different antenna structures for the evaluation. Four different scenarios are examined to analyze the impact of these configurations on wireless communication performance:

Scenario 1 Commercial dipole antennas: This is the baseline scenario. Both the TxU and RxU are attached with standard commercial dipole antennas.

Scenario 2 LaYUA and commercial dipole: This scenario includes the LaYUA, attached to the TxU, and the commercial dipole antenna, attached to the RxU.

Scenario 3 Commercial dipole and TuSA: This scenario utilizes the commercial dipole antenna, attached to the TxU, and the TuSA, attached to the RxU.

Scenario 4 LaYUA and TuSA: This scenario uses the LaYUA, attached to the TxU, and the TuSA, attached to the RxU.

3. Results

This section presents the results obtained in each phase of this experimental study.

3.1. Antenna Measurements

The radiation pattern measurements were carried out in an anechoic chamber. Figure 3 shows photographs of the custom-designed antennas LaYUA and TuSA, which were implemented, respectively, on FR-4 substrates and mounted in the anechoic chamber. Figure 4 illustrates the simulated and measured reflection coefficient's magnitude and normalized pattern at 2.4 GHz, when both the antennas are radiating into free space. There is a good agreement between the simulations and measurements. In this way, the simulated gain of the LaYUA and TuSA are about 9 dBi and 3 dBi, respectively.

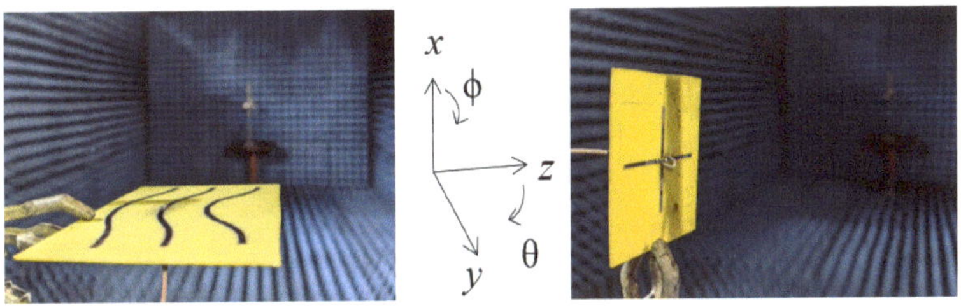

Figure 3. Photographs of the antenna prototypes mounted in an anechoic chamber, along with the definition of their spherical coordinates. (**Left**) LaYUA. (**Right**) TuSA.

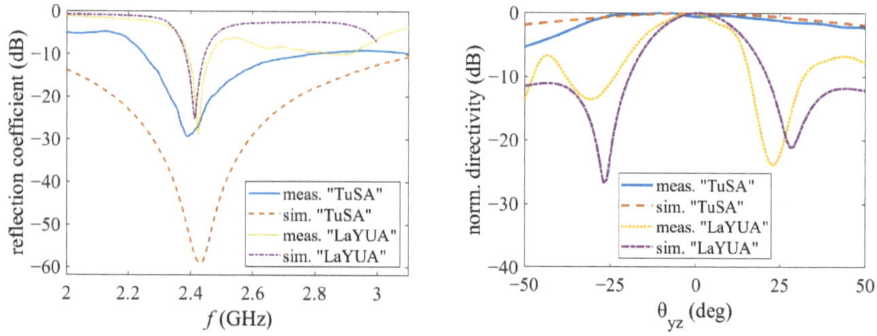

(a) (b)

Figure 4. Measured (solid blue and dotted yellow) and simulated (dashed red and dash–dot violet) antenna results. (**a**) Reflection coefficient magnitude. (**b**) Normalized radiation pattern on the azimuth plane (ϕ = 90°) at 2.4 GHz. The results are invalid in a sector between 50° and 180° due to the measurement setup.

3.2. System Performance Comparison

Figure 5 presents a comparative analysis of the four scenarios described above, using the three antennas, that tests their performance in terms of the bit error rate versus distance when a wireless communication takes place. These results are also very useful for understanding how capable different antenna configurations/devices are of preserving the signal quality as transmission distances get longer.

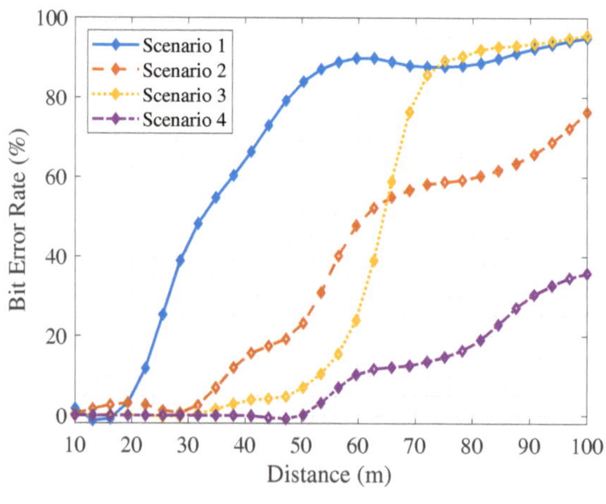

Figure 5. BER obtained from measurements. Scenario 1 (solid blue): commercial dipole (TxU)–commercial dipole (RxU). Scenario 2 (dashed red): LaYUA (TxU)–commercial dipole (RxU). Scenario 3 (dotted yellow): commercial dipole (TxU)–TuSA (RxU). Scenario 4 (dash–dot violet): LaYUA (TxU)-TuSA (RxU).

Scenario 1, utilizing commercial dipole antennas for both TxU and RxU, serves as the baseline and shows the poorest performance. The bit error rate rapidly increases to approximately 90% at 60 m. In contrast, Scenario 4, employing the LaYUA for TxU and the TuSA for RxU, demonstrates a remarkable improvement. At 60 m, Scenario 4 maintains a bit error rate of only about 10%, representing a significant improvement over Scenario 1. This better performance extends to the maximum tested distance of 100 m, where the Scenario 4 error rate remains below 40%, while Scenario 1's approaches 100%, indicating an approximately 60% improvement in signal integrity at long range.

Scenario 2 and Scenario 3 offer intermediate improvements. At 60 m, Scenario 2 shows a bit error rate of 50%, an approximately 40% improvement over Scenario 1. Scenario 3 performs better that Scenario 2 up to 60 m but degrades rapidly beyond this point. At 80 m, Scenario 3 exhibits a bit error rate of about 90%, while Scenario 2 maintains one around 60%, demonstrating the 30% improvement of Scenario 2 over Scenario 3 at this distance. Notably, Scenario 4 outperforms all others consistently, showing a 40% lower error rate compared to Scenario 2 and a 70% improvement over Scenario 3 at the 80-meter mark. These quantitative comparisons underscore the substantial impact of antenna selection on wireless communication system performance, with the LaYUA-TuSA combination (Scenario 4) offering the most significant improvements in reliability and range.

This comparative analysis underscores the significant impact that antenna selection has on wireless communication system performance. Scenario 4, employing the LaYUA-TuSA configuration, demonstrates the most robust performance in terms of reliability and range, offering the lowest bit error rate, even at long distances. This makes it the most suitable for applications requiring strong signal integrity over large distances.

3.3. Discussion

The results presented in Figure 4a show that the measured reflection coefficients for both the LaYUA and TuSA antennas align well with the simulated results. This suggests that the designs of these antennas are robust and their performance characteristics in real-world conditions closely match theoretical expectations. Furthermore, the bandwidth of approximately 300 MHz for the LaYUA antenna is particularly advantageousfor various applications within the 2.4 GHz frequency range, as it ensures stable communication across a wide spectrum. In contrast, the TuSA exhibits a much wider bandwidth, which could contribute to its more stable performance over varying environmental conditions and distances.

The simulated and measured radiation patterns on the azimuth plane ($\phi = 90^\circ$) shown in Figure 4b agree very well within the half-power beamwidth for both antennas, thus, the simulated antenna characteristics are verified.

The high directivity of the LaYUA antenna plays a critical role in its superior performance, particularly in scenarios requiring long-distance communication. Directivity, in this case, refers to an antenna's ability to focus the signal in a specific direction, thus minimizing energy loss and enhancing the effective signal strength at the receiving end. This is clearly demonstrated in Scenario 4, where the use of the LaYUA at the transmission unit (TxU) significantly improves the overall system performance compared to the baseline configuration (Scenario 1). The reduced bit error rate (BER) across increasing distances demonstrates the advantage of this high-gain antenna for long-range applications.

In contrast, the commercial dipole antennas used in Scenario 1 are omnidirectional, meaning they radiate energy equally in all directions. While this provides coverage over a wider area, it also dilutes the signal strength in any specific direction, leading to a higher BER as the distance increases. At 60 m, the BER for Scenario 1 reaches approximately 90%, whereas Scenario 4 (the LaYUA and TuSA combination) maintains a BER of only 10%. This drastic improvement highlights the importance of using antennas with focused radiation patterns for long-distance wireless communication.

The benefits of using the LaYUA in combination with the TuSA include the high directivity of the quasi Yagi-Uda structure, which enables long-distance communication and has the best performance, as shown in Figure 5. Additionally, the circular polarization characteristic of the TuSA reduces the need for exhaustive polarization calibration to ensure a clear line of sight.

The better performance of Scenario 4 in Figure 5 can be attributed to the combined characteristics of the utilized radiators. Both Scenarios 2 and 4 employ the LaYUA, known for its high directivity, which significantly enhances signal focus. However, Scenario 4 is further enhanced by the inclusion of a turnstile antenna, which is polarization-independent due to its circular polarization [15]. This combination leverages the high directivity of the

LaYUA at the TxU and the polarization versatility of the TuSA at the RxU, resulting in the best bit error rate performance across all tested scenarios.

In Scenarios 2 and 3, where only one custom antenna is used, the performance is still markedly better than the baseline, but not as good as when both custom antennas are used. This suggests that while each antenna independently offers performance improvements, the combination of both the high-directivity LaYUA and the polarization-independent TuSA offers the most significant benefits. At 80 m, Scenario 4 maintains a BER of about 40%, whereas Scenario 1 approaches 100% and Scenario 3 reaches 90%. This underscores the complementary nature of the LaYUA and TuSA antennas in enhancing communication's reliability and range.

4. Conclusions

In this work, a performance comparison, in terms of the bit error rate versus distance of an IoT communication system, has been evaluated using four scenarios with three antennas each attached to the Tx and Rx RF modules of NRF24L01. For the evaluation, a commercial dipole of the brand Siretta, a previously designed frequency-scaled Landstorfer Yagi-Uda antenna (LaYUA), and a turnstile antenna (TuSA) were utilized.

The LaYUA, with a directivity of about 9 dBi and horizontal polarization, was optimized with parameterized elements to enhance its performance, along with the TuSA, which features a directivity of about 3 dBi and circular polarization to provide consistent omnidirectional reception. The results illustrated that the combination of the LaYUA for the TxU and the TuSA for the RxU (Scenario 4) shows the best performance. At a distance of 60 m, this scenario maintained a bit error rate (BER) of approximately 10%, representing a 60% improvement over the baseline Scenario 1 (commercial dipole antennas), which exhibited a 90% BER. Even at 100 m, Scenario 4 kept the BER below 40%, significantly outperforming the baseline, which approached a 100% BER. This represents a substantial reduction in transmission errors, making it the most reliable and effective solution for long-range wireless communication. Therefore, the results suggest that high-performance antennas like those in Scenario 4 are essential for applications requiring extended transmission distances and reliable data integrity.

The results demonstrated that using non-commercial antennas like the custom radiators designed herein significantly enhanced wireless communication performance, particularly by reducing bit error rates over longer distances. This approach highlighted the substantial benefits of employing specialized antennas to optimize wireless systems, particularly in environments where standard commercial antennas may be insufficient. Using the LaYUA in combination with the TuSA enables many IoT devices to function effectively in remote environments such as agriculture, energy grids, and environmental monitoring. This combination facilitates robust communication between nodes, reducing data loss and ensuring reliable monitoring over larger areas.

Author Contributions: Conceptualization, J.G.A. and M.V.B.; methodology, J.G.A., M.V.B. and A.R.N.; validation, J.G.A., M.V.B., and A.R.N.; investigation, J.G.A., M.V.B. and C.Y.V.; writing—original draft preparation, J.G.A., M.V.B. and C.Y.V.; writing—review and editing, A.R.N. and H.B.M. All authors have read and agreed to the published version of the manuscript.

Funding: This research received no external funding.

Institutional Review Board Statement: Not applicable

Informed Consent Statement: Not applicable

Data Availability Statement: Data are contained within the article.

Conflicts of Interest: The authors declare no conflicts of interest.

Abbreviations

The following abbreviations are used in this manuscript:

FR-4	Flame retardant-4: woven glass-reinforced epoxy resin
RF	Radio Frequency
TxU	Transmission unit
RxU	Reception unit
SPI	Serial Peripheral Interface
LaYUA	Landstorfer Yagi-Uda Antenna
TuSA	Turnstile antenna
BER	Bit error rate

References

1. Rappaport, T.S. *Wireless Communications: Principles and Practice*; Cambridge University Press: Cambridge, UK, 2024.
2. Hanes, D.; Salgueiro, G.; Grossetete, P.; Barton, R.; Henry, J. *IoT Fundamentals: Networking Technologies, Protocols, and Use Cases for the Internet Of Things*; Cisco Press: Indianapolis, IN, USA, 2017.
3. Rathod, N.; Jain, P.; Subramanian, R.; Yawalkar, S.; Sunkenapally, M.; Amrutur, B.; Sundaresan, R. Performance analysis of wireless devices for a campus-wide IoT network. In Proceedings of the 2015 13th International Symposium on Modeling and Optimization in Mobile, Ad Hoc, and Wireless Networks (WiOpt), Mumbai, India, 25–29 May 2015; IEEE: Piscataway, NJ, USA, 2015; pp. 84–89.
4. Kulkarni, P.; Hakim, Q.O.A.; Lakas, A. Experimental evaluation of a campus-deployed IoT network using LoRa. *IEEE Sens. J.* **2019**, *20*, 2803–2811. [CrossRef]
5. Shen, P.; Qi, Y.; Yu, W.; Fan, J.; Li, F. OTA measurement for IoT wireless device performance evaluation: Challenges and solutions. *IEEE Internet Things J.* **2018**, *6*, 1223–1237. [CrossRef]
6. *IEEE Standard for Information Technology—Telecommunications and Information Exchange Between Systems—Local and Metropolitan Area Networks—Specific Requirements—Part 11: Wireless LAN Medium Access Control (MAC) and Physical Layer (PHY) Specifications—Amendment 4: Enhancements for Very High Throughput for Operation in Bands below 6 GHz*; IEEE: Piscataway, NJ, USA, 2013; IEEE Standard No. 802.11ac-2013.
7. Vakaliuk, T.A.; Andreiev, O.V.; Dubyna, O.F.; Korenivska, O.L.; Andreieva, Y.O. Wireless technologies in IoT projects with distributed computing. In Proceedings of the Doors 2024, 4th Edge Computing Workshop, Zhytomyr, Ukraine, 5 April 2024; pp. 4–13.
8. De, B.; Karmakar, A. Arduino based RF Compact Module for Short Range Wireless Communication using nRF24L01. *Int. J. Eng. Res. Technol. (IJERT)* **2022**, *10*. PANE-2021.
9. Anchidin, L.; Lavric, A.; Mutescu, P.M.; Petrariu, A.I.; Popa, V. The Design and Development of a Microstrip Antenna for Internet of Things Applications. *Sensors* **2023**, *23*, 1062. [CrossRef] [PubMed]
10. Thiruvenkadam, S.; Parthasarathy, E.; Mythali, N.R.; Jevalikar, R.; Agarwal, A. Design and analysis of UWB antenna for IoT applications. *AIP Conf. Proc.* **2023**, *2946*, 060003. [CrossRef]
11. Reyes Narváez, A.; Barba Molina, H. Landstorfer Antenna Structure Shaping Based on Parameterized Parallel Curves. In *Proceedings of the Emerging Research in Intelligent Systems;* Olmedo Cifuentes, G.F., Arcos Avilés, D.G., Lara Padilla, H.V., Eds. Springer Nature: Cham, Switzerland; Sangolquí, Ecuador, 2023; pp. 61–71. [CrossRef]
12. V, D.K.; Bhardwaj, A.; Mishra, D. Investigation of a turnstile nanoantenna. *Micro Nano Lett.* **2011**, *6*, 94–97. [CrossRef]
13. Kraus, J.D.; Marhefka, R.J. Antennas for all applications. In *Antennas for all Applications;* McGraw-Hill Professional: New York, NY, USA, 2002.
14. Semiconductor, N. nRF24L01 Single Chip 2.4GHz Transceiver. 2017. Available online: https://www.nordicsemi.com/Products/nRF24-series (accessed on 29 August 2024).
15. Balanis, C.A. *Antenna Theory: Analysis and Design*, 3rd ed.; John Wiley & Sons, Inc.: Hoboken, NJ, USA, 2005.

engineering proceedings

MDPI

Proceeding Paper

Real-Time Cyber–Physical Power System Testbed for International Electrotechnical Commission 61850 Generic Object-Oriented Substation Event Transfer Time Measurements †

Le Nam Hai Pham [1,*], Veronica Rosero-Morillo [2], Anup Shukla [3], Francisco Gonzalez-Longatt [1,4] and Viviana Meza-G [5]

1 Digital Energy Systems Laboratory, University of South-Eastern Norway, 3918 Porsgrunn, Norway; fglongatt@fglongatt.org
2 Institute of Electrical Energy IEE, National University of San Juan, San Juan 5400, Argentina; vrosero@iee.unsj.edu.ar
3 Electrical Engineering Department, IIT Jammu, Jammu 181121, India; anup.shukla@iitjammu.ac.in
4 Centre for Renewable Energy Systems Technology, Loughborough University, Loughborough LE11 3TU, UK
5 Quito Electric Company, Quito 170802, Ecuador; vmeza@eeq.com.ec
* Correspondence: le.pham@usn.no

† Presented at the XXXII Conference on Electrical and Electronic Engineering, Quito, Ecuador, 12–15 November 2024.

Abstract: Towards the decarbonisation of the power system, digital substations have gradually increased in smart grids, where Ethernet cables have replaced large quantities of copper wires. With this transition, the standardised communication protocols through the LAN network play a central role in exchanging information and data between the physical power system and the control centres. One of the well-known protocols in the digital substations is IEC 61850 GOOSE (Generic Object-Oriented Substation Event), which is used to share time-critical information related to protection, automation, and control. The transmission time of this protocol affects power system operation and raises various issues, such as communication latencies and incorrect information. Therefore, it is necessary to consider the protocol transmission time for further protection and control mechanisms to ensure the stability and efficiency of the power system. For this purpose, this paper contributes the implementation of a cyber–physical power system (CPPS) testbed to measure the transfer time of IEC 61850 GOOSE under the real-time domain using the real-time simulator, Typhoon HIL, and its toolchains. This paper can benefit scholars and researchers in the relevant domains in implementing a CPPS testbed and an approach for transfer time measurement of communication protocols within the laboratory, eliminating the need for real-world substation devices.

Keywords: cyber–physical power system; communication protocol; GOOSE; IEC 61850; real-time simulation; Typhoon HIL

Citation: Pham, L.N.H.; Rosero-Morillo, V.; Shukla, A.; Gonzalez-Longatt, F.; Meza-G, V. Real-Time Cyber–Physical Power System Testbed for International Electrotechnical Commission 61850 Generic Object-Oriented Substation Event Transfer Time Measurements. *Eng. Proc.* **2024**, *77*, 17. https://doi.org/10.3390/engproc2024077017

Academic Editor: Jaime Cepeda

Published: 18 November 2024

1. Introduction

The design of upcoming power systems has been experiencing significant challenges as it moves from the traditional centralised method of electricity generation to a decentralised system [1,2]. This transition enables the development of flexible, low-carbon energy and power systems [3]. In addition, the use of numerous interactive distributed units creates new technical difficulties with the implementation of various smart grid concepts, such as virtual power plants (VPPs) [4], microgrids [5], and digital substations [6]. The consistent improvement in communication technologies, in terms of speed, reliability, and adaptability, has facilitated the conversion of conventional substation wiring into a fully digital environment.

Digital substations offer significant advantages compared to their traditional counterparts [7]. The availability of high-bandwidth communication enables the transmission of large amounts of data over a single Ethernet cable, reducing the large amount of copper wiring in the substation [8]. To facilitate this transition to digital substations, there is a need for a universal communication standard that enhances interoperability and ensures the system's sustainability. As a result, the IEC 61850 standard was established to provide comprehensive models for power system devices across different vendors.

The IEC 61850 communication includes time-critical services that use GOOSE messages to transfer data and is usually associated with protective relaying activities. The popularity of this protocol has expanded in recent years [9]. For example, in [10], the authors proposed the conventional protective relay testing and the performance testing of IEC 61850-based relays to evaluate GOOSE message performance over traditional hardwired testing. Accordingly, functional testing has proved that GOOSE provides a very flexible, fast, high-priority, and reliable method for exchanging substation events among IEDs for interlocking and protection purposes. Similarly, another paper [11] proposed a methodology for the real-time implementation of a communication-dependent, logically selective, adaptive protection algorithm for AC microgrids using hardware-in-the-loop IEC 61850 GOOSE protocol testing. The authors in this paper agreed that the whole process of fault detection, isolation, and adaptive setting using Ethernet communication is possible within the standard low-voltage ride-through curve, which maintains the seamless transition to the islanded mode. Furthermore, in [12], Sidhu et al. provided a step-by-step configuration process comprising IEC 61850 data modelling, datasets of GOOSE within individual IEDs, and system integration of GOOSE.

In the aforementioned papers, the authors emphasised the importance of the IEC 61850 GOOSE communication protocol in the control and protection of digital substations. Therefore, it is necessary to consider the time transmission of GOOSE to ensure the accuracy of the proposed control and protection mechanisms. Since conventional power systems have transformed into cyber–physical systems (CPS), the cyber–physical power system (CPPS) testbed is considered a valuable solution. Indeed, the CPPS has the ability to integrate communication protocols and advanced control and protection algorithms [11]. For example, the authors in [13] used a CPPS testbed to verify the monitoring and wide-area control. Additionally, with the advent of simulation modelling, the appearance of real-time simulators such as RTDS, Typhoon HIL, or OPAL-RT enhanced the simulation capabilities to replicate the realistic power system behaviours and characteristics [14]. Leveraging these real-time simulators, the authors in [15] proposed a CPPS testbed for optimal power flow study using Typhoon HIL. The communication protocols were also involved in this research paper, demonstrating practical scenarios in grid operations. Similarly, other researchers have leveraged CPPS to advance power system security and control, highlighting its versatility across different areas of power system research [16]. A key advantage emphasised in these studies is CPPS's ability to operate in real-time environments, effectively simulating real-world scenarios without the need for costly infrastructure (such as RTUs, PMUs, or digital protection relays), making it a cost-effective and scalable approach for power system experimentation and development [2,17]. Therefore, there is an urgent need to deploy the CPPS testbed for further research and studies in control and protection domains.

To address this need, this paper proposes a CPPS testbed to implement an IEC 61850 GOOSE communication protocol under real-time simulation, aiming to measure its transfer time within a laboratory environment. The main contributions of this paper can be summarised as follows:

- A construction methodology for the CPPS testbed is proposed using a real-time simulator, Typhoon HIL. This testbed is used to mimic the actual operations of power systems such as a manual open/close circuit breaker or an automatic tripping protection relay under fault scenarios.

- The communication protocol, IEC 61850 GOOSE, is integrated into the CPPS testbed. The deployment of this protocol within the laboratory environment opens a new horizon for the further testing of control and protection mechanisms.
- The methodology of transmission time estimation of IEC 61850 GOOSE is provided in this paper. One-way and round-trip transfer time experiments are conducted to verify the working conditions of the laboratory for further studies in the control and protection domains.

The remaining sections of the paper are organised as follows. Section 2 provides the theoretical background of the IEC 61850 GOOSE communication protocol. In this section, the structure of the GOOSE message and the transfer time measurement based on international standards are provided. Then, the methodology of CPPS testbed implementation and the time measurement approach are provided in Section 3. Section 4 describes the experimental cases with relevant hardware and software in a laboratory environment. Finally, the last section, Section 5, summarises the main contribution of the work that was carried out.

2. Theory Background

2.1. IEC 61850 GOOSE

The IEC 61850 device model begins with a physical device connected to the network and is defined by its network address [18]. Within each physical device, one or more logical devices may provide a common interface between various physical devices in the substation and the SCADA system. Each logical device contains one or more logical nodes and functional units within a logical device [19]. A logical node is a named grouping of data and associated services and is logically related to the power system function. The IEC 61850 defines a standardised way of addressing data objects. Figure 1 shows the default addressing scheme that includes the following:

- LDName—stands for logical device name.
- LNRef—stands for Logical Node name.
- DataObjectName—the data objects within a logical node represent the information that is exchanged between devices.
- DataAttributeName—describes the properties of the data objects. These attributes may include data type, measurement range, and accuracy.
- FC—stands for function code defining the type of message being sent, such as a control command or status update.

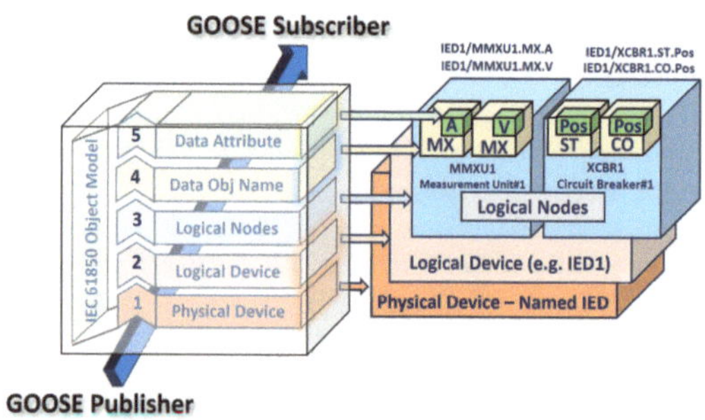

Figure 1. Default naming scheme of IEC 61850 GOOSE.

The IEC 61850 GOOSE protocol specifies that engineers and vendors use Substation Configuration Language (SCL), which was developed to configure the settings and format the functions within a particular IED [20].

Manufacturer-specific IED configuration tools are used to convert the functionality, communication mechanism, and parameters of an entire IED into a hierarchy of SCL system files. The ".scd" file extension is used for SCL files, which contain the description of the substation configuration data in XML format. The ".scd" file defines the logical nodes, data classes, and data objects within the substation, along with their attributes and relationships. In addition, this type of format serves as a blueprint for the configuration of the communication network and system in the substation, and the engineering tools and IEDs use it to understand and interpret the data exchanged in the system.

The IEC 61850 GOOSE protocol specifies transmission times based on various operation applications. As outlined in [21], these transmission time requirements are summarised in Table 1. For critical operations like interlocking or trip signals, which necessitate a rapid response, the protocol mandates a maximum transmission time of 10 milliseconds. This ensures the reliable and timely exchange of fast messages, essential for the proper functioning of time-sensitive protection and control operations.

Table 1. Transmission time requirements for substation operations.

Type	Performance Class	Application Type	Max Transmission Time
Type 1A—Trip	P1	Interlocking, inter trip, logic discrimination	10 ms
Type 1B—"Others"	P1	Interaction of the automation system	100 ms
Type 2	P2	Status monitoring, control commands	100 ms
Type 3	P1	Historical data, event logging	500 ms

2.2. IEC 61850 GOOSE Transfer Time

The idea of transfer time, as described in IEC 61850-5 [21], is used to ensure adequate transmission time between two devices [22]. Part 5 of this standard provides a diagram, as shown in Figure 2a, that depicts the transfer time for transmitting a message from one device to another.

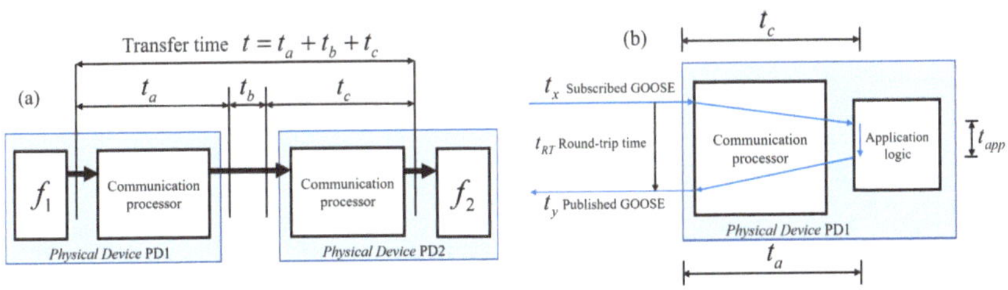

Figure 2. (**a**) GOOSE transfer time and (**b**) GOOSE round-trip time.

The measurement of data transfer is calculated from the moment the sender (physical device 1, PD1) sends the data to the point that the receiver (physical device 2, PD2) receives it. This duration is the transfer time, which includes t_a and t_c, the time taken by the communication processors at both ends to encode and decode the data, and t_b, the time taken for the data to travel over the network.

2.3. IEC 61850 GOOSE Round-Trip Time

The internal data of commercial devices are typically inaccessible to end users, which makes it challenging to measure transfer time according to IEC 61850-5 specifications [22]. However, by following the transfer time definition provided in the standard, various tests can be performed to assess the performance of devices that comply with IEC 61850.

The round-trip time specified in IEC 61850-10 [23], as shown in Figure 2b, can be used to measure the transfer time. According to this standard, the time between receiving a subscribed GOOSE message (t_x) and publishing a GOOSE message (t_y) on the network determines the round-trip time, t_{RT}. It is noticed that while this definition is similar, it deviates from the standard use of round-trip time in the networking community.

2.4. Ping-Pong Test

As shown in Figure 3, the ping-pong test is performed by two devices that publish and subscribe to GOOSE messages to measure the round-trip time between the two devices. In this test, the Test Set (TS) device, device A, publishes a message to the Device Under Test (DUT), device B. Device B then responds by publishing a message that device A subscribes to. According to IEC 61850-10, this test focuses on measuring the round-trip time and is intended as a benchmark for comparing the performance of different IEDs.

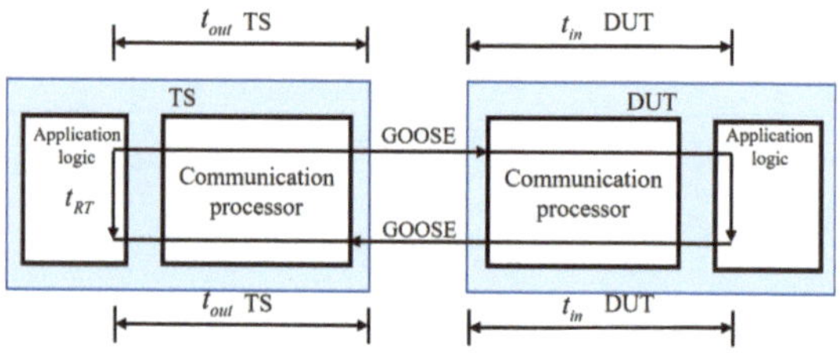

Figure 3. Ping-pong test.

3. Implementation of Cyber–Physical Power System Testbed and Methodology of Transfer Time Measurement

3.1. CPPS Testbed Implementation

Figure 4 illustrates the architecture of the CPPS testbed that relies on communication between two real-time simulators. The real-time simulator has the capability to communicate with external devices via LAN Ethernet [24]. In recent years, there have been many studies utilising real-time simulators to transmit communication signals to external devices [2]. For example, the author in [25] proposed an approach consisting of using the OPAL-RT platform to interact with protection relays. In the same manner, the authors in this paper utilise two real-time simulators to develop a CPPS testbed to simulate the behaviour of the real-world substations in the virtual environment.

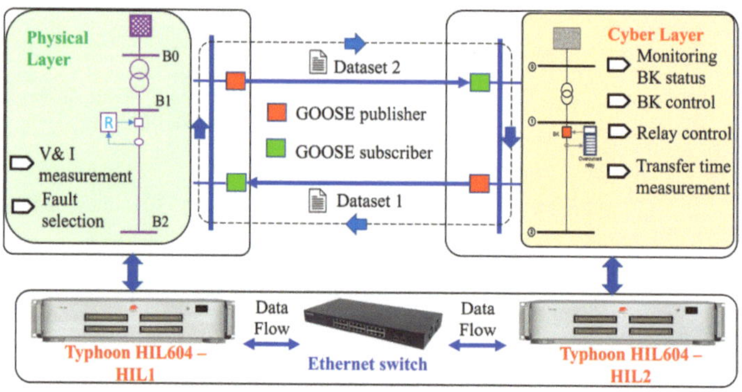

Figure 4. CPPS framework.

The CPPS testbed contains two layers: the physical layer and the cyber layer. The physical layer sends the information and data of physical devices, while the cyber layer provides the control feedback. In this setup, the CPPS is designed to replicate the specific operations in substations regarding the protection relay and circuit breaker (BK), as detailed in Table 2.

Table 2. Operations in CPPS testbed.

Operation	GOOSE Publisher	GOOSE Subscriber	Transfer Type
When a fault occurs, the tripping signal from the relay causes the BK to open, and the BK status is sent from the substation to the control centre.	Physical layer	Cyber layer	One-way transfer
The control centre sends a reset signal for the protection relay in the test system, which clears the tripping signal and takes the BK return to the closed state. In addition, the BK can be manually opened or closed by command from the control centre.	Cyber layer	Physical layer	Round-trip transfer
The control centre updates the status of the BK accordingly.	Physical layer	Cyber layer	

3.2. Transfer Time Measurement

To measure the transmission time of GOOSE messages within the communication system, the authors employed a timer with a specified execution time, T_S. This execution time represents the duration required to solve a system of complex equations and produce the output results, and is synchronised with the real-world clock. The operating principle of detecting logic state changes in the transmitted signal is straightforward; the timer is activated upon the sender's detection of a state change. It remains active until the signal is successfully recorded at the receiver, marking the end of the transmission process. The transfer time is equal to the product of the T_S and n points at the counter.

As illustrated in Figure 5a, the one-way transfer time needs to be precisely measured between two points: point 1 and point 4. Point 1 is the sender transmission stack and point 2 is after the execution of the receiver transmission stack, when the GOOSE message is transmitted onto the network. This timing ensures accuracy, as point 2 aligns with the moment the message is effectively placed onto the network. Similarly, for the client, measurement occurs between point 3, before the receiver transmission stack is initiated, and point 4, after the stack has been fully executed.

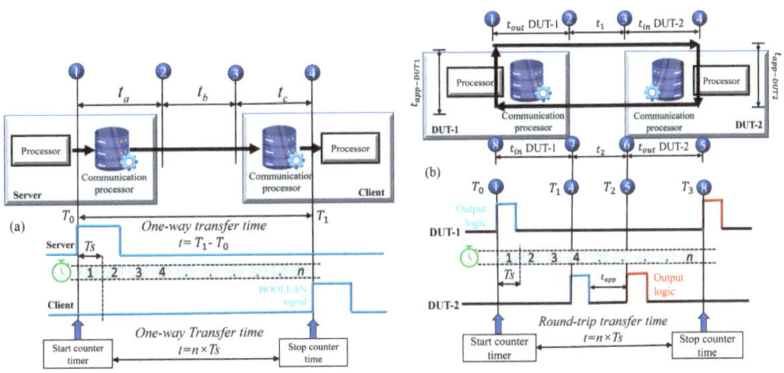

Figure 5. IEC 61850 GOOSE transfer time measurement using timer counter. (**a**) One-way transfer and (**b**) round-trip transfer.

For the round-trip transmission time depicted in Figure 5b, the evaluation includes measuring the time interval between two distinct points: point 1 and point 8. This includes two primary segments: points 1 to 4, reflecting the transfer time from DUT-1 to DUT-2, and points 5 to 8, which represent the reverse journey from DUT-2 back to DUT-1. These segments encapsulate the complete transmission process between two devices. Additionally, points 4 to 5 denote the time required for application communication within the devices themselves. In addition, Algorithm 1 is provided to measure round-trip time based on commands of open and close BK at the cyber layer.

Algorithm 1: GOOSE round-trip time measurement

```
Inputs:
Ts ← Execution rate
Co ← Open BK = 0/Close BK = 1              [BK control = GOOSE send]
SBC← BK opened = 0/BK closed = 1           [BK status = GOOSE receive]
Clock ← Clock off = 0/Clock on = 1         [Enable/Disable Clock]
en ← Encounter
t_Rt ← GOOSE round-trip time
for i in n = 1, 2...N do:                  [N point of measurement]
   if (Clock = 1) and (Co = 1):            [Condition to enable the counter]
      en = 1;
   else
      en = 0;
   end if
   count = 0;
   if en = 1 and SBC = 1:                  [Condition to measure t_Rt]
      count = count + 1;
      t_Rt = count * Ts;
   else
      count = 0;
   end if
   Max = 0;
   if t_Rt(t − 1) > t_Rt(t)
      c = 1;
   else
      c = 0;
   end if
   Max = Max + t_Rt(t − 1)*c               [Maintain t_Rt]
end
Outputs:
N measured values of GOOSE round-trip time
```

4. Experimental Case

The experimental cases of the CPPS testbed are conducted at DIgEnSys-Lab at the Department of Electrical Engineering, Information Technology, and Cybernetics at the University of South-Eastern Norway (USN) at the campus in Porsgrunn, Norway (https://fglongattlab.fglongatt.org). The laboratory setup is shown in Figure 6. The experiment involved the use of two digital real-time simulators, Typhoon HIL version HIL 604s, each with eight computing cores, two ARM cores, and a combination of digital and analogue input/output. The CPPS testbed is modelled and visualised on two corresponding PCs. One PC was responsible for managing the physical layer, while the other PC was responsible for the cyber layer. Both PCs operate on Windows 10. All devices were connected through a LAN in the laboratory. GOOSE block [16] executes the communication between two layers in the Typhoon HIL environment.

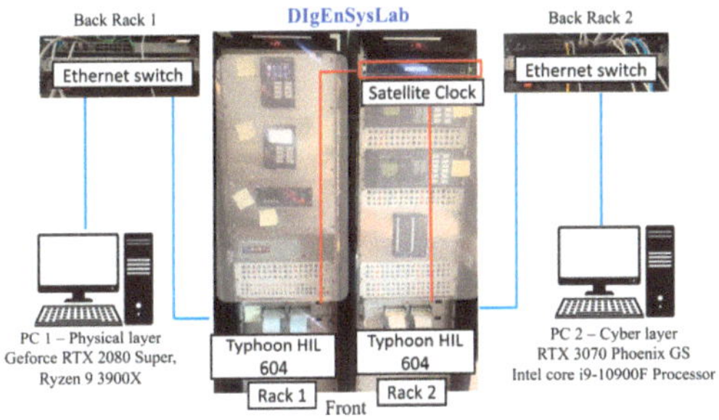

Figure 6. Laboratory setup for CPPS with blue line as Ethernet cable and red line for coaxial cable.

4.1. Test System

The physical layer includes the test model, which is a modified model of the three-phase radial Feeder 1 of the distribution benchmark network, and the CIGRE European Medium Voltage, which was proposed by CIGRE Task Force C6.04.02 [8], as shown in Figure 7. The nondirectional overcurrent protection relay located at transmission line B1–B2 is the instantaneous overcurrent protection relay. The configuration of the relay is provided in [17,26].

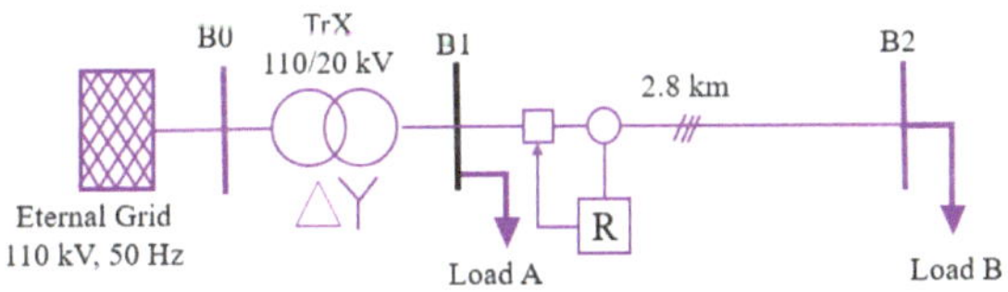

Figure 7. Test system.

4.2. GOOSE Configuration

To implement communication between the two layers, it is necessary to configure the dataset for data transfer. In this paper, the authors used SEL acSELerator Architect Software, Version 2.3.11.2125 [27], which is a Schweitzer Engineering Laboratories (SEL) product, to configure the GOOSE message. The GOOSE files are then exported in a .scd format and imported into GOOSE blocks in the two models. The configuration of the GOOSE files is shown in Table 3.

Table 3. GOOSE configuration.

GOOSE dataset 1	
appID	0 × 1000
VLAN ID	0 × 000
Min Time	4 ms
Max Time	1000 ms
Dataset	PIOCSTq (Relay reset signal) BKRCSWSTstVal (BK Open/Close)
GOOSE dataset 2	
appID	0 × 1001
VLAN ID	0 × 001
Min Time	4 ms
Max Time	1000 ms
Dataset	BKRCSWISTstVal (BK status)

4.3. Results and Discussion

The GOOSE package is typically transferred in the LAN with its timestamps, and relevant information can be captured using network analyser software. The well-known software Wireshark (Version 3.6.1) is a network sniffer tool that can capture network traffic through the host network interface card. Therefore, the GOOSE package can be captured using Wireshark, as shown in Figure 8.

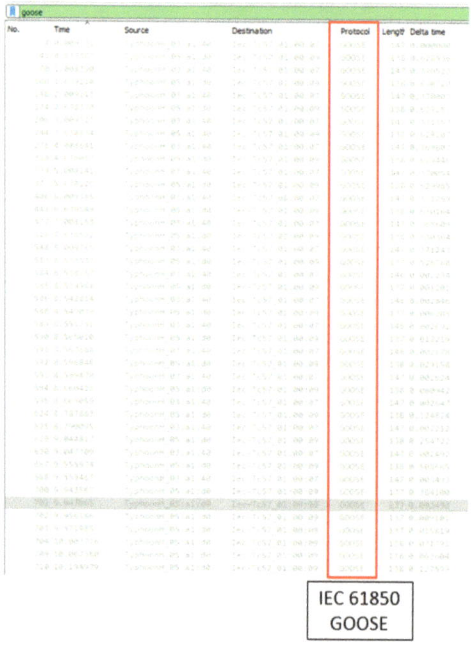

Figure 8. IEC 61850 GOOSE captured by Wireshark during the experiments.

However, it is important to note that Wireshark timestamps in Windows environments are sourced from the WinPcap library, which is a Microsoft Windows package capture for software installed in Microsoft Windows computers. WinPcap consists of a driver for low-level network access and a library for accessing low-level network layers [28]. Consequently, utilising Wireshark to capture files and analyse timestamps could potentially result in inaccurate outcomes [29].

The transmission time is measured according to the CPPS testbed operations outlined in Table 2. The one-way transmission time follows the BK state sent from the physical layer to the cyber layer when the fault occurs. On the other hand, the ping-pong test based on BK control and status signal is conducted to correctly obtain the round-trip time of the GOOSE communication protocol between two real-time simulators. The acquisition time from the transmitted and received GOOSE signal is based on Algorithm 1. This algorithm is simulated through logical and computation blocks in Typhoon HIL and connected to the communication part in the control centre model. The execution time for the counter is selected as 20 µs, and the transmission time is measured at 1000 points.

The GOOSE time measurement consists of 1000 samples over a duration of 25 min. These values are used for analysis to obtain the mean value of the transmission time as well as maximum and minimum values. The measurement time includes the processing time of the communication processor in a real-time simulator under LAN traffic in a laboratory environment.

The measurement results from the experiments conducted in the CPPS testbed are presented and analysed using a histogram and box plot, as illustrated in Figure 9. Specifically, Figure 9a and 9b display the one-way and round-trip transmission times of GOOSE messages, respectively. For one-way transmission, the analysis indicates a mean transmission time of 7.28 ms. The recorded maximum and minimum transmission times are 10.6 ms and

5 ms, respectively. However, the most probable one-way transmission time is 6.5 ms, with a probability density of 0.29. For round-trip transmission, the analysis reveals an average transmission time of 19.68 ms, with a maximum of 32.32 ms and a minimum of 7.65 ms. The most probable round-trip time is 17.3 ms, with a probability density of 0.08.

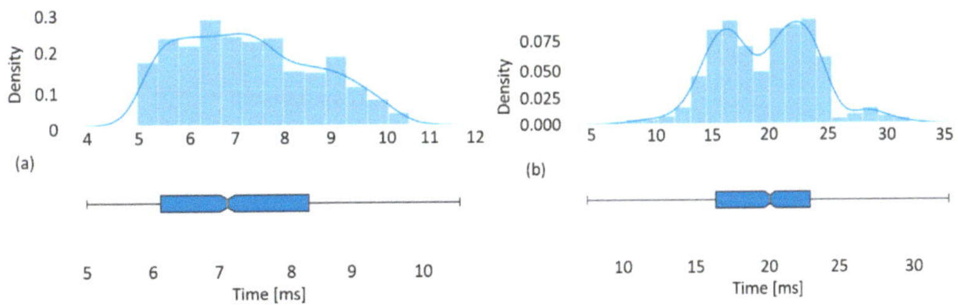

Figure 9. Histogram and box plot of IEC 61850 GOOSE transfer time. (**a**) One-way transfer and (**b**) round-trip transfer.

When compared to Table 1, which outlines the required transmission times for operational applications in the power system, the results show that both the mean and most probable one-way transmission times are under 10 ms. Similarly, for round-trip transmission, the mean value of 19.68 ms and the most probable value of 17.3 ms for the outbound and return journeys remain within acceptable limits (below 20 ms for round-trip transfer). These results demonstrate that the laboratory environment satisfies the requirements for researching communication protocols, as well as for control and protection systems in the power grid, particularly for fast message processing and other time-sensitive applications.

However, the observed maximum transmission times for both one-way and round-trip transmissions indicate that there is room for improvement. Increasing the transmission bandwidth and speed in the laboratory environment would help improve the transmission speed and ensure more consistent performance across all scenarios.

5. Conclusions

The shift towards digitalisation in power systems has led to an increase in digital substations, where traditional copper wiring is gradually being replaced by Ethernet cables. This transition has made standardised communication via Ethernet LAN an essential requirement. IEC 61850 is a commonly used communication protocol in these digital substations. The transfer time of this protocol is crucial because it directly affects the operation of substation components and can impact the wider area network. Therefore, it is essential to consider the transfer time of this protocol for further advance control and protection algorithm development. Within the Typhoon HIL framework associated with its toolchains, the lab-based environment is set up for the CPPS testbed to perform the specific operation in the power substation. The CPPS testbed can be used for further research in power system domains without the actual high-cost devices.

The transfer time of the IEC 61850 GOOSE message exchange between layers is measured using 1000 points over approximately 25 min. The presented results show mean transfer times of 7.28 ms and 19.68 ms for one-way and round-trip transfers, respectively. The measurement values obtained from both one-way and round-trip transmission experiments meet the fast message requirements essential for substation operations. This confirms that the laboratory environment is well-suited for conducting experiments and research related to further control and protection system development. Given the capabilities of the CPPS testbed and the potential for improving transmission times, the following suggestions for future work are proposed:

- Transmission time improvement: enhance Ethernet bandwidth and speed to ensure consistent and reliable communication protocols.

- In-depth research on communication latencies: conduct comprehensive studies to further analyse and understand the factors contributing to communication latencies within the system.
- Development of advanced control and protection strategies: utilise the CPPS testbed to explore and implement innovative control and protection strategies, leveraging its capabilities for improved performance and reliability in power system operations.

Author Contributions: Conceptualization, L.N.H.P. and F.G.-L.; methodology, L.N.H.P. and F.G.-L.; software, L.N.H.P.; validation, L.N.H.P. and F.G.-L.; writing—original draft preparation, L.N.H.P., V.R.-M. and A.S.; writing—review and editing, L.N.H.P., F.G.-L., V.R.-M., A.S. and V.M.-G.; visualization, L.N.H.P.; supervision, F.G.-L. All authors have read and agreed to the published version of the manuscript.

Funding: This research was funded by the Research Council of Norway.

Institutional Review Board Statement: Not applicable.

Informed Consent Statement: Not applicable.

Data Availability Statement: Data are contained in the article.

Acknowledgments: Le Nam Hai Pham wants to express his thanks to the University of South-Eastern Norway, Porsgrunn campus, for support during his PhD. F. Gonzalez-Longatt recognises the collaboration and support from Typhon HIL and Schweitzer Engineering Laboratories (SEL).

Conflicts of Interest: Except where specifically indicated, the authors declare no conflicts of interest. V.M.-G. is an employee of Electrica Quito, which could be considered a potential conflict of interest related to the subject of this manuscript. However, all necessary measures have been taken to ensure that this connection has not inappropriately influenced or biased the results presented in this work.

References

1. Gonzalez-Longatt, F.; Rueda Torres, J.L. Introduction to smart grid functionalities. *Green Energy Technol.* **2018**, 1–18. [CrossRef]
2. Wagle, R.; Pham, L.N.H.; Tricarico, G.; Sharma, P.; Rueda, J.L.; Gonzalez-Longatt, F. Experiences in a Cyber-Physical Co-Simulation Testbed Development for a Smart-er Distribution Network. In Proceedings of the 2023 IEEE PES Conference on Innovative Smart Grid Technologies-Middle East (ISGT Middle East), Abu Dhabi, United Arab Emirates, 12–15 March 2023; pp. 1–5. [CrossRef]
3. Wagle, R.; Pham, L.N.H.; Tricarico, G.; Sharma, P.; Rueda, J.L.; Gonzalez-Longatt, F. Co-simulation-based optimal reactive power control in smart distribution network. *Electr. Eng.* **2023**, *106*, 2391–2405. [CrossRef]
4. Sanchez Gorostiza, F.; Gonzalez-Longatt, F. Optimised TSO–DSO interaction in unbalanced networks through frequency-responsive EV clusters in virtual power plants. *Wiley Online Libr.* **2020**, *14*, 4908–4917. [CrossRef]
5. Acosta, M.N.; Gonzalez-Longatt, F.; Topić, D.; Andrade, M.A. Optimal microgrid–interactive reactive power management for day–ahead operation. *Energies* **2021**, *14*, 1275. [CrossRef]
6. Melo, A.; Ulrich, L.; Netto, U.; González-Longatt, F. High Impedance Fault Detection Distributed Scheme Using Modern Auto-reclosers and IEC 61850 | PAC World. Available online: https://www.pacw.org/high-impedance-fault-detection-distributed-scheme-using-modern-auto-reclosers-and-iec-61850 (accessed on 2 October 2024).
7. Juriić, G.; Havelka, J.; Capuder, T.; Sučić, S. Laboratory test bed for analyzing fault-detection reaction times of protection relays in different substation topologies. *Energies* **2018**, *11*, 2482. [CrossRef]
8. Pham, L.N.H.; Gonzalez-Longatt, F. *Distance Protection Relay Testing Using Virtual Hardware-in-the-Loop Device*; Springer Nature: Singapore, 2023; pp. 379–424. [CrossRef]
9. Sanchez-Garrido, J.; Jurado, A.; Medina, L.; Rodriguez, R.; Ros, E.; Diaz, J. Digital Electrical Substation Communications Based on Deterministic Time-Sensitive Networking over Ethernet. *IEEE Access* **2020**, *8*, 93621–93634. [CrossRef]
10. Daboul, M.; Orsagova, J.; Bajanek, T.; Wasserbauer, V. Testing protection relays based on IEC 61850 in Substation Automation Systems. In Proceedings of the 2015 16th International Scientific Conference on Electric Power Engineering (EPE), Kouty nad Desnou, Czech Republic, 20–22 May 2015; pp. 335–340. [CrossRef]
11. Memon, A.A.; Kauhaniemi, K. Real-Time Hardware-in-the-Loop Testing of IEC 61850 GOOSE-Based Logically Selective Adaptive Protection of AC Microgrid. *IEEE Access* **2021**, *9*, 154612–154639. [CrossRef]
12. Sidhu, T.; Kanabar, M.; Parikh, P. Configuration and performance testing of IEC 61850 GOOSE. In Proceedings of the 2011 IEEE International Conference on Advanced Power System Automation and Protection (APAP), Beijing, China, 16–20 October 2011; Volume 2, pp. 1384–1389. [CrossRef]
13. Cui, H.; Li, F.; Tomsovic, K. Cyber-physical system testbed for power system monitoring and wide-area control verification. *IET Energy Syst. Integr.* **2020**, *2*, 32–39. [CrossRef]

14. Phan, L.N.H.; Melo, A.F.S.; Gonzalez-Longatt, F. Testing a Non-directional Overcurrent Protection Relay: Power-Hardware-in-The Loop Approach. In Proceedings of the 2024 4th International Conference on Smart Grid and Renewable Energy (SGRE), Doha, Qatar, 8–10 January 2024. [CrossRef]
15. Pham, L.N.H.; Wagle, R.; Tricarico, G.; Melo, A.F.S.; Rosero-Morillo, V.; Shukla, A.; Gonzalez-Longatt, F. Real-Time Cyber-Physical Power System Testbed for Optimal Power Flow Study using Co -Simulation Framework. *IEEE Access* **2024**, *12*, 1. [CrossRef]
16. Poudel, S.; Ni, Z.; Malla, N. Real-time cyber physical system testbed for power system security and control. *Int. J. Electr. Power Energy Syst.* **2017**, *90*, 124–133. [CrossRef]
17. Pham, L.N.H.; Wagle, R.; Gonzalez-Longatt, F.; Acosta, M. Non-directional Overcurrent Protection Relay Testing Using Virtual Hardware-in-the-Loop Device. In *Real-Time Simulation and Hardware-in-the-Loop Testing Using Typhoon HIL*; Springer Nature: Singapore, 2023; pp. 283–337. [CrossRef]
18. *IEC 61850-7-1*; Communication Networks and Systems in Substations-Part 7-1: Basic Communication Structure for Substation and Feeder Equipment—Principles and Models. IEC: London, UK, 2011.
19. Melo, A.F.S.; Roldan-Fernandez, J.M.; Gonzalez-Longatt, F.; Burgos-Payan, M.; Netto, U.C. Novel IEC 61850-based off-site engineering and validation methodology for protection, automation, and control systems. *Electr. Power Syst. Res.* **2024**, *232*, 110409. [CrossRef]
20. *IEC 61850-6*; Communication Networks and Systems in Substations-Part 6: Configuration Description Language for Communication in Electrical Substations Related to IEDs. IEC: London, UK, 2009.
21. *IEC 61850-5*; Communication Networks and Systems in Substations—Part 5: Communication Requirements for Functions and Device Models. IEC: London, UK, 2013.
22. Gonzalez-Redondo, M.J.; Moreno-Munoz, A.; Pallares-Lopez, V.; Real-Calvo, R.J.; Lopez, M.A.O.; Moreno-Garcia, I.M. IEC 61850 GOOSE transfer time measurement in development stage. In Proceedings of the 2013 IEEE 22nd International Symposium on Industrial Electronics (ISIE), Taipei, Taiwan, 28–31 May 2013. [CrossRef]
23. *IEC 61850-10*; Communication Networks and Systems in Substations-Part 10: Conformance Testing. IEC: London, UK, 2022.
24. Riquelme-Dominguez, J.M.; Gonzalez-Longatt, F.; Rueda, J.L.; Palensky, P.; Felipe, A.; Melo, S. Cyber-physical testbed co-simulation real-time: System frequency response. In Proceedings of the 2022 IEEE International Conference on Environment and Electrical Engineering and 2022 IEEE Industrial and Commercial Power Systems Europe (EEEIC/I&CPS Europe), Prague, Czech Republic, 28 June–1 July 2022. [CrossRef]
25. Avalos, A.; Zamora, A.; Escamilla, O.; Paternina, M.R.A. Real-Time Hardware-in-The-loop Implementation for Power Systems Protection. In Proceedings of the 2018 IEEE PES Transmission & Distribution Conference and Exhibition-Latin America (T&D-LA), Lima, Peru, 18–21 September 2018. [CrossRef]
26. Pham, L.N.H.; Wagle, R.; Gonzalez-Longatt, F. Concise Definition of the Overcurrent Protection System for CIGRE European Medium Voltage Benchmark Network. In Proceedings of the 2023 IEEE PES Conference on Innovative Smart Grid Technologies-Middle East (ISGT Middle East), Abu Dhabi, United Arab Emirates, 12–15 March 2023; pp. 1–5. [CrossRef]
27. SEL-5032 ACSELERATOR Architect Software | Schweitzer Engineering Laboratories. Available online: https://selinc.com/products/5032/ (accessed on 22 March 2023).
28. Wang, H.; Ma, R. Design of network protocol analyzers using WinPcap. *Open Cybern. Syst. J.* **2014**, *8*, 779–783. [CrossRef]
29. Melo, A.F.S.; Riquelme-Dominguez, J.M.; Gonzalez-Longatt, F.; Rueda, J.L.; Palensky, P. Sampled Values ROCOF performance methodology breakdown. In Proceedings of the 2022 IEEE International Conference on Environment and Electrical Engineering and 2022 IEEE Industrial and Commercial Power Systems Europe, EEEIC/I and CPS Europe 2022, Prague, Czech Republic, 28 June–1 July 2022; pp. 1–5. [CrossRef]

engineering proceedings

MDPI

Proceeding Paper

Development and Evaluation of Proportional-Derivative, Proportional-Derivative with Friction Compensation, Inverse-Dynamics, and Sliding-Mode Control Strategies for Trajectory-Tracking in Robotic Manipulators †

David Robles [1], Ney Medrano [1], Yuliana Chicay [1], Marjorie Pilatasig [1], Gabriela M. Andaluz [1,2] and Paulo Leica [1,*]

1 Departamento de Automatización y Control Industrial, Facultad de Ingeniería Eléctrica y Electrónica, Escuela Politécnica Nacional, Quito 170525, Ecuador; ney.medrano@epn.edu.ec (N.M.); gabriela.andaluz@epn.edu.ec (G.M.A.)
2 Department of Electronic Engineering and Communications, Universidad de Zaragoza, 44003 Zaragoza, Spain
* Correspondence: paulo.leica@epn.edu.ec
† Presented at the XXXII Conference on Electrical and Electronic Engineering, Quito, Ecuador, 12–15 November 2024.

Abstract: In this paper, four control strategies are developed and evaluated for the trajectory-tracking of a two-degree-of-freedom SCARA-type robotic manipulator: (i) a proportional-derivative controller (PD), (ii) a proportional-derivative controller with friction compensation (PD + G), (iii) an inverse-dynamics controller and (iv) a sliding-mode controller with a dynamic model (SMCD). These controllers are implemented in a dynamic model of a manipulator robot, and their performance is assessed based on trajectory-tracking accuracy and robustness against disturbances. Robustness tests are conducted by varying the parameters of the dynamic model of the robot. The performance of each controller is analyzed using the Integral Squared Error (ISE) and the Integral of Time-weighted Squared Error (ITSE) indexes to compare their effectiveness. This study offers a comprehensive evaluation of each control strategy, demonstrating that the SMCD achieves the optimal balance between accuracy and disturbance robustness.

Keywords: inverse dynamics; manipulator; trajectory-tracking; PD controller; robustness; SMC

Citation: Robles, D.; Medrano, N.; Chicay, Y.; Pilatasig, M.; Andaluz, G.M.; Leica, P. Development and Evaluation of Proportional-Derivative, Proportional-Derivative with Friction Compensation, Inverse-Dynamics, and Sliding-Mode Control Strategies for Trajectory-Tracking in Robotic Manipulators. *Eng. Proc.* **2024**, *77*, 18. https://doi.org/10.3390/engproc2024077018

Academic Editor: Pablo Proaño

Published: 18 November 2024

1. Introduction

Robotic manipulators are fundamental components of industrial automation and play a critical role in tasks that require precision, repeatability, and flexibility. However, the precise control of these systems remains a complex challenge and of interest to researchers, especially when faced with model uncertainties, external disturbances, and dynamic motions [1]. Several control strategies have been developed to control a robotic manipulator [2,3].

There are controllers as in [4], which employ improved classical techniques such as adaptive PID that do not depend on the model and are robust to load variations. However, nonlinear controllers, such as sliding-mode controllers (SMCs), efficiently improve the tracking of the robotic manipulator [5]. The proposal of [6] focuses on the improvement of chattering by using three SMCs with the sign function, saturation, and hyperbolic tangent function—the SMC with saturation had the best results. In [7], an SMC + PD is implemented with an algorithm that does not require knowledge of the robot model, so the SMC is induced to consider the robot dynamics as an uncertainty and avoid using the model. It is worth noting the superiority in the efficiency of SMCs over classical controllers such as PDs, whose steady-state error is notoriously higher compared to an SMC-type controller [8]. SMCs have also been combined with fuzzy techniques to improve controller performance [9]. One of the challenges of SMCs is implementing techniques to mitigate the chattering effect through controller stability analysis using Lyapunov theory [10] and other SMC controls with backstepping [11]. In the work presented in [12], it is observed that the determination of an unsteady sliding surface improves the implemented control law for the sliding-mode controller; however, the control is applied to the robot joints and not to the trajectory generation. In [13–16], emphasis is placed on the implementation of a robust controller based on the sliding mode for the trajectory-tracking of a robotic manipulator; however, these works do not present a robustness analysis of the variation present in the model.

The main contributions of this work are that the controller works on the end-effector error and not on the joints, allowing us to take advantage of the redundancy of the manipulator. Another contribution corresponds to the design of an SMC that is based on the robot model, thus obtaining a controller robust to model variations, unlike other works where the perturbations are considered in the references or at the output and not with regard to the model. Also, in the design, the discontinuous component has been considered as a variant of the sigmoid to improve the chattering of the system. Another proposal of this work is the comparison through the development and evaluation of four control strategies for the trajectory-tracking of a SCARA-type robotic manipulator: (i) a proportional-derivative (PD) controller; (ii) a PD controller with friction compensation (PD + G); (iii) an inverse-dynamics controller; and (iv) a sliding-mode controller with a dynamic model (SMCD). This includes the formulation of the cinematic and dynamic model of the manipulator robot and the design of the SMCD, which includes an optimized sliding surface and a control law based on Lyapunov's theory, ensuring superior robustness to system perturbations. The validation of the proposed approach is performed by a simulation of the robotic manipulator, including trajectory algorithms, and the implementation of several test scenarios that allow for a comprehensive evaluation of the position error. The use of the Integral Squared Error (ISE) and Integral of Time-weighted Squared Error (ITSE) indexes allows for measuring the performance of the controllers, providing a quantitative comparison.

This paper is organized as follows: Section 2 presents the modeling of the SCARA-type manipulator robot; Section 3 describes the four implemented controllers; Section 4 outlines the simulation level tests of the controllers and the results of the system under perturbations; and finally, Section 5 provides the conclusions obtained from the results of the development and evaluation of the controllers.

2. Modeling

The manipulator robot used in this work is of an industrial type and corresponds to a SCARA-type structure (model BOSCH SR-800 of German origin), which has two degrees of freedom. The implementation of the four control strategies requires knowledge of its kinematic and dynamic models, which are described below.

2.1. Kinematic Model of the Manipulator Robot

The kinematic model of the two-degree-of-freedom SCARA manipulator shown in Figure 1 relates the position of the operating end in the (x, y) plane to the variables of the two joints q_1, q_2. (x, y) is the referential plane of the end-effector position, and q_1, and q_2, are the angular positions of joint 1 and 2, respectively. The robot parameters are the length of the first link, $l_1 = 0.445$ m, and the length of the second link, $l_2 = 0.355$ m, [17]. The kinematic model of the robot is defined by

$$x = l_1 cos(q_1) + l_2 cos(q_1 + q_2); \; y = l_1 sin(q_1) + l_2 sin(q_1 + q_2) \tag{1}$$

The inverse kinematics of the robot can be obtained by means of geometric analysis:

$$q_1 = tan^{-1}\left(\frac{y}{x}\right) - tan^{-1}\left(\frac{l_2 - sin(q_2)}{l_1 + l_2 cos(q_2)}\right) \tag{2}$$

$$q_2 = cos^{-1}\left(\frac{x^2 + y^2 - l_1^2 - l_2^2}{2l_1 l_2}\right) \quad (3)$$

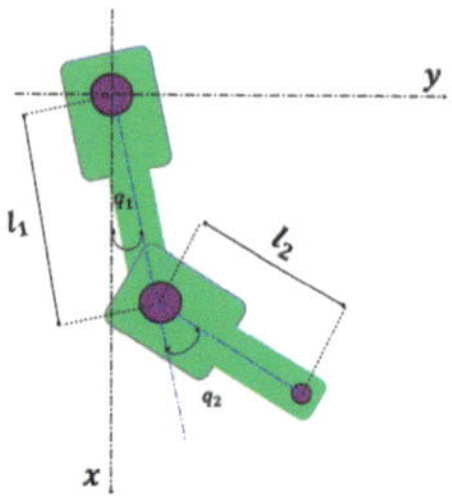

Figure 1. SCARA robotic manipulator.

From the derivation of (1), the kinematic model can be represented as

$$\begin{bmatrix} \dot{\tilde{x}} & \dot{\tilde{y}} \end{bmatrix}^T = J(q)\begin{bmatrix} \dot{\tilde{q}}_1 & \dot{\tilde{q}}_2 \end{bmatrix}^T \quad (4)$$

$\begin{bmatrix} \dot{\tilde{x}}\dot{\tilde{y}} \end{bmatrix}'$ is the derivative of the error in x and y, respectively; $\begin{bmatrix} \dot{\tilde{q}}_1\dot{\tilde{q}}_2 \end{bmatrix}^{\mathrm{T}}$ is the derivative of the angular error of joint 1 and 2; and $J(q)$ is the Jacobian matrix, which is given by

$$J(q) = \begin{bmatrix} -(l_1 sin(q_1) + l_2 sin(q_1 + q_2)) & -l_2 sin(q_1 + q_2) \\ (l_1 cos(q_1) + l_2 cos(q_1 + q_2)) & l_2 cos(q_1 + q_2) \end{bmatrix} \quad (5)$$

2.2. Robot Dynamic Model

The dynamic model for the SCARA robotic arm with two degrees of freedom is defined as

$$\ddot{q} = M^{-1}(q)\left(\tau - C(q,\dot{q})\dot{q} - f(\dot{q})\right) \quad (6)$$

τ is the vector of torques of joints 1 and 2, respectively; $M(q)$ is the inertia matrix of the manipulator robot; $C(q,\dot{q})$ is the matrix representing the centrifugal and Coriolis forces; and $f(\dot{q})$ is the vector representing the effects of viscous friction. The dynamic parameters of the Bosh SR-800 manipulator robot are obtained from [17].

$$M(q) = \begin{bmatrix} 1.7277 + 0.1908cos(q_2) & 0.0918 + 0.0954cos(q_2) \\ 0.0918 + 0.0954cos(q_2) & 0.9184 \end{bmatrix} \quad (7)$$

$$C(q,\dot{q}) = \begin{bmatrix} 31.8192 - 0.0954sin(q_2)\dot{q}_2 & -0.0954sin(q_2)(\dot{q}_1 + \dot{q}_2) \\ 0.3418sin(q_2)\dot{q}_1 & 12.578 \end{bmatrix} \quad (8)$$

$$f(\dot{q}) = \begin{bmatrix} 1.0256\, sig(\dot{q}_1) \\ 1.7842\, sig(\dot{q}_2) \end{bmatrix} \quad (9)$$

3. Controllers

Four control strategies for trajectory-tracking the SCARA-type manipulator robot are presented: (i) a proportional-derivative (PD) controller, (ii) a PD controller with friction compensation (PD + G), (iii) an inverse-dynamics controller, and (iv) a sliding-mode controller with a dynamic model (SMCD).

3.1. Dynamic Sliding Mode Controller (SMCD)

The controller in sliding mode with a dynamic model allows us to improve the performance of the system. Since it considers the dynamics of the manipulator, it will allow us to reduce the position error in the steady state, which means $\tilde{p} \to 0$ as time tends to infinity $\lim_{t\to\infty} \tilde{\rho} = 0 \in R^n$. The sliding-mode control structure with a dynamic model (SMCD) is defined as the following:

$$\tau_{SMCD} = \tau_{cD} + \tau_{dD} \tag{10}$$

where τ_{SMCD} describes the SMCD control action; τ_{cD} represents the continuous part; and τ_{dD} represents the discontinuous part of the control action. Figure 2 shows the SMCD controller scheme, where ε_d represents the robot's desired trajectory and ε represents the trajectory described by the robot.

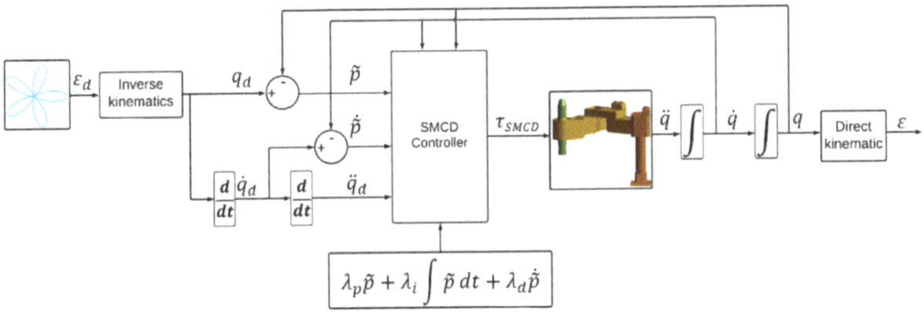

Figure 2. SMCD control scheme.

The sliding surface is defined by

$$s = \lambda_p\tilde{p} + \lambda_i \int \tilde{p}dt + \lambda_d\dot{\tilde{p}} \tag{11}$$

where $\tilde{p} = [q_d - q]$ is the end-effector position error; q_d represents the desired end-effector positions; q is the current position of the robot end effector; λ_p, λ_i, and λ_d are the coefficients of the proportional, integral, and derivative parts of the sliding surface, respectively. $\dot{\tilde{p}} = [\dot{q}_d - \dot{q}]$ is the error of the end-effector velocity; $\dot{q}_d$ is the desired velocity of the end effector; $\ddot{p} = [\ddot{q}_d - \ddot{q}]$ is the acceleration error; $\ddot{q}_d$ is the desired end-effector acceleration; and $\ddot{q}$ is the acceleration of the robot end-effector. Deriving (11) and substituting $\ddot{p} = \ddot{q}_d - \ddot{q}$, we obtain (12):

$$\dot{s} = \lambda_p\dot{\tilde{p}} + \lambda_i\tilde{p} + \lambda_d(\ddot{q}_d - \ddot{q}) \tag{12}$$

Substituting (6) in (12), we obtain the following expression:

$$\dot{s} = \lambda_p\dot{\tilde{p}} + \lambda_i\tilde{p} + \lambda_d\left(\ddot{q}_d - M^{-1}(q)\,(\tau - C(q,\dot{q})\dot{q} - f(\dot{q}))\right) \tag{13}$$

To determine τ_{cD}, we assume $\tau_{dD} = 0$ and set $\dot{s} = 0$, obtaining (14):

$$\tau_{cD} = M(q)\left[\frac{\lambda_p\dot{\tilde{p}}}{\lambda_d} + \frac{\lambda_i\tilde{p}}{\lambda_d} + \ddot{q}_d\right] + C(q,\dot{q})\dot{q} + f(\dot{q}) \tag{14}$$

For τ_{dD}, the Lyapunov candidate function is defined as $V = \frac{1}{2}s^Ts$. Deriving V with respect to time, $\dot{V} = s^T\dot{s}$ is obtained. Substituting (12) and in a closed loop, $\tau = \tau_{SMCD} = \tau_{cD} + \tau_{dD}$, and replacing (14), we obtain the following:

$$\dot{V} = s^T\left(-M^{-1}\lambda_d\tau_{dD}\right) \tag{15}$$

To ensure that $\dot{V} < 0$, τ_{dD} is defined as

$$\tau_{dD} = \lambda_d^{-1} M\, \delta \frac{s}{|s| + \beta} \tag{16}$$

where δ and β represent a positive scalar; $sign(s)$ is a sign function. We substitute (16) in (15).

$$\dot{V} = -s^T \delta \frac{s}{|s| + \beta} \tag{17}$$

This guarantees that $\dot{V} < 0$; therefore, $s \to 0$ as $t \to \infty$. Deriving (11) and setting it equal to zero, we obtain the following:

$$0 = \lambda_d\, \ddot{\tilde{p}} + \lambda_p\, \dot{\tilde{p}} + \lambda_i\, \tilde{p} \tag{18}$$

The roots of the second-order system can be defined as $r = \frac{-\lambda_p - \sqrt{\lambda_p^2 - 4\lambda_d \lambda_i}}{2\lambda_d}$. In this second-order system, to have different real roots, it must be satisfied that $\lambda_p^2 > 4\lambda_d \lambda_i$; then, $\tilde{p} \to 0$ with $t \to \infty$.

3.2. Inverse-Dynamics Controller

The control law used to implement the model is taken from [18], and is defined as follows:

$$\tau = M(q)\left(\ddot{q}_d + K_v\, \dot{\tilde{q}} + K_p \tilde{q}\right) + C(q, \dot{q})\dot{q} + f(\dot{q}) \tag{19}$$

where $\dot{\tilde{q}} = [\dot{q}_d - \dot{q}]$ is the end-effector velocity error; $\dot{q}_d$ is the desired end-effector speed; $\dot{q}$ is the manipulator robot end-effector speed; $\tilde{q} = [q_d - q]$ is the end-effector position error; and K_v and K_p represent the velocity and position gain matrices, respectively.

3.3. Controller PD + G

For the PD + G controller, the following control law given in [19] is used, described by

$$\tau = K_v\, \dot{\tilde{q}} + K_p\, \tilde{q} + f(\dot{q}) \tag{20}$$

3.4. Controller PD

For the PD controller, we use the control law given in [19], which is defined as follows:

$$\tau = K_v\, \dot{\tilde{q}} + K_p\, \tilde{q} \tag{21}$$

4. Results

In order to compare the four controllers, the tests of all the controllers consist of the manipulator robot tracking a trajectory of a five-petal flower. The parameters of the desired trajectory are as follows: $r = 0.1 + 0.1\, cos(0.25t)$; $x_d = r\, cos(0.05t) + 0.4$; and $y_d = r\, sin(0.05t) + 0.4$, where $\varepsilon_d = [x_d\ y_d]$ and $\varepsilon = [x\ y]$. The parameters of the proportional and derivative actions of the PD controller are $K_p = \begin{bmatrix} 40 & 0 \\ 0 & 40 \end{bmatrix}$ and $K_v = \begin{bmatrix} 10 & 0 \\ 0 & 10 \end{bmatrix}$, and those of the PD + G controller are $K_p = \begin{bmatrix} 40 & 0 \\ 0 & 20 \end{bmatrix}$ and $K_v = \begin{bmatrix} 20 & 0 \\ 0 & 14 \end{bmatrix}$. The parameters of the proposed SMCD are $\lambda_p = 0.1$, $\lambda_i = 70$, $\lambda_d = 0.5$, $\beta = 0.6$; and $\delta = 1.2$. To analyze the behavior of the controllers, it is proposed to develop two experiments. A time of 130 s is considered the simulation duration of Experiment 1, and 260 s is the simulation duration of Experiment 2. In Experiment 1, the desired trajectory is implemented without any disturbance. In Experiment 2, the desired trajectory is considered with disturbance at time

$t = 130$ s. For both experiments, two error-tracking indices are evaluated: ISE and ITSE. The simulation platform used in this work is Matlab version R2022B.

4.1. Experiment 1

Using the dynamic model from (6), each of the four controllers was implemented. Figure 3 shows the trajectory-tracking performance achieved by each controller. The results show that the trajectory-tracking signals described by the PD controller exhibit the highest error, while the SMCD achieves the most precise trajectory-tracking. Figure 4 shows the position errors for each controller. From the results obtained, the SMCD performs better than the others, since it has a lower amplitude and the error is visually negligible compared to the PD, PD + G, and inverse-dynamics controllers.

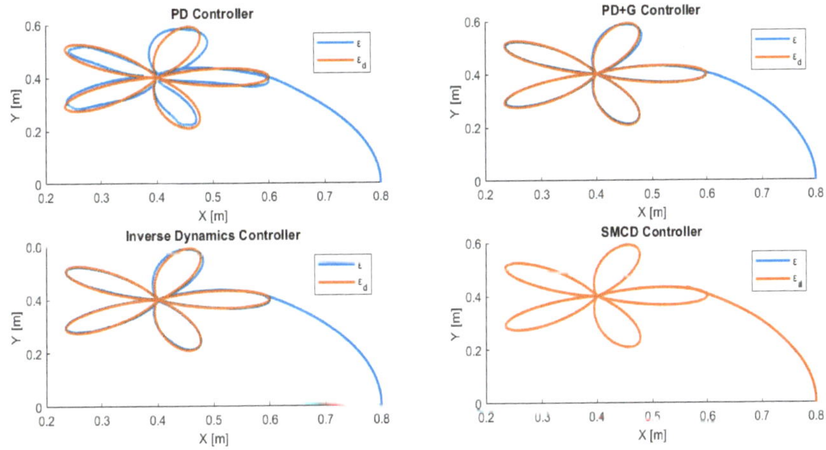

Figure 3. Trajectories achieved by each controller without disturbance conditions.

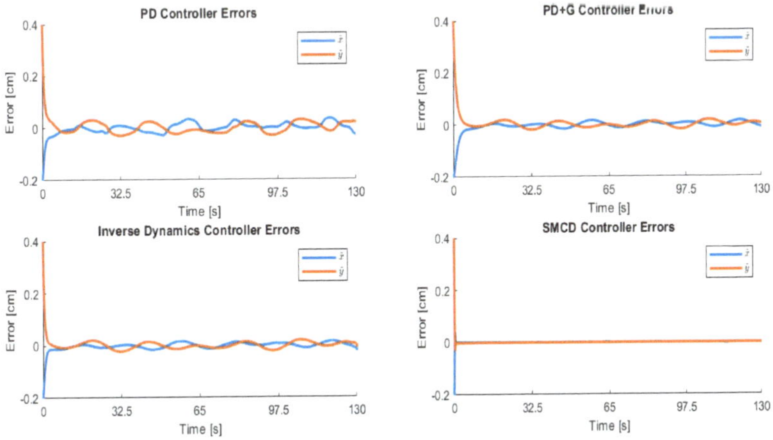

Figure 4. Position errors were obtained for each controller without disturbance.

Table 1 shows the error-based performance indices (ISE and ITSE) for the four controllers. From the results shown, it is observed that the values obtained for the SMCD have the lowest indices, so it is verified that it is the best controller in terms of trajectory-tracking without disturbance. On the other hand, the PD controller is the least accurate due to its high error rates.

Table 1. Performance index based on trajectory-tracking error with each type of controller.

Error-Based Performance Index	PD		PD + G		Inverse Dynamics		SMCD	
	X-axis	Y-axis	X-axis	Y-axis	X-axis	Y-axis	X-axis	Y-axis
ISE	0.05944	0.1115	0.0584	0.1068	0.0334	0.0624	0.00503	0.0125
ITSE	1.913	2.99	0.6313	0.8451	0.5995	0.9947	0.00150	0.00083

4.2. Experiment 2

For the robustness analysis, a perturbation is incorporated into the matrix $M(q)$ obtained in (7); this disturbance factor is referred to as F_P. For the analysis with perturbation, each controller has a different perturbation factor, F_P. However, the disturbance in all controllers is considered to occur at the time instant of 130 s.

$$M(q) = \begin{bmatrix} 1.7277 + 0.1908cos(q_2) & 0.0918F_P + 0.0954cos(q_2) \\ 0.0918F_P + 0.0954cos(q_2) & 0.9184F_P \end{bmatrix} \tag{22}$$

Figure 5 displays the trajectories obtained by incorporating the maximum allowed disturbance magnitude for each controller without compromising their execution capability. The results indicate that the SMCD exhibits superior performance in trajectory-tracking under disturbance conditions. In contrast, the other controllers—PD, PD + G, and inverse dynamics—experience significant destabilization due to the disturbance, and they deviate notably from the desired trajectory.

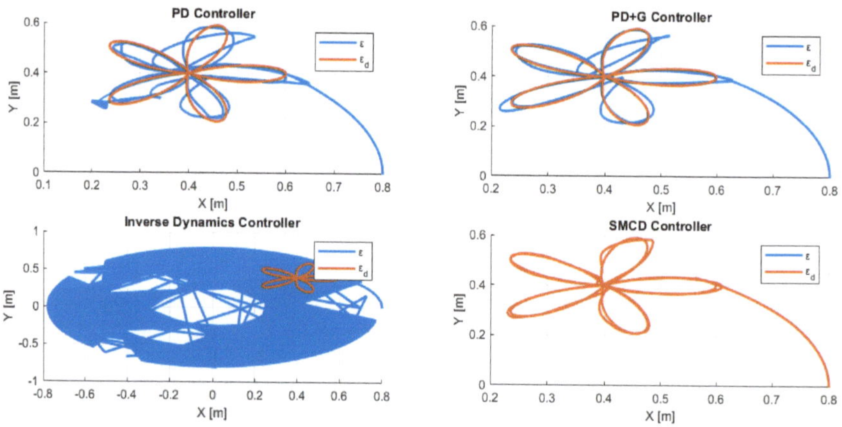

Figure 5. Trajectories achieved by each controller under disturbance conditions.

Finally, Figure 6 presents the position error values obtained by introducing the disturbance into the system. An analysis of the different results reveals that the error magnitude increases at the moment of disturbance occurs; despite the control actions of the different controllers, they fail to minimize the position error. The ability of each controller to handle the maximum disturbance without losing control and becoming unstable was evaluated. As a result, we found that the PD controller can tolerate a disturbance of $F_P = 110$, the PD + G controller can tolerate $F_P = 90$; the inverse-dynamics controller can tolerate $F_P = 82$, and the SMCD can tolerate $F_P = 163$, making the SMCD the most robust against disturbances.

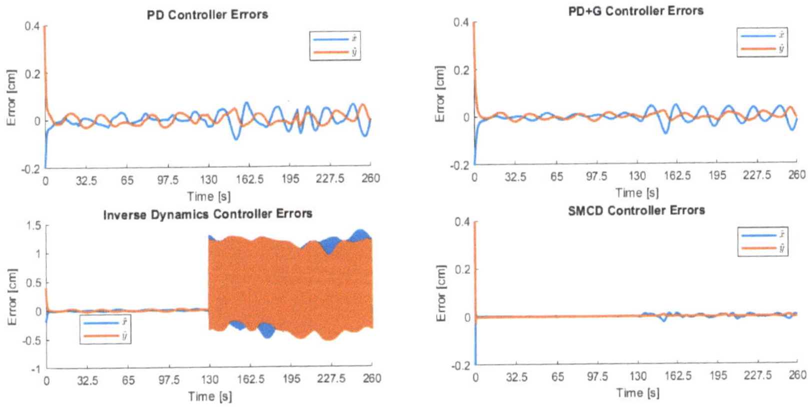

Figure 6. Position errors were obtained for each controller with disturbance at $t = 130$ s.

5. Conclusions

The SMCD demonstrated significant superiority in trajectory-tracking the robotic manipulator, outperforming the other evaluated controllers both under normal conditions and with disturbances. This controller exhibited the lowest error indices (ISE and ITSE) and maintained good accuracy, even in the presence of external disturbances, highlighting its robustness and effectiveness for applications in dynamic environments. In contrast, the inverse-dynamics controller showed good performance under normal conditions but displayed notable limitations in handling disturbances, suggesting that it may not be suitable for robotic systems that require high precision and operate in dynamic environments. Meanwhile, the PD and PD + G controllers experienced considerable increases in position error when faced with disturbances, resulting in inferior performance compared to the SMCD. These results underscore the importance of considering robustness and disturbance-recovery capability in the controller design of robotic systems intended to operate in non-ideal environments.

Author Contributions: This work presents the development and evaluation of PD, PD + G, inverse-dynamics, and sliding-mode control strategies for trajectory-tracking in a robotic manipulator. For the development, the contribution was as follows: Conceptualization: G.M.A. and P.L.; Investigation and Methodology: G.M.A. and P.L.; Software and Validation: D.R., N.M., Y.C. and M.P.; Writing—original draft preparation: D.R., N.M., Y.C. and M.P.; Writing—review and editing: G.M.A. and P.L.; Supervision: P.L. All authors have read and agreed to the published version of the manuscript.

Funding: This research received no external funding.

Institutional Review Board Statement: Not applicable.

Informed Consent Statement: Not applicable.

Data Availability Statement: Data are contained within the article.

Acknowledgments: The authors would like to thank the GIECAR group and ARCI for the technical support for the work carried out. Also, the ESPE-LA and Escuela Politécnica Nacional that through the PIEX-DACI-ESPE-24 project, Autonomous control of Aerial Manipulator Robots, and PIS-23-09 have provided research hours for the development of the project.

Conflicts of Interest: The authors declare no conflicts of interest.

References

1. Nuhel, A.K.; Sazid, M.M.; Bhuiyan, N.M.; Arif, A.I. Designing and Performance-Analysis of a 3 DOF Robotic Manipulator Arm and its Higher Order Integration of 7 DOF Robotic Arm. In Proceedings of the 2022 4th International Conference on Sustainable Technologies for Industry 4.0 (STI), Dhaka, Bangladesh, 17–18 December 2024. [CrossRef]
2. Khan, G.D. Adaptive Neural Network Control Framework for Industrial Robot Manipulators. *IEEE Access* **2024**, *12*, 63477–63483. [CrossRef]
3. Dai, L.; Yu, Y.; Zhai, D.-H.; Huang, T.; Xia, Y. Robust Model Predictive Tracking Control for Robot Manipulators with Disturbances. *IEEE Trans. Ind. Electron.* **2021**, *68*, 4288–4297. [CrossRef]
4. Lee, J.; Chang, P.H.; Yu, B.; Jin, M. An Adaptive PID Control for Robot Manipulators Under Substantial Payload Variations. *IEEE Access* **2020**, *8*, 162261–162270. [CrossRef]
5. Truong, T.N.; Vo, A.T.; Kang, H.-J. A Backstepping Global Fast Terminal Sliding Mode Control for Trajectory Tracking Control of Industrial Robotic Manipulators. *IEEE Access* **2021**, *9*, 31921–31931. [CrossRef]
6. Animesh, A.; Ohri, J. Trajectory Tracking in a 3-DOF Robotic Manipulator using Sliding Mode Controller. In Proceedings of the 2020 First IEEE International Conference on Measurement, Instrumentation, Control and Automation (ICMICA), Kurukshetra, India, 24–26 June 2020. [CrossRef]
7. Ouyang, P.; Acob, J.; Pano, V. PD with sliding mode control for trajectory tracking of robotic system. *Robot. Comput. Manuf.* **2014**, *30*, 189–200. [CrossRef]
8. Ashagrie, A.; Salau, A.O.; Weldcherkos, T. Modeling and control of a 3-DOF articulated robotic manipulator using self-tuning fuzzy sliding mode controller. *Cogent Eng.* **2021**, *8*, 1950105. [CrossRef]
9. Obando, C.; Chávez, D.; Leica, P.; Camacho, O. Sliding Mode Controller Based on a Hybrid Surface for Tracking Improvement of Non-Linear Processes. *IFAC-Pap.* **2020**, *53*, 11747–11752. [CrossRef]
10. Zhao, R.; Yang, J.; Li, X.; Mo, H. Adaptive Variable Universe Fuzzy Sliding-Mode Control for Robot Manipulators with Model Uncertainty. *IEEE J. Radio Freq. Identif.* **2024**, *8*, 658–664. [CrossRef]
11. Li, Z.; Zhai, J. Event-Triggered Based Sliding Mode Asymptotic Tracking Control of Robotic Manipulators. *IEEE Trans. Circuits Syst. II: Express Briefs* **2023**, *71*, 1266–1270. [CrossRef]
12. Ma, Z.; Sun, G. Dual terminal sliding mode control design for rigid robotic manipulator. *J. Frankl. Inst.* **2018**, *355*, 9127–9148. [CrossRef]
13. Azeez, M.I.; Abdelhaleem, A.M.M.; Elnaggar, S.; Moustafa, K.A.F.; Atia, K.R. Optimized sliding mode controller for trajectory tracking of flexible joints three-link manipulator with noise in input and output. *Sci. Rep.* **2023**, *13*, 12518. [CrossRef] [PubMed]
14. Naik, P.R.; Samantaray, J.; Roy, B.K.; Pattanayak, S.K. 2-DOF robot manipulator control using fuzzy PD control with SimMechanics and sliding mode control: A comparative study. In Proceedings of the 2015 International Conference on Energy, Power and Environment: Towards Sustainable Growth (ICEPE), Shillong, India, 12–13 June 2015; pp. 1–6. [CrossRef]
15. Vijay, M.; Jena, D. Optimal GA based SMC with adaptive PID sliding surface for robot manipulator. In Proceedings of the 2014 9th International Conference on Industrial and Information Systems (ICIIS), Gwalior, India, 15–17 December 2014; pp. 1–6. [CrossRef]
16. Azeez, M.I.; Elnaggar, S.; Abdelhaleem, A.M.M.; Moustafa, K.A.F.; Atia, K.R. Enhancing robustness and noise rejection in flexible joint manipulators: An optimized sliding mode controller with enhanced gray wolf optimization for trajectory tracking. *J. Braz. Soc. Mech. Sci. Eng.* **2023**, *45*, 4–7. [CrossRef]
17. Roberti, F.; Soria, C.; Slawinski, E.; Mut, V.; Carelli, R. Open Software Structure for Controlling Industrial Robot Manipulators. In *Robot Manipulators Trends and Development*; INTECH Open Access Publisher: Rijeka, Croatia, 2010; pp. 497–520. [CrossRef]
18. Orozco-Soto, S. Fuzzy PD+G Type 2 Interval Fuzzy PD+G Control for Position Control of Robot Manipulators. *Technol. Mag.* **2015**, *2*, 899–909.
19. Lozano, R.; Valera, A.; Albertos, P.; Arimoto, S.; Nakayama, T. PD control of robot manipulators with joint flexibility, actuators dynamics and friction. *Automatica* **1999**, *35*, 1697–1700. [CrossRef]

engineering proceedings

MDPI

Proceeding Paper

An Integrated System of Industrial Robotics and Machine Vision for the Automation of the Assembly and Packaging Process of Industrial Hinges †

Carlos Calderon-Cordova *, David Castillo, José Fernandez, Roger Sarango and Raúl Castro

Department of Computer Science and Electronics, Universidad Técnica Particular de Loja, Loja 1101608, Ecuador; dicastillo1@utpl.edu.ec (D.C.); jgfernandez4@utpl.edu.ec (J.F.); rasarango1@utpl.edu.ec (R.S.); jrcastro@utpl.edu.ec (R.C.)

* Correspondence: cacalderon@utpl.edu.ec or cacalderon@ieee.org

† Presented at the XXXII Conference on Electrical and Electronic Engineering, Quito, Ecuador, 12–15 November 2024.

Abstract: The manufacturing, assembly, and packaging processes of industrial hinges must improve their productivity. To address this challenge, a technological system based on a robot arm and vision machine was developed, integrated, and evaluated. The devices that make up the hardware architecture are the following: the Epson VT6L six-degrees-of-freedom industrial robot arm, the Compact Vision CV2-HA image-processing device, the Basler acA1600-20gc camera, and the end effector based on 3 kg holding force electromagnets. The software elements used for the development of this automated system are the SPEL+ programming language and VisionGuide tool, both of which are integrated into the Epson RC+ robotic system development environment. For the performance evaluation of the hinges and packaging identification algorithms, 80 hinges of each type were used as input to the robotic system workstation. As a result, the following results were obtained: (1) the hinge identification algorithms were 100% correct, (2) the primary packaging identification algorithm was 100% correct, and (3) the hinge assembly algorithm was 92.5% correct because it had six errors. These results provide evidence of the effectiveness of the developed system. In addition, the motion times of the robot arm were analyzed in detail to identify opportunities for improving the automated production process.

Keywords: robotic arm; machine vision; object recognition; robot motion times; motion control algorithms

Citation: Calderon-Cordova, C.; Castillo, D.; Fernandez, J.; Sarango, R.; Castro, R. An Integrated System of Industrial Robotics and Machine Vision for the Automation of the Assembly and Packaging Process of Industrial Hinges. *Eng. Proc.* **2024**, *77*, 19. https://doi.org/10.3390/engproc2024077019

Academic Editor: Pablo Proaño

Published: 18 November 2024

1. Introduction

According to [1], the metal-mechanical sector is a priority sector because it is the origin of other industries such as food, construction, and textiles, among others, as it provides the necessary machinery for the production of their goods. This industrial sector is one of the engines of progress in the economy of Ecuador, the essential activity of which is the manufacture of base metals. Evidently, essential changes are required for the development and improvement of applied technologies [2].

This sector represents 10% of the total non-oil manufacturing GDP, generates more than 80 thousand jobs, and is among the most sectorially interrelated industries in the economy. It shows high growth potential because of its capacity to produce not only final products, but also supplies for other industries or economic sectors [3], which makes it an attractive field for the application of new technologies to improve and optimize its processes.

The manufacturing, assembly, and packaging processes of industrial hinges need to improve their productivity, that is, the number of units produced and packaged per hour. This is because of a constant increase in product demand. Industrial tubular hinges are made up of two parts, male and female, and their application is the manufacture of doors and windows of industrial characteristics. The technological system coupled with the process automation objective is a robotic system complemented by a machine vision system for the identification of objects to be manipulated.

Based on the preliminary elements of the system, related works that contribute to the research and development of robotic systems operated with artificial vision have been identified. In [4], a machine vision system was implemented for a pick-and-place application of the ABB IRB120 industrial robotic arm, which is an Intel RealSense D435 camera that captures the color image and depth information. Finally, it uses pattern recognition algorithms for object location. In [5], the authors developed a collection and manipulation system based on the Kawasaki RS03N robotic arm, which consists of a Basler acA1300 camera and the OpenCV image processing library. The machine vision system performs object capture, pre-processing, segmentation, and recognition. The relevant operations performed on the image included Gaussian filtering, geometric transformation, and thresholding.

In [6], a positioning system for a six-degrees-of-freedom robotic arm was developed using artificial neural networks. Four neural networks that control the movements of the joints were used to move the position of the end effector. In [7], a positioning system for a robotic arm was developed using two cameras, and processing was performed using a computer vision model called YOLOv8. On the other hand, in [8], the positioning of a robot arm was based on RGB-D images, that is, color images with depth information. The authors used an Iterative Closest Point (ICP) algorithm.

Now, focusing on the algorithms related to the present project, the effectiveness of pattern recognition algorithms in industrial manipulator robots can be assessed through key factors: accuracy, speed, and adaptability. Recent advancements in deep learning and computer vision have notably strengthened these algorithms, enabling them to handle complex industrial tasks with greater precision and efficiency. The following sections highlight essential characteristics of the most effective algorithms.

Regarding accuracy: Deep learning models, especially those employing convolutional neural networks (CNNs), have shown high accuracy, reaching up to 94% in object recognition tasks with robotic manipulators [9]. Additionally, simpler real-time methods have achieved 100% accuracy in certain image classification tasks, demonstrating that high accuracy can also be attained with less complex algorithms [10].

Regarding speed: Algorithms such as SIFT and SURF, implemented via the OpenCV library, offer fast feature detection and matching, which is essential for real-time applications in industrial environments [11]. Leveraging deep learning models, especially on powerful GPUs, can further enhance processing speed, enabling efficient handling of complex tasks [12].

Regarding adaptability: Learning algorithms empower robots to adapt to diverse tasks without the need for explicit programming, greatly enhancing their flexibility in dynamic environments [13]. Furthermore, the capability to recognize and respond to variations in object features, such as shape and color, underscores the adaptability of these systems [9].

Although deep learning and advanced algorithms provide substantial advantages, simpler methods can also deliver high accuracy and speed. This highlights a potential trade-off between algorithmic complexity and performance, depending on the specific application requirements.

Based on the above, the objective of the present project is to develop a system to identify, assemble, and pack industrial hinges using an industrial robotic arm and machine-vision algorithms. The outline of this study consists of Section 2, Robotic System Design and Machine Vision, which deals with the description of the production process and the technological architecture of the system; Section 3, Development of Object Identification Routines, deals with the main elements of the machine vision routines; Section 4, Results and Discussion, deals with the evaluation of the motion control algorithms and the identification algorithms, and Section 5 summarizes the conclusions of this study.

2. Robotic System Design and Artificial Vision

2.1. Description of the Production Process

The production process of industrial hinges consists of five stages: turning, drilling, assembly, primary packaging, and secondary packaging (Figure 1). In the turning stage, male (M25) and female (H25) hinges were cut and manufactured from round black steel rods. The dimensions obtained were a diameter of 25 mm, length of 55 mm (female hinge), and length of 65 mm (male hinge). The hinge types were manufactured separately on different machines, and the boxes were then transported to the punching stage.

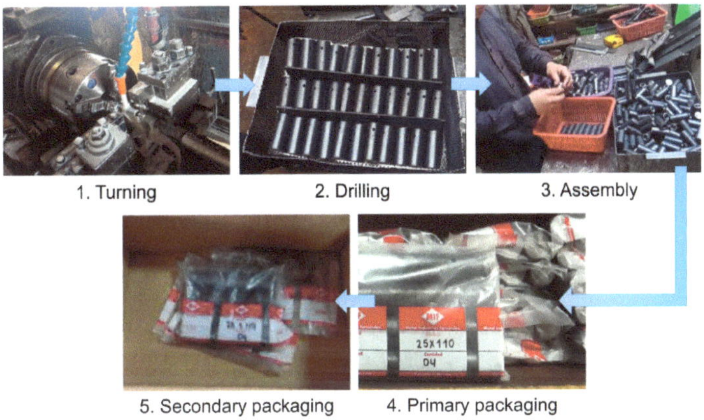

Figure 1. Graphic description of the stages of the original production process.

The drilling stage consists of drilling a 4 mm diameter hole in the female hinge for lubrication of the mechanism. Subsequently, they were placed in a metal basket with partitions. The assembly stage receives the male and female hinge baskets, couples them, and continues to the primary packaging stage, in which four assembled hinges are grouped in a plastic envelope. Finally, the secondary packaging consists of storing 20 primary packages in cardboard boxes distributed in two rows, two columns, and five levels.

The average monthly production of this process is approximately 2800 hinges; however, it is desirable to increase these values using an automation system. Based on the analysis of the production process, a robotic system was designed to perform assembly and secondary packaging tasks.

2.2. Robotic System Architecture

The robotic system to perform the assembly and secondary packaging tasks of the industrial hinges consists of the following devices: the Epson VT6L industrial robot arm, Compact Vision CV2-HA device (Nagano, Japan), Basler acA1600-20gc camera (Ahrensburg, Germany), and electromagnet-based end effector (see Figure 2).

The EPSON VT6L robot arm has six degrees of freedom, a reach of 900 mm, and a payload of 6 kg. The robot controller is built into the base [14]. The Compact Vision CV2-HA device is a high-speed image processor that supports Gigabit Ethernet (up to four) and USB (up to two) cameras, high-resolution cameras up to 5MP, and is compatible with Epson RC+ software (RC+ 7.5 R2) and the VisionGuide (7.5 R2) tool for image processing applications [15]. The Basler GigE acA1600-20gc camera is a color camera located on the wrist link of the robot arm with a resolution of 2 megapixels 1628 × 1236, frame rate of 20 fps, pixel depth of 12 bits, and ICX274 image sensor (Sony Semiconductor Solutions, Kanagawa, Japan) [16].

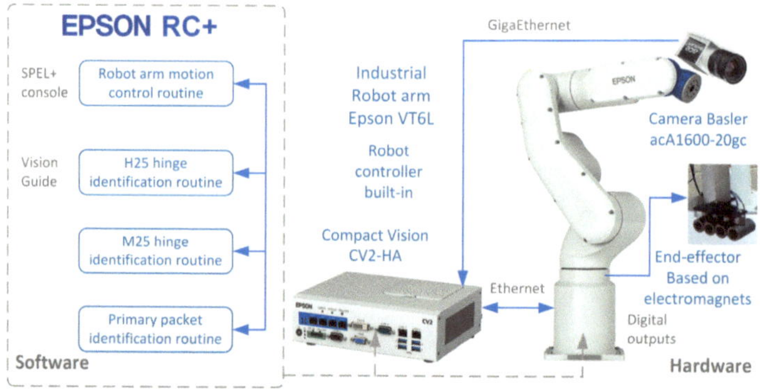

Figure 2. System hardware and software architecture.

The end effector was designed based on the requirements of the problem and consisted of a physical structure built with Grilon thermoplastic, the four electromagnets of model KK P20/15 with a holding force of 3 kg, an operating voltage of 12 V, a maximum current of 170 mA, dimensions of 20 mm × 15 mm [17], and an electronic board that energizes the electromagnets when it receives a digital signal from the robot controller.

The software tool used in this project was Epson RC+, which is used to design and implement applications with robotic systems. The programming language used was SPEL+, which is based on text instructions. In addition, it has an integrated 3D simulator, in which the movements of the robot can be tested. The library of functions used for processing the images captured by the camera is VisionGuide, which includes camera calibration, edge detectors, object position detectors, color detectors, and object geometric characteristic meters [15].

The robot workstation consisted of the following distribution of spaces: male hinge basket, female hinge basket, assembly ramp, hinge storage area with primary packaging, and secondary packaging area (see Figure 3). The workstation was illuminated by a 100 W LED spotlight with white light (6000 K).

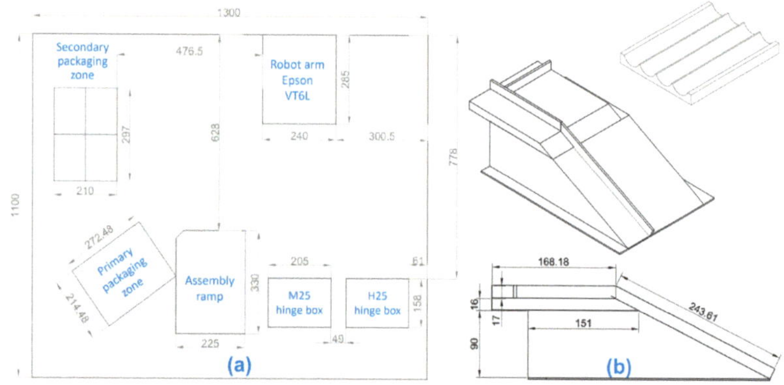

Figure 3. Epson VT6L Robot Workstation Geometry: (**a**) workstation elements and (**b**) assembly ramp design.

3. Development of Object Identification Routines

3.1. Main Program Routine

The tasks possessed by the main routine are as follows: (1) turn on the actuators of the robot arm, (2) position it over the M25 hinges to search and identify them, (3) if it detects hinges, it picks them up and arranges them on the assembly ramp, (4) the robot proceeds to position itself over the H25 hinges, (5) if it detects hinges, it picks them up and proceeds to assemble them with the H25 hinges on the ramp, (6) once the hinges are assembled, the robot moves them towards the operator who performs the primary packaging, and (7) the operator places the primary packaging in a specific position and then the robot identifies the presence of them and proceeds to pick them up and place them in the secondary packaging.

3.2. M25 and H25 Hinges Identification and Collection Routine

The M25 hinge identification routine was performed using VisionGuide. The primary functions used were ImageOp, Blob, and Point. The ImageOp function allows morphological operations to be performed on the region of interest (ROI) of an image. The Blob function calculates geometric and topological characteristics to determine the presence/absence, position, size, and orientation of an object. The Point function allows the reference positions within the workspace of the robot to be defined.

Specifically in the developed routine, the ImageOp function is used to erode the image so that only the objects of interest remain, then the Blob function is used to search for the objects of interest, determine and compare the area within a defined range, and then obtain the coordinates. The Point operation determines the central coordinates of a set of four M25 hinges.

Once the coordinates of the M25 hinge groups have been identified, they are called from the main routine such that the robot performs the pick-and-place function. To connect the robot arm motion control program and the image processing program, the VRun command is used to execute the specific hinge identification routine, and the VGet command is used to obtain the resulting coordinates of the identified hinges.

With respect to the H25 hinge identification routine, the same ImageOp, Blob, and Point functions are used; the parameters that vary between the two routines are due to the differences in length of this type of hinge.

3.3. Primary Packet Identification Routine

The routine for the identification of primary packages is simple. The main element is the Blob function that allows to determine the area of the object found and compare it within a specific range. After the position of the object is identified, the robot picks it up and places it in the space where the secondary package is located, considering that the objects are arranged in layers of two rows and two columns.

Once the identification and collection routines for the M25 hinges, H25 hinges, and primary packages were developed, we proceeded with offline validation of the algorithms. The results of the preliminary validation of the algorithms are shown in Figure 4. In addition, it can be seen that the program stores and visualizes the execution times of the movements to analyze them later.

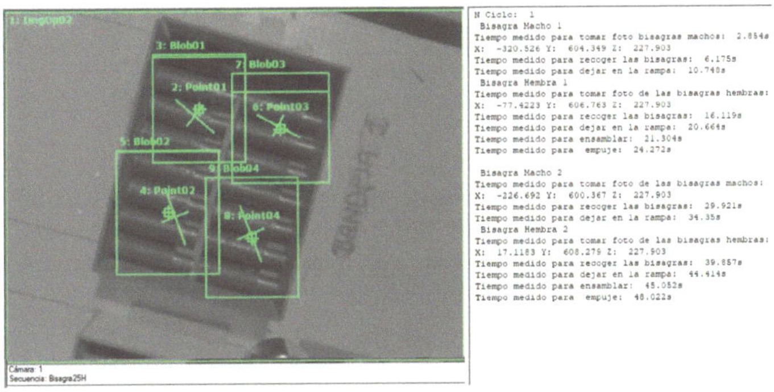

(a)

Figure 4. *Cont.*

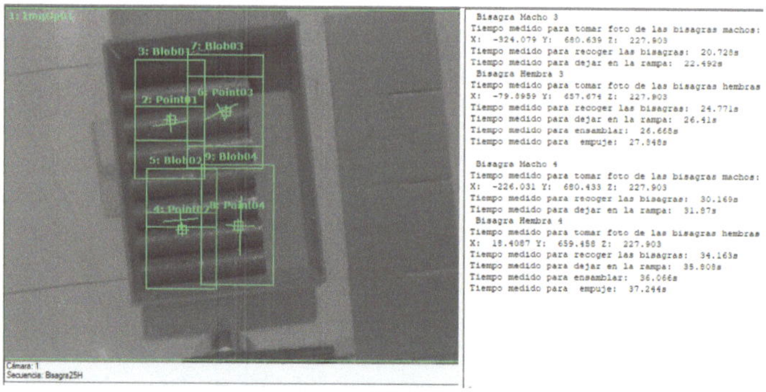

(**b**)

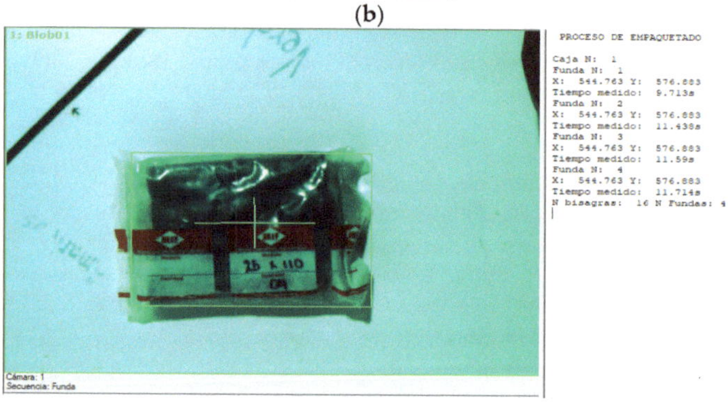

(**c**)

Figure 4. Validation of identification and collection routines for (**a**) M25 hinges, (**b**) H25 hinges, and (**c**) primary packages.

4. Results and Discussion

4.1. Evaluation of Robot Arm Motion Control Algorithms

Photographs of the instances of the object identification and robot arm motion control algorithms can be seen in Figure 5. The instances in the figure are as follows: (a) The algorithm identifies the M25 hinges. (b) The robot arm collects the M25 hinges by means of the electromagnetic type gripper. (c) The robot arm deposits them on the assembly ramp. (d) The robot arm leaves them on the assembly ramp. (e) The algorithm identifies the H25 hinges by means of the electromagnetic type gripper. (f) The robot arm collects the H25 hinges. (g) The robot arm takes them to the assembly ramp. (h) The robot arm assembles the M25 and H25 hinges. (i) The robot arm pushes the assembled hinges towards the operator, at this moment the robot will repeat the identification and control routines until there are no more M25 and H25 hinges on the workstation. While these routines are running, the operator groups the hinges into primary packaging and deposits them in a specific area in the workspace. When no more hinges are to be assembled, the robotic system proceeds with the following algorithms. (j) The algorithm identifies the group of hinges with the primary package. (k) The robot arm collects the primary packages, and (l) the robot arm arranges the primary packages in a 2 $\times$ 2 matrix for subsequent secondary packaging.

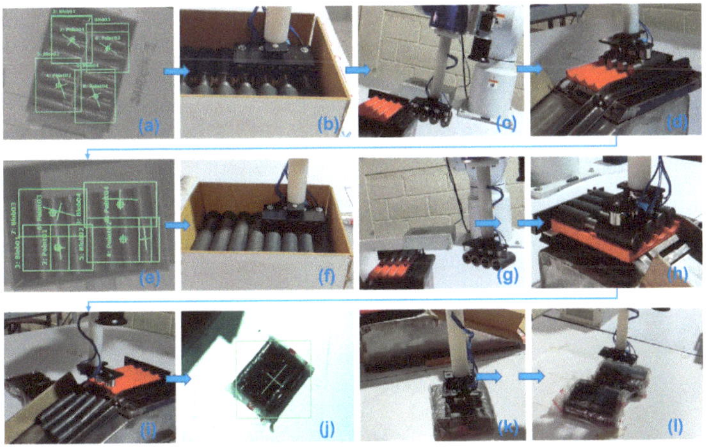

Figure 5. Photographs of the object identification and motion control algorithms of the robot arm. (**a**) identify M25 hinges, (**b**) collect M25 hinges, (**c**) carry hinges to assembly ramp, (**d**) leave M25 hinges on ramp, (**e**) identify H25 hinges, (**f**) collect H25 hinges, (**g**) carry hinges to assembly ramp, (**h**) assemble the M25 and H25 hinges, (**i**) slide the assembled hinges toward the operator, (**j**) identify the primary package of hinges, (**k**) collect the primary package, and (**l**) arrange the primary packages in a matrix.

4.2. Evaluation of Identification Algorithms

For the performance evaluation of the M25 and H25 hinges and packaging identification algorithms, 80 hinges of each type were used as inputs to the robotic system workstation. After executing all algorithms, the following results were obtained: (1) the M25 and H25 hinge identification algorithms were 100% correct, (2) the primary packaging identification algorithm was 100% correct, and (3) the hinge assembly algorithm was 92.5% correct because it had six errors. In conclusion, the effectiveness of the identification algorithms is very good, whereas the assembly process has errors because it depends on the mechanical factors of the gripper and workstation.

4.3. Analysis of Process Time Results

To analyze the times of the hardware and software systems in executing the developed functions, the program labeled the time segments for each of the robot's movements. Table 1 lists the labels of the time segments and their descriptions.

Table 1. Time segments of the VT6L robot arm movements.

Segment	Description
A	Movement to the position for capturing and processing images of M25 hinges
B	Movement to pick up the M25 hinges
C	Movement to deposit the M25 hinges on the assembly ramp
D	Movement to the position for capturing and processing images of H25 hinges
E	Movement to pick up H25 hinges
F	Movement to deposit the H25 hinges on the assembly ramp
G	Movement to assemble the M25 and H25 hinges
H	Movement to transfer the assembled hinges to the operator

The experiment to measure the process times involved assembling 80 industrial hinges, for which a box of 80 M25 hinges and a box of 80 H25 hinges were entered. The robotic system identifies, picks up, and assembles the hinges into groups of four; therefore, there are 20 cycles of movements A, B, C, D, E, F, G, and H, as described in Table 1.

Figure 6 shows the time required for 20 movement cycles to assemble the hinges. The following analyses were extracted from the data:

- The average time for each movement cycle was 24.94 s. This is the time to identify, pick up, and assemble the four industrial hinges. The total time required to assemble the 80 hinges was 8 min 19 s. Consequently, a time of 6.24 s was determined for each hinge assembled.
- The time used by the robotic system to assemble the hinges is similar to the time of the operator; however, two aspects are emphasized: the robotic system maintains its productivity throughout the working day while the operator constantly decreases its productivity owing to various factors, and it is also emphasized that these productivity values can be increased when the design of the electromagnetic gripper is improved, and the operating speed of the robot is 30% of its nominal speed to ensure the correct grip of the manipulated hinges. When the speed was higher than 416 mm/s, the parts started to detach from the gripper.
- The longest cycle times are movements C and F, which correspond to the movements to deposit the M25 and H25 hinges on the assembly ramp, because in these movements, the speed is reduced so that the hinges fit into the assembly guides.
- It is identified that the total cycle time has a repeating pattern every four cycles because the trays of the M25 and H25 hinges are in layers of 16; therefore, each layer of the trays is divided into four quadrants because the robot picks up the hinges in groups of 4. The central position of each quadrant causes a repeating pattern in the total time curve.
- Finally, the relationship between the number of assembled hinges and the time taken was determined using (1). With this function, it is estimated that the robotic system with the current conditions would assemble 577 hinges in 1 h, and 4621 hinges on a working day of 8 h.

$$Assembly\ Time = 6.231 * Number\ Of\ Hinges + 0.412 \quad (1)$$

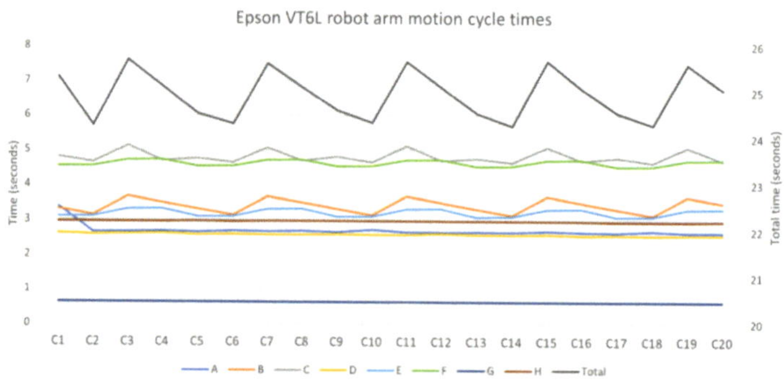

Figure 6. Motion cycle times for assembling the hinges using the VT6L robot arm. The graphs of time segments A, B, C, D, E, F, G and H use the left Y-scale. The graph of total time uses the right Y-scale.

Concerning the packing time, we also proceeded to automatically capture the times at which it identified, picked up, and deposited the primary packages in the secondary package box. Twenty packing cycles were measured because each primary package contained four assembled hinges, resulting in 80 hinges available for testing.

Based on the tests, the average time to pack each primary package of the four hinges was 11.96 s. Taking into account the data obtained, it is estimated that the robotic system with the current conditions would pack 299 packages in one hour and 2392 packages in an 8 h working day.

5. Conclusions

A system was designed and implemented to identify, pick, assemble, and pack the industrial hinges. The system is based on an industrial robotic system and machine vision system. The devices that comprise the hardware architecture are as follows: the Epson VT6L 6 DoF industrial robot arm, the Compact Vision CV2-HA image processing device, the Basler acA1600-20gc camera, and the end effector based on 3 kg force electromagnets. The software elements used for the development of this automated system were integrated into an Epson RC+ development environment, and the motion control algorithms were developed using the SPEL+ programming language. In addition, with the VisionGuide tool, routines for the identification and collection of hinges and packages were developed.

For the performance evaluation of the M25 and H25 hinge and packaging identification algorithms, 80 hinges of each type were used as inputs to the robotic system workstation. Consequently, the following was determined: (1) the M25 and H25 hinge identification algorithms were 100% correct, (2) the primary packaging identification algorithm was 100% correct, and (3) the hinge assembly algorithm was 92.5% correct, as it had six errors.

Finally, concerning the analysis of the process times, the following was determined: (1) the average time to assemble a hinge is 6.24 s, (2) using the relationship found, it is estimated that the robotic system would assemble 577 hinges per hour, (3) the average time to pack each primary package of four hinges is 11.96 s, and (4) it is estimated that the robotic system would pack 299 packages in one hour.

In future work, we plan to improve the gripper mechanism to increase the speed of movement of the joints of the VT6L robot arm. In addition, improvements will be made to the workstation layout to optimize the times and collaborate with the company's operator. Finally, the algorithms will be improved to be less dependent on the type of manipulator robot and the illumination level of the workstation. One option is to use a Reinforcement Learning-based system [18,19].

Author Contributions: Conceptualization, methodology, C.C.-C., D.C. and J.F.; software, D.C. and J.F.; validation, D.C. and J.F.; writing—original draft preparation, C.C.-C. and R.S.; writing—review and editing, C.C.-C., R.S. and R.C.; supervision, C.C. C.; project administration and funding acquisition, C.C.-C. and R.C. All authors have read and agreed to the published version of the manuscript.

Funding: This research was funded by the Universidad Tecnica Particular de Loja, grant number PROY_ARTIC_CE_2022_3667.

Institutional Review Board Statement: Not applicable.

Informed Consent Statement: Not applicable.

Data Availability Statement: Data are contained within the article.

Conflicts of Interest: The authors declare no conflicts of interest.

References

1. Criollo Segura, K.V.; Ramírez Rázuri, M.d.C. Análisis de la Inversión en Actividades Exportadoras para las Empresas del Sector Metalmecánico Ecuatoriano Asociadas a FEDIMETAL Hacia los Países de la CAN. Bachelor's Thesis, Salesian Polytechnic University, Cuenca, Ecuador, 2014.
2. Loor Sandoval, B.I. Estudio de la Evolución del Sector Metalmecánico Cuya Actividad es la Fabricación de Metales Comunes en el Ecuador en el Período 2010–2015. Master's Thesis, Universidad Andina Simón Bolívar—Sede Ecuador, Quito, Ecuador, 2018.
3. Zabala, V. Situación del Sector Metalmecánico y su Importancia en la Economía Ecuatoriana. Available online: https://ekosnegocios.com/articulo/situacion-del-sector-metalmecanico-y-su-importancia-en-la-economia-ecuatoriana (accessed on 31 October 2024).
4. Ortiz de Zuñiga Mingot, I. Optimización del Sistema de Visión Artificial de un Robot Industrial para una Aplicación de Pick and Place. Bachelor's Thesis, Universidad Pontificia Comillas, Madrid, Spain, 2020.
5. Valencia Armijos, Á.J.; Idrovo Urgilés, R.M. Diseño e implementación de un sistema de reconocimiento y manipulación de frutas utilizando visión artificial y brazo robótico industrial. Bachelor's Thesis, Escuela Superior Politécnica del Litoral, Guayaquil, Ecuador, 2017.

6. Borja, M.G. Sistema de posicionamiento con visión artificial para un brazo robótico articulado de seis grados mediante redes neruonales artificiales. In Proceedings of the 16th LACCEI International Multi-Conference for Engineering, Education, and Technology: "Innovation in Education and Inclusion", Lima, Peru, 19–21 July 2018.
7. Wei, S.; Zhu, X. Spatial recognition and positioning system of robotic arm based on binocular vision. In Proceedings of the 2024 IEEE 6th Advanced Information Management, Communicates, Electronic and Automation Control Conference (IMCEC), Chongqing, China, 24–26 May 2024; Volume 6, pp. 998–1002.
8. Yang, C.; Li, Z.; Cai, Z.; Gao, Y.; Xu, T.; He, G.; Yan, F.; Shao, C. Target position and posture recognition based on RGB-D images for autonomous grasping robot arm manipulation. In Proceedings of the 2020 10th International Conference on Information Science and Technology (ICIST), Bath, London, and Plymouth, UK, 9–15 September 2020; pp. 65–70.
9. Anh, M.N.; Bien, D.X. A Solution of Pattern Recognition Based on Deep Learning for Robotic Manipulator to Pick Up and Drop Objects. *Int. J. Emerg. Technol. Adv. Eng.* **2023**, *13*, 114–120. [CrossRef] [PubMed]
10. Voleti, V.; Mohan, P.; Gupta, S.; Iqbal, J. Simple Real-Time Pattern Recognition for Industrial Automation. In Proceedings of the 2017 International Conference on Industrial Design Engineering, Dubai, United Arab Emirates, 29–31 December 2017; pp. 107–111.
11. Mukhanov, S.B.; Uskenbayeva, R. Pattern Recognition with Using Effective Algorithms and Methods of Computer Vision Library. In *Optimization of Complex Systems: Theory, Models, Algorithms and Applications. WCGO 2019*; Advances in Intelligent Systems and Computing; Springer: Cham, Switzerland, 2020; Volume 991, pp. 810–819.
12. Miyajima, R. Deep Learning Triggers a New Era in Industrial Robotics. *IEEE MultiMedia* **2017**, *24*, 91–96. [CrossRef]
13. Cipriani, G.; Bottin, M.; Rosati, G. Applications of Learning Algorithms to Industrial Robotics. In *Advances in Italian Mechanism Science: Proceedings of the 3rd International Conference of IFToMM Italy*; Mechanisms and Machine Science; Springer: Cham, Switzerland, 2021; Volume 91, pp. 260–268.
14. EPSON. 6-Axis Robots VT Series Manipulator Manual. Rev.9. 2018; p. 316. Available online: https://files.support.epson.com/far/docs/epson_vt6l_robot_manual_(r9).pdf (accessed on 31 October 2024).
15. EPSON. Robot Specifications 2023 Catalog. 2023. Available online: https://files.support.epson.com/far/docs/epson_robot_specification_2023_catalog_cpd-54833r3.pdf (accessed on 31 October 2024).
16. Baslerweb. Basler ace Classic acA1600-20gc. Available online: https://www.baslerweb.com/en/shop/aca1600-20gc/ (accessed on 31 October 2024).
17. Robu.in. DC 12V KK-P20/15 3KG Lifting Solenoid Electromagnet. Available online: https://robu.in/product/electric-sucker-electromagnet-kk-p20-15-12v/ (accessed on 31 October 2024).
18. Calderón, C.; Sarango, R.; Castillo, D.; Lakshminarayanan, V. A Deep Reinforcement Learning Framework for Control of Robotic Manipulators in Simulated Environments. *IEEE Access* **2024**, *12*, 103133–103161. [CrossRef]
19. Calderon-Cordova, C.; Sarango, L.; Chamba, D.; Sarango, R.; Castro, R. Performance Analysis of Motion Control Algorithms of an Industrial Robot Arm Applied to 3D Concrete Printing Systems. *Eng. Proc.* **2023**, *47*, 9. [CrossRef]

engineering proceedings

MDPI

Proceeding Paper

Energy Production Assessment of Intermittent Renewable Generation Projects from a Stochastic Perspective †

Nilo Quirola * and Jaime Cepeda

Faculty of Electrical and Electronic Engineering, Escuela Politécnica Nacional, Quito 170525, Ecuador; cepedajaime@ieee.org

* Correspondence: nilo-esteban@hotmail.com; Tel.: +593-98-409-6499

† Presented at the XXXII Conference on Electrical and Electronic Engineering, Quito, Ecuador, 12–15 November 2024.

Abstract: In response to Ecuador's ongoing energy crisis, it is essential to explore solutions such as incorporating renewable energy projects into the system. This article presents a novel methodology to evaluate the energy impact of integrating intermittent renewable energy sources into the operation of power systems from a stochastic perspective, applied to the Yanahurcu wind farm project of 52.8 MW installed capacity. Using historical wind data from reanalysis databases, the wind resource is modeled using a Histogram-Based Gaussian Space Correlation Model. Afterwards, the wind farm modeling enables simulations to be run for a variety of case studies within SimSEE software, version 127.263. The results indicate that the methodology is effective, showing that the wind farm could match Ecuador's current annual wind energy production.

Keywords: wind energy; wind resource stochasticity; Histogram-Based Gaussian Space Correlation Model (CEGH); power system simulation

1. Introduction

The growth of the world population, coupled with the accelerated depletion of fossil fuel reserves, which, at the current rate of exploitation, are expected to be completely exhausted within a few centuries [1], has motivated researchers to evaluate the impact of integrating sources of clean and environmentally friendly energy, such as solar and wind, into electric power systems [2]. In line with this, the United Nations' 2030 Agenda established 17 Sustainable Development Goals, one of which is Affordable and Clean Energy. This goal encourages countries to invest in renewable energy, setting targets such as increasing the share of renewable energy in the global energy mix and promoting international cooperation to facilitate access to clean energy research and technology [3].

Wind energy represents an inexhaustible natural resource with the potential to combat climate change [2]. By 2023, the total installed global wind capacity reached 1021 GW, reflecting a growth of 13% compared to 2022 [4]. In Ecuador, the current regulatory framework includes the utilization of renewable energy sources, with a short-term feasible wind potential of 884 MW. To develop new renewable generation projects, a policy was established that allows the implementation of power and energy blocks from various primary sources, to be covered by projects proposed by the private sector [5] through Public Selection Processes (PPSs). Within this context, the PPS for the Non-Conventional Renewable Energy Block (ERNC I) was launched and fully awarded in 2023. The block comprises six photovoltaic solar projects, three hydroelectric projects, and one wind farm project, totaling 511 MW with private investment exceeding USD 800 million [6]. One of the awarded projects is the Yanahurcu wind farm project, with 52.8 MW installed capacity and 44.81 MW effective capacity, to be installed in the province of Loja, at a price of 60.63 USD/MWh [7].

As regards wind farm projects, the estimation of energy production is critically important for the successful planning and operation of renewable energy systems, and stochastic

Citation: Quirola, N.; Cepeda, J. Energy Production Assessment of Intermittent Renewable Generation Projects from a Stochastic Perspective. *Eng. Proc.* **2024**, *77*, 20. https://doi.org/10.3390/engproc2024077020

Academic Editor: Walter Vargas

Published: 18 November 2024

analysis plays a key role in enhancing the accuracy of these estimates. Wind power is inherently variable and influenced by several unpredictable factors, such as meteorological conditions and grid topology availability, which introduces significant uncertainty into production forecasts. Traditional deterministic models fall short of capturing this variability, often leading to mismatches between expected and actual output. Stochastic methods, on the other hand, generate multiple scenarios of wind power production based on probability distributions of wind speeds, wind directions, and other variables, offering a more realistic and comprehensive understanding of possible outcomes. By incorporating stochastic analysis, decision-makers can better manage the risks associated with wind farm production variability, optimize system performance, and improve grid integration. Moreover, it allows for the assessment of different weather scenarios and their impact on supply–demand balance, enhancing the resilience and reliability of power systems.

A method that integrates wind power and temperature correlations to generate stochastic scenarios for wind power production is presented in [8]. It emphasizes capturing both regular and extreme conditions, such as multi-day periods without wind, to perform supply–demand balance analysis. This method helps in improving the accuracy of long-term wind power forecasts and optimizing planning under stochastic conditions. In [9], a hybrid method using Vector Autoregressive Moving Average (VARMA) models and copula techniques to generate wind power scenarios with spatiotemporal correlations is proposed. This approach effectively captures both the spatial and temporal dependencies between multiple wind farms, offering a robust solution for short-term wind power forecasting. A comprehensive review of various probabilistic forecasting models used in wind power generation is stated in [10]. It discusses the differences between deterministic and probabilistic models, with a particular focus on how the latter can capture the inherent uncertainty in wind energy production. The review also identifies future research directions in the field of stochastic wind. All these contributions allow researchers to face the problem of stochastic wind farm energy production forecasting, showing important advances in this field; however, most of the approaches are not yet included in mass-market commercial software capable of being used for performing confident enough studies related to the connection of energy renewable projects. In this regard, the Ecuadorian entity responsible for planning the operation of power generation, the National Electricity Operator CENACE, optimizes economic dispatch in medium- and long-term scenarios using the SDDP (Stochastic Dual Dynamic Programming) tool, a modularly licensed software that requires significant investment [11]. As a free software alternative, the Instituto de Energía Eléctrica in Uruguay has developed the SimSEE platform for the Simulation of Electric Power Systems, which obtains an Optimal Operation Policy through the resolution of a Stochastic Dynamic Programming (SDP) problem [12]. Whereas SDDP allows researchers to manage the problem of sizing large volumes of decision variables and restrictions, SimSEE uses an SDP algorithm and offers the possibility of implementing user models, so the versatility of modeling new energy sources is greater. Although these two packages have a great modeling and simulation capacity, SimSEE surpasses SDDP in three important aspects: (i) it is an open-source free software, (ii) it is capable of accepting user-defined models, and (iii) it is more flexible for including intermittent renewable generation such as wind farms and photovoltaic plants, based on its Histogram-Based Gaussian Space Correlation (CEGH) model. In this connection, SimSEE is the selected software to develop the proposed methodology, since this software has the capabilities and enough references to be accepted by the Electricity Regulation and Control Agency in Ecuador (ARCONEL) as a replacement of SDDP.

In [11], a database has been created containing all the power plants of the National Interconnected System (SNI), modeled with parameters provided by CENACE to simulate a single-node dispatch over a 10-year horizon.

Based on the aforementioned, this paper evaluates the power generation of the Yanahurcu wind project by integrating the plant into the SimSEE database. A CEGH model is generated from wind data to analyze the wind resource in the project's installation area. For this aim, a data reanalysis strategy is first applied to obtain corrected data. Subsequently, simulations are carried out with different time horizons, thus obtaining the estimated production of the plant, which will finally be analyzed from a probabilistic perspective. The main contribution of this paper is the integration of a free tool like SimSEE with an open-access meteorological database like MERRA2. This enables researchers to conduct renewable energy studies even when measurement data are not available, orienting the forecasting to the development of actual wind farm projects since SimSEE can be accepted by ARCONEL as a replacement of SDDP.

2. Wind Energy Production Based on Histogram-Based Gaussian Space Correlation (CEGH) Model

The main challenge with wind power generation is the stochastic nature of the wind resource. This makes it crucial for energy producers to accurately model the stochastic behavior of wind in order to make reliable predictions [13]. Two types of forecasts can be distinguished: short-term forecasts, which range from seconds to minutes and are useful for controlling turbine operation; and long-term forecasts, which range from hours to days and are used for planning the integration of additional plants into the grid [14].

The CEGH model allows the generation of synthetic series that correspond to the outputs of a stochastic process. This often requires the historical realization of the variables involved in the stochastic process since the modeling is based on an available historical realization from which estimations are generated [15]. In electric power systems, the processes involved usually have values with a certain continuity, which allows the process outputs to be represented in the form of a future cone, as shown in Figure 1.

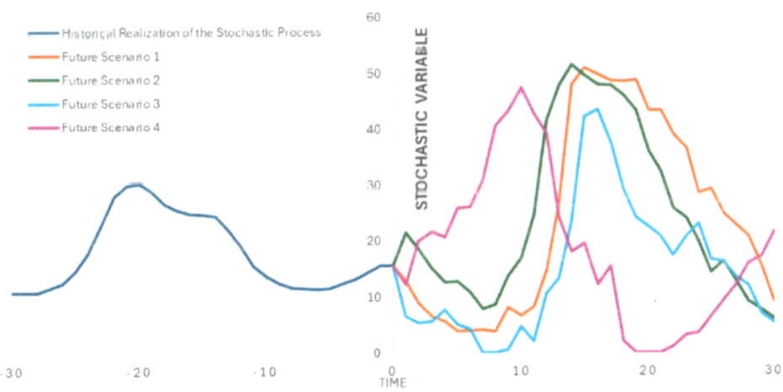

Figure 1. Output of a stochastic process—future cone [15].

The CEGH model uses nonlinear transformations to bring historical series into Gaussian space, where a filter or linear system is identified that generates time series with correlation and autocorrelation functions, like the historical ones. These synthetic series have amplitudes corresponding to Gaussian variables, so inverse nonlinear transformations are used to obtain the original amplitude histograms [16].

3. Materials and Methods

To account the stochastic nature of the wind in simulations, it is necessary to define a CEGH model created from measurement data, or, in this case, historical data obtained through reanalysis tools, which have been pre-processed to improve result quality. Then, a model of the wind farm must be generated in SimSEE, with two options: one representing wind speed as a module and another representing the two components of wind speed, allowing the representation of its direction. For simulations where the dispatch problem is solved using SDP, two databases are generated in SimSEE in order to represent different case studies that aim to emulate the dispatch stages performed by the system operator [17]. Finally, a probabilistic analysis of the results is performed to evaluate the energy impact of the wind project's connection. This methodology is described in the flowchart illustrated in

Figure 2. It is important to highlight that this comprehensive methodology constitutes an important practical contribution to the current practices of evaluating wind farm power forecasts using mass-market commercial software capable of being used for performing confident enough studies related to the connection of energy renewable projects when measured data are not available. For this aim, a CEGH model is generated based on a proposed data reanalysis strategy that is first applied to obtain corrected data. Thus, the main contribution is defining a methodology to integrate a free tool like SimSEE with an open-access meteorological database like MERRA2, ensuring confident enough data correction. With this, the power forecasting of actual wind farm projects is also allowed since SimSEE might be accepted as a replacement of SDDP or other commercial software.

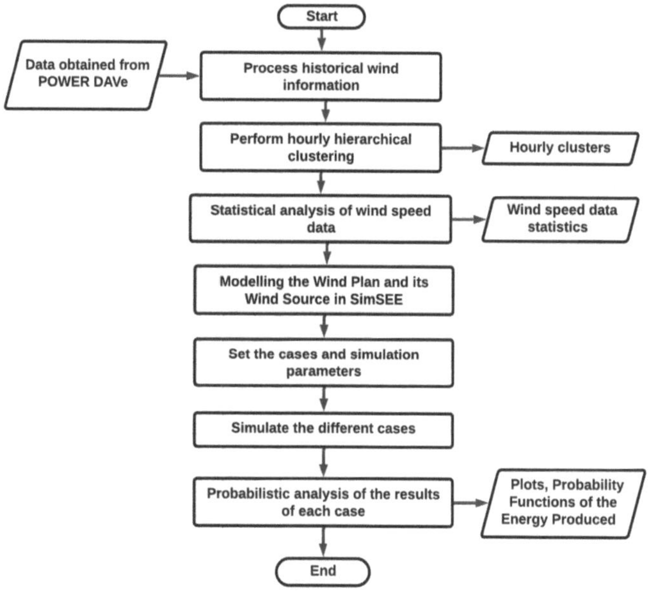

Figure 2. Proposed methodology.

3.1. Analysis of Historical Wind Data

Wind measurement data at the installation site were not available for modeling the wind resource, so the POWER DAVe platform was used to access meteorological information from the MERRA-2 reanalysis database. With the approximate location of the project as the known datum, it is possible to obtain wind speed and direction data, using the POWER DAVe website. The available data correspond to wind speed, in meters per second, at 50 m above the ground level and wind direction, in degrees. These data are first reanalyzed to define the input data for the stochastic wind farm power production forecast. To carry out reliable energy analyses, it is necessary to have data series spanning 25 to 30 years. However, since this historical information was unavailable, the recommendation is to have data for periods ranging from 5 to 10 years as a minimum [18]. Therefore, it was established that the historical data used would be from 2015 to 2023.

Subsequently, three corrections are applied to the historical data, following the procedure described in [19]:

- Vertical Wind Speed Profile Correction (1), related to the height above the ground at which the wind speed value was obtained and the height at which the wind turbine hub will be located.

$$v_2 = v_1 \frac{ln\left(\frac{h_2}{z_0}\right)}{ln\left(\frac{h_1}{z_0}\right)} \quad (1)$$

where $h_1 = 50$ is the height at which MERRA-2 data were obtained; $h_2 = 86$ is the turbine hub height; $z_0 = 0.4$ is the surface roughness coefficient; v_1 is the wind speed at height h_1; and v_2 is the wind speed at height h_2.

- Air Density Correction (2), which adjusts the atmospheric pressure at the wind farm height.

$$v_{corrected} = v_{original}\left(\frac{\rho_2}{\rho_1}\right)^{a_h} \tag{2}$$

where ρ_1 is the air pressure at the reanalysis data height $h_1 = 1871.5$; ρ_2 is the air pressure at the actual project site height $h_2 = 2700$; and a_h is a factor calculated as indicated in Equation (3).

$$a_h = \begin{cases} 1/3 & for\ v_{original} >= 7.5 \\ 1/15 \cdot v_{original} - 1/6 & for\ 7.5 < v_{original} < 12.5 \\ 2/3 & for\ v_{original} \geq 12.5 \end{cases} \tag{3}$$

- Bias Correction or Downscaling (4), which corrects the bias of reanalysis data by adjusting them around a known mean [20].

$$v_{downscaling} = \alpha \cdot v_{reanalysis} + \beta \tag{4}$$

Parameters α and β are obtained as follows:

$$\alpha = 0.6 \cdot \frac{v_{mean}}{\alpha \cdot v_{reanalysis}} + 0.2 \tag{5}$$

$$\beta = v_{mean} - \alpha \cdot v_{reanalysis} \tag{6}$$

where v_{mean} is the known average speed for the project installation site, and $v_{reanalysis}$ corresponds to each wind speed value obtained from MERRA-2.

Additionally, it is necessary to decompose the wind into the Vx and Vy components that SimSEE uses as the wind source for the actor considering wind direction.

With the corrected data, the next phase is to analyze them statistically. Two classifications are distinguished: hourly analysis and monthly analysis. For the hourly analysis, clusters were identified to group hours with similar behavior using a hierarchical clustering procedure, resulting in the dendrogram shown in Figure 3. The statistical measures calculated for each hourly cluster, for the hourly analysis, and for each month in the monthly analysis were mean, median, standard deviation, kurtosis, skewness coefficient, shape, and scale parameters for the Weibull distribution.

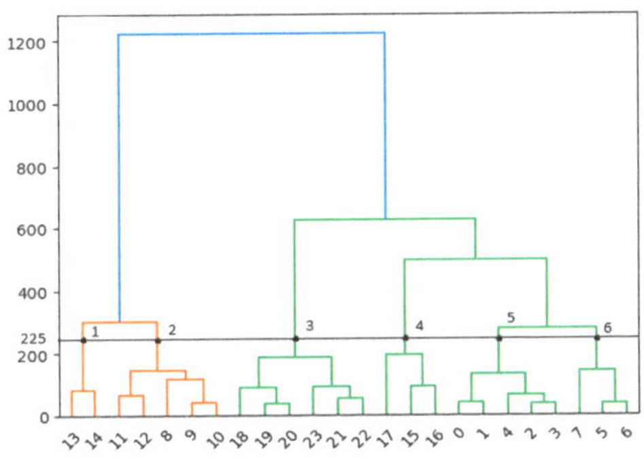

Figure 3. Hierarchical clustering dendrogram by hour.

3.2. Modeling in SimSEE

To simulate the behavior of the wind farm, it is necessary to have a wind source. In this case, the source will be of the CEGH Synthesizer type, associated with a CEGH model generated from the historical data. The Serial Analysis tool is used to create the wind source, generating a .txt file with the wind CEGH model from the historical data. Once this file is obtained, it is loaded into the SimSEE Edit database, and the corresponding wind source is generated. Since the wind farm was modeled using two different actors, two sources had to be created, considering the input variables for each actor. First, the SimSEE wind farm actor allows for the incorporation of a source with two terminals: wind speed [m/s] and temperature [°C]. On the other hand, the SimSEE wind farm Vxy actor allows for the incorporation of a source with two terminals, Vx and Vy, which are the wind speed components in [m/s].

3.3. *Definition of Case Studies*

Considering the stages of operation scheduling, four case studies have been defined. For each case study, specific parameters will be considered, such as different time steps and time horizons, which allow each planning stage to be better represented.

3.3.1. Case 1: Daily Dispatch

Simulating the daily production of a wind farm can be useful for observing resource behavior throughout the day. Although intermittent energy plants do not participate in solving the economic dispatch problem, it is still valuable to estimate their expected production during peak and valley hours. The results were analyzed for days with maximum and minimum resource availability in a year. This case should be simulated with an hourly time step, so the simulation was carried out in a simplified database where only the wind farm and a demand are modeled. The simulation time horizon was set to one year (2026), and an additional year was considered for the optimization of the time horizon.

3.3.2. Case 2: Weekly Dispatch

This type of dispatch is used for Unit Commitment. In this case, the wind farm's production was analyzed over a week of high wind resource availability and another week of low availability. It was also developed in the simplified database. Additionally, daily simulation blocks were established. The duration of the posts was set according to the hourly clustering of the wind data. The same optimization and simulation horizons as in the previous case were used.

3.3.3. Case 3: Monthly Dispatch for One Year

For this case, the annual production was analyzed with an appropriate simulation time step. The results allow the identification of production periods throughout the year and estimate the annual energy output of the wind farm. Additionally, this type of study helps determine the fuel requirements for thermal power plants and establish the maintenance schedule for power system components. The simulations were carried out in the SimSEE database that contains the entire National Interconnected System (SNI). Weekly simulation steps were considered, where the hourly duration of each post was multiplied by 7 to represent the 7 days of the week. The same optimization and simulation horizons as in the previous cases were used.

3.3.4. Case 4: Long-Term Dispatch

Long-term dispatch was used to analyze the wind farm's production over 25 years with an appropriate time step. This long-term planning helps determine the needs to satisfy demand and to implement system expansion plans. As in the previous case, this case was developed in the complete SNI database, where monthly simulation posts were established. For the time horizons, 25 years were considered for the simulation and 29 years for the optimization.

3.4. *Simulation in SimSEE*

Once the simulation parameters for each case study have been established, the simulation is run from the SimSEE Edit Simulator window. First, the optimization process is executed to obtain the Optimal Operation Policy. Then, the simulation is executed, which has a shorter duration than the previous process.

3.5. Probabilistic Analysis of Results

After obtaining the simulation results for each of the case studies, a specific treatment of the output data from each case is necessary.

3.5.1. Determination of Days and Weeks of Maximum and Minimum Energy Production

For case studies 1 and 2, it is necessary to determine, based on the results for a year, the corresponding days and weeks of highest and lowest energy production.

3.5.2. Obtaining Monthly Clusters

The hierarchical clustering procedure was used to determine two periods in the year: one of high wind resource availability and another of low availability. Considering that planning in Ecuador distinguishes between two semiannual periods in the year (rainy and dry), the aim was to establish something similar for the wind resource in the project installation area, based on the results of case study 4.

3.5.3. Exceedance Probability

The exceedance probability refers to the probability of exceeding a certain value over a given period [21], and it can be obtained using Equation (7).

$$P(X > x) = 1 - F(X \leq x) \tag{7}$$

where $P(X > x)$ is the probability that the random variable X exceeds the value x, and $F(X \leq x)$ is the cumulative probability distribution function.

Thus, for each case study, the exceedance probability of the energy produced during each analyzed period was assessed, obtaining the adjusted curves based on the results provided by SimSEE.

4. Results

The statistical analysis of the historical wind data provided useful information about the behavior of this resource in the project installation area. In Figure 4a, the hourly behavior throughout the day is illustrated using the clusters obtained according to Figure 3, and monthly behavior throughout the year is also illustrated in Figure 4b. This plot allows for a better appreciation of the statistical measures than simply presenting the numerical results.

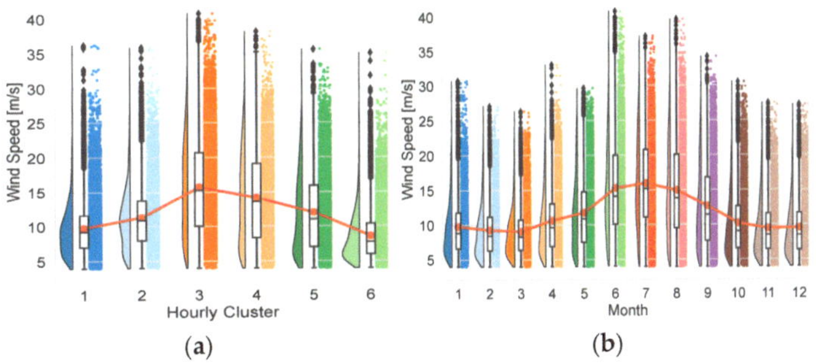

Figure 4. Historical wind speed raincloud plot: (**a**) analysis by hourly cluster; (**b**) analysis by month.

Therefore, the results of this analysis provide insights into wind behavior in the project area and allow researchers to assess whether the conditions are favorable for installing a wind power plant. This constitutes an important contribution to generation expansion planning in Ecuador.

Subsequently, the energy production results obtained through the SimSEE simulation are presented. First, case study 1 is presented, where the exceedance probability for the daily energy of the Wind farm Vxy model is shown in Figure 5.

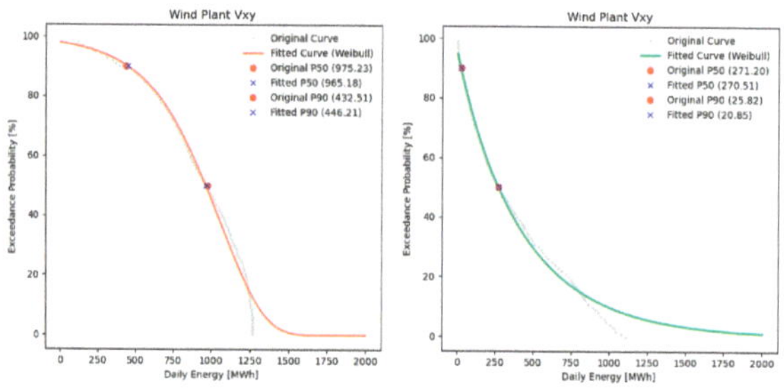

(**a**) (**b**)

Figure 5. Daily energy exceedance probability curves: (**a**) day of maximum production; (**b**) day of minimum production.

The values obtained from the original and fitted curves were extracted and are shown in Table 1, where the energy corresponding to the 50% and 90% exceedance probabilities for the days of maximum and minimum production, respectively, is provided.

Table 1. Daily energy values for 50% and 90% exceedance probabilities.

Condition	Exceedance Probability	Daily Energy Original Curve [MWh]	Daily Energy Fitted Curve [MWh]
Maximum Production	P50	975.23	965.18
	P90	432.51	446.21
Minimum Production	P50	271.20	270.51
	P90	25.82	20.85

These daily energy values provide power system operators with the estimated daily production limits for the wind plant, which should be considered into the daily dispatch.

With the results from case 4, hierarchical clustering was performed, whose dendrogram is illustrated in Figure 6. Since the results show two clear wind production periods per year, a suitable cutoff line was chosen, resulting in one cluster formed by the months of May to September and the other by the months of October to April.

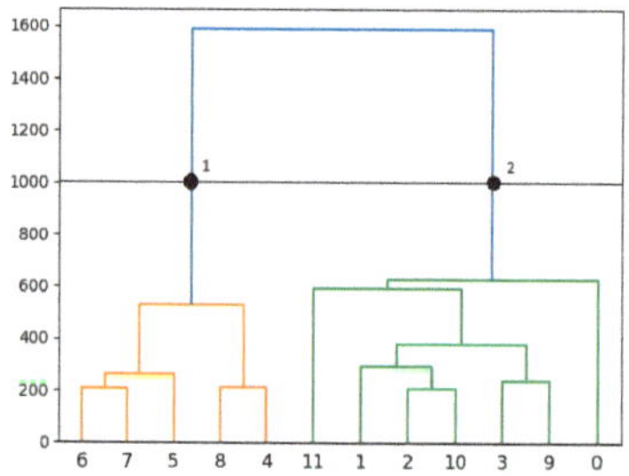

Figure 6. Hierarchical clustering dendrogram by month.

The identified periods closely resemble those of the country's hydrology. Therefore, maintenance planning for this wind plant should align with the maintenance schedules of hydroelectric plants. In addition, during these periods of maintenance, fuel for thermal plants should be secured to ensure a reliable energy supply.

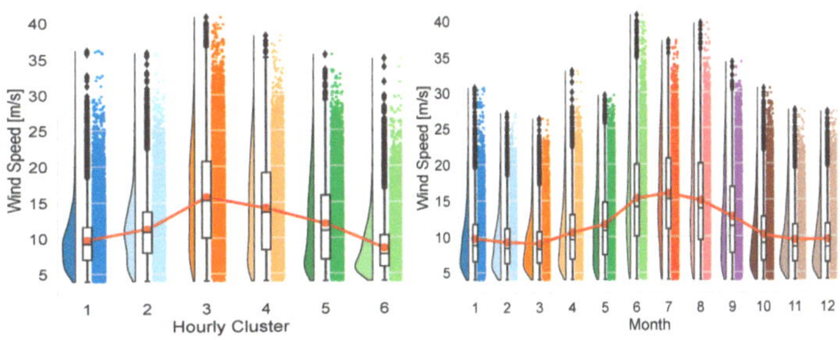

Thus, for case 3, the annual production as well as the production for each of the monthly periods was analyzed. In Figure 7, the adjusted exceedance probability curves for these three periods are presented, both for the wind farm model and for the wind farm Vxy model.

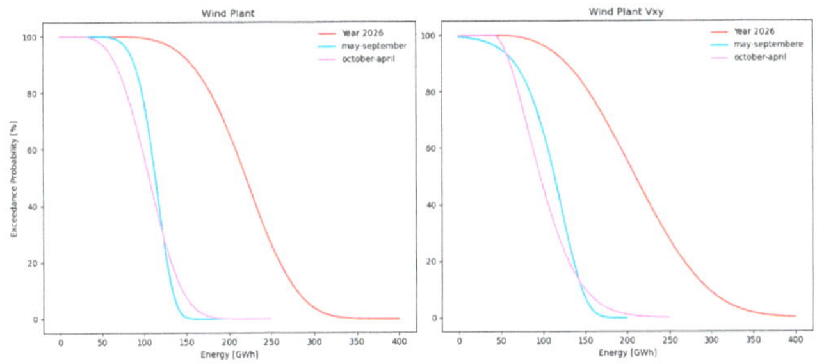

(a) (b)

Figure 7. Annual energy exceedance probability curves: (**a**) wind farm model; (**b**) wind farm Vxy model.

In Figure 8, the expected production for the entire life of the plant is compared between the two SimSEE wind farm models.

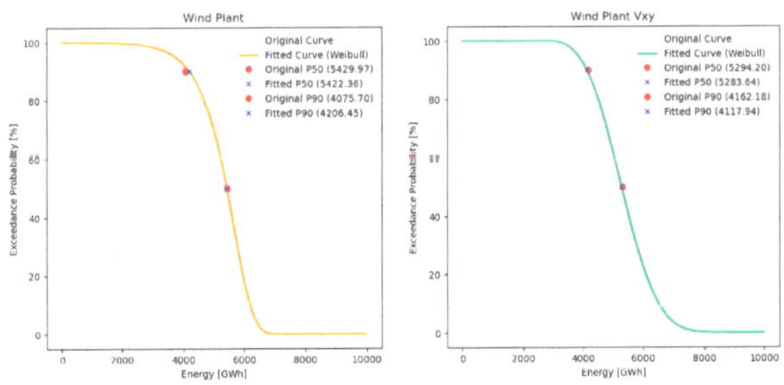

(a) (b)

Figure 8. Total energy exceedance probability curves: (**a**) wind farm model; (**b**) wind farm Vxy model.

With the results of the entire life of the plant, the expected annual energy, full load equivalent hours, and capacity factor, which are important parameters when evaluating power generation projects, can be calculated, as shown in Table 2.

Table 2. Metrics to assess power generation projects.

Exceedance Probability	Energy [GWh/Year]	Full Load Equivalent Hours	Capacity Factor [%]
P50	211.35	4003	45.7
P90	164.72	3120	35.6

This information should be factored into system expansion plans to forecast future energy requirements, considering the estimated energy production from existing plants with a certain level of confidence.

5. Discussion and Conclusions

This paper provides a comprehensive solution for conducting energy impact studies of intermittent renewable energy production when measurement data are unavailable, using free software tools, which is a significant advantage that allows for broad application.

To this end, a methodology has been developed that allows for the estimation of expected renewable generation output considering the stochasticity of the primary resource, demonstrating good performance for the analyzed wind farm project. The main contribution is defining a methodology to integrate a free tool like SimSEE with an open-access meteorological database like MERRA2, ensuring confident enough data correction. With this, the power forecasting of actual wind farm projects is allowed since SimSEE might be accepted as a replacement of SDDP or other commercial software. In this connection, it is important to mention that, although there are other novel proposals for wind farm stochastic power forecasting, such as the application of VARMA or copula, the proposed methodology is focus on using a well-known software since the purpose is to define a procedure capable of being accepted for assessing actual renewable projects in Ecuador.

As a main result, it has been demonstrated that the connection of the Yanahurcu wind farm project will have a considerable impact on the Ecuadorian electric power system, since this plant will deliver 211.35 GWh/year, which equals the wind energy production in Ecuador in 2023; this means that through the implementation of this project, the current annual production from wind sources would be doubled.

The proposed methodology has been developed as part of the project PII-DEE-2023-04, "Analysis of steady and dynamic state microgrid operation considering the implementation of Battery Energy Storage Systems (BESS)". Further research is under development for evaluating the performance of this methodology with other types of renewable sources.

Author Contributions: N.Q. and J.C. contributed equally to this paper. All authors have read and agreed to the published version of the manuscript.

Funding: This research received no external funding.

Institutional Review Board Statement: Not applicable.

Informed Consent Statement: Not applicable.

Data Availability Statement: The data presented in this study are available on request from the corresponding authors. The data are not publicly available due to being the intellectual property of universities.

Conflicts of Interest: The authors declare no conflicts of interest.

Nomenclature

ARCONEL	Ecuadorian Electricity Regulation and Control Agency
CEGH	Histogram-Based Gaussian Space Correlation Model
CENACE	Ecuadorian Electricity National Operator
ERNC	Non-Conventional Renewable Energy
PPS	Public Selection Process
SDDP	Stochastic Dual Dynamic Programming
SDP	Stochastic Dynamic Programming
SNI	National Interconnected System
WECS	Wind Energy Conversion Systems

References

1. Mostafaeipour, A.; Sedaghat, A.; Dehghan-Niri, A.; Kalantar, V. *Wind Energy Feasibility Study for City of Shahrbabak in Iran*; Elsevier: Amsterdam, The Netherlands, 2011. [CrossRef]
2. Kollu, R.; Rayapudi, S.R.; Narasimham, S.V.L.; Pakkurthi, K.M. Mixture probability distribution functions to model wind speed distributions. *Int. J. Energy Environ. Eng.* **2012**, 3, 37. [CrossRef]
3. United Nations. Energía—Desarrollo Sostenible. Available online: https://www.un.org/sustainabledevelopment/es/energy/ (accessed on 4 February 2024).
4. Lee, J.; Zhao, F. *Global Wind Energy Report 2024*; Global Wind Energy Council (GWEC): Brussels, Belgium, 2024.
5. Ministerio de Energía y Minas. Plan Maestro de Electricidad. Available online: https://www.recursosyenergia.gob.ec/plan-maestro-de-electricidad/ (accessed on 19 November 2023).
6. Review Energy. Ecuador Adjudica la Totalidad del Bloque de 500 MW Renovables. Available online: https://www.review-energy.com/otras-fuentes/ecuador-adjudica-la-totalidad-del-bloque-de-500-mw-renovables (accessed on 1 December 2023).

7. Ministerio de Enegía y Minas. 10 Empresas Presentaron sus Ofertas Económicas Para el Desarrollo del «Bloque de Energías Renovables no Convencionales de 500 Megavatios. Available online: https://www.recursosyenergia.gob.ec/10-empresas-presentaron-sus-ofertas-economicas-para-el-desarrollo-del-bloque-de-energias-renovables-no-convencionales-de-500-megavatios/ (accessed on 1 December 2023).
8. Li, F.; Liu, D.; Sun, K.; Hong, S.; Peng, F.; Zhang, C.; Tao, T.; Qin, B. Stochastic and Extreme Scenario Generation of Wind Power and Supply–Demand Balance Analysis Considering Wind Power–Temperature Correlation. *Electronics* **2024**, *13*, 2100. [CrossRef]
9. Sharma, K.C.; Gupta, A.; Bhakar, R. Wind Power Scenario Generation Considering Spatiotemporal Correlations: A Distribution Free Hybrid VARMA-Copula Approach. *Smart Grids Sustain. Energy* **2023**, *8*, 17. [CrossRef]
10. Bazionis, I.K.; Georgilakis, P.S. Review of Deterministic and Probabilistic Wind Power Forecasting: Models, Methods, and Future Research. *Electricity* **2021**, *2*, 13–47. [CrossRef]
11. Benítez, J.F.; Jaramillo, C.P. *Análisis del Impacto de las Energías Renovables no Convencionales en la Planificación Operativa de Largo Plazo del Sistema Nacional Interconectado Utilizando la Plataforma SimSEE*; Escuela Politécnica Nacional: Quito, Ecuador, 2022.
12. "SimSEE," Introducción. Available online: https://simsee.org/simsee/introduccion.html (accessed on 12 February 2024).
13. Letcher, T. *Wind Energy Engineering a Handbook for Onshore and Offshore Wind Turbines*, 2nd ed.; Charlotte Cockle: London, UK, 2023.
14. Peddapelli, S.K.; Virtic, P. *Wind and Solar Energy Applications; Technological Challenges and Advances*, 1st ed.; CRC Press: Boca Raton, FL, USA, 2023. [CrossRef]
15. Chaer, R.A. Fundamentos del Modelado CEGH de Procesos Aleatorios; Montevideo, Uruguay. 2011. Available online: https://www.researchgate.net/publication/279515834_Fundamentos_del_modelado_CEGH_de_procesos_aleatorios (accessed on 29 August 2024).
16. Chaer, R. Manual SimSEE Volumen 5: AnálisisSerial; Montevideo, Uruguay. 2019. Available online: https://simsee.org/db-docs/Docs secciones/comunes/ManualSimSEE_Vol5_Analisisserial.pdf (accessed on 12 February 2024).
17. Agencia de Regulación y Control de Energía y Recursos Naturales No Renovables. *REGULACIÓN Nro. ARCERNNR-004/20 (Codificada): Planificación Operativa, Despacho y Operación del Sistema Eléctrico de Potencia*; Agencia de Regulación y Control de Energía y Recursos Naturales No Renovables: Quito, Ecuador, 2023.
18. Villarubia, M. *Ingeniería de la Energía Eólica*, 1st ed.; Alfaomega Mexico: Mexico City, Mexico, 2013.
19. PSR. *Time Series—Lab Manual de Metodología*; PSR: Mysore, India. Available online: https://timeserieslab.com/manual (accessed on 1 August 2024).
20. Staffell, I.; Pfenninger, S. Using bias-corrected reanalysis to simulate current and future wind power output. *Energy* **2016**, *114*, 1224–1239. [CrossRef]
21. Frangopol, D.M.; Kim, S. Prognosis and life-cycle assessment based on SHM information. In *Sensor Technologies for Civil Infrastructures*; Elsevier: Amsterdam, The Netherlands, 2014; Volume 1, pp. 145–171. [CrossRef]

engineering proceedings

MDPI

Proceeding Paper

Estimation of the Current Uncertainty in the Dielectric Shoe Test According to the ISO/IEC 17025 Standard in the High Voltage Laboratory LABAV of the Escuela Politécnica Nacional

Juan D. Ramírez [1,*], Darwin Pozo [1], Edison Novoa [1], Jorge Medina [2], William O. Chamorro [2], Dolores V. Ramírez [3], Victoria Paca [4] and Alex Valenzuela [5]

1 Department of Electric Energy, Faculty of Electrical and Electronic Engineering, Escuela Politécnica Nacional, Ladrón de Guevara, Quito 170143, Ecuador; darwin.pozo@epn.edu.ec (D.P.); edison.novoa@epn.edu.ec (E.N.)
2 Facultad de Ingeniería Eléctrica y Electrónica, Departamento de Automatización y Control Industrial, Escuela Politécnica Nacional, Quito 170525, Ecuador; jorge.medinaj@epn.edu.ec (J.M.); william.chamorro@epn.edu.ec (W.O.C.)
3 Colegio Isaac Newton, Quito 170124, Ecuador; dramirez@isaacnewton.edu.ec
4 Facultad de Ciencias Administrativas, Escuela Politécnica Nacional, Quito 170124, Ecuador; victoria.paca@epn.edu.ec
5 Maestría en Electricidad, Universidad Tecnica de Cotopaxi, Latacunga EC050101, Ecuador; alex.valenzuela8644@utc.edu.ec
* Correspondence: juan.ramirezd@epn.edu.ec

Abstract: The High Voltage Laboratory (LABAV) at the Escuela Politécnica Nacional conducts dielectric tests on safety shoes in accordance with the ASTM F2412-18 standard. Additionally, as per the NTE INEN ISO 17025 standard, the laboratory must estimate the uncertainty of its measurements. Despite the scarcity of examples in the existing literature, this work provides a real-world example to assist other laboratories in replicating the uncertainty estimation process. In this article, we systematically present the calculation of leakage current uncertainty in shoes using both the traditional "Guide to the Expression of Uncertainty in Measurement" (GUM) method and the Monte Carlo method (MCM) for validation. The results from both approaches yield a similar uncertainty value of u = 0.0733 mA. Finally, we highlight the advantages that the MCM method offers in this context.

Keywords: High Voltage Laboratory; dielectric test; uncertainty estimation; Monte Carlo method; safety shoes; GUM

Citation: Ramírez, J.D.; Pozo, D.; Novoa, E.; Medina, J.; Chamorro, W.O.; Ramírez, D.V.; Paca, V.; Valenzuela, A. Estimation of the Current Uncertainty in the Dielectric Shoe Test According to the ISO/IEC 17025 Standard in the High Voltage Laboratory LABAV of the Escuela Politécnica Nacional. *Eng. Proc.* **2024**, 77, 21. https://doi.org/10.3390/engproc2024077021

Academic Editor: Jaime Cepeda

Published: 7 November 2024

1. Introduction

In the field of personnel safety for those working with electricity, dielectric shoes are a vital element in safeguarding workers' lives. Consequently, standards ASTM F2412-18a and ASTM F2413-18 [1,2] establish the methods and requirements that various types of safety shoes must meet. For dielectric shoes, three randomly selected shoes from a production lot must withstand a voltage of 18,000 volts for 1 min, with the current passing through the shoe not exceeding 1 mA. If the shoe burns, punctures, or exceeds a current of 1 mA, it does not meet the standard's requirements, and the entire associated production lot is deemed non-compliant with safety requirements.

Measuring the current in the shoe is crucial during testing because it is a relatively low value, meaning that measurement uncertainty can affect the compliance criterion of 1 mA. For instance, if the test had an uncertainty of 0.5 mA, the laboratory could only guarantee compliance for shoes that reach a value of 0.5 mA, as the current measurement would be 0.5 mA ± 0.5 mA. For higher measurement values, the uncertainty would exceed the compliance limit, and a conformity statement could not be issued. Hence, accurately estimating and achieving low uncertainty is paramount.

To ensure the reliability of laboratory test results, the guidelines of the NTE INEN-ISO/IEC 17025:2018 standard "General requirements for the competence of testing and calibration laboratories" [3] are followed. This standard serves as the foundation for implementing and maintaining a quality management system (QMS) in a testing laboratory. Section 7.2.1.1 of this standard states, "The laboratory shall use appropriate methods and procedures for all laboratory activities and, where appropriate, for the evaluation of measurement uncertainty, as well as statistical techniques for data analysis". This requirement is also harmonized with the Ecuadorian Accreditation Service (SAE) in its document "PL02-Policy for the estimation of measurement uncertainty" [4].

For estimating uncertainty, the "Guide to the Expression of Uncertainty in Measurement" (GUM) or alternative methods accepted by the International Bureau of Weights and Measures (BIPM) are available [5]. The GUM defines measurement uncertainty as the inability to know the exact value of the measurand. Uncertainty is affected by numerous factors or sources and is generally categorized into two types based on the evaluation method: "Type A" and "Type B". Type A uncertainty evaluation is determined in most cases by a statistical analysis of a series of observations, obtaining the experimental standard deviation of the measurement, usually from an averaging procedure or regression analysis. In contrast, Type B uncertainty uses non-statistical means, applying scientific judgment based on available information and experience regarding the potential variability of the variable under study.

The High Voltage Laboratory (LABAV) of the Escuela Politécnica Nacional conducts dielectric shoe testing for the public. Based on the presented background, this article aims to present a real example of uncertainty estimation in dielectric testing of safety shoes according to GUM guidelines, validated through Monte Carlo simulations. The goal is to provide a bibliographic source for other laboratories requiring similar uncertainty estimations, as no articles or reference sources with procedures or numerical examples to guide LABAV were found during the estimation process.

2. Materials and Methods

The applied methodology is divided into two phases:

Phase 1: Uncertainty estimation using the GUM Guide [5].

Phase 2: Monte Carlo simulation estimation [6].

2.1. Phase 1: Identification of Sources of Uncertainty

In the testing of dielectric shoes, the main sources of uncertainty affecting the current measurement are as follows:

- Resolution: The resolution of the equipment is a type B uncertainty and is obtained from the equipment's catalog. In this case, the shoe tester is a GESTER GT-KB42 model whose ammeter resolution is 0.01 mA.
- Current meter error: The uncertainty of the current meter is obtained from the calibration certificate issued by an accredited laboratory. This uncertainty is considered Type B, as it is based on expert judgment rather than direct measurement. The calibration certificate was obtained from a metrological laboratory accredited under ISO 17025, with an uncertainty of 0.007 mA.
- Repeatability and reproducibility: The repeatability of the dielectric shoe test is primarily ensured by the use of an automated testing system, which minimizes operator influence. In metrology, repeatability refers to the variability in results when the same procedure is performed under identical conditions. In this case, the shoe tester (GESTER GT-KB42) is fully automated, ensuring high repeatability as the operator's role is limited to preparing the test setup. The operator follows simple and standardized steps: they fill the shoe with a conductive material according to the [2] method, place the shoe inside the tester, attach the voltage electrodes, and close the testing chamber. Once the chamber is sealed, the operator initiates the test through a digital interface, which is pre-programmed to follow the standardized testing procedure. The

system then conducts the test autonomously, applying the voltage and measuring the resulting current without further human intervention.

This automation eliminates variability due to operator handling, which is a critical factor in ensuring consistent and repeatable results. To quantify the repeatability of the test, ten repetitions were performed, yielding a standard deviation of 0.107 mA. This result highlights the precision and stability of the automated system when conducting the dielectric test under controlled conditions.

In addition to ensuring repeatability through the use of an automated testing system, reproducibility is also a key consideration in dielectric shoe testing. Reproducibility refers to the consistency of test results when conducted under varying conditions, such as by different operators, using different equipment, or in different laboratories. In this case, reproducibility is supported by adherence to standardized test methods, specifically [2]. Since the procedure is well-defined and the equipment is programmed to follow this standard without manual adjustments, variations due to human intervention are minimized.

While reproducibility testing often involves comparing results across different locations or setups, the use of an internationally recognized standard and an automated system in this test significantly reduces variability between tests conducted in different environments. Moreover, the calibration of the equipment in accordance with ISO 17025 ensures that the system operates consistently across different setups. This calibration, alongside the use of the same ASTM test method, enhances the reproducibility of the test, even when conducted by different laboratories. The system's capacity for automated, standardized procedures ensures that tests performed in different settings can yield comparable results, providing confidence in the reproducibility of the dielectric shoe test.

- Temperature: Ambient temperature can affect the characteristics of both the measuring equipment and the shoe under test. This uncertainty is considered Type A and is estimated from temperature measurements taken during the test. Ten repetitions were performed, resulting in a standard deviation of 0.000979 mA/°C. Temperature fluctuations can influence the test because voltage is applied to a dielectric material; in this case, the shoe and insulation materials may change their conductivity with temperature. However, it is important to note that according to [2], the shoe must rest for 24 h in the testing area, which must maintain a temperature of 22 ± 2 °C. During the test, the temperature should remain between 22 ± 2 °C, and the test itself lasts only one minute, meaning that temperature fluctuations are minimal when following the standardized method.
- Voltage variability: The electric tester can exhibit voltage fluctuations that affect the measurement. This uncertainty is considered Type A and is estimated from voltage and current measurements in the laboratory during the test. Ten repetitions were performed, resulting in a standard deviation of 0.000979 mA/°C.

These uncertainty parameters are presented in Equation (1) and in Table 1.

Table 1. Estimated values of the uncertainty components of the proposed model.

Source of Uncertainty	Estimated Value	Units	Type	Distribution	Divisor
Resolution Δ_{Res}	0.01	mA	B	Rectangular	$2\sqrt{3}$
Error ε	0.007	mA	B	Normal	$k;\ k = 2$
Repeatability Δ_{Rep}	0.107	mA	A	Normal	$\sqrt{n};\ n = 10$
ΔI_t Temperature Correction	0.000979	mA/°C	A	Normal	$\sqrt{n};\ n = 10$
ΔI_v Voltage Correction	0.0424	mA/V	A	Normal	$\sqrt{n};\ n = 10$

2.2. Phase 1: Selection of the Measurement Model

The measurement model that relates the measured current I to the influencing variables can be represented by Equation (1):

$$I_{fc} = I_m + \Delta_{Res} + \Delta_{Rep} - (\Delta I_t + \Delta I_v) - \varepsilon, \quad (1)$$

where

I_m is the leakage current measured;
I_{fc} is the leakage current corrected;
Δ_{Res} is the deviation of the current due to the resolution of the instrument;
Δ_{Rep} is the deviation of the current due to measurement repeatability;
ΔI_t is the deviation of the current due to temperature variation;
ΔI_v is the deviation of the current due to voltage variation;
ε is the error of the instrument due to calibration.

2.3. Phase 1: Estimation of Uncertainty Components

2.3.1. Type A Uncertainty

Type A uncertainty is obtained from a series of repeated measurements. The experimental standard deviation s is calculated by Equation (2):

$$s = \sqrt{\frac{\sum_{i=1}^{n}(m_i - \bar{m})^2}{n-1}}, \quad (2)$$

where

m_i is the value in the i th measurement;
n is the number of measurements;
$\bar{m}$ is the average value.

Type A uncertainty (u_A) is calculated by Equation (3):

$$u_A = \frac{s}{\sqrt{n}}. \quad (3)$$

2.3.2. Type B Uncertainty

Type B uncertainty is estimated from the available information about the sources of uncertainty, such as calibration certificates, previous studies, or expert knowledge. An appropriate probability distribution is used to represent the variability of each source of uncertainty.

2.4. Phase 1: Combination of Uncertainty Components

The combined standard uncertainty u_c is calculated by combining the Type A and Type B uncertainty components using Equation (4):

$$u_c = \sqrt{\sum_{i=1}^{N} u_A^2 + \sum_{j=1}^{z} u_B^2} = \sqrt{\sum_{i=1}^{M} (C_i.u_i)^2}, \quad (4)$$

where

N is the number of Type A uncertainty sources;
μ_A is the Type A uncertainty of the i-th source;
z is the number of Type B uncertainty sources;
μ_B is the Type B uncertainty of the j-th source;
M is the number of total uncertainty sources;
u_i are the individual standard uncertainties.

To determine u_c in Equation (4), we need to find a sensitivity coefficient C_i by taking the partial derivative of the mathematical model from Equation (1) with respect to the variable for which the contribution is sought according to Equation (5).

$$C_i = \frac{\partial y}{\partial x_i}, \tag{5}$$

where

C_i represents the sensitivity coefficient for the input variable x_i;
y is the output of the model;
x_i is the specific input variable;
$\partial y / \partial x_i$ represents the partial derivative of output y with respect to input x_i.

The contributions $C_i.u_i$ from Equation (4) are calculated individually by multiplying each uncertainty by its sensitivity coefficient obtained using Equation (5).

Expression of Uncertainty

The total uncertainty is expressed with a confidence interval and a confidence level. The confidence interval represents the range within which the true value of the measurement is found with a certain probability. The confidence level indicates the probability that the true value lies within the confidence interval. This value is known as expanded uncertainty in Equation (6):

$$u = u_c \cdot t_{0.95,n-1}, \tag{6}$$

where

$t_{0.95,n-1}$ is the value of the Student's t-distribution with a 95% confidence level and $n-1$ degrees of freedom. This value will be known as coverage factor k.

To determine the total degrees of freedom of the uncertainty, we must calculate the contributions of the degrees of freedom for each uncertainty u_i. For Type A uncertainties, the degrees of freedom are equal to $n-1$, with n representing the number of measurements taken to estimate that uncertainty. In this study, for Type B uncertainties, which originate from the calibration certificate and the instrument resolution, a very high value is used; in this case, 200.

Once the individual uncertainties u_i are known and the combined standard uncertainty u_c is calculated from Equation (4), the total degrees of freedom are determined using the Welch–Satterthwaite equation, as in Equation (7).

$$v_{\text{eff}} = \frac{u_c^4}{\sum_{i=1}^{N} \left(\frac{(C_i \cdot u_i)^4}{v_i} \right)}, \tag{7}$$

where

v_{eff} is the effective degrees of freedom;
u_c is the combined standard uncertainty;
u_i are the individual standard uncertainties;
v_i are the degrees of freedom associated with each u_i;
N is the number of uncertainty components.

With the total degrees of freedom, based on the Student's t-distribution, we determine the coverage factor k for a 95% confidence level.

2.5. Phase 2: Monte Carlo Simulation Estimation

The Monte Carlo method (MCM) is a widely used numerical method for estimating measurement and calibration uncertainties. This method generates random values of the measured variable (in this case, current), considering the probability distributions of influencing variables and their correlations. By repeating the simulation a considerable number of times, a distribution of the measured current is obtained, allowing for uncertainty calculation and comparison with the estimation obtained using the GUM Guide [6].

To calculate the uncertainty estimation using the GUM Guide, the free application MCM Alchimia was used, which implements the Monte Carlo method to estimate measurement and calibration uncertainties according to the JCGM 101 Guide [6]. This version includes a complete uncertainty budget for the GUM framework and allows for the handling of correlated quantities and regression curves.

Steps of Monte Carlo Simulation

Definition of influencing variables: Identify all variables that can affect the measurement of the current, such as equipment calibration, electric network variability, temperature, and humidity.

Selection of probability distributions: Assign an appropriate probability distribution to each influencing variable, considering available information such as calibration certificates, previous studies, or expert knowledge. For example, a normal distribution can be used for equipment calibration, and uniform distributions can be used for temperature and humidity.

Generation of random values: Use Monte Carlo simulation software to generate random values of influencing variables according to selected probability distributions.

Calculation of current: For each set of random values of influencing variables, calculate the corresponding current value using the defined measurement model.

Repetition of simulation: Repeat steps 3 and 4 a large number of times (e.g., 10,000 or 100,000) to obtain a representative distribution of the measured current.

Analysis of distribution: Analyze the distribution of the measured current, including the mean, standard deviation, and other measures of dispersion.

Estimation of uncertainty: Calculate the total uncertainty u_c by combining Type A and Type B uncertainty components, as described in Section 2.4 of this article.

Comparison with GUM estimation: Compare the estimated total uncertainty from the Monte Carlo simulation with the estimation obtained using the GUM Guide. If the estimations are consistent, the methodology used is validated.

3. Results

3.1. Uncertainty Results Using the GUM Method

The variables from the utilized model contributing to uncertainty are presented in Table 1. The estimated resolution value comes from the instrument's certificate, while the error originates from the instrument's calibration certificate. For repeatability and temperature corrections, ten repetitions are performed, and the standard deviation is obtained. The same table specifies the type of uncertainty, the distribution function they follow, and the divisor used to calculate the standard uncertainties u_i, which is defined as the division of the estimated value by the divisor.

In Table 2, the sensitivity coefficients C_i calculated from the mathematical model and the standard uncertainties u_i are presented.

Table 2. Sensitivity coefficients of each uncertainty component of the proposed model.

Source of Uncertainty	Sensitivity Coefficient Calculation	C_i	Standard Uncertainties, u_i
Resolution Δ_{Res}	$\frac{\partial I_{fc}}{\partial \Delta Res}$	1	0.002886751
Error ε	$\frac{\partial I_{fc}}{\partial \varepsilon}$	−1	0.003500000
Repeatability Δ_{Rep}	$\frac{\partial I_{fc}}{\partial \Delta Rep}$	1	0.033836371
ΔI_t Temperature Correction	$\frac{\partial I_{fc}}{\partial \Delta It}$	−1	0.000309587
ΔI_v Voltage Correction	$\frac{\partial I_{fc}}{\partial \Delta Iv}$	−1	0.013408057

Finally, in Table 3, the contributions $C_i \cdot u_i$ from each uncertainty variable are determined, and degrees of freedom are assigned.

Table 3. Contributions and degrees of freedom of each uncertainty component of the proposed model.

Source of Uncertainty	Contribution of Uncertainty $C_i \cdot u_i$	$(C_i \cdot u_i)^2$	$(C_i \cdot u_i)^4$	Degrees of Freedom v_i
Resolution Δ_{Res}	0.002886751	8.33333×10^{-6}	6.94444×10^{-11}	9
Error ε	−0.003500000	1.2225×10^{-5}	1.50063×10^{-10}	200
Repeatability Δ_{Rep}	0.033836371	1.1449×10^{-3}	1.3108×10^{-6}	200
ΔI_t Temperature Correction	−0.000309587	9.58771×10^{-8}	9.18609×10^{-15}	9
ΔI_v Voltage Correction	−0.013408057	1.79776×10^{-4}	3.23194×10^{-8}	9

With the calculations from Table 3, we obtain a combined standard uncertainty of $u_c = 0.036679084$.

We apply (7) and obtain a total of 12 effective degrees of freedom.

Using the Student's t-distribution, we determine factor k for 95% confidence, resulting in $k = 2.179$. However, for testing laboratories, it is recommended to use factor $k = 2$.

Therefore, the expanded uncertainty is $u = k \cdot u_c$, which results in $\boldsymbol{u} = \mathbf{0.0733581673}$ **mA** for this analysis.

3.2. Uncertainty Results Using the MCM Method

For this section, we utilize the MCM Alchimia program. The procedure is straightforward:

1. Input the set of equations for the model (the same model as in Equation (1));
2. Define each variable of the model along with its units and values;
3. Specify the distribution function for each variable (as shown in Table 1);
4. Assign infinite degrees of freedom to each variable;
5. Finally, run 500,000 simulations for this case.

The software generates a result of $\boldsymbol{u} = \mathbf{0.07336}$ **mA** with a coverage factor of $\boldsymbol{k} = \mathbf{2}$ and a 95.45% confidence interval.

MCM Alchimia proposes in Figure 1 a comparison of the results between the MCM method and the GUM method.

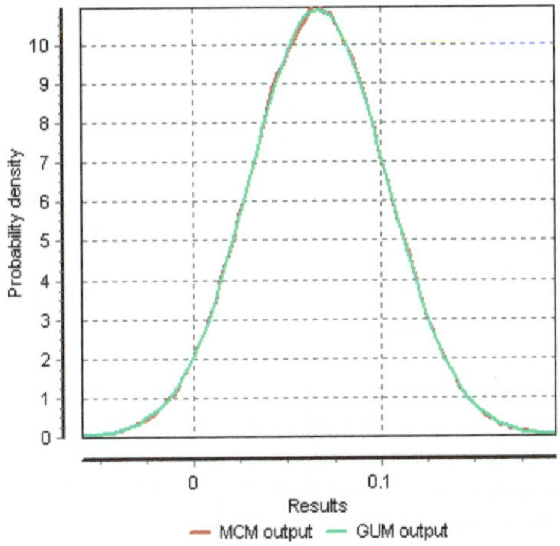

Figure 1. Results of the Monte Carlo simulation for uncertainty estimation using the MCM and GUM methods.

3.3. Expression of the Measurement

The expression of the measurement should include uncertainty. For example, if the measurement in the test yields 0.84 mA, it should be expressed as $I = \mathbf{0.84}$ mA with uncertainty $u = 0.733$ mA and a coverage factor of $k = 2$ at a confidence level of 95%.

In the classical form, the current would be expressed as I = (0.84 ± 0.733) mA.

4. Discussion

4.1. On Uncertainty

When observing the results from Sections 3.1 and 3.2, it is evident that the uncertainty is practically the same for both methods. This validation allows us to use the result for laboratory test reports in accordance with the NTE INEN ISO 17025 standard.

4.2. Advantages of Monte Carlo Simulation

When calculating uncertainty using the GUM method, the traditional approach involves creating a set of tables in a spreadsheet and following all the steps described in Section 2.1. However, when using the MCM method, there is significant ease in the calculation process. Essentially, the problem reduces to determining the mathematical model of the measurement and its variables in the best possible way. The rest of the simulation is rapid, and the results are practically the same as those indicated by the GUM.

Additionally, for more complex measurements with intricate partial derivatives, the MCM method simplifies the process even further. In general, we recommend using this method due to the following advantages:

- Flexibility: Monte Carlo simulation allows for handling a wide variety of measurement models and probability distributions. This flexibility is beneficial in modeling complex scenarios and diverse sources of uncertainty.
- Consideration of correlations: It can incorporate correlations between influencing variables, providing a more accurate estimation of uncertainty. This feature is crucial as it reflects real-world conditions where variables often influence each other.
- Validation of GUM estimation: Monte Carlo simulation serves as a tool to validate uncertainty estimations obtained using the GUM Guide. By comparing simulated results with theoretical estimations, it helps us to ensure the robustness and reliability of uncertainty assessments.

- Visualization of distribution: It enables the visualization of the distribution of the measured current, offering valuable insights into measurement variability. This visualization aids in understanding the range of potential outcomes and the likelihood of different measurement results under varying conditions. These advantages make Monte Carlo simulation a powerful tool for uncertainty analysis in measurement and calibration, particularly in contexts where complex interactions and diverse sources of uncertainty need to be considered.

5. Conclusions

The uncertainty estimation for current measurement in dielectric shoes has been carried out according to the NTE INEN ISO 17025 standard. The overall estimation result indicates an uncertainty of $u = \mathbf{0.0733}$ mA with a coverage factor of $k = \mathbf{2}$ at a confidence level of 95%. It can be observed from Table 3 that the largest contribution to the uncertainty $C_i \cdot u_i$ comes from the repeatability variable. In other words, the equipment applying voltage to the shoe experiences some difficulty in reproducing test conditions.

Considering that, according to the ASTM F2413-18 standard, the current limit for a dielectric shoe subjected to 18,000 V for one minute is 1 mA, the most reliable measurement ensuring compliance with the standard would be 0.9267 mA.

Author Contributions: Conceptualization, J.D.R., D.P., E.N. and A.V.; methodology, J.D.R., D.V.R., E.N. and D.P.; software, D.P.; validation, W.O.C., J.M. and V.P.; formal analysis, J.D.R.; investigation, J.D.R. and D.P.; resources, J.D.R. and D.P.; data curation, J.D.R. and D.P.; writing—original draft preparation, J.D.R.; writing—review and editing, J.D.R. and D.V.R.; visualization, J.D.R.; supervision, A.V. All authors have read and agreed to the published version of the manuscript.

Funding: This research received no external funding.

Institutional Review Board Statement: Not applicable.

Informed Consent Statement: Not applicable.

Data Availability Statement: All data are included in the article.

Conflicts of Interest: The authors declare no conflicts of interest.

Abbreviations

The following abbreviations are used in this manuscript:

ASTM	American Society for Testing and Materials
NTE INEN	Norma Técnica Ecuatoriana Servicio Ecuatoriano de Normalización
ISO	International Organization for Standardization
IEC	International Electrotechnical Commission
QMS	Quality management system
SAE	Ecuadorian Accreditation Service
GUM	Guide to the Expression of Uncertainty in Measurement
BIPM	International Bureau of Weights and Measures
MCM	Monte Carlo method

References

1. ASTM International. F2413-18 Standard Specification for Performance Requirements for Protective (Safety) Toe Cap Footwear. 2018. Available online: https://www.astm.org/f2413-18.html (accessed on 3 July 2024).
2. ASTM International. F2412-18a Standard Test Methods for Foot Protection. 2018. Available online: https://www.astm.org/f2412-18a.html (accessed on 3 July 2024).
3. INEN. NTE INEN-ISO/IEC 17025 Requisitos Generales para la Competencia de los Laboratorios de Ensayo y Calibración (ISO/IEC 17025:2017, IDT). 2017. Available online: https://www.gob.ec/regulaciones/norma-tecnica-ecuatoriana-nte-inen-isoiec-17025-requisitos-generales-competencia-laboratorios-ensayo-calibracion-isoiec-170252017-idt (accessed on 3 July 2024).
4. SAE. PL02 Política para la Estimación de la Incertidumbre de la Medición. 2021. Available online: https://www.acreditacion.gob.ec/wp-content/uploads/downloads/2018/02/PL02-R02-Politica-incertidumbre.pdf (accessed on 3 July 2024).

5. JCGM. Evaluation of measurement data—Guide to the expression of uncertainty in measurement. *Int. Organ. Stand. Geneva ISBN* **2008**, *50*, 134. [CrossRef]
6. JCGM. Evaluation of Measurement Data—Supplement 1 to the 'Guide to the Expression of Uncertainty in Measurement'—Propagation of Distributions Using a Monte Carlo Methodr. 2008. Available online: https://www.bipm.org/en/doi/10.59161/JCGM101-2008 (accessed on 3 July 2024).

engineering proceedings

MDPI

Proceeding Paper

Influence of Optimal Charging Station Integration on Electric Power Distribution Grid: Case of Electric Bus-Based Transport System †

Daniel Orbe *, Luis Salazar, Paúl Vásquez, William Chamorro and Jorge Medina

Escuela Politécnica Nacional, Quito 170525, Ecuador; luis.salazar@epn.edu.ec (L.S.); paul.vasquez@epn.edu.ec (P.V.); william.chamorro@epn.edu.ec (W.C.); jorge.medinaj@epn.edu.ec (J.M.)

* Correspondence: daniel.orbej@epn.edu.ec

† Presented at the XXXII Conference on Electrical and Electronic Engineering, Quito, Ecuador, 12–15 November 2024.

Abstract: Electric mobility is one of the main pillars of the global energy transition towards a more sustainable and environmentally responsible model. Greenhouse gas emissions could be drastically reduced with electric mobility massification. Public transport systems represent the first step of this massification due to government policies, but these electromobility projects should optimize their resources to be cost-effective. Furthermore, the implementation of charging stations could cause negative impacts on electrical distribution networks, which should be evaluated beforehand for the adequate planning of power grids. A methodology was developed and implemented herein for the technical and economic evaluation of electric bus-based transport systems.

Keywords: electric mobility; electric power distribution; electric public transportation; optimization

Citation: Orbe, D.; Salazar, L.; Vásquez, P.; Chamorro, W.; Medina, J. Influence of Optimal Charging Station Integration on Electric Power Distribution Grid: Case of Electric Bus-Based Transport System. *Eng. Proc.* **2024**, 77, 22. https://doi.org/10.3390/engproc2024077022

Academic Editor: Jaime Cepeda

Published: 18 November 2024

1. Introduction

The incorporation of electric buses (EBs) into the public transportation system is a valuable asset in the energy transition process, as they reduce environmental pollution with low noise levels and zero CO_2 emissions [1]. Due to the advantages of electric mobility, the incorporation of EBs is growing worldwide. According to Bloomberg New Energy Finance's Electric Vehicle Outlook, it is projected that by 2025, the number of EBs worldwide will reach 1.2 million, which would imply that 47% of buses will be electric [2]. In 2024, Ecuador approved the "Energy Competitiveness Law", which established that by 2030, all vehicles incorporated into the country's public transportation system must be zero-emission or electric [3].

EB-based transport systems have some limitations, for example their limited driving range and extended charging time, much longer than a fuel-powered bus. Therefore, it is essential to implement charging strategies for EBs to maintain their daily operations properly in accordance with planning schedules. The choice of these strategies will influence not only the number of buses available to be deployed on a transit route but also the energy consumption level and battery life expectancy. As studied in [1], to implement charging strategies for EBs, it is necessary to accurately know the energy consumption of each transport unit along the operational route. This depends not only on the mechanical and electrical characteristics of each unit but also on the topographical features of the route and traffic conditions [2]. The conventional charging strategy consists of recharging the battery when it is almost depleted until it is fully charged. However, based on [2], it has been shown that bringing the vehicle battery to its minimum charge level significantly reduces its lifespan. Therefore, to mitigate this depreciation, it is advisable to maintain battery operation at an adequate charge level (ACL). However, this condition would require increasing the size of the bus fleet. Considering battery characteristics, an intelligent option is to maximize downtime during off-peak hours to recharge EBs. This will increase their driving range, reduce energy consumption, and avoid the early need to expand the fleet size due to charging demands, maintaining the ACL at a reasonable level and prolonging battery life. The duration and number of downtimes for each bus during daily operations directly depend on the trips it must make, which are organized by schedules [2].

So far, several investigations emphasize planning the charging strategy from a battery, vehicle, and route perspective. Besides the charger deployment, how to schedule the bus fleet in an efficient way is also a critical issue for the BEB system. It is demonstrated that the reasonable scheduling of the bus fleet has great potential to reduce the fleet size and corresponding investment costs [4]. Simulation optimization approaches have been used to evaluate energy consumption and determine optimal charging station placement [5]. Bi-level optimization models have been developed to design efficient transit routes and locate charging infrastructure simultaneously [6]. Operational challenges, such as limited range and long charging times, can be addressed through optimization models that assign buses to routes, allocate parking spaces, and optimize charging schedules [7]. Several techniques such as genetic algorithms have been employed to solve this combinatorial optimization problem, considering factors like battery autonomy, charging opportunities, user demand, and vehicle capacity to determine optimal timetables and fleet composition [8]. These previous studies demonstrate the potential of optimization techniques to overcome the limitations of electric buses and support the design of sustainable public transportation systems.

Likewise, it is necessary to analyze the impact of incorporating charging stations for EBs into the electric power distribution grid (EPDG) [9]. Charging stations connect to the EPDG, and due to their consumption characteristics, they can generate negative impacts on the grid, such as transformer overloading, distribution line overloading, increased power and energy losses, voltage disturbances, and power quality issues [10]. Therefore, a key challenge in station planning is to find an optimal sizing for new charging stations that meets transportation demand and is cost-effective from an electrical grid perspective. Considering both perspectives will not only reduce total costs from a systems perspective, thus increasing social welfare, but will also accelerate the incorporation of EBs into transportation companies [11]. Over the years, efforts toward the construction and management of electrified transit systems have been made by both the industrial and academic communities [12].

This article focuses primarily on analyzing the impact of incorporating charging stations for electric buses into the EPDG. The analysis will be conducted for the route studied in [3], starting with a description of the case study, and then studying EB energy consumption and incorporating battery charging restrictions. The optimization problem will be described, as it is necessary to find an optimal sizing for the charging stations. The second part of this work is focused on vehicle charging stations' impact on the EPDG. It begins with a demand analysis, incorporating energy consumption characteristics of BEBs into a database modeled in specialized software for electrical studies in distribution power systems such as CYME 9.3. Once the charging station demand is incorporated into the feeder that provides electricity in the operational area of the case study route, quasi-steady-state simulations will be performed to evaluate distribution grid performance. Finally, the results obtained will be presented and analyzed, followed by the conclusions drawn from this work.

2. Study Case Description

2.1. Public Transportation Route Model

For this study, the "Marín–Ciudadela Tarqui" route was used, which is currently a public transportation route in Quito–Ecuador. This route covers a total distance of 10.67 km. The selection of this route was due to the significant elevation changes along its path and the traffic levels it experiences. It is considered that the altitude difference is a determining parameter for electric vehicle consumption in Quito. Topographic information was obtained from the United States Geological Survey (USGS), which provides the necessary SRTM files to determine the elevation of any point on Earth with 30 m resolution [3].

2.2. Electric Bus Parameters

Technical specifications of electric buses used in this study were obtained from manufacturer catalogs and are summarized in Table 1. The selected bus manufacturer has units operating in several cities across Latin America, including Ecuador, and according to the e-bus radar platform [13], their vehicles account for 45% of the buses operating in the region.

Table 1. Technical specifications of electric urban transport buses.

Parameter	Value
Brand	BYD
Model	C9
Engine Power	360 kW
Battery Capacity	438 kWh
Autonomy	300 km
Height	3.55 m
Length	12.9
Width	2.55
Weight	18,000 kg
Capacity	53 p

Hence, an optimization model was developed to determine the optimal strategy for the electric bus charging system, aiming to minimize annual operational costs while ensuring the proper operation of the BEB transportation system. The most critical considerations for the optimization problem are detailed in Section 4, including the objective function and problem constraints.

3. Energy Consumption for Electric Buses

Electric bus battery (EBB) performance depends on environmental variables such as temperature and humidity, construction factors like manufacturing tolerances, and other variables that are more difficult to determine precisely, such as the battery's age and previous usage patterns. The route model presented in Section 2.1 was implemented using SUMO 1.21.0 (Simulation of Urban Mobility) software. It is one of the most efficient and frequently adopted microscopic traffic simulation tools. An energy estimation model was implemented based on vehicle and road characteristics [14]. BEBs were also modeled based on the parameters shown in Table 1, considering traffic demand.

For this study, five operational scenarios were analyzed to evaluate battery consumption. Two scenarios assessed the battery state of charge (SOC) for the "Marín–Ciudadela Tarqui" route, considering the topographic profile presented in Figure 1. The remaining three scenarios incorporated levels of service (LOSs) related to vehicular traffic, considering three levels of service: light, moderate, and heavy. Each scenarios was also associated with an occupancy level ranging from 20% to 100% of the total passenger capacity [3]. Based on the conclusions reached in [3], battery charge has a moderate dependency on traffic levels, but its consumption behavior significantly changes when considering the elevation profile. This is evident in Figure 1a, which shows the evolution of the SOC along the route at 100% occupancy, considering both flat and hilly routes. In the present study, the battery consumption model considered a hilly route with elevation changes, medium traffic levels, and 100% occupancy, to provide an energy performance very close to the real operating conditions performance. With these considerations, the battery consumption is shown in Figure 1b.

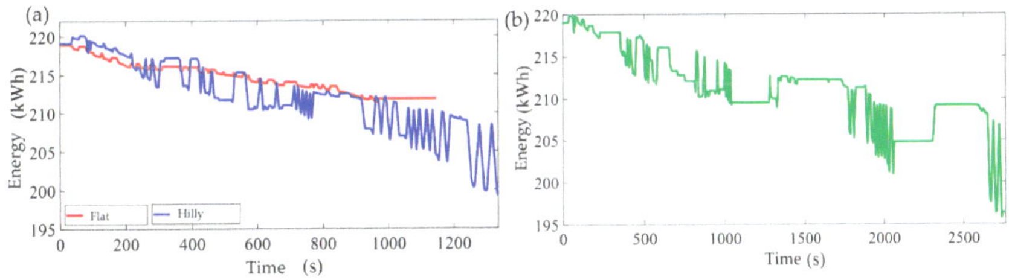

Figure 1. BEB SOC considering (**a**) elevation and (**b**) elevation and traffic level [3].

4. Optimization Problem

The case case entails a transport system, which is composed of a departure terminal (the starting point) and a main terminal (the terminal point). The departure station has a plug-in charging infrastructure and a bus depot. A service loop is considered to be the path from where the BEB departs from the starting point and operates along the route until it returns to the starting point. The BEB one-line transport system considered in this study is illustrated in Figure 2. This section aims to determine the optimal scheduling of the BEB charging system, minimizing the total costs while ensuring the normal operation of the system [15]. Therefore, some assumptions were considered to achieve the optimization problem formulation:

- Each bus has a predetermined schedule that specifies the arrival and departure frequencies to terminals.
- In every loop, the charging times are the same for all buses in the route.
- All loops have the same length, duration, and energy consumption, and all buses were produced by the same manufacturer and have the same driving range.
- There is only one charging station in the transportation system located at the starting point.
- All chargers are identical, and each charger is equipped with only one outlet.

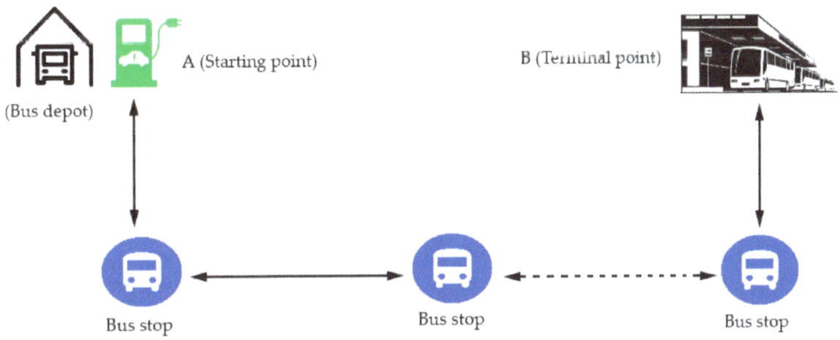

Figure 2. BEB transportation system architecture.

4.1. Problem Formulation

Considering the assumptions declared above, the objective function is structured as follows:

$$min\ C = \sum_{b\in B}\sum_{i\in S} c_m d_{ib} D + \sum_{i\in S}\sum_{n\in N} (c_e p_c u_n) D X_i^n + \sum_{n\in N} 12\ p_c c_d + \sum_{n\in N} \alpha c_{f1} + \sum_{n\in N} \left(\alpha c_{f2} + D c_{m2}\right) \quad (1)$$

where c_m is unit cost of maintenance ($/km), d_{ib} is each loop driving distance (km), c_e is the electric energy cost ($/kWh), c_d is the electric demand cost ($/kW), p_c is the charger power (kW), u_n is the charging duration (h), D is the number of operating days in a year, x_i^n is a binary variable which represents whether each bus recharges its battery in trip i at charging station n, α is the annualized factor, c_{f1} is the fixed cost per BEB (purchase investment, $), c_{f2} is the charging station fixed cost (construction investment, $), and c_{m2} is the charging station maintenance costs. N indexed n is the set of chargers in charging station, and S is the set of scheduled loops [16]. B indexed b is the set of BEBs operating on the route.

The constraints associated with the study problem are shown below [16]:

$$\sum_{n\in N} x_i^n \leq 1 \qquad \forall\, i\, \in S \tag{2}$$

$$E_i + \sum_{n\in N} (T_c u_n)\, x_i^n \leq\, E_{max} \quad \forall\, i\, \in S \tag{3}$$

$$E_j = E_i + \sum_{n\in N} (T_c u_n)\, x_i^n \quad \forall\, i\, \in S \tag{4}$$

$$E_i \geq E_{min} \qquad \forall\, i\, \in S \tag{5}$$

$$E_{min} = 0.2\ E_{cap},\ E_{max} = 0.8\, E_{cap} \qquad \forall\, i\, \in S \tag{6}$$

$$x_i^n = \{0, 1\}, \qquad \forall\, i\, \in S, \forall\, n \in N \tag{7}$$

where E_i is the remaining energy at the end of trip i (kW), T_c is the recharging rate, E_{max} is the electric bus battery maximum useful charging capacity (kW), E_{min} is the electric bus battery minimum useful charging capacity (kW), E_j is the remaining energy at the beginning of the trip j (kW), and E_{cap} is the electric bus maximum battery capacity declared by the manufacturer. The constraint in Equation (2) prevents two buses from charging on the same charger at the same time. The constraint in Equation (3) does not allow the maximum percentage of battery charge to be exceeded during charge cycles. The constraint in Equation (4) refers to energy conservation. The constraint in Equation (5) ensures that energy is not consumed beyond the minimum battery capacity. Constraint (6) establishes the minimum and maximum recommended SOC values. Constraint (7) is a binary requirement.

4.2. Route Optimization Results

The problem formulated above corresponds to a combinatorial optimization problem, where the decision variables are discrete (p_c and N). In the optimization method choice process, rhw objective function evaluation computational cost was analyzed. Since the assumptions considered simplify the problem substantially, it is possible to perform multiple evaluations of objective function in a relatively short time with low computational cost, which makes the optimization problem quite tractable. The problem was solved using a combinatorial search algorithm under different scenarios of BEB prevalence. For example, a 20% prevalence rate will mean that 20% of the route units are electric. Considering incentives and regulations issued by the Ecuadorian government, an exponential growth of electric bus prevalence in public transportation is expected. Under a "Maximum effort towards energy transition" (MEET) scenario, 85% of the bus fleet being electric would be reached by the year 2050 by considering the aggressive growth in the number of electric buses, while under a "Conservative effort towards energy transition" (CEET) scenario, the electric bus number would reach 34% in 2040. Conversely, in a "Business as Usual" (BAU) scenario, electric bus prevalence is negligible [17].

Route frequencies are adapted to different levels of BEB prevalence. Figure 3a,b show the distance covered as a function of time in each loop completed by BEBs under 10% and 50% prevalence scenarios, respectively.

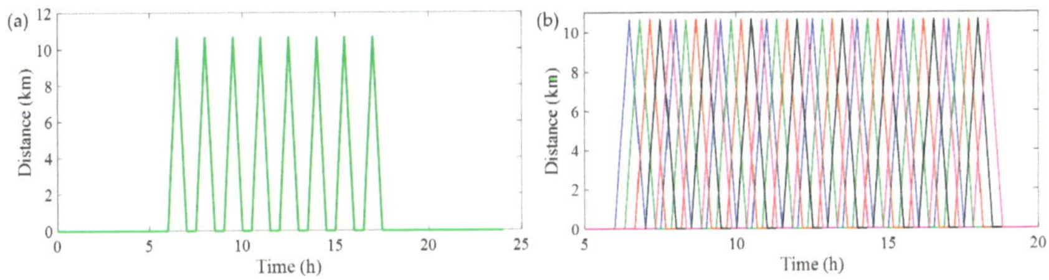

Figure 3. BEB daily position; (**a**) prevalence of 10%, (**b**) prevalence of 50%.

One of the most important results of the optimization problem is the operational programming of charging time intervals in accordance with technical and logistical constraints. The BEB public transport system operation time per day is divided into Y periods, and travel and charging periods are interleaved equally between all transport units. The charging models used in this work to calculate the battery SOC were obtained experimentally by the Ecuadorian National Energy Regulatory Agency ARCERNNR, based on measurements at various charging station models. Figure 4 shows the charging models obtained.

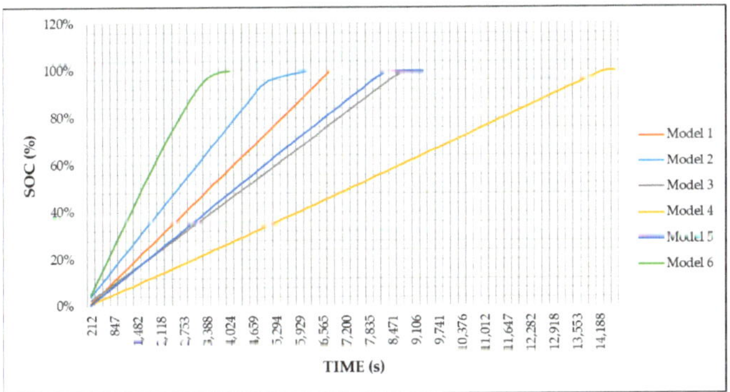

Figure 4. Electric vehicle charging models.

Optimal results were obtained for the scenarios shown in Table 2, considering 80 kW chargers.

Table 2. Optimal solutions summary.

Scenario (Prevalence)	Projected Year	Charging Stations	Minimum Total Costs (USD)
20%	2030 [1]	1	85,464.55
50%	2037 [1]	1	204,626.64
100%	2055 [1]	2	413,614.68

[1] Foresight scenarios based on [17].

Figure 5 shows the battery SOC of each electric bus on the route in the highest prevalence scenario. It denotes the constraints fulfillment considering that batteries can only safely operate without a detrimental impact on their health and longevity at between a 20 and 80% SOC level [18].

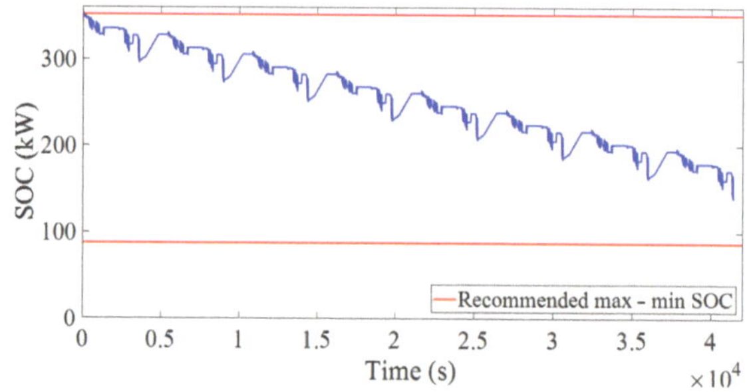

Figure 5. BEBs daily SOC—prevalence of 100%.

5. Demand Analysis

A feeder labeled as 5D in "Chilibulo´s" substation is where the charging station would be connected, and it belongs to the "Empresa Eléctrica Quito" electric utility company. This feeder is located in southern Quito–Ecuador, and it is characterized by predominantly residential customers. It supplies power to 268 distribution transformers at a 22.8 kV voltage level. To further analyze the impact of implementing the charging stations on the case study feeder, it is necessary to estimate the customer and charging station demand curves. The customer demand curves correspond to typical curves obtained from stratified customer cluster analyses [19]. This methodology allows for the obtaining of typical customer power demand curves, which adapt to the monthly consumption of any feeder customer. Using estimated daily power load profiles, quasi-dynamic load flows can be carried out on the feeder.

Electric Chargers Demand Profile

Based on optimal charging scenarios obtained in the previous section, the charger demand curves corresponding to each prevalence level are established and included in the 5D feeder power flow analysis. Figure 6a shows the daily power demand curve of the charging station corresponding to the 20% prevalence level. The charging time and intervals allow the electric bus fleet to adequately maintain its charge level and minimize transport system operating costs. Figure 6b shows the daily power demand curve for the 50% prevalence level, which represents the maximum level at which a single charging station can supply all the system. Note that hourly tariffs have not been considered in the energy cost calculation, which means that low-demand nighttime hours are not prioritized for charging.

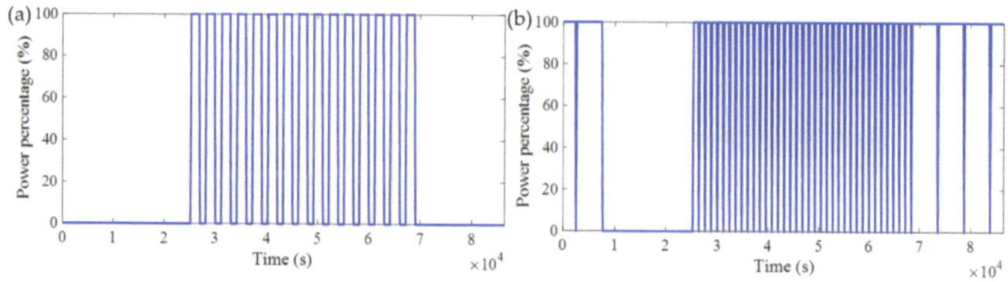

Figure 6. Charging station demand curve; (**a**) prevalence of 20%, (**b**) prevalence of 50%.

Figure 7 displays the daily power demand curve in a 60% prevalence scenario where one single charging station does not meet the BEB system's charging needs, even if it operates 24 h a day. This prevalence scenario requires the installation of a second charging station. Finally, the daily power demand profile for two charging stations required in a 100% prevalence scenario is shown in Figure 8.

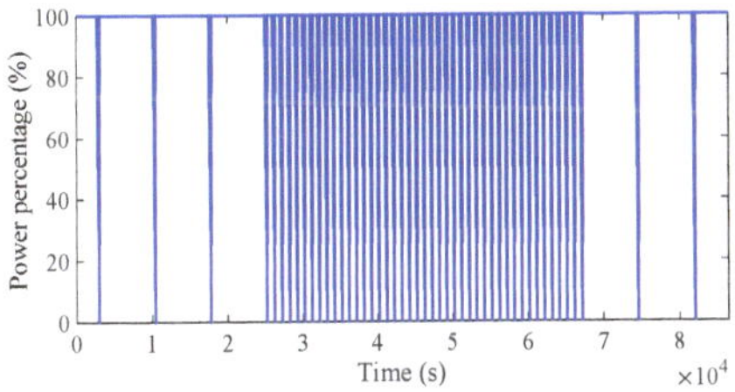

Figure 7. Charging station demand curve—prevalence of 60%.

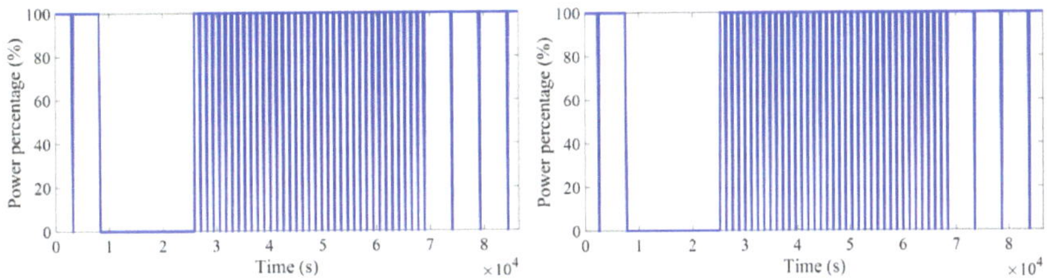

Figure 8. Charging stations demand curves—prevalence of 100%.

6. Impact Evaluation

The CYME software was used to analyze the impact of installing charging stations on the studied feeder during this work. In Ecuador, the CYME software is widely used among electric utility companies to perform several studies in primary distribution networks, as in the present work. Figure 9 shows the current power demand daily profiles of the feeder under study. The three curves displayed represent the power demand for the respective phases of the feeder. Demand data at 15 min intervals are considered for the power flow analysis.

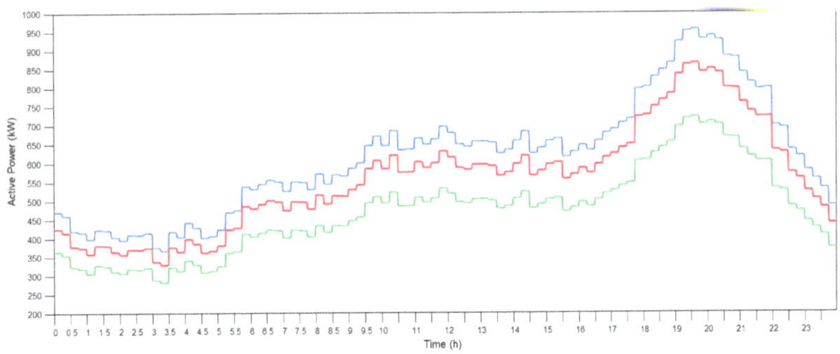

Figure 9. Current 5D Chilibulo's feeder daily power demand curves.

Electric utilities are now facing the challenge of managing new energy uses, such as electromobility and distributed renewable sources of energy interconnected with the EPDG. Both cases should be analyzed using power demand and energy production curves, allowing for the conducting of quasi-dynamic power flows. The CYME software can perform time-series simulations to study the impact of load variations and intermittency of renewable sources, preparing the EPDG for future higher prevalence levels. Figure 10 shows the feeder power demand daily curves for a 100% electric bus prevalence level. A 6% increase in the maximum demand can be noted, considering that the charging times coincide with the system's peak hours. This results in a significant increase in power losses of 8% during peak hours. The increase in demand is consistent across all phases, attributable to the installation of three-phase charging stations.

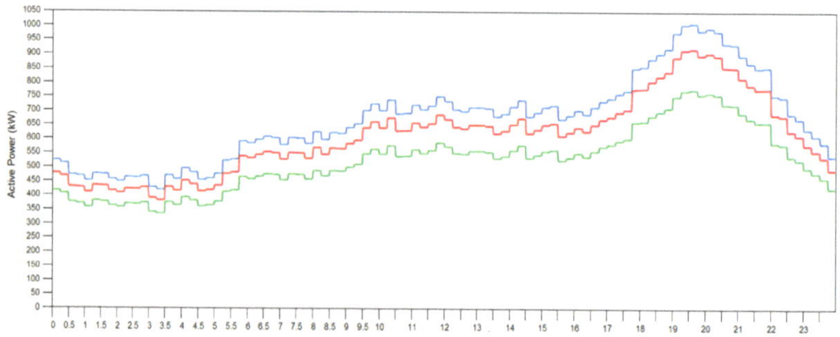

Figure 10. The 5D Chilibulo's feeder daily power demand curves—prevalence of 100%.

7. Conclusions

The operation of a BEB transport system requires the optimal management of financial and energy resources, considering technical and operational constraints. In the present work, a methodology for solving the optimization problem was successfully proposed and implemented, with a focus on a small case study. However, modern transportation systems include multiple routes, with variable frequencies adapted to passenger demand. Although these issues increase the complexity of the optimization problem (increased extension with multiple solutions), the problem formulation does not change significantly.

The results indicate that BEB purchase costs, energy consumption, and demand charges have the most significant impacts on total system annualized costs (approx. 65%).

Although the investment and maintenance costs of charging stations are not significant in contrast to total system annualized costs, in more demanding routes, system efficiency and BEB autonomy could be improved through opportunity charging implementation, where pantograph or induction are typically used.

The impact degree of charging stations on the power distribution grid is highly dependent on the BEB prevalence level; high prevalence levels will cause high occupancy rates in charging stations; therefore, the coincident peak demand of the system may be significantly increased, producing possible negative effects on grid such as feeder overloading, voltage drops, and an increase in power losses.

Author Contributions: Conceptualization, D.O. and P.V.; methodology, D.O. and L.S.; software, D.O., L.S. and W.C.; validation, L.S., W.C., and J.M.; formal analysis, P.V., D.O. and L.S.; investigation, D.O. and L.S.; data curation, J.M. and W.C.; writing—original draft preparation, D.O. and L.S.; writing—review and editing, P.V., J.M. and W.C.; project administration, P.V. All authors have read and agreed to the published version of the manuscript.

Funding: This research received no external funding.

Institutional Review Board Statement: Not applicable.

Informed Consent Statement: Not applicable.

Data Availability Statement: The data will be available on request.

Conflicts of Interest: The authors declare no conflicts of interest.

References

1. Skuza, A.; Jurecki, R.S. Analysis of Factors Affecting the Energy Consumption of an EV Vehicle—A Literature Study. *IOP Conf. Ser. Mater. Sci. Eng.* **2022**, *1247*, 012001. [CrossRef]
2. Bie, Y.; Ji, J.; Wang, X.; Qu, X. Optimization of Electric Bus Scheduling Considering Stochastic Volatilities in Trip Travel Time and Energy Consumption. *Comput.-Aided Civ. Infrastruct. Eng.* **2021**, *36*, 1530–1548. [CrossRef]
3. Orbe, D.; Salazar, L.; Vásquez, P. Estimación y Análisis de Sensibilidad del Consumo Energético de Buses Eléctricos mediante Simulaciones Microscópicas en líneas de Transporte Público. *Rev. Téc. Energ.* **2024**, *21*, 105–113. [CrossRef]
4. Ibarra-Rojas, O.J.; Delgado, F.; Giesen, R.; Muñoz, J.C. Planning, Operation, and Control of Bus Transport Systems: A Literature Review. *Transp. Res. Part B Methodol.* **2015**, *77*, 38–75. [CrossRef]

5. Sebastiani, M.T.; Lüders, R.; Fonseca, K.V.O. Evaluating Electric Bus Operation for a Real-World BRT Public Transportation Using Simulation Optimization. *IEEE Trans. Intell. Transp. Syst.* **2016**, *17*, 2777–2786. [CrossRef]
6. Iliopoulou, C.; Tassopoulos, I.; Kepaptsoglou, K.; Beligiannis, G. Electric Transit Route Network Design Problem: Model and Application. *Transp. Res. Rec.* **2019**, *2673*, 264–274. [CrossRef]
7. Messaoudi, B.; Oulamara, A. Electric Bus Scheduling and Optimal Charging. In Proceedings of the Computational Logistics; Paternina-Arboleda, C., Voß, S., Eds.; Springer International Publishing: Cham, Switzerland, 2019; pp. 233–247.
8. Peña, D.; Dorronsoro, B.; Tchernykh, A.; Ruiz, P. Public Transport Timetable and Charge Optimization Using Multiple Electric Buses Types. In Proceedings of the Genetic and Evolutionary Computation Conference Companion, Boston, MA, USA, 9–13 July 2022; Association for Computing Machinery: New York, NY, USA, 2022; pp. 751–754.
9. Unterluggauer, T.; Rich, J.; Andersen, P.B.; Hashemi, S. Electric Vehicle Charging Infrastructure Planning for Integrated Transportation and Power Distribution Networks: A Review. *eTransportation* **2022**, *12*, 100163. [CrossRef]
10. Alamatsaz, K.; Hussain, S.; Lai, C.; Eicker, U. Electric Bus Scheduling and Timetabling, Fast Charging Infrastructure Planning, and Their Impact on the Grid: A Review. *Energies* **2022**, *15*, 7919. [CrossRef]
11. Liu, K.; Yamamoto, T.; Morikawa, T. Impact of Road Gradient on Energy Consumption of Electric Vehicles. *Transp. Res. Part Transp. Environ.* **2017**, *54*, 74–81. [CrossRef]
12. Li, X.; Castellanos, S.; Maassen, A. Emerging Trends and Innovations for Electric Bus Adoption—A Comparative Case Study of Contracting and Financing of 22 Cities in the Americas, Asia-Pacific, and Europe. *Res. Transp. Econ.* **2018**, *69*, 470–481. [CrossRef]
13. E-BUS RADAR. Available online: https://ebusradar.org/es/ (accessed on 27 August 2024).
14. Validi, A.; Morales-Alvarez, W.; Olaverri-Monreal, C. Analysis of the Battery Energy Estimation Model in SUMO Compared with Actual Analysis of Battery Energy Consumption. In Proceedings of the 2021 16th Iberian Conference on Information Systems and Technologies (CISTI), Chaves, Portugal, 23–26 June 2021; pp. 1–6.
15. He, Y.; Liu, Z.; Song, Z. Optimal Charging Scheduling and Management for a Fast-Charging Battery Electric Bus System. *Transp. Res. Part E Logist. Transp. Rev.* **2020**, *142*, 102056. [CrossRef]
16. Wang, Y.; Huang, Y.; Xu, J.; Barclay, N. Optimal Recharging Scheduling for Urban Electric Buses: A Case Study in Davis. *Transp. Res. Part E Logist. Transp. Rev.* **2017**, *100*, 115–132. [CrossRef]
17. Estudio de Análisis y Prospectiva de La Electro-Movilidad En Ecuador y El Mix Energético al 2030, Instituto de Investigación Geológico y Energético, Ministerio de Energía y Minas. Available online: https://www.geoenergia.gob.ec/wp-content/uploads/downloads/2023/08/Estudio-de-analisis-y-prospectiva-de-la-electromovilidad-en-Ecuador-y-el-mix-energetico-al-2030.pdf (accessed on 3 August 2023).
18. Juan, A.; Marugan, C.; Ahsini, Y.; Fornes, R.; Panadero, J.; Martin, X. Using Reinforcement Learning to Solve a Dynamic Orienteering Problem with Random Rewards Affected by the Battery Status. *Batteries* **2023**, *9*, 416. [CrossRef]
19. Orbe Játiva, D.; Salazar Yépez, G.; Vásquez Miranda, P. Estimación de Curvas Diarias de Demanda Eléctrica En Transformadores de Distribución Mediante Redes Neuronales Artificiales. In Proceedings of the 2022 IEEE Biennial Congress of Argentina (ARGENCON), San Juan, Argentina, 7–9 September 2022; pp. 1–5.

engineering proceedings

MDPI

Proceeding Paper

On the Performance Comparison of Fuzzy-Based Obstacle Avoidance Algorithms for Mobile Robots †

José Zúñiga, William Chamorro, Jorge Medina *, Pablo Proaño, Renato Díaz and César Chillán

Departamento de Automatización y Control Industrial, Facultad de Ingeniería Eléctrica y Electrónica, Escuela Politécnica Nacional, Quito 170143, Ecuador; jose.zuniga01@epn.edu.ec (J.Z.); william.chamorro@epn.edu.ec (W.C.); pablo.proano@epn.edu.ec (P.P.); victor.diaz@epn.edu.ec (R.D.); cesar.chillan@epn.edu.ec (C.C.)

* Correspondence: jorge.medinaj@epn.edu.ec

† Presented at the XXXII Conference on Electrical and Electronic Engineering, Quito, Ecuador, 12–15 November 2024.

Abstract: One of the critical challenges in mobile robotics is obstacle avoidance, ensuring safe navigation in dynamic environments. In this sense, this work presents a comparative study of two intelligent control approaches for mobile robot obstacle avoidance based on a fuzzy architecture. The first approach is a neuro-fuzzy interface that combines neural networks' learning capabilities with fuzzy logic's rule-based reasoning, offering a flexible and adaptable control strategy. The second is a classic Mamdani fuzzy system that relies on human-defined fuzzy rules, providing an intuitive approach to control. A key contribution of this work is the development of a fast comprehensive, model-based dataset for neural network training generated without the need for real sensor data. The results show the evaluation of these two systems' performance, robustness, and computational efficiency using low-cost ultrasonic sensors on a Pioneer 3DX robot within the Coppelia Sim environment.

Keywords: ANFIS; Mamdani; obstacle avoidance; fuzzy; mobile robots; intelligent control systems; IoT; ultrasonic sensors; synthetic dataset; CoppeliaSim

Citation: Zúñiga, J.; Chamorro, W.; Medina, J.; Proaño, P.; Díaz, R.; Chillán, C. On the Performance Comparison of Fuzzy-Based Obstacle Avoidance Algorithms for Mobile Robots. *Eng. Proc.* **2024**, *77*, 23. https://doi.org/10.3390/engproc2024077023

Academic Editor: Jackeline Abad

Published: 18 November 2024

1. Introduction

Mobile robots have become increasingly prevalent in various applications, from industrial automation to personal assistance. One of the critical challenges in mobile robotics is obstacle avoidance, which ensures the safe navigation of the robot in its environment [1]. Conventional control techniques have limitations in dealing with the inherent uncertainties and complexities of real-world environments [2].

Conventional, model-based control techniques often struggle to handle the inherent uncertainties and complexities of real-world navigation. These methods often rely on precise environmental models and struggle to adapt to unexpected obstacles or changing conditions [3]. To address these challenges, intelligent control systems, such as fuzzy logic and adaptive neuro-fuzzy inference systems, have emerged as promising solutions [4–6]. Fuzzy logic, inspired by human reasoning, excels in dealing with vagueness and linguistic information [7], while neural networks offer powerful learning capabilities for adapting to dynamic environments [8].

In this work, we present a comparative study of two intelligent control approaches for mobile robot obstacle avoidance: the adaptive neuro-fuzzy inference system (ANFIS) and the Mamdani fuzzy system. The ANFIS combines the learning capabilities of neural networks with the rule-based reasoning of fuzzy logic, offering a more flexible and adaptable control strategy [4]. Adaptive neuro-fuzzy inference system leverages the advantages of fuzzy logic to handle uncertainty in sensor data, ensuring smooth and intuitive control actions, while simultaneously employing the adaptive learning capabilities of neural networks to optimize performance over time [9]. The Mamdani fuzzy system, on the other hand, relies on a set of human-defined fuzzy rules to make decisions, providing a more intuitive approach to control [5]. Both intelligent control methods have been extensively studied and applied in the field of mobile robotics, and each approach has its own strengths and limitations [10]. This comparative study aims to evaluate the performance, robustness, and computational efficiency of these two control systems in the context of mobile robot obstacle avoidance, providing valuable insights for researchers and engineers working in this field.

Interestingly, while the adaptive neuro-fuzzy inference system offers greater adaptability and learning capabilities, the Mamdani fuzzy system may be more suitable for scenarios where the underlying system dynamics are well understood, and the control objectives can be easily translated into a set of fuzzy rules [11]. The choice between the adaptive neuro-fuzzy inference system and the Mamdani fuzzy system for mobile robot obstacle avoidance depends on the specific requirements of the application, the available data, and the level of uncertainty in the environment [12]. Autonomous robots are quite commonly used but the scale is limited to repeated tasks or indoor applications in most cases. For autonomous vehicles, the sensor technology must be made more accommodating since a diversity of possibilities occurs which populates into a huge dimensional problem [4]. In this context, the adaptive neuro-fuzzy inference system and the Mamdani fuzzy system can play a crucial role in enhancing the obstacle avoidance capabilities of mobile robots, paving the way for more reliable and versatile autonomous systems.

In this work, we assess the performance of ANFIS and classic Mandami fuzzy controllers for obstacle avoidance in unknown scenarios using the measurements given by low-cost ultrasonic sensors as sensing data. Both controllers use as a base trajectory controller a fuzzy cluster arrangement that handles the heading angle towards the desired goal and moves the robot with a constant velocity. It is worth mentioning that this work deals with static obstacle avoidance within a controlled environment. The development of a fuzzy neural network for the avoidance of moving (dynamic) obstacles, as well as a performance comparison with other types of controllers, is part of future research and will be presented in future works. A key contribution of this work lies in the development of a comprehensive, model-based dataset for training the ANFIS controller. This dataset is generated with random data within a range that emulates the low-cost sensor measurements, a combination of avoidance rules, and the model of the robot. The resultant is a synthetic dataset that does not require real sensor data, enhances the ANFIS training, and can be adapted to the robot's geometry and sensor models.

The performance assessment of both fuzzy topologies was carried out in simulation using the Pioneer DX3 robot within Coppelia Sim. The fuzzy systems were assembled and trained in Matlab which communicates with Coppelia Sim in real time. For the benefit of the community, the dataset generation along with the simulation files are available at https://github.com/WChamorro/Neuro-Fuzzy-Obstacle-Avoidance.git (accessed on 7 November 2024).

The rest of the paper is organized as follows: Methodology presents a detailed overview of the adaptive neuro-fuzzy inference system and the Mamdani fuzzy system, including their key components and decision-making processes applied to a robot to avoid obstacles considering data from ultrasonic sensors. Then, the results are presented with a discussion about them, and finally, relevant conclusions are presented.

2. Methodology

Adaptive neuro-fuzzy inference system (ANFIS) and classic Mamdani fuzzy systems represent two distinct approaches to modeling and control. ANFIS integrates the learning capabilities of neural networks with the interpretability of fuzzy logic, providing a powerful tool for handling complex, non-linear problems through adaptive learning [13]. In contrast, Mamdani fuzzy systems rely on a rule-based approach, where human expertise is encoded into a set of linguistic rules, offering intuitive and easily understandable solutions. The following sections bring details about the fuzzy architecture and the dataset generation for training the ANFIS.

2.1. Neuro-Fuzzy Inference System

The base robot's behavior is managed by a Sugeno fuzzy clustering approach, as shown in Figure 1a,c. This controller returns three sets of velocity: low, zero, and high based on the heading angle (HA). It is worth mentioning that the robot knows where the goal position is, and its 2D relative pose (to the starting point). The fuzzy clustering will be active when the sensors do not detect a closer obstacle [14].

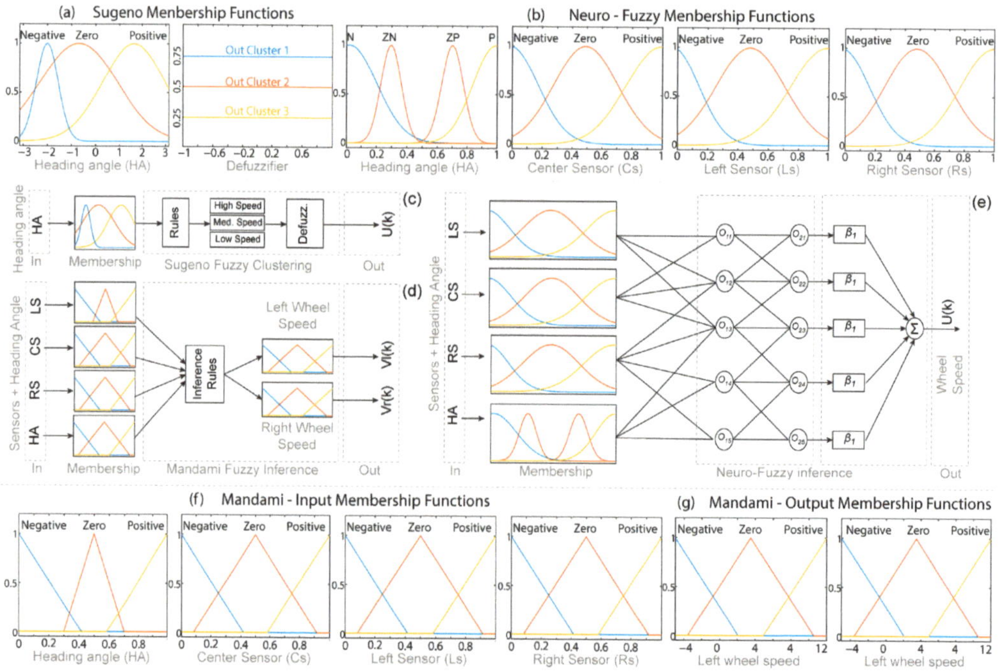

Figure 1. Fuzzy systems architecture: (**a**) Sugeno membership functions. (**b**) Neuro-fuzzy membership functions. (**c**) Sugeno clustering architecture. (**d**) Mandami architecture. (**e**) Neuro-Fuzzy architecture. (**f**) Mandami input membership functions. (**g**) Mandami output membership functions.

The ANFIS architecture shown in Figure 1b,e drives the robot's wheel speeds based on the inputs from each ultrasonic sensor: Right Sensor (RS), Central Sensor (CS), Left Sensor (LS), and the heading angle. This scheme was replicated for both wheels, and it was built with three sigmoidal membership functions for the sensors, and four for the heading angle for a more precise control. The Mamdani architecture in Figure 1d will be detailed in the following sections.

Model-Based Dataset Generation

The training dataset was created with uniformly generated random values representing the sensor readings. Low-cost ultrasonic sensors were used in this study to measure distance to the obstacles in meters. The shorter this distance, the more significant the control action required from the robot, whether it involves continuing to move forward, turning right, or turning left. In this work, we use the model of the sensor HC-SR04, which has a range from 2 to 400 cm and a 15-degree detection angle. The detection range was limited between d_{min} and d_{max} as shown in Figure 2a. This configuration ensures the robot can anticipate and avoid obstacles in advance, it also helps to prevent the sensors from receiving noisy data or errors due to distant objects. The desired heading angle is computed as $\theta = atan2(target_y, target_x)$, since we know the goal position noted with coordinates $(target_x,\ target_y)$.

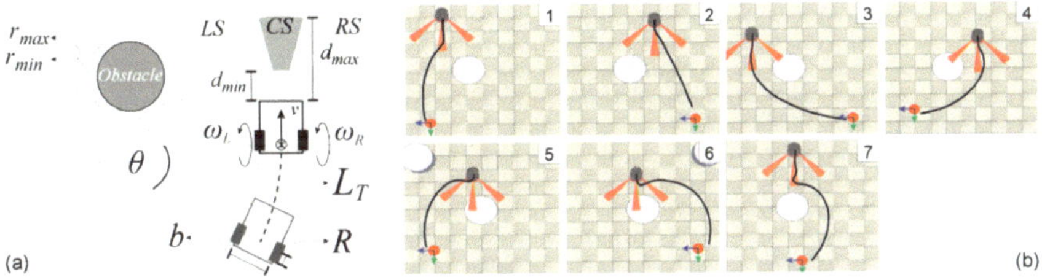

Figure 2. Tools for the dataset generation: (**a**) Robot geometry. (**b**) Avoidance rules visualization.

The dataset was generated in Python and requires only a specific set of parameters that depend on the geometry of the problem. The dataset is composed of the sensor data the heading angle, and the estimated right and left wheel velocities. The sensor data were randomly generated with a 25% bias focused on shorter distances and small alignment angles, as shown in Figure 3. The biased data are intended to enhance robot training for navigating tight spaces and handling multiple obstacles. It is worth mentioning that the data generated is normalized within their respective ranges to have an approach as general as possible.

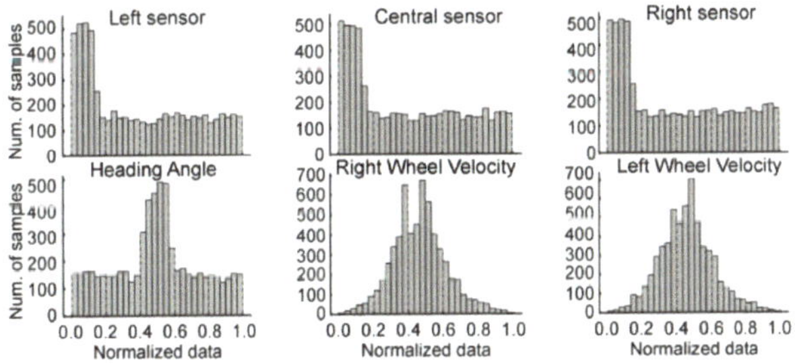

Figure 3. Training data distributions.

The wheel speeds in the dataset are estimated considering the robot's geometry and the sensor data following [15], as

$$\omega_{R/L} = \left(1 \pm \frac{b\theta}{2L_T}\right)\frac{v}{R}\ [\text{rad/s}] \tag{1}$$

where b is the distance between the robot's wheels, and R the radius of each wheel. L_T is the normalized robot's curvature length computed as follows:

$$L_T = d \cdot (L_{T_{max}} - L_{T_{min}}) + L_{T_{min}}\ [\text{m}] \tag{2}$$

recall that L_T is a normalized expression between L_{Tmin} and L_{Tmax}, which depends on the geometric characteristics of the robot as illustrated in Figure 2a. The remaining term v, is the estimated robot's linear velocity computed as follows:

$$v = v_{max} - \left(G_q \cdot \left(\frac{v_{mean}}{2}\right)\right)\ [\text{m/s}]\ \text{with}\ G_q = 1 - 4d(1-d) \tag{3}$$

where G_q is a quadratic function that uses the sensor-measured distance d. G_q models a desired velocity boost if the sensors detect an obstacle closer to d_{min} or d_{max}, or a moderate speed in the mid-range. This behavior allows us to reach the goal as fast as possible if we have a clear path or produce faster reactions in the presence of an obstacle.

In the presence of an obstacle, the robot should decide which direction must be followed. These decisions are predefined by providing the knowledge with the statement of the possible avoidance rules pictured in Figure 2b. Cases 1 and 2 are the simplest to interpret because the robot must move away from the obstacle while approaching the target, without compromising trajectory. In cases 3 and 4, the target is in the same orientation as the obstacle, so a change to the desired angle is made to ensure the robot does not move too far from the obstacle, thereby avoiding trajectory compromise. For cases 5 and 6, the front sensor will detect the closest distance, so the position of the target will determine the robot's movement. Case 7 applies to the remaining cases, when the robot is aligned with the target, involving a real-time condition that consists of changing the orientation angle, forcing the wheels to make a sharper turn, and preventing the robot from colliding with the obstacle.

The dataset was generated using a Python script yielding synthetic adjustable data to train the ANFIS. Figure 4 shows the result of the training process that was carried out with the neuro-fuzzy designer in MATLAB 2024A [16].

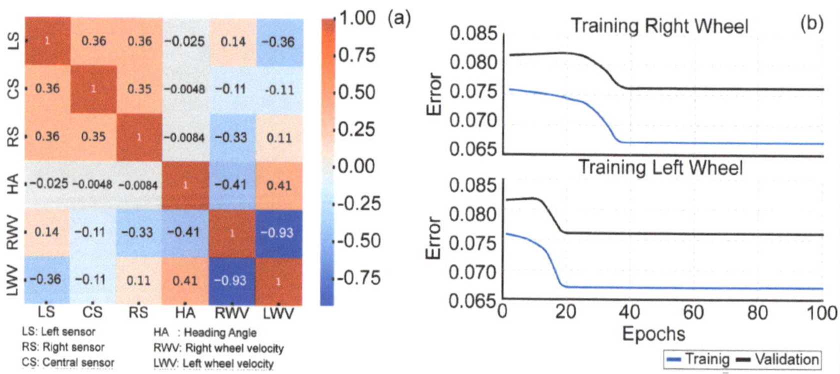

Figure 4. ANFIS training results: (**a**) Correlation Matrix. (**b**) Training error.

The generated dataset for the neuro-fuzzy scheme was split into 80% for training and 20% for validation. The training error stabilizes after 30 epochs approximately, as shown in Figure 4b. The validation error is slightly higher than the training error, indicating that the ANFIS is capable of generalizing with a lower error. As shown in Figure 4a, LS, CS, and RS sensors exhibit moderate positive correlations with each other (0.36), indicating that they typically detect similar features in the environment. The heading angle shows a very low correlation with the sensors, suggesting minimal direct influence from sensor readings. The right wheel velocity (RWV) and left wheel velocity (LWV) display a strong negative correlation (−0.93), reflecting the expected inverse relationship during differential steering maneuvers. The heading angle also moderately correlates with wheel velocities, showing a negative correlation with RWV (−0.41) and a positive correlation with LWV (0.41), indicating the significant role of wheel velocities in determining the robot's heading.

2.2. Mamdani Fuzzy Inference System

The Mamdani fuzzy inference system (FIS) pictured in Figure 1d,f,g, enables inference based on the evaluation of one or more membership sets according to a predetermined number of rules proposed by an expert system designer [17]. Unlike a Sugeno-type fuzzy inference system, Mamdani allows for multiple inputs and outputs [18]. Consequently, it is not necessary to separate the calculations for the speeds of the left and right wheels of the motor. The following sections describe the structure of this fuzzy approach.

2.2.1. Membership Functions for Inputs and Outputs

The membership function of each input variable (sensors and heading angle) was built with three triangular fuzzy sets, which are uniformly distributed, as shown in Figure 1f,g. For the variables related to the distance sensors, the following fuzzy sets are defined with triangular functions with these parameters: Close: [−0.42, 0, 0.42], Medium: [0.08, 0.5, 0.92], Far: [0.6, 1, 1.42]. The output membership functions are similarly constructed with triangular fuzzy sets: Slow: [−0.42, 0, 0.42], Normal: [0.08, 0.5, 0.92], Fast: [0.6, 1, 1.42]. These intervals were set based on the desired robot's speed limitations.

2.2.2. Inference and Defuzzification

To evaluate a Mamdani-type fuzzy set, it is necessary to develop rules, which are also defined by the designer. The base rules allow handling the avoidance cases displayed in Figure 2b. Other rules were added based on the possible combinations between input and output membership functions. For the defuzzification of the FIS output, the centroid method (COG) is applied, as it provides balanced, smooth, and natural results, which enhance the performance of the mobile robot control [9].

3. Results and Discussion

The fuzzy strategies for obstacle avoidance were evaluated through simulation using Coppelia Sim, where we assembled virtual scenarios of different complexity with obstacles of different shapes. Whitin Copelia the scenarios were configured with the Bullet engine as a physics computation motor due to its high performance in collision detection and rigid body dynamics, making it suitable for real-time applications for obstacle avoidance.

The Pioneer 3DX used in the experiments was set with geometry constants based on the real robot specifications to: $b = 0.381$ [m], $R = 0.195$ [m]. The robot's maximum velocity was set to v_{max} = 1.2 [m/s] which leads to considering an average velocity of v_{mean} = 0.6 [m/s]. The path's curvature is constrained to $L_{T_{min}}$ = 0.3 [m] and $L_{T_{min}}$ = 1.2 [m]. The low-cost sensor model mounted in the virtual robot is the HC-SR04 from which we use constrained distance measurements from d_{min} = 0.3 [m] to d_{max} = 1 [m]. Recall that the sensing data and wheel speed were normalized from 0 to 1 to generate the training dataset for the ANFIS.

Avoidance Experiments

The ANFIS and Mamdani fuzzy strategies were assessed in two complex scenarios. Scenario A was built with scattered obstacles with rounded and squared shapes. In the center of the scenario, we place an elongated object that will test the avoidance capabilities due to the curve that the robot should make to reach the goal. The results in scenario A are pictured in Figure 5, where the ANFIS scheme handles all the obstacles efficiently. As expected, the robot navigates closely to the large obstacle until it can turn towards the goal. On the other hand, the Mamdani approach produces an unnecessary loop while trying to avoid the large obstacle and detecting another obstacle on its path. Note that in Figures 5 and 6, the ultrasonic beams are shown in red when no obstacle is detected and in yellow when an obstacle is present.

Scenario B emulates a narrow corridor where the robot should navigate closely to the walls. This experiment includes a closed curve towards the goal as displayed in Figure 6. The ANFIS results in this scenario, shown in Figure 6a, show a clean trajectory while the robot navigates closely to the walls even during the closed curve, where the robot follows the shape of the wall. In Figure 6b the Mandami approach tries to face the goal all the time, yielding an oscillatory behavior in the presence of a persistent obstacle. This issue is related to the constant set of membership functions that were designed to address generic avoidance cases. The distribution of membership functions in the Mamdani-type system was not subjected to a tuning process; instead, a uniform distribution was directly applied across the entire evaluation range of the term set. This approach may require a fine tuning since the robot responses aggressively to an obstacle which may produce large deviations from the goal.

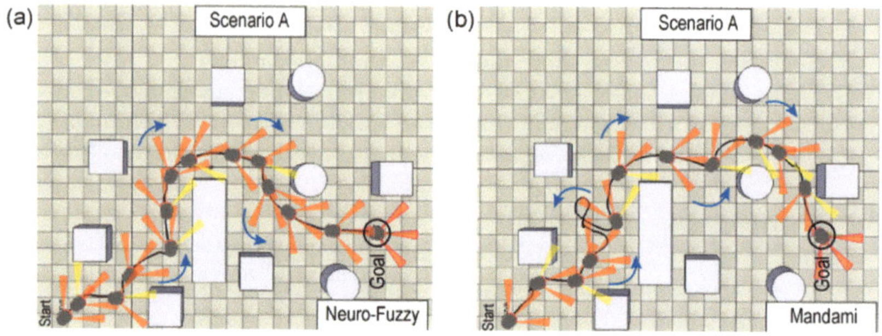

Figure 5. Experiments in scenario A: (**a**) ANFIS performance. (**b**) Mandami performance.

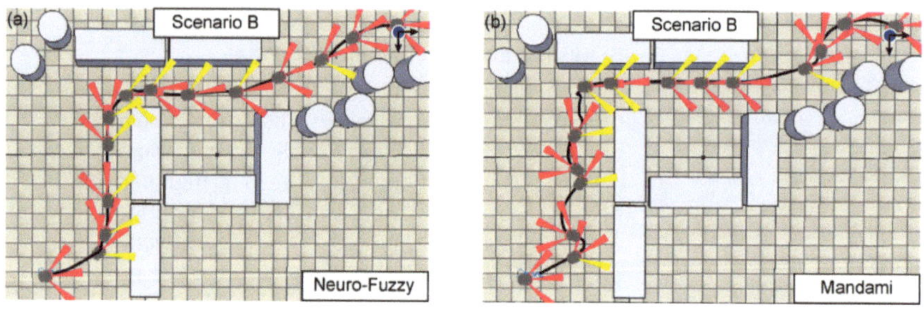

Figure 6. Experiments in scenario B: (**a**) ANFIS performance; (**b**) Mandami performance.

Evaluating obstacle avoidance schemes in robotics and autonomous systems can be approached through both qualitative and quantitative methods. Qualitatively, one can assess the effectiveness of the scheme by observing the robot's ability to navigate around obstacles without collisions, maintaining smooth and natural movement patterns. Quantitative evaluation involves measuring specific metrics such as velocity and acceleration. By analyzing these parameters, one can determine the efficiency and responsiveness of the avoidance scheme, ensuring that the robot moves at an optimal speed while minimizing abrupt changes in acceleration, which can indicate inefficient or unsafe maneuvers. In this sense, the ANFIS produced smooth displacements with controlled accelerations as shown in Figure 7b-bottom. This yields controlled movements and avoids unnecessary oscillations, especially while sensing large walls. In both scenarios, the ANFIS shows constant velocities with minimal acceleration especially when the robot navigates closely to the walls, see Figure 7b-top. On the other hand, the Mandami approach tends to oscillate while avoiding an obstacle. In this case, large accelerations are observed in Figure 7a-bottom, which may cause deviations or loops as the one observed in scenario A.

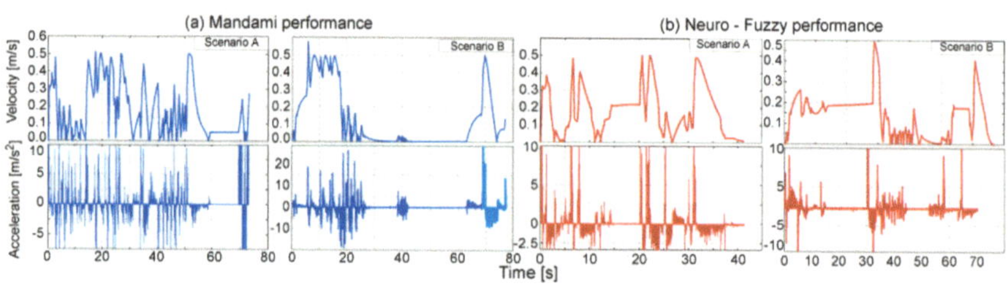

Figure 7. Velocity and Acceleration performance results: (**a**) Mandami; (**b**) ANFIS.

As shown in Figure 7, the acceleration frequency is lower in the neuro-fuzzy control compared to the Mamdani control. This smoothness allows the robot to move more fluidly, avoiding getting stuck when evading an obstacle. Additionally, a lower acceleration frequency has the advantage of preventing mechanical wear and fatigue, particularly in experiments with a real robot.

One of the efficiency parameters used to determine the efficacy between the two controllers is the performance ratio. This metric is defined as the ratio between the shortest distance the robot can travel, which is the straight line the robot would take in an obstacle-free environment, and the distance actually traveled by the robot in each experiment. In this way, efficiency in identical scenarios can be compared using a dimensionless parameter, where the perfect distance (completely straight trajectory) would result in a performance ratio of 1. The lower this value, the poorer the performance of the controller analyzed in a given scenario.

Table 1 gives a clear overview of the metrics from each experiment. It includes the Root Mean Square (RMS) values for both speed and acceleration. When using the Mamdani-type fuzzy controller, there are significant changes in acceleration during orientation shifts. However, with the neuro-fuzzy controller, these peaks are greatly reduced, suggesting that this controller aims to keep the speed constant. In Scenario B, the values are quite similar and stay below one, mainly because the presence of multiple obstacles forces the robot to move more slowly. However, the time it takes for the robot to reach its target is significantly reduced when using the neuro-fuzzy control.

Table 1. Performance statistics in the test scenarios.

Statistics	Scenario A		Scenario B	
	ANFIS	Mamdani	ANFIS	Mamdani
Trajectory time [s]	41.45	29.65	70.45	77.17
Traveled path [m]	14.47	13.21	18.02	20.46
RMS velocity [m/s]	0.24	0.24	0.19	0.20
RMS acceleration [m/s^2]	1.09	2.07	0.74	0.68
Performance ratio	0.87	0.85	0.82	0.74
Computational time per cycle [ms]	1.3	0.56	1.3	0.56

4. Conclusions

The results show the effectiveness of using intelligent control systems, specifically the adaptive neuro-fuzzy inference system (ANFIS) for obstacle avoidance in mobile robots. Findings highlight several key insights: ANFIS controller exhibited superior performance in complex obstacle avoidance scenarios compared to the Mamdani fuzzy system. This is primarily due to ANFIS's ability to adapt and learn from the environment, resulting in smoother and more efficient navigation. In scenarios with scattered and complex obstacles, ANFIS managed to navigate without unnecessary loops or deviations, unlike the Mamdani system which sometimes struggled with large obstacles. The ability of ANFIS to adjust its parameters dynamically allowed it to handle unexpected changes and obstacles more effectively. Overall, the neuro-fuzzy controller demonstrates significant advantages in maintaining smoother and more efficient control, particularly in more complex scenarios with obstacles. In Scenario B, where multiple obstacles are present, the neuro-fuzzy controller not only reduces the trajectory time from 77.17 s (Mamdani) to 70.45 s but also maintains a favorable balance between speed and acceleration. This improvement in trajectory time suggests that the neuro-fuzzy controller can navigate obstacles more effectively, achieving faster goal attainment. Despite the slightly higher RMS acceleration in Scenario B, the neuro-fuzzy controller still manages to maintain efficient control, as indicated by a higher performance ratio of 0.82 compared to 0.74 for the Mamdani controller. This ratio highlights the neuro-fuzzy controller's ability to find a more optimal path, staying closer to the ideal straight-line trajectory even in complex environments.

In narrow corridor scenarios, ANFIS maintained a clean trajectory closely following the walls, even during sharp turns. In contrast, the Mamdani system exhibited a sinusoidal trajectory due to aggressive responses to obstacles, indicating the need for fine-tuning to improve performance in such environments. Quantitative analysis of velocity and acceleration showed that ANFIS provided controlled movements with minimal unnecessary oscillations, especially when navigating close to walls. The Mamdani approach, however, led to larger accelerations and deviations, which can be inefficient and potentially unsafe. A significant contribution of this work is the development of a comprehensive, model-based dataset for training the ANFIS controller. This synthetic dataset, generated without real sensor data, enhances the training process and can be adapted to various robot geometries and sensor models, providing a versatile tool for future research.

Author Contributions: Conceptualization, J.Z. and W.C.; methodology, J.Z., W.C. and J.M.; software, J.Z., W.C. and J.M.; validation, P.P., R.D., C.C. and J.Z.; formal analysis, J.Z. and W.C.; investigation, J.Z., J.M., W.C., P.P., R.D. and C.C.; resources, J.Z.; data curation, W.C. and J.M.; writing—original draft preparation, J.Z., W.C. and J.M.; writing—review and editing, J.Z., W.C., J.M., P.P., R.D. and C.C.; visualization, W.C.; supervision, W.C. and J.M. All authors have read and agreed to the published version of the manuscript.

Funding: This research received no external funding.

Institutional Review Board Statement: Not applicable.

Informed Consent Statement: Not applicable.

Data Availability Statement: Data are contained within the article.

Conflicts of Interest: The authors declare no conflicts of interest.

References

1. Thrun, S.; Burgard, W.; Fox, D. *Probabilistic Robotics*; MIT Press: Cambridge, MA, USA, 2005; Intelligent Robotics and Autonomous Agents.
2. Siciliano, B.; Sciavicco, L.; Villani, L.; Oriolo, G. *Robotics: Modelling, Planning and Control*; Springer Science & Business Media: Berlin/Heidelberg, Germany, 2010.
3. Zohaib, M.; Pasha, M.; Riaz, R.A.; Javaid, N.; Ilahi, M.; Khan, R.D. Control Strategies for Mobile Robot with Obstacle Avoidance. *arXiv* **2013**. [CrossRef]
4. Menghal, P.M.; Laxmi, A.J. Adaptive Neuro Fuzzy based dynamic simulation of induction motor drives. In Proceedings of the 2013 IEEE International Conference on Fuzzy Systems, Hyderabad, India, 7–10 July 2013; pp. 1–8. [CrossRef]
5. Yet, W.; Qidwai, U. Intelligent sensor network for obstacle avoidance strategy. In Proceedings of the SENSORS, 2005 IEEE, Irvine, CA, USA, 30 October–3 November 2005; p. 4. [CrossRef]
6. Xiao, X.; Liu, B.; Warnell, G.; Stone, P. Motion Planning and Control for Mobile Robot Navigation Using Machine Learning: A Survey. *arXiv* **2022**, arXiv:2011.13112. [CrossRef]
7. Garg, S.; Sharma, D. A Critical Study of Fuzzy Logic Systems and Its Applications. Available online: https://www.ijtsrd.com/engineering/computer-engineering/105/a-critical-study-of-fuzzy-logic-systems-and-its-applications/deepak-sharma (accessed on 26 July 2024).
8. Chiou, K.-C.; Huang, S.-J. An adaptive fuzzy controller for robot manipulators. *Mechatronics* **2005**, *15*, 151–177. [CrossRef]
9. Driankov, D.; Hellendoorn, H.; Reinfrank, M. Introduction. In *An Introduction to Fuzzy Control*; Driankov, D., Hellendoorn, H., Reinfrank, M., Eds.; Springer: Berlin/Heidelberg, Germany, 1993; pp. 1–36. [CrossRef]
10. Castillo, O.; Soria, J.; Arias, H.; Morales, J.B.; Inzunza, M. Intelligent Control and Planning of Autonomous Mobile Robots Using Fuzzy Logic and Multiple Objective Genetic Algorithms. In *Analysis and Design of Intelligent Systems using Soft Computing Techniques*; Melin, P., Castillo, O., Ramírez, E.G., Kacprzyk, J., Pedrycz, W., Eds.; Springer: Berlin/Heidelberg, Germany, 2007; pp. 799–807. [CrossRef]
11. Titov, A.P. Analysis of models of adaptive neuro-fuzzy systems. *Vestn. RGGU Seriâ Inform. Inf. Bezop. Mat.* **2024**, 21–35. [CrossRef]
12. Medina-Santiago, A.; Camas-Anzueto, J.L.; Vazquez-Feijoo, J.A.; León, H.R.H.; Mota-Grajales, R. Neural Control System in Obstacle Avoidance in Mobile Robots Using Ultrasonic Sensors. *J. Appl. Res. Technol.* **2014**, *12*, 104–111. [CrossRef]
13. Jang, J.-S.R. ANFIS: Adaptive-network-based fuzzy inference system. *IEEE Trans. Syst. Man Cybern.* **1993**, *23*, 665–685. [CrossRef]
14. Chiu, S.L. Fuzzy Model Identification Based on Cluster Estimation. *J. Intell. Fuzzy Syst.* **1994**, *2*, 267–278. [CrossRef]
15. Maulana, E.; Muslim, M.A.; Zainuri, A. Inverse kinematics of a two-wheeled differential drive an autonomous mobile robot. In Proceedings of the 2014 Electrical Power, Electronics, Communicatons, Control and Informatics Seminar (EECCIS), Malang, Indonesia, 27–28 August 2014; pp. 93–98. [CrossRef]

16. MathWorks Design, Train, and Test Sugeno-Type Fuzzy Inference Systems—MATLAB—MathWorks América Latina. Available online: https://la.mathworks.com/help/fuzzy/neurofuzzydesigner-app.html (accessed on 16 August 2024).
17. Wondosen, A.; Shiferaw, D. *Fuzzy Logic Controller Design for Mobile Robot Outdoor Navigation*; Cornell University Press: Ithaca, NY, USA, 2024; Volume 1, pp. 1–20.
18. Pop, M.-D.; Pescaru, D.; Micea, M.V. Mamdani vs. Takagi–Sugeno Fuzzy Inference Systems in the Calibration of Continuous-Time Car-Following Models. *Sensors* **2023**, *23*, 8791. [CrossRef] [PubMed]

engineering proceedings

MDPI

Proceeding Paper

Optimal Sizing of Hybrid Generation Systems (Photovoltaic System and Energy Storage System) for Off-Grid Applications †

Jaime Guamangallo [1,*,‡], Jefferson Porras [1,‡], Carlos Quinatoa [1,‡], Jimmy Vaca [1,‡] and Luis Chiza [2,*,‡]

1 Dirección de Posgrados, Facultad de Ciencias de la Ingeniería y Aplicadas, Universidad Técnica de Cotopaxi, Latacunga 050150, Ecuador; jefferson.porras0449@utc.edu.ec (J.P.); carlos.quinatoa7864@utc.edu.ec (C.Q.); jimmy.vaca1539@utc.edu.ec (J.V.)

2 Departamento de Energía Eléctrica, Facultad de Ingeniería Eléctrica y Electrónica, Escuela Politécnica Nacional, Quito 170525, Ecuador

* Correspondence: jaime.guamangallo4058@utc.edu.ec (J.G.); luis.chiza@epn.edu.ec (L.C.)

† Presented at the XXXII Conference on Electrical and Electronic Engineering, Quito, Ecuador, 12–15 November 2024.

‡ These authors contributed equally to this work.

Abstract: This paper presents an optimal sizing strategy for a hybrid generation system combining photovoltaic (PV) and energy storage systems. To achieve this, the optimization problem is solved using the simplex method for linear programming, implemented through Python. The model considers test data on electrical energy demand and solar irradiation, alongside battery operating conditions such as state of charge (SOC) and upper and lower charge limits as key decision variables. Conventional PV system sizing serves as a benchmark to assess the effectiveness of the optimization, with particular attention given to the computational resources required for problem solving.The results obtained from the optimization method demonstrate a substantial improvement in the utilization of energy resources, both from the photovoltaic system and the energy storage system. This approach enabled the design of an optimized system based on the proposed model, which was further refined using Matlab/Simulink.

Keywords: hybrid systems; photovoltaic system; battery system; optimization; Python; linear programming; simplex method

Citation: Guamangallo, J.; Porras, J.; Quinatoa, C.; Vaca, J.; Chiza, L. Optimal Sizing of Hybrid Generation Systems (Photovoltaic System and Energy Storage System) for Off-Grid Applications. *Eng. Proc.* **2024**, *77*, 24. https://doi.org/10.3390/engproc2024077024

Academic Editors: Jaime Cepeda and Walter Vargas

Published: 18 November 2024

1. Introduction

Currently, the development of electrical generation systems based on renewable energies is of great importance in addressing the challenges posed by the reduction of CO_2 emissions and the search for sustainable and environmentally friendly alternative energies. Energy, along with resources such as water and transportation, among other factors, has a significant impact on development. These resources often become essential services in remote areas, thereby limiting the development of rural communities. The costs associated with conventional energy resources for remote areas, such as gas, coal, and other fossil-fuel-based resources, are generally much higher than in urbanized areas due to their challenging accessibility in most cases [1].

Considering the cost of electrical service, particularly for lighting, it can be more expensive in sparsely populated rural areas compared with those with more frequent access to the conventional electrical grid. In light of this issue, renewable energy generation systems have become a crucial resource within energy systems, offering a solution to environmental problems and reducing dependence on fossil resources. Additionally, an important aspect to emphasize is that they can be highly useful in the electrification of locations and areas that do not have easy access to energy services [2].

The issues associated with the use of renewable energy generation systems relate to the unavailability of resources 24 h a day, due to their stochastic nature and factors such as weather, which limit their use in isolated areas considering their inability to meet energy demand at all times. Considering these aspects, there arises a need to develop hybrid systems based on renewable energies, in which the power system operates with two or more energy resources, with at least one being a renewable energy source. Additionally, these systems can be complemented with conventional generation systems such as a diesel generator and/or energy storage systems, in order to supply energy demand during critical moments. This ensures the continuity and quality of service [3].

Considering the development of hybrid systems, various studies have been found in the relevant state of the art. Initially, hybrid systems were considered for early applications in telecommunications stations and economic activities. They were later implemented in various countries with a focus on rural electrification for areas lacking interconnection, where the extension of different distribution and transmission networks incurred high costs, as well as in distributed generation systems [4–9].

Applications such as water pumping systems using autonomous hybrid photovoltaic–wind systems in remote rural communities are proposed in [10]. A system enhancement is proposed in [11], which involves a hybrid photovoltaic–wind power system supplemented with a fuel cell, serving as a backup for a mobile dwelling. Based on this type of system, ref. [12] presented an adaptation of a battery system for this type of configuration, allowing for battery charging during periods of excess energy in the hybrid system, as well as execution of supplementary functions.

In the context of managing hybrid generation systems, various studies have been found in the literature, for example, in [13]. It encompassed an analysis related to the main benefits and motivations regarding the implementation of hybrid generation systems, considering the various opportunities associated with different renewable resources, in contrast with the different simulation and optimization tools available for the optimal management and development of diverse energy resources. To provide a broader perspective on the development of hybrid generation systems, the study in [14] presented a comprehensive review of the state of the art, considering aspects such as system architecture, energy storage systems, auxiliary generation components, software, algorithms, and economic and reliability criteria related to the optimization of these systems. Additionally, [15] offers an analysis of various structural types, addressing design opportunities and simulation based on optimization models for hybrid generation systems utilizing renewable energy sources.

As can be observed, hybrid energy systems have a wide range of applications in addition to their development potential. Given that these are electrical power generation systems, they may interact directly or indirectly with the traditional electrical grid. Consequently, two technology categories are identified: on-grid (grid-connected) and off-grid (standalone or non-grid connected) [16].

The simplex algorithm is a well-established and efficient method for solving linear programming problems, making it highly suitable for the optimal design of hybrid generation systems involving photovoltaic (PV) and energy storage components. These systems typically require the optimization of multiple decision variables, such as the sizing of PV panels and battery storage, while adhering to constraints related to energy demand, solar irradiance, and battery performance (e.g., state of charge limits). The simplex method is particularly adept at handling linear relationships, ensuring that the system is optimized for both performance and cost-effectiveness [17].

In the context of hybrid systems, where renewable energy generation and storage must be carefully balanced to meet demand and maximize resource efficiency, the simplex algorithm offers a robust solution by efficiently navigating feasible regions of the problem space to find an optimal solution [18].

Its computational efficiency is especially valuable in real-time decision-making or iterative processes, such as those required in dynamic environments with varying solar irradiance and energy demand. By employing the simplex algorithm, designers can ensure the system is both technically and economically optimized, a crucial factor for the widespread adoption of renewable energy technologies in modern power systems [19].

The optimal design of hybrid energy systems becomes a complex task due to the difficulty of accurately predicting the performance of these energy systems. This complexity arises for various reasons. Firstly, the optimization problem involves a large number of variables. Additionally, conflicting objectives are present, making the optimization problem complex, such as cost, performance, supply and demand management, network constraints, and other factors. Furthermore, another reason arises due to coupled nonlinearities, non-convexities, and mixed-type variables, which often eliminate the possibility of using conventional optimization methods to solve this type of problem [20].

2. Hybrid Electrical Power Generation Systems

Hybrid power generation systems can be defined as the combination of two or more energy conversion elements, such as electrical power generators or energy storage elements, or two or more fuels for the same element, which, when integrated, have the capability to overcome the limitations inherent to any of them [21].

The proposed definition of hybrid systems encompasses a wide range of possibilities, characterized by the integration of multiple energy conversion systems. This includes stationary energy systems, where at least one energy conversion element leverages renewable energy sources. A critical analysis of conventional energy systems, which are typically employed, is necessary to identify potential applications where a hybrid system could be effectively utilized as a substitute, thereby enhancing the overall efficiency and sustainability.

Based on their characteristics, hybrid electric power systems have a wide range of applications. The main applications include [20]:

- Remote area AC electrical grids, where traditional grid extension is not feasible.
- Integration of distributed energy resources in existing distribution infrastructure.
- Power supply solutions for isolated, rural, or special-purpose electrical loads.

In the context of hybrid generation systems, a typical scheme is shown in Figure 1. Here, the interaction between three different generation systems based on renewable energies is illustrated, using an energy storage system to supply a load.

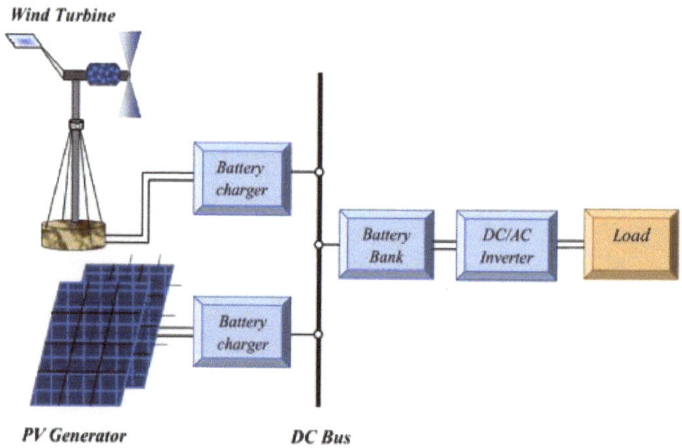

Figure 1. Typical schematic of a hybrid power generation system [13].

3. Simplex Method Algorithm for Solving Linear Programming Optimization

This linear programming method provides a tool to systematically assess or examine the vertices of a feasible region, so in order to estimate and calculate the optimal value of the objective function [22]. The simplex optimization method is an iterative technique that geometrically corresponds to moving from one so-called feasible corner point to another, until the optimal feasible point is located. Slack variables are introduced to ensure that the corner points are feasible and remain within the solution region. Algebraically, the method involves transitioning from one feasible corner point to another by repeatedly identifying the pivot column, pivot row, and consequently, the pivot element within a sequence of matrix tableaux. Upon identifying the pivot element, a new tableau is created by pivoting (using the Gauss–Jordan method) around this element. This section considers the use of slack variables and pivoting within the context of the standard maximization problem [23]. The following characteristics are considered in the optimization model:

- The objective function is linear and maximized.
- The variables are non-negative.
- Structural constraints, all of the form:

$$ax + by + \ldots \leq c, wherec \geq 0 \quad (1)$$

4. Methods

4.1. Hybrid System Modeling

The proposed methodology utilizes linear programming techniques to determine the optimal size of the photovoltaic generation system and energy storage system for an off-grid system, ensuring minimal costs and maximal efficiency. To achieve this, historical solar irradiance data and test energy consumption profiles will be utilized as inputs. The flowchart in Figure 2 provides a step-by-step illustration of the process to be developed, outlining the key stages and decision points involved in the optimization process.

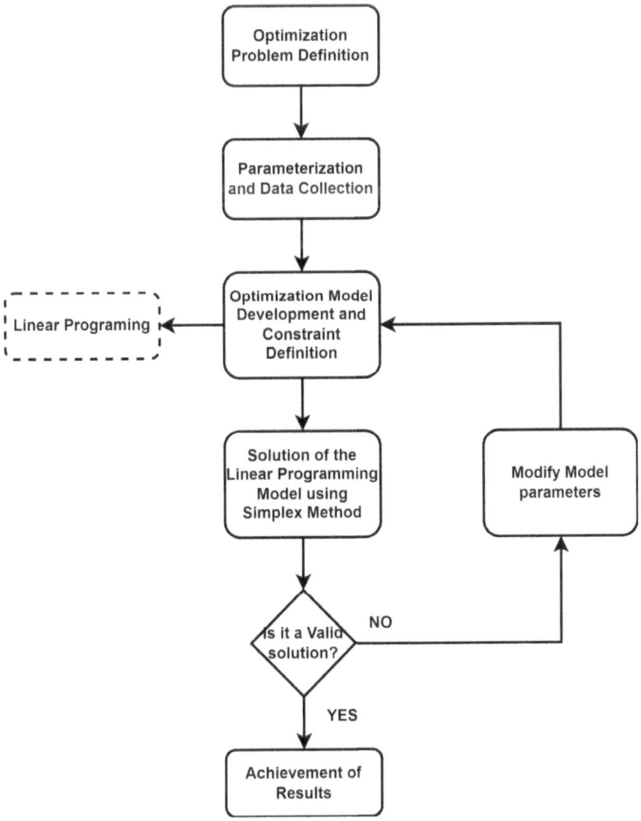

Figure 2. Diagram of the proposed methodology.

Referring to the flowchart in Figure 2, the following processes are involved:

- *Formulation of the optimization problem:* the problem is defined, and in this practical scenario, a hybrid power generation system consisting of a photovoltaic system and an energy storage system is considered.
- *Model parameterization and data acquisition:* key parameters of the photovoltaic system and energy storage system models will be identified and collected, including solar irradiance data, energy storage capacity, and electrical load profiles.

The next steps of the proposed scheme are detailed in the following section.

4.2. Hybrid Power Generation System Design and Optimization Methodology

The electrical energy generated by the photovoltaic system (PV_{size}) exhibits a direct linear relationship with the system's size, indicating that an increase in system size will result in a corresponding increase in energy production. For the purpose of designing the hybrid energy system, a total system efficiency of 20% is assumed, taking into account losses in the photovoltaic array, energy storage, and power conversion components.

The design of the hybrid energy system is based on the following key considerations:

- The design of the hybrid energy system must ensure that the sum of the photovoltaic generation (PV_{gen}) and battery discharge ($P_{discharge}$) is greater than or equal to the electrical load, guaranteeing a reliable and continuous power supply.
- In scenarios where the photovoltaic generation capacity exceeds the electrical energy demand, the excess energy can be effectively harnessed to charge the battery system (P_{charge}), thereby optimizing energy storage and reducing energy waste.
- If the excess photovoltaic generation exceeds the available capacity in the battery system, it can be assumed that the battery is fully charged, and it is necessary to reduce the photovoltaic generation. This defines the maximum amount of charge delivered.
- The amount of energy that can be discharged by the battery system must be less than the available capacity in the battery system.
- The state of charge of the battery system (SOC) is defined as $B_{state}(t)$, and the maximum charge is defined as B_{max}.
- The objective function is defined with the intention of minimizing the system cost, which is defined as the size of the photovoltaic system multiplied by the cost of the photovoltaic system, plus the cost of the battery system, which is related to the battery capacity and its cost, in addition to a penalty for the cumulative discharge of the battery system.

$$min_{PV-Bat_{size}}(PV_{size} \times cost_{PV} + B_{max} \times cost_{bat} + penalty_{value}) \quad (2)$$

For the solution of the linear programming model, Python programming tools have been utilized. In this context, the use of programming tools such as Python becomes highly valuable for addressing and solving optimization problems. Additionally, as an open-source software, it generally does not impose a significant computational burden in terms of performance.

5. Results and Discussion

The proposed scenario comprises a comprehensive dataset of 24 h solar irradiation and electrical energy demand profiles, carefully selected to facilitate in-depth analysis, testing, and validation of the proposed system.

The demand profile illustrated in Figure 3 represents a typical residential load pattern, characterized by a peak demand of approximately 1.4 kW at 21:00 p.m., and a minimum demand of around 0.2 kW, showing residential energy consumption characteristics, considering as a reference the load indexes used in the study carried out in [24].

The test solar irradiance profile depicted in Figure 4 showcases a representative diurnal cycle, with maximum irradiance levels attained during the daylight period, specifically between approximately 09:00 a.m. and 16:00 p.m., thereby illustrating the typical daily variation in solar radiation.

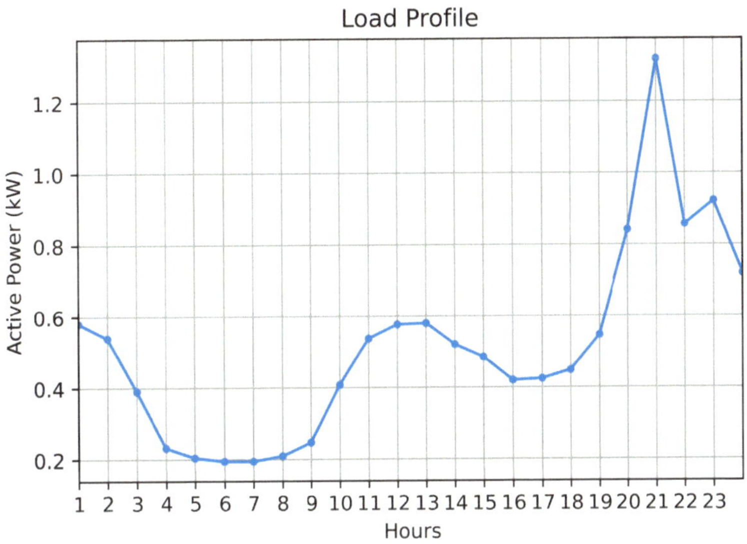

Figure 3. Load profile. Test data.

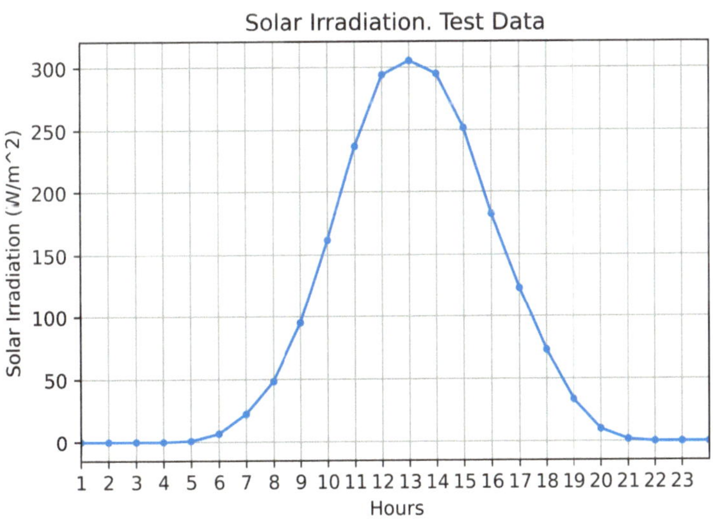

Figure 4. Solar irradiation. Test data.

The analysis of the irradiance profile indicates that there are specific hours of the day when photovoltaic energy generation is not viable, requiring the utilization of energy storage systems to ensure a stable energy supply. Notably, during critical periods such as from 18:00 p.m. onwards, the absence of irradiance coincides with peak electrical energy demand, as depicted in Figure 3, highlighting the need for energy storage solutions to mitigate this mismatch.

By applying Equation (2), a comprehensive estimation of the photovoltaic system's capacity is conducted, taking into account the monthly energy demand, to estimate the annual energy yield and ensure a reliable and efficient energy supply.

$$P_{PV} = \frac{Energy_{monthly} * 12}{fp * 8760} \quad (3)$$

By analyzing the demand profile presented in Figure 3 and applying a plant factor of 0.2, which is representative of photovoltaic systems, the estimated power rating of the photovoltaic system is determined to be 25.48 kW, ensuring a reliable and efficient energy supply.

For the battery system, a 0% State of Charge (SOC) has been set as the discharge limit, resulting in the following outcomes.

The operation of the hybrid generation system, as illustrated in Figure 5, shows a balanced interaction between energy generation and demand. During periods of maximum solar irradiance, the battery system charges, while during peak energy demand, the stored energy is discharged. This results in a reliable and resilient energy supply that meets the expected performance standards.

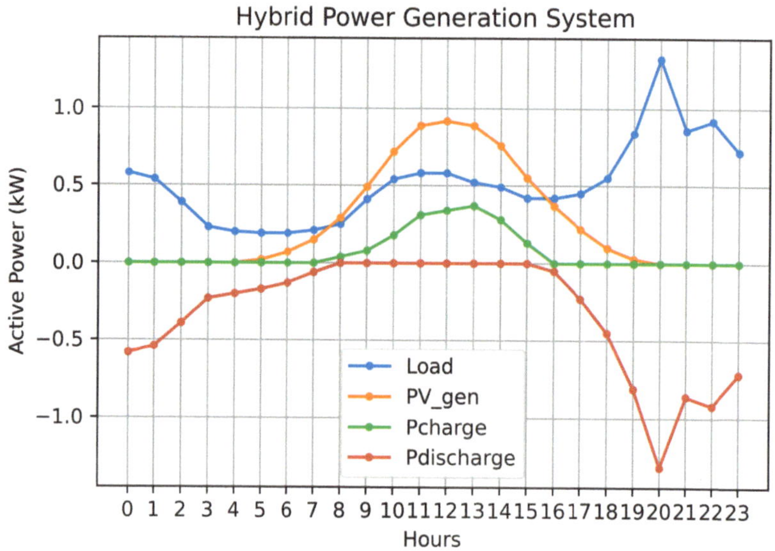

Figure 5. Results obtained. Hybrid system optimization.

Table 1 presents a comprehensive summary of the hybrid generation system sizing results, tailored to the specific conditions and requirements outlined in the analysis.

Table 1. Hybrid Generation System Sizing Results.

System	Value [kW]
PV system	15.1
Energy storage system	5.9

The results presented in Table 1 demonstrate that the optimized hybrid generation system design achieves a considerable reduction of approximately 40% in the initial photovoltaic system size, while also accurately determining the required battery system capacity. Furthermore, as depicted in Figure 5, the battery system exhibits the capability to reliably supply the demand during the specific time instants when it is required, thereby ensuring a robust and efficient energy supply.

To validate the obtained results, the optimization model developed in Simulink has been adapted from [25] (Figure 6). Using input data such as energy demand, solar resource availability, and a loss factor, the model estimates the design of the photovoltaic system. This process outputs key parameters, including the system's capacity, the required number of solar panels, and the battery system capacity in ampere-hours (Ah).

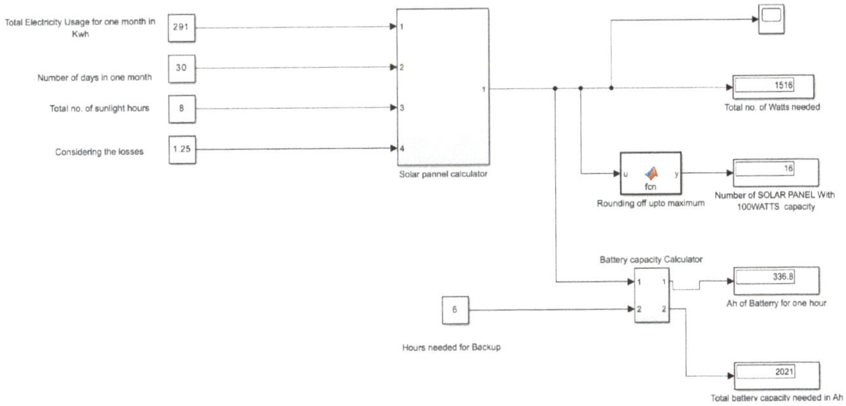

Figure 6. Solar panel calculator. Adapted from [25].

The operating conditions considered include a 30 day period, 8 h of total daily solar irradiation, and a loss factor of 1.25. Additionally, for the battery system, a backup duration of 6 h is taken into account. The obtained results are detailed in Table 2.

Table 2. Solar panel results.

Parameter	Value
Active Power of PV system	15.16 [kW]
Number of Solar panel	16
Ah of Battery per hour	336.8 [Ah]
Total battery capacity in Ah	2021 [Ah]

The results indicate that a photovoltaic (PV) system with 16 panels is required. For the energy storage system, the battery capacity must be at least 336.8 Ah.

6. Conclusions

An optimization model based on linear programming was developed using the simplex method. This approach does not demand significant computational resources and readily conforms to the characteristics of the mathematical model proposed for the hybrid generation system. Given the optimization model's low computational demands, it has been employed both for estimating the energy production of the photovoltaic system relative to solar resource availability (solar irradiation) and for the design of the generation system. The latter includes considerations for the number of solar panels and the energy storage system's capacity in ampere-hours (Ah). In conclusion, the adoption of this linear programming-based optimization model, particularly with the simplex method, demonstrates its efficacy in balancing computational efficiency with the accuracy of energy estimation and system design. This approach not only streamlines the modeling process, but also enhances the feasibility of implementing hybrid generation systems in scenarios where computational resources are limited, thereby promoting broader adoption of renewable energy solutions.

Author Contributions: Methodology, software, investigation: J.G. and L.C.; Conceptualization: J.P.; Project administration: C.Q. and J.V.; Supervision, review, and editing: L.C. All authors have read and agreed to the published version of the manuscript.

Funding: This research received no external funding.

Institutional Review Board Statement: Not applicable.

Informed Consent Statement: Not applicable.

Data Availability Statement: The data presented in this study are available upon request from the corresponding author. The data are not publicly available due to intellectual property of universities.

Conflicts of Interest: The authors declare no conflicts of interest.

Abbreviations

The following abbreviations are used in this manuscript:

fp	plant factor
$Energy_{monthly}$	Monthly energy consumption
P_{PV}	PV system power capacity
Ah	Ampere-hour

References

1. Bouhier,A. Guía Para la Optimización de Sistemas Híbridos de Energía, Dimensionamiento, Operación y Control. Master's Thesis, Universidad Nacional del Rosario, Rosario, Argentina, 2017.
2. REN21. *Renewables Global Status Report. Technical Report, Renewable Energy Policy Network for the 21st Century*; REN21: Paris, France, 2016.
3. Nallolla, C.A.; Perumal, V. Optimal Design of a Hybrid Off-Grid Renewable Energy System Using Techno-Economic and Sensitivity Analysis for a Rural Remote Location. *Sustainability* **2022**, *14*, 15393. [CrossRef]
4. Naikodi, A. Solar-wind hybrid power for rural Indian cell sites. In Proceedings of the 2010 IEEE International Energy Conference, Manama, Bahrain, 18–21 December 2010; pp. 69–72. [CrossRef]
5. Elhassa, Z.A.M.; Moh Zain, M.F.; Sopian, K.; Awadalla, A. Design of Hybrid Power System of Renewable Energy for Domestic Used in Khartoum. *J. Appl. Sci.* **2011**, *11*, 2270–2275. [CrossRef]
6. Saheb-Koussa, D.; Haddadi, M.; Belhamel, M. Economic and technical study of a hybrid system (wind–photovoltaic–diesel) for rural electrification in Algeria. *Appl. Energy* **2009**, *86*, 1024–1030. [CrossRef]
7. Nandi, S.K.; Ghosh, H.R. Prospect of wind—PV-battery hybrid power system as an alternative to grid extension in Bangladesh. *Energy* **2010**, *35*, 3040–3047. [CrossRef]
8. Al-Badi, A.H. Hybrid (solar and wind) energy system for Al Hallaniyat Island electrification. *Int. J. Sustain. Energy* **2011**, *30*, 212–222. [CrossRef]
9. Dalwadi, P.; Shrinet, V.; Mehta, C.R.; Shah, P. Optimization of solar-wind hybrid system for distributed generation. In Proceedings of the 2011 Nirma University International Conference on Engineering, Ahmedabad, India, 8–10 December 2011; pp. 1–4. [CrossRef]
10. Vick, B.D.; Neal, B.A. Analysis of off-grid hybrid wind turbine/solar PV water pumping systems. *Sol. Energy* **2012**, *86*, 1197–1207. [CrossRef]
11. Eroglu, M.; Dursun, E.; Sevencan, S.; Song, J.; Yazici, S.; Kilic, O. A mobile renewable house using PV/wind/fuel cell hybrid power system. *Fuel Energy Abstr.* **2011**, *36*, 7985–7992. [CrossRef]
12. Yazici, M.S.; Yavasoglu, H.A.; Eroglu, M. A mobile off-grid platform powered with photovoltaic/wind/battery/fuel cell hybrid power systems. *Int. J. Hydrogen Energy* **2013**, *38*, 11639–11645. [CrossRef]
13. [CrossRef] Mohammed, Y.S.; Mustafa, M.W.; Bashir, N. Hybrid renewable energy systems for off-grid electric power: Review of substantial issues. *Renew. Sustain. Energy Rev.* **2014**, *35*, 527–539, ISSN 1364-0321. [CrossRef]
14. León Gómez, J.C.; De León Aldaco, S.E.; Aguayo Alquicira, J. A Review of Hybrid Renewable Energy Systems: Architectures, Battery Systems, and Optimization Techniques. *Eng* **2023**, *4*, 1446–1467. [CrossRef]
15. Lazarov, V.D.; Notton, G.; Zarkov, Z.; Bochev, I. Hybrid Power Systems with Renewable Energy Sources—Types, Structures, Trends for Research and Development. In Proceedings of the International Conference ELMA2005, Sofia, Bulgaria, 15–16 September 2005.
16. Li, J.; Liu, P.; Li, Z. Optimal design of a hybrid renewable energy system with grid connection and comparison of techno-economic performances with an off-grid system: A case study of West China. *Comput. Chem. Eng.* **2022**, *159*, 107657. [CrossRef]
17. Dantzig, G.B. *Linear Programming and Extensions*; Princeton University Press: Princeton, NJ, USA, 1963
18. Bertsimas, D.; Tsitsiklis, J.N. *Introduction to Linear Optimization*; Athena Scientific: Nashua, NH, USA, 1997
19. Bazaraa, M.S.; Sherali, H.D.; Shetty, C.M. *Nonlinear Programming: Theory and Algorithms*; John Wiley & Sons: Hoboken, NJ, USA, 2010.
20. Bourennani, F.; Rahnamayan, S.; Naterer, G.F. Optimal Design Methods for Hybrid Renewable Energy Systems. *Int. J. Green Energy* **2013**, *12*, 148–159. [CrossRef]
21. Manwell, J.F. *Hybrid Energy Systems, en Encyclopedia of Energy*; Cleveland, C.J., Ed.; Elsevier: New York, NY, USA, 2004; pp. 215–229. [CrossRef]
22. Velinov, A.; Gicev, V. Practical application of simplex method for solving linear programming problems. *J. Appl. Math. Inform.* **2018**, *1*, 7–15.

23. Linear Programming: The Simplex Method (2020). Purdue University Northwest. Available online: https://www.pnw.edu/wp-content/uploads/2020/03/attendance5-1.pdf (accessed on 15 October 2024).
24. Chiza Segovia, L.L. Evaluación de la Seguridad dinámica en línea en Base a una Interfaz de Protocolo de Comunicación OPC de Arquitectura Unificada. 127 Páginas. Master's Thesis, National Polytechnic School, Quito, Ecuador, 2023
25. Sabareeshwaran. Solar Panel and Battery Capacity needed calculator. Matlab Central File Exchange. 2024. Available online: https://www.mathworks.com/matlabcentral/fileexchange/43449-solar-panel-and-battery-capacity-needed-calculator (accessed on 5 August 2024).

engineering proceedings

Proceeding Paper

Optimal Load Shedding Scheme Considering the Dynamic Frequency Response †

Carlos Lozada [1,*], Walter Vargas [2], Nelson Granda [2] and Marlon Chamba [1]

1 Operador Nacional de Electricidad (CENACE), Quito 17211991, Ecuador; mchamba@cenace.gob.ec
2 Department of Electrical Engineering, Faculty of Electrical and Electronic Engineering, Campus José Rubén Orellana Ricaurte, Escuela Politécnica Nacional, Quito 170525, Ecuador; walter.vargas@epn.edu.ec (W.V.); nelson.granda@epn.edu.ec (N.G.)
* Correspondence: clozada@cenace.gob.ec
† Presented at the XXXII Conference on Electrical and Electronic Engineering, Quito, Ecuador, 13–15 November 2024.

Abstract: A power system is never in a steady state due to continuous load variations, disturbances, maneuvers, and the operation of protection systems. A generation deficit causes a frequency drop in the system that must be controlled. If this frequency deviation is not properly managed, it can result in the loss of synchronism between generators and, eventually, lead to a partial or even total system collapse. This article presents a load shedding scheme applied to the IEEE 39-bus New England system. The scheme considers an N-1 contingency space to evaluate the dynamic frequency response, aiming to determine the appropriate settings for low-frequency relays activated by the rate of change of frequency (*ROCOF*).

Keywords: load shedding; optimal load shedding scheme; dynamic frequency response; rate of change of frequency; *ROCOF*

Citation: Lozada, C.; Vargas, W.; Granda, N.; Chamba, M. Optimal Load Shedding Scheme Considering the Dynamic Frequency Response. *Eng. Proc.* **2024**, *77*, 25. https://doi.org/10.3390/engproc2024077025

Academic Editor: Jaime Cepeda

Published: 18 November 2024

1. Introduction

Power system stability is defined as the ability of a power system to remain in an operational equilibrium state under normal operating conditions and to evolve to an acceptable equilibrium state after a disturbance [1]. A power system can be vulnerable to instability problems when operating near its physical limits; these problems must be controlled, or they can lead to partial or even total system collapse. Traditionally, the following types of stabilities exist: angular stability, voltage stability, and frequency stability.

Frequency stability is associated with the balance between power generation and electrical demand. When there is a variation in power generation, it causes a deviation in the system frequency, which can result in values outside the safe operating ranges. To maintain stability, power–frequency controllers are incorporated to regulate the balance between generation and demand, ensuring that the frequency remains within appropriate operating ranges [2].

Power–frequency control can be organized into three levels: primary, secondary, and tertiary. The main characteristics of each control level are their specific operational time ranges and associated variables. Primary control aims to limit the frequency deviation during a contingency, restoring the balance between power generation and electrical demand by bringing the system to a new operating point where the frequency differs from the nominal value. This control operates within a time range of 2 to 30 s. The primary frequency response results from the interaction of generator inertia, load damping, speed regulators, and other devices that supply energy to the system, such as battery energy storage systems (BESSs) [3].

Secondary control operates in a time range of 30 s to 10 min. It operates within the control area, considering the frequency and power exchange with neighboring areas, and it is implemented by Automatic Generation Control (AGC). Finally, tertiary control operates in a time margin greater than 10 min. It acts in the scope of a large electrical system, seeking optimized load sharing to ensure sufficient energy reserves [3]. During the operation of the EPS, situations may arise in which imbalances between generated power and consumed power are significantly pronounced. In these circumstances, the mechanical valves controlled by the governors may be too slow to react in time before the frequency crosses acceptable operating limits. This may violate safe operating parameters, which could result in damage to the generating units [1,4]. In these cases, remedial strategies consisting of under-frequency load shedding or generation tripping are designed to prevent possible damage to the generating machines and the collapse of the system.

Load shedding strategies can be classified into the following categories: conventional, computational, and adaptive [5]. These categories identify the value of frequency and *ROCOF* for the operation of their algorithms. Strategies involving *ROCOF* have been investigated to achieve better results by identifying the minimal amount of load that needs to be disconnected in critical scenarios.

Reference [6] proposes a scheme that estimates the rate of change of frequency (*ROCOF*) at the center of inertia (CoI) and, consequently, the size of the generation loss using local frequency measurements. An innovative turning point detection technique is presented to eliminate the effect of local frequency oscillations.

Reference [7] presents a methodology based on time-domain simulations to obtain the frequency evolution in response to generation loss across various operating scenarios. These scenarios are classified based on the rate of change of frequency (*ROCOF*). A prioritized list of loads for disconnection is generated. Using these results, a load shedding scheme is proposed, and its performance is compared through dynamic simulations.

Reference [8] proposes sizing the storage capacity for virtual energy contribution based on *ROCOF*. In this article, *ROCOF* measurements are identified locally, and calculations are performed using center of inertia parameters. This technique is applied to specific areas of the system; however, it does not provide a characterization of the system under different contingencies.

Reference [9] describes a methodology for identifying characteristic *ROCOF*s (rate of change of frequency) in a power system under N-1 contingencies. This methodology is applied to the IEEE 39-bus New England system to determine the number of *ROCOF*s that may arise during dynamic frequency behavior. Based on these results, appropriate load shedding measures can be established.

2. Methodology

The rate of change of frequency (*ROCOF*) is a critical indicator of the robustness of an electrical power system. Analyzing the dynamic behavior of this parameter enables the estimation of power imbalances within the system, which can be calculated using the following equation.

$$ROCOF = \frac{\Delta P}{S} \cdot \frac{f}{2H} \tag{1}$$

where ΔP represents the power imbalance caused by an event, f denotes the nominal frequency of the EPS, H is the system's total inertia constant after the event, and S refers to the system's nominal power [10]. The *ROCOF* calculation, as part of the proposed methodology, is performed using a 0.5-s window, following the recommendations provided in [11–13].

The proposed methodology employs a reduced first-order model to represent the system's frequency response. This model facilitates the calculation of the dynamic frequency response to an imbalance between generation and load, providing an effective tool for stability analysis. Comprising a turbine, a speed governor, a synchronous generator, and a load, the model allows for a more efficient evaluation of the influence of these elements on frequency stability. The process flow of the methodology is illustrated in Figure 1.

Figure 1. Stages of the proposed methodology.

2.1. Representative Events of the Contingency Space

An equivalent first-order model is implemented to characterize the dynamic frequency response using the following variables: load imbalance (*ΔPo*), system control regulation (*R*), and load damping (*D*). It is assumed that the dynamic capacity available is a fraction (*F*) of the immediate reserve capacity, while the complementary fraction (1 − *F*) is represented as a first-order lag with a time constant (*T*). Additionally, (*Km*) is a gain constant for the spinning reserve of the generators. The proposed model considering N generators is illustrated in Figure 2 [14].

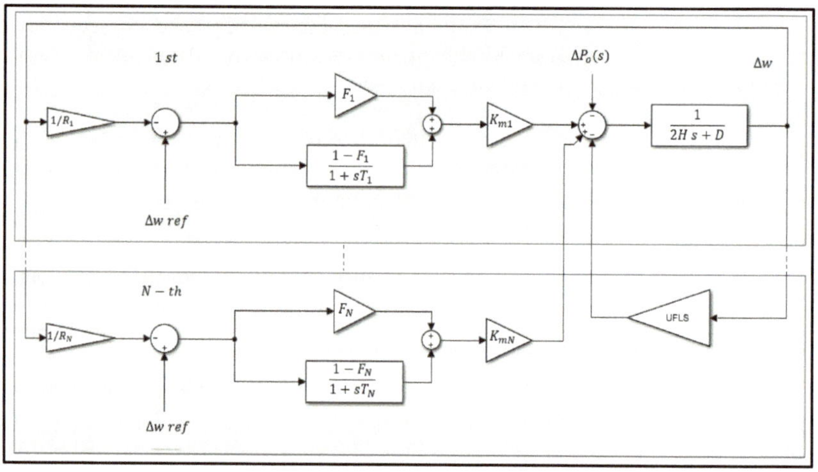

Figure 2. First-order model for dynamic frequency response.

Reference [9] details the methodology used to develop a first-order model for the dynamic frequency response. Monte Carlo simulations are used to generate possible operational scenarios, which are then subjected to N − 1 contingencies to evaluate their behavior. Subsequently, a database is created that includes parameters such as the rate of change of frequency (*ROCOF*), the system's equivalent inertia before and after the contingency, the generation loss power during the event, and the NADIR. Using this database, an unsupervised classification algorithm known as Clustering is applied. The results obtained with this methodology are shown in Table 1 and Figure 3.

Table 1. Representative events of the contingency space.

Groups	*ROCOF* [Hz/s]	Hpre	Hpos	Pout	NADIR
Cluster 1	−0.38	4.89	4.57	403.75	58.72
Cluster 2	−0.30	4.89	4.63	322.99	59.08
Cluster 3	−0.25	4.88	4.66	270.70	59.31

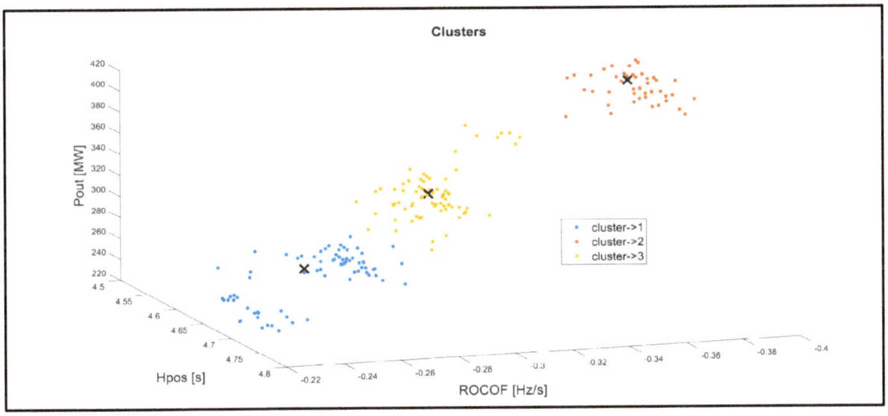

Figure 3. Correlation of variables *ROCOF*, Hpos, and Pout.

2.2. Optimization

After identifying the events that represent the behavior of the established database, load shedding optimization is performed using the "surrogateopt" algorithm available in MATLAB 2021a. This algorithm aims to minimize the percentage of load shed, with the objective function specifically designed to achieve this minimization.

2.2.1. Mathematical Modeling of the Optimization Problem

Problem definition:

$$FO : min\left\{\sum_{i=1}^{n} C_i(\Delta t_i) \times P_{load_i}\right\} \tag{2}$$

where i: representative scenario; P_{load_i}: system demand in scenario I; Δt_i: variation in relay actuation time and measurement; $C_i(\Delta t_i)$: percentage of load shed; n: number of representative scenarios.

Decision Variables:

$$C_i(\Delta t_i) : \text{ Percentage of load shed.} \tag{3}$$

Constraints:

$$F_{ss_i} > 59.2 \text{ Hz} \tag{4}$$

$$t_{i+1} - t_i < 5 \text{ ms} \tag{5}$$

$$0 < C_i < 1 \tag{6}$$

where F_{ss}: Stabilization Frequency; t_i: Action Time of Scenario i.

ROCOF calculation is performed using a 0.5 s time window. The database must be restructured by excluding scenarios where the system returns to safe levels after a contingency. The selection criterion is based on the *NADIR*: scenarios in which the frequency does not drop below the 59.4 Hz threshold are considered safe and removed from the database. The 59.4 Hz value corresponds to the first stage of under-frequency load shedding, as recommended in [15].

2.2.2. Definition of *ROCOF* Settings for Relays

As a result of implementing the previous stages, the *ROCOF* values for the centroids of each characteristic group within the system's N − 1 contingency universe are obtained and denoted by the symbol x in Figure 3.

These values provide statistical insights into potential events within the system. For the optimization model, it is assumed that the *ROCOF* threshold required to activate the protection relay has already been identified.

In this context, the frequency derivative value for each centroid must be adjusted to cover the associated cluster elements. To achieve this, the risk management criterion based on the mean and standard deviation proposed in [16] is used. Figure 4 illustrates the effect of this adjustment; the solid black lines represent the mean of the *ROCOF* values for the associated group; the dashed lines represent the standard deviation of the *ROCOF* values for the associated group; the colored solid lines represent the *ROCOF* set for each associated group. Additionally, the statistical analysis for the *ROCOF* values of group 3 (blue) is shown in the box.

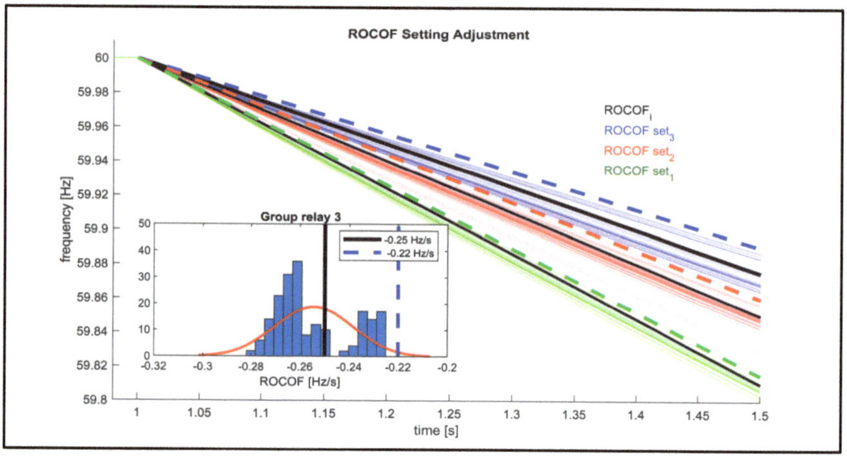

Figure 4. *ROCOF* setting adjustment.

Criterion Based on the Mean and Standard Deviation:

$$ROCOF_{set} = u - 2\sigma \tag{7}$$

where $ROCOF_{set}$: the variable considered for optimization; u: the mean of the *ROCOF* values for the associated group; and σ: the standard deviation of the *ROCOF* values for the associated group.

For the implementation of the heuristic optimization algorithm, the stochastic variation of the percentage of load that must be disconnected based on the demand at that moment is considered. The frequency behavior is evaluated for each of the defined *ROCOF* settings, given that three distinct settings must be coordinated to ensure the selectivity of the protections.

In this context, a model of a *ROCOF*-activated low-frequency protection relay is implemented for each characteristic group. The objective is to have a stochastic parameter that determines the delay time each relay must consider before sending the trip signal, thereby ensuring the selectivity of the protection settings.

3. Application of the Methodology and Analysis of Results

The test system used is an IEEE 39 bus bar system. This system has nineteen loads, ten generators, twelve transformers, and thirty-five transmission lines, whose data can be found in reference [17]. In the present study, a specific modification was made that involves dividing the original generators into eighteen to increase the sensitivity in the analysis of N-1 contingency events. After applying the proposed methodology, the results are presented in Table 2.

Table 2. Settings of each low-frequency relay based on representative groups.

Groups	*ROCOF* Set [Hz/s]	Load Shed [%]	Delay Δt [s]
Relay 1	−0.37	5.8	0.21
Relay 2	−0.28	2.2	0.51
Relay 3	−0.22	1.1	0.84

To evaluate the methodology, all scenarios generated in the *MCS* were analyzed by placing the protection relays with the obtained settings. Figure 5 shows a histogram comparing the stabilization frequency (*FSS*) values and the lowest frequency point (*NADIR*) before and after implementing the protection relays. This comparison demonstrates a shift towards safer frequency values, thereby validating the implemented methodology.

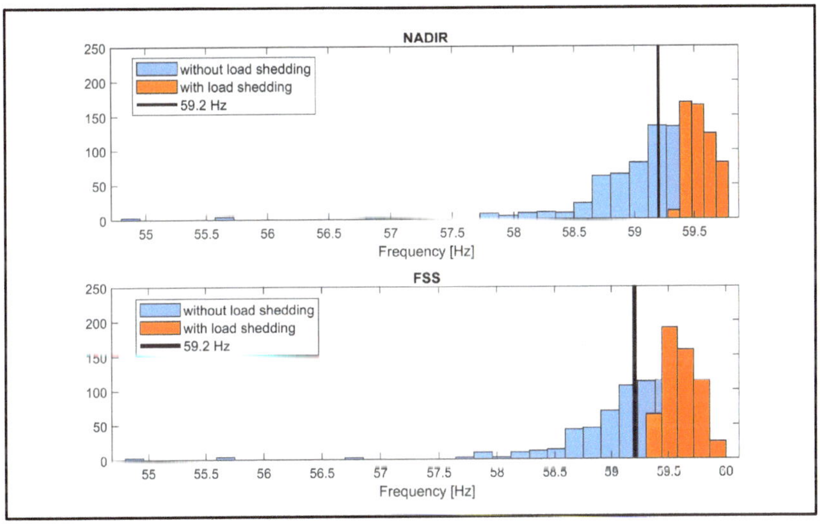

Figure 5. Comparison histograms of the methodology's impact on frequency.

4. Conclusions and Recommendations

The application of the present methodology assumes that the speed regulators are functioning correctly and that there is an accurate identification of machine variables such as machine inertia, power limits, and regulator droop, among others explained in the methodology. Therefore, a preliminary parameter validation stage must be conducted before applying this methodology.

The present methodology has the characteristic of observing different conditions of the electrical system as long as the input variables, such as generation loss power, *ROCOF*, and post-contingency equivalent inertia, are identified. These variables can be easily obtained by reading the current state of the system. This characteristic justifies its application in systems with a high penetration of non-inertial and low-inertia generation, such as photovoltaic and wind power plants, and even considering synthetic inertia responses provided by batteries.

In the present methodology, equivalents of the generator-governor groups were implemented to simplify the analysis. These equivalents can be used to perform a zonal analysis in systems with large generation parks. This requires a preliminary analysis to identify zones with very similar frequency behavior. However, for a more detailed analysis, it is recommended to use the entire network, although this would require significantly higher computational capacity.

Author Contributions: Methodology, software, and investigation: C.L.; Conceptualization: W.V.; Project administration: N.G.; Supervision, review, and editing: M.C. All authors have read and agreed to the published version of the manuscript.

Funding: This research received no external funding.

Institutional Review Board Statement: Not applicable.

Informed Consent Statement: Not applicable.

Data Availability Statement: Data are contained within the article.

Conflicts of Interest: The authors declare no conflicts of interest.

Abbreviations

ROCOF	Rate of Change of Frequency
EPS	Electric Power System
UFLS	Under Frequency Load Shedding
Governors	Speed Regulators
BESS	Battery Energy Storage System
NADIR	Lowest Point of the Frequency
AGC	Automatic Generation Control
MCS	Monte Carlo Simulation
Hpre	Equivalent Inertia before the Contingency
Hpos	Equivalent Inertia after the Contingency
Pout	Lost Generation Power
ΔP_0	Load Power Variation
Δw	Speed Variation
Km_i, F_i, T_i, and R_i	Meters of the First-Order Reduced Model
H_{eq}	Equivalent System Inertia
D	Load Damping

References

1. Kundur, P.; Balu, N.J. *Power System Stability and Control*; McGraw-Hill: New York, NY, USA, 1994; ISBN 978-0-7803-3463-2.
2. Vargas, W.; Chamba, S.; Torre, A.D.L.; Echeverría, D. Protocolo de pruebas y validación de reguladores de velocidad—Aplicación práctica en la central hidroeléctrica Delsitanisagua. *Rev. Téc. Energ.* **2022**, *19*, 22–33. [CrossRef]
3. Chamba, S.; Vargas, W.; Echeverría, D.; Riofrio, J. Regulación Primaria de Frecuencia Mediante Sistemas de Almacenamiento de Energía con Baterías en el Sistema Eléctrico Ecuatoriano. *Rev. Téc. Energ.* **2022**, *19*, 13–21. [CrossRef]
4. Rudez, U.; Mihalic, R. Analysis of Underfrequency Load Shedding Using a Frequency Gradient. *IEEE Trans. Power Deliv.* **2011**, *26*, 565–575. [CrossRef]
5. Laghari, J.A.; Mokhlis, H.; Bakar, A.H.A.; Mohamad, H. Application of Computational Intelligence Techniques for Load Shedding in Power Systems: A Review. *Energy Convers. Manag.* **2013**, *75*, 130–140. [CrossRef]
6. Sun, M.; Liu, G.; Popov, M.; Terzija, V.; Azizi, S. Underfrequency Load Shedding Using Locally Estimated *ROCOF* of the Center of Inertia. *IEEE Trans. Power Syst.* **2021**, *36*, 4212–4222. [CrossRef]
7. Granda, N.; Jácome, V. Esquema Automático de Alivio de Carga para Sistemas Eléctricos que sirven a Plataformas Petroleras. *Rev. Téc. Energ.* **2023**, *19*, 58–68. [CrossRef]
8. Alonso Sørensen, D.; Vázquez Pombo, D.; Torres Iglesias, E. Energy Storage Sizing for Virtual Inertia Contribution Based on *ROCOF* and Local Frequency Dynamics. *Energy Strategy Rev.* **2023**, *47*, 101094. [CrossRef]
9. Lozada, C.X.; Vargas, W.A.; Granda, N.V.; Chamba, M.S. Methodology for Identifying Representative Rates of Change of Frequency (*ROCOFs*) in an Electric Power System against N-1 Contingencies. *Eng. Proc.* **2023**, *47*, 8. [CrossRef]
10. Egido, I.; Fernandez-Bernal, F.; Centeno, P.; Rouco, L. Maximum Frequency Deviation Calculation in Small Isolated Power Systems. *IEEE Trans. Power Syst.* **2009**, *24*, 1731–1738. [CrossRef]
11. Wu, L. Power System Frequency Measurement Based Data Analytics and Situational Awareness. Ph.D. Thesis, University of Tennessee, Knoxville, TN, USA, 2018.
12. Wright, P.S.; Davis, P.N.; Johnstone, K.; Rietveld, G.; Roscoe, A.J. Field Measurement of Frequency and *ROCOF* in the Presence of Phase Steps. *IEEE Trans. Instrum. Meas.* **2019**, *68*, 1688–1695. [CrossRef]
13. ENTSOE. *Rate of Change of Frequency (ROCOF) Withstand Capability: ENTSO-E Guidance Document for National Implementation for Network Codes on Grid Connection*; ENTSOE: Brussels, Belgium, 2017.
14. Aik, D.L.H. A General-Order System Frequency Response Model Incorporating Load Shedding: Analytic Modeling and Applications. *IEEE Trans. Power Syst.* **2006**, *21*, 709–717. [CrossRef]

15. Centro Nacional de Control de Energía. *Determinación Del Esquema De Alivio De Carga Por Baja Frecuencia Para El Sistema Nacional Interconectado*; CENACE: Quito, Ecuador, 2022.
16. Chamba, M.; Vargas, W.; Cepeda, J. Evaluación probabilística de la estabilidad transitoria considerando la incertidumbre de la demanda y gestión del riesgo. *Rev. Téc. Energ.* **2018**, *15*, 1–10. [CrossRef]
17. Athay, T.; Podmore, R.; Virmani, S. A Practical Method for the Direct Analysis of Transient Stability. *IEEE Trans. Power Appar. Syst.* **1979**, *PAS-98*, 573–584. [CrossRef]

engineering proceedings

MDPI

Proceeding Paper

Directional Overcurrent Protection Design for Distribution Network: CIGRE European Medium-Voltage Benchmark Network †

Le Nam Hai Pham [1,*], Veronica Rosero-Morillo [2] and Francisco Gonzalez-Longatt [1,3]

1 Digital Energy Systems Laboratory, University of South-Eastern Norway, 3918 Porsgrunn, Norway; fglongatt@fglongatt.org
2 Institute of Electrical Energy IEE, National University of San Juan, San Juan 5400, Argentina; vrosero@iee.unsj.edu.ar
3 Centre for Renewable Energy Systems Technology, Loughborough University, Loughborough LE11 3TU, UK
* Correspondence: le.pham@usn.no

† Presented at the XXXII Conference on Electrical and Electronic Engineering, Quito, Ecuador, 12–15 November 2024.

Abstract: Overcurrent protection is a fundamental aspect of power system protection and is widely utilised in distribution networks. The increasing integration of renewable energy sources (RESs) into the conventional power system has introduced operating challenges due to the variability in fault directions. As a result, protection engineers must not only adjust basic parameters such as pickup current or time delay, but also carefully evaluate the directional protection to align with specific protection objectives and the devices being protected. The complexity of considering multiple aspects in the protection system design can pose challenges for operators in configuring their settings. Therefore, it is necessary to have a systematic approach for protection system design. For this purpose, this paper proposes a methodology for protection system design focusing on directional overcurrent protection setting configuration with detailed implementation. A well-known distribution network, the CIGRE European (EU) medium-voltage (MV) benchmark network, is used to test and validate the proposed methodology with the support of DIgSILENT PowerFactory version 2023 SP1. This article provides a useful document for the configuration of overcurrent protection systems in order to prepare for the challenges arising from the high integration of RESs in the future grid.

Keywords: CIGRE EU MV network; distribution network; directional; overcurrent protection; RES; testing; validation

Citation: Pham, L.N.H.; Rosero-Morillo, V.; Gonzalez-Longatt, F. Directional Overcurrent Protection Design for Distribution Network: CIGRE European Medium-Voltage Benchmark Network. *Eng. Proc.* **2024**, 77, 26. https://doi.org/10.3390/engproc2024077026

Academic Editor: Jaime Cepeda

Published: 18 November 2024

1. Introduction

The transition toward clean energies has led to the widespread integration of renewable energy sources (RESs) in many sectors, particularly power systems [1]. This transformation offers substantial advantages across economic [2], technological [3], and environmental [4] domains. Nevertheless, alongside these benefits, the power system is facing many challenges in management and operation due to the inherent uncertainty and intermittency of power generation from RESs [5]. RESs such as wind and solar are dependent on weather conditions, leading to fluctuating power generation. This variability complicates the task of protection coordination, where devices like relays, circuit breakers, and fuses must distinguish between normal operational changes and fault conditions [6]. In addition, RESs contribute another source of fault current to the network, potentially increasing the total fault level while changing the magnitude and direction of fault currents [7]. Indeed, many RESs, especially those connected via power electronics, have different fault current characteristics. They can rapidly change the magnitude of their fault contribution and, in some cases, even reverse the direction of fault current flow [8]. Therefore, this requires the protection system to be adapted to this transformation.

In recent years, there has been much research on RES penetration influencing the protection system [9,10]. The literature review in this field can be subdivided into two main categories. The first category delves into the short circuit analysis under the influences of RES penetration within the system. Accordingly, the authors in [11] investigated the transient response of short circuit current fed from non-synchronous generators. The short circuit currents in the high integration of RESs witnessed a significant change in magnitude and direction compared to the observed in conventional power systems without integration. Similarly, the authors in [12] believe that the increase in short circuit level can burden the protection system and require solutions to address this issue. Based on that, some groups of researchers are focusing their research on the second category, protection system configuration according to the short circuit characteristic. Research on this category can be highlighted at [13,14]. In [15], the combination of circuit breakers and protection relays is proposed for the CIGRE MV EU benchmark distribution network. Similarly, in paper [16], in the context of overcurrent protection, the authors defined the relay settings based on power system protection theory. However, the validation of the proposed protection system design is still missing. To address the lack of protection relay testing before actual implementation, there are some proposed testing methods such as real-time simulation [17] and hardware-in-the-loop [18]. The relevant devices and hardware are emphasised in these methods. The presented literature shows a lack of general overview of protection system configuration settings. Therefore, it is essential to have a systematic approach to configure protection systems in a future-extendable manner in the scenario of more RES integration.

To meet this need for a comprehensive protection system design, this paper presents a methodology for configuring a protection system that adapts to rapid changes in the future power system. Accordingly, an overcurrent protection system, which is one of the most popular features in power system protection, is implemented in this paper. The design of overcurrent protection is executed for a well known test system, the CIGRE MV EU benchmark distribution system, under different operating conditions. With the support of DIgSILENT PowerFactory, the short circuit is analysed and used to configure the overcurrent protection. The proposed protection system is validated with the relay model inside the software.

The remainder of the article is divided into the following sections. Section 2 provides an overview of overcurrent protection with two features, non-directional and directional. Section 3 gives the methodology of protection system configuration in step-by-step implementation. Section 4 gives the case study with the test system description, the design of overcurrent protection, and its configuration. The validation of the proposed protection system is also given in this section. Finally, Section 5 concludes this paper with a summary and suggestions for further studies.

2. Overcurrent Protection

Overcurrent protection is one of the most popular features of protection systems in distribution networks. The overcurrent protection relay works on a very basic principle; it generates and sends the trip signal to the circuit breaker (BK) by comparing the current flowing through the secondary winding of the current transformer (CT) with a predetermined threshold (pickup current). The overcurrent protection relay can be divided into two categories: (a) non-directional and (b) directional.

2.1. Non-Directional Overcurrent Protection

Non-directional overcurrent protection relays are applied widely in radial distribution feeders, the most common worldwide type of distribution system. In addition to comparing current flowing through it with preset thresholds, it also stands out with the ability of operating time coordination by time versus current magnitude characteristic curves (*I-t*).

Let us take an example of its time-operating coordination. A system with one source and parallel transmission line with two protection relays located at both ends is illustrated in Figure 1a,b. In case of a fault occurring in the middle of Line A, as shown in Figure 1a, relays R1 and R2 must be tripped before relays R3 and R4 to ensure a continuous system (Line B is still in service). This can be achieved by setting the operating time of relays R1 and R2 faster than relays R3 and R4. Similarly, as shown in Figure 1b, in case of a fault occurring at bus 2, relays R1 and R2 must be tripped primarily, while relays R3 and R4 in this case are backup protection.

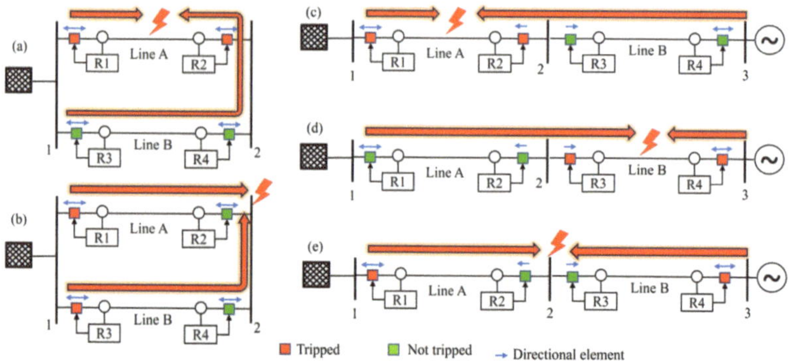

Figure 1. Typical applications of overcurrent protection relays in transmission system. (**a**,**b**). A system with one source and parallel transmission line. (**c**–**e**). Double-end-fed radial system.

Therefore, depending on the protection purposes and fault scenarios, the operating time of overcurrent protection can be coordinated. The time coordination can become simple with three current-time characteristics: (i) instantaneous (ANSI 50), (ii) time-dependent, defined time and inverse time (ANSI 51), (iii) combination of instantaneous and time-dependent. The operating time according to inverse-time characteristics can be defined as:

$$t_{trip} = TMS\frac{a}{PSM^b - c} \quad (1)$$

$$PSM = \frac{I_r}{PS} \quad (2)$$

In Equations (1) and (2), *TMS* is the time-multiplier setting of the relay, *PSM* is the plug-setting multiplier, *PS* is the plug setting of the relay, I_r is the pickup current, and the parameters of characteristic curves a, b, and c follow IEC 60255 [19] and IEEE C37.112 [20].

2.2. Directional Overcurrent Protection

The directional overcurrent protection relay is essentially non-directional overcurrent with the additional feature of a preset direction of protection (forward or reverse), as shown in Figure 2.

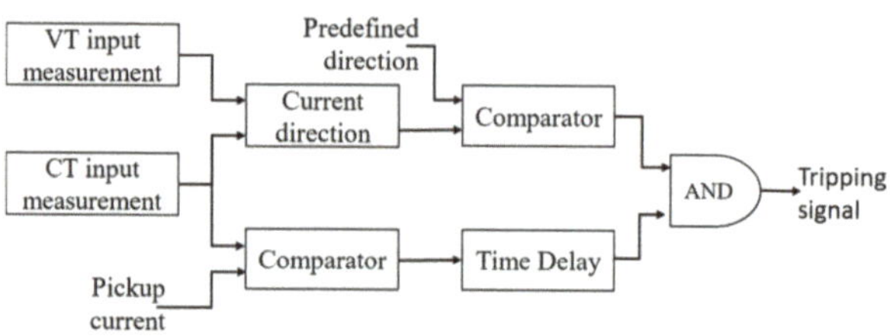

Figure 2. Block diagram of directional overcurrent protection.

The directional overcurrent protection relay is widely used on the transmission side in distribution networks. Taking its application illustrated in Figure 1c–e, a double-end-fed radial system is used, with two relays located at both ends of the transmission line. In case of a fault occurring in the middle of Line A, as shown in Figure 1c, the fault current flows in both directions, from source at bus 1 and from source at bus 3. Relay R1 is a non-directional overcurrent relay; therefore, it operates in this case without considering the current direction. In contrast, relay R2 is a forward directional overcurrent relay, which can 'see' the current flowing in front of it (from the source at bus 1 to relay R2). In this case, relay R2 can operate because the current direction is the same as its preset direction. Similarly, in Figure 1d, relays R3 and R4 can 'see' the fault current. However, in case of a fault occurring at bus 2, as shown in Figure 1e, relays R2 and R3 do not operate since their preset directions are not matched with the fault current direction. Therefore, setting the direction for directional overcurrent protections in different fault scenarios is extremely important.

The directional decision of the relay can be made using the relay characteristic angle (RCA). RCA is the angle referred to as the maximum torque angle (MTA) with polarising voltage as shown in Figure 3. It defines the operating sectors of the relay, and the forward and reverse zones. By evaluating the phase angle between the operating current and the polarising voltage, the relay determines whether it falls within the predetermined forward or reverse region specified by RCA, thereby enabling or disabling fault directional trip. Additionally, the flexible directional relay is designed to guarantee the right working zone and prevent relay mis-operations. It allows the phase angle operating region to be stretched or retracted through its minimum/maximum forward and reverse angle settings. In certain popular networks, the minimum/maximum forward and reverse angles of the protection system are set to ±86 degrees and ±94 degrees, respectively, while RCA is commonly set to 90 degrees.

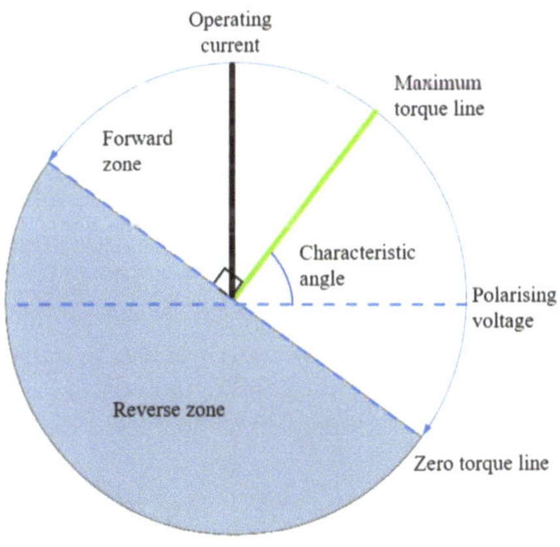

Figure 3. Directional overcurrent relay operating characteristics.

3. Protection System Design Methodology

Figure 4 shows the methodology for protection device setting configuration. The authors believe that this methodology can be applied in every type of protection system, such as fuse, overcurrent protection relay, and distance protection relay. To obtain the effective protection device configuration, these steps must be followed:

- **Identification of protection devices**: During this phase, the selection of protection devices, such as fuses, circuit breakers, and protection relays, is determined according to voltage level, current rating, specific standards, or system requirements. Some standards outline specific requirements for protection devices corresponding to each voltage level within power systems. For example, at low-voltage levels, IEC 60947-2 requires the use of circuit breakers for a maximum operational voltage of 1 kV AC or 1.5 kV DC [21]. In medium-voltage applications, standards such as IEC 60255 provide requirements for protective relays designed for voltage levels ranging from 1 kV to 36 kV [22].

- **Definition of protection zones**: The selection of protection region, which is isolated by protection devices in case of fault, is established in this phase. This involves identifying distinct areas of the power system that require protection, such as feeders, buses, transformers, and generators. Once the regions are selected, clear boundaries for each protection zone must be established, corresponding to the protective devices selected in the initial step. The possibility of overlap between protection devices is also considered in this phase.
- **System analysis**: The analysis is a comprehensive analysis of the power system to understand potential fault scenarios and normal operating conditions. This analysis examines various fault types that could occur, including three-phase faults, single-line-to-ground faults, and line-to-line faults. During this phase, it is essential to assess the operational scenarios in which these faults might arise, taking into consideration variations in load and generation, or network topologies. Additionally, normal operating currents are evaluated to establish baseline conditions. This analysis provides critical data necessary for setting the protection parameters, ensuring that devices can effectively detect and respond to abnormal conditions while remaining operational during normal operational scenarios.
- **CT (current transformer) and VT (voltage transformer) ratio selection**: The selection of CT and VT ratios is performed based on the findings from the analysis in the second step. The appropriate CT ratio is determined by the maximum expected load current and potential fault currents in the system. It is crucial to select a ratio that allows CT to accurately reproduce current signals without saturation during fault conditions, ensuring reliable protection device operation. Similarly, the VT ratio is chosen based on the system voltage levels to ensure that the voltage measurements accurately reflect the operating conditions.
- **Threshold configurations**: Threshold configurations involve establishing precise threshold values for the protection devices to ensure that they detect and respond to abnormal conditions accurately. Each protection device, such as relays and circuit breakers, has specific operational thresholds for parameters like overcurrent, undercurrent, and voltage levels. During this phase, these threshold values are defined based on the results of the system analysis and normal operating conditions. For example, overcurrent thresholds should be set above the maximum load current but below the expected fault current to ensure the timely detection of faults while preventing nuisance tripping during normal fluctuations.
- **Time setting configurations**: This step entails defining specific time delays for the operation of protection devices, ensuring timely and effective responses to fault conditions. In this phase, protection engineers establish coordination settings to ensure that devices operate in a sequential manner, allowing only the nearest device to a fault to trip first. The time settings are determined based on the characteristics of the protection devices, the type of fault, and the specific protection zone layout. Additionally, engineers consider the selectivity of the protection system during this phase, ensuring that devices are coordinated properly across different zones.
- **Other configurations**: Other necessary parameters, such as directional elements and zone interlocking, are defined to complete the protection setting configuration. Additionally, configurations may include settings for features of protection devices such as harmonic filtering, under-voltage protection, and frequency protection, ensuring comprehensive coverage for various fault scenarios.
- **Protection system validation**: The design of the protection system must be validated before actual implementation. The protection system needs to meet the criteria and standards to ensure that it provides adequate protection for the power system under various faults and operating conditions. Offline simulation, real-time simulation, and hardware-in-the-loop are popular methods that can be used to test the protection configuration.

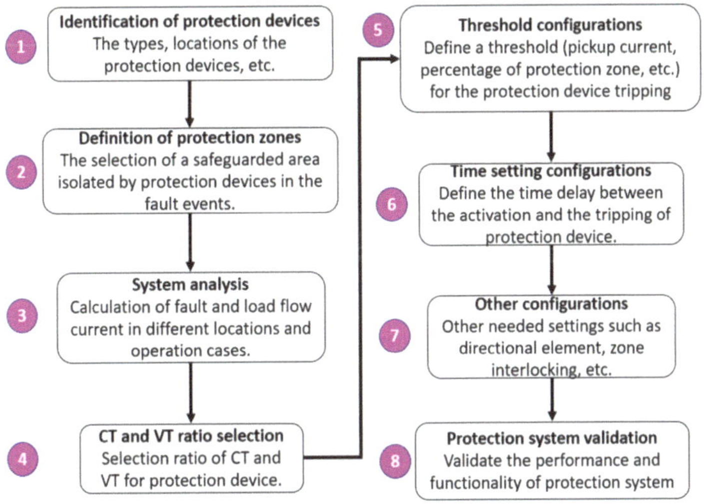

Figure 4. Methodology of protection system design.

4. Case Study

4.1. Test System

In this paper, the CIGRE EU MV benchmark network is used as the test system for the overcurrent protection system design. This benchmark system is a scaled-down representation of a medium-voltage network in southern Germany and is designed to reflect the 50 Hz power system commonly used across European countries. This test system is recommended by CIGRE for studying behaviours of power systems due to the high integration of RESs [23].

This network consists of 14 nodes with a nominal based voltage of 20 kV and connects 18 loads, which are divided into residential, commercial, and industrial types. The benchmark has two feeders. The network topology can be configured through these feeders by using the three isolation switches (S1, S2, and S3). More detailed information on this test system is available in [23]. Further variety may be introduced through eight operating cases of means of configuration switches S1, S2, and S3 as shown in Table 1. If these switches are open, then both feeders are radial. Closing S2 and S3 in feeder 1 would create a loop or mesh. With the given location of S1, it can either be assumed that both feeders are fed by the same substation or by different substations, and closing S1 interconnects the two feeders through a distribution line.

Table 1. Operating cases with switch states. 0: Open; 1: Closed.

Case	S1	S2	S3	Case	S1	S2	S3
1	0	0	0	5	1	0	0
2	0	0	1	6	1	0	1
3	0	1	0	7	1	1	0
4	0	1	1	8	1	1	1

4.2. Overcurrent Protection System Design

Following the methodology detailed in Section 3, the subsequent steps for protection system design in the test system are executed as follows:

- **Identification of protection devices**: In this case study, the overcurrent protection system is used, including both non-directional (at the beginning of the transmission line) and directional overcurrent protection relays (at the end of the transmission line) in conjunction with circuit breakers to meet the MV requirements. The arrangement of relays and circuit breakers is located at both ends of transmission lines, as illustrated in Figure 5.

- **Definition of protection zones**: Protection zones are selected to cover the entirety of the transmission line. In this setup, the primary protection is assigned to the relay positioned at the end of the transmission line (protection relay Bx), while the backup protection is assigned to the relay situated at the beginning of the transmission line (protection relay Ax).
- **System analysis**: The short circuit analysis is illustrated in Figure 6. The two types of short circuit which are used in this analysis are a three-phase short circuit and a single-line-to-ground short circuit. Both cases are executed in maximum short circuit calculations. As shown in Figure 6, the magnitude of the short circuit current varies due to different operating conditions of the test system. Notably, case 8, where all three switches are closed, exhibits the highest short circuit current values for both the three-phase and single-line-to-ground cases. This is due to the interconnected topology, where the short circuit current is the sum values of the currents from the two feeders. However, the phase angle of the short circuit current does not exhibit substantial variation. Additionally, normal operating current levels were evaluated during this phase, revealing significant variations resulting from the different network topologies.
- **CT and VT ratio selection**: The calculation of the primary value of CT is based on the largest operating current across all the operational scenarios involving switches. To ensure the reliability of the relays, the CT ratio is considered with the overload situation as including more than 25%, or, in other words, it is equal to 125% of the largest operating current magnitude. The primary value of VT is selected based on the normal-based voltage of the system. The secondary values of CT and VT are selected as 1 A and 5 V, respectively.
- **Threshold configurations**: The overcurrent protection uses a combination of inverse-time and instantaneous features. Following [16], as shown in Table 2, for inverse-time features, the pickup current is selected as 2.0 to 3.0 times the maximum operating current, while for instantaneous features, the pickup current for the primary protection is set at 1.5 times the rated thermal current of the feeder line, while the pickup current for the backup protection is established as 1.5 times the minimum three-phase short circuit current.
- **Time setting configurations**: Regarding the inverse-time characteristic, the ANSI extremely inverse curve is used for the entire overcurrent protection relay with TMS as 0.05. The delay time of the instantaneous feature is selected as 300 milliseconds and 50 milliseconds corresponding to backup and primary protection.
- **Other configurations**: In this case study, protection relay Ax is of the non-directional type. On the other hand, the directional element is configured for relay Bx to operate in the forward direction, enabling the protection devices to effectively "see" and safeguard the area directly in front of them. Accordingly, MTA in the setting of the directional overcurrent relay in the network is set to 90 degrees.
- **Protection system validation**: For validation, the authors in this paper use offline simulation in DIgSILENT PowerFactory 2023 SP1 software. The model of real-world relay REF 630 from the ABB company is selected to test the relay configuration. The validation study is implemented in case 8, interconnected topology. In this case, as shown in Figure 7a, in the scenario of a three-phase short circuit occurring in the middle of Line 1–2, both relays can enable the trip signal. In contrast, in the case of a three-phase short circuit occurring at bus 2, relay B2 with a forward direction setting fails to operate the trip signal, as shown in Figure 7b.

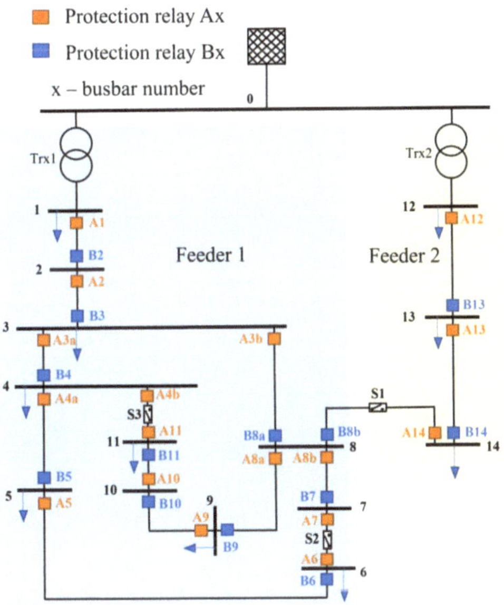

Figure 5. Protection relay location in the test system.

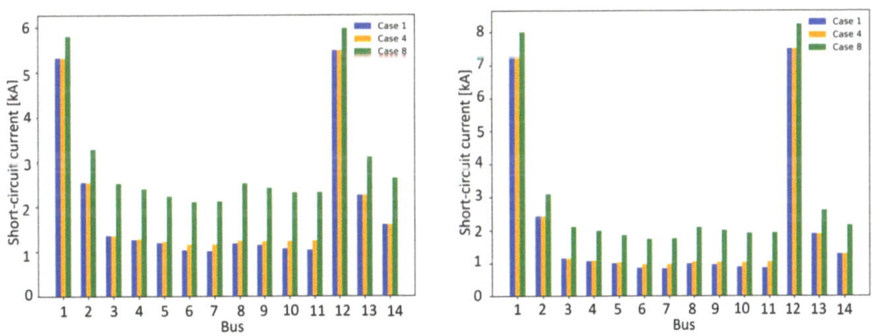

(**a**) (**b**)

Figure 6. Short circuit current magnitude obtained at buses. (**a**) Three-phase short circuit, and (**b**) single-line-to-ground short circuit.

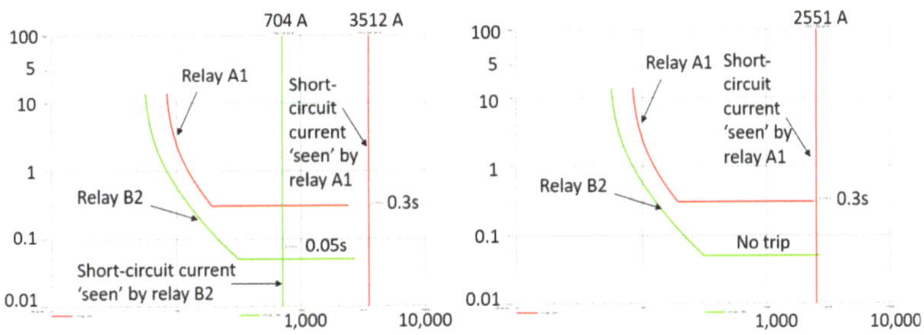

(**a**) (**b**)

Figure 7. Relay A1 and B2 characteristic. (**a**) Short circuit occurs in the middle of Line 1–2, and (**b**) short circuit occurs at bus 2.

Table 2. Protection relay configuration.

Relay	CT Ratio	Ioc [A]	ANSI 51	ANSI 50		Relay	CT Ratio	Ioc [A]	ANSI 51	ANSI 50	
			Ir [kA]	Ir [kA]	Delay (ms)				Ir [kA]	Ir [kA]	Delay (ms)
A1	200:1	170	0.40	3.80	300	A3b	100:1	84	0.19	1.77	300
B2	200:1	169	0.27	2.94	50	B8	100:1	85	0.13	1.06	50
A2	200:1	169	0.40	3.63	300	A8a	50:1	59	0.14	1.72	300
B3	200:1	170	0.27	2.26	50	B9	75:1	59	0.08	1.02	50
A3a	75:1	69	0.16	1.89	300	A9	50:1	34	0.08	1.59	300
B4	75:1	69	0.11	1.21	50	B10	50:1	34	0.05	0.95	50
A4a	50:1	49	0.12	1.78	300	A10	50:1	13	0.03	1.54	300
B5	50:1	49	0.08	1.13	50	B11	50:1	13	0.02	0.92	50
A5	50:1	21	0.05	1.54	300	A12	100:1	84	0.19	3.37	300
B6	50:1	21	0.03	1.07	50	B13	100:1	84	0.13	2.02	50
A8b	75:1	23	0.05	1.51	300	A13	100:1	83	0.19	2.14	300
B7	50:1	20	0.03	0.92	50	B14	100:1	83	0.13	1.43	50

5. Conclusions

The widespread integration of RES into power systems offers numerous benefits; however, it also presents operational challenges; for example, it impacts on power system protection. Indeed, RESs contribute to increased short circuit magnitude values and alter the direction of current flow, necessitating a systematic approach to power system protection design to mitigate these impacts. This paper introduces a methodology to address this need. Accordingly, the design of an overcurrent protection system that incorporates both directional and non-directional features is implemented in a well-known test system, the CIGRE EU MV distribution network.

This paper can be regarded as a valuable resource for protection engineers who are seeking an aid document in preparation for further RES integration in the future. The presented methodology in this paper employs software-based validation methods and theoretical calculations for protection system design. Therefore, specialised analysis tools and advanced validation methods, such as real-time simulations and hardware-in-the-loop testing, can be utilised for further research. Moreover, leveraging this methodology in a larger test system can significantly advance wide-area monitoring, control, and protection capabilities in future power systems.

Author Contributions: Conceptualization, L.N.H.P. and F.G.-L.; methodology, L.N.H.P. and F.G.-L.; software, L.N.H.P.; validation, L.N.H.P. and F.G.-L.; writing—original draft preparation, L.N.H.P. and V.R.-M.; writing—review and editing, L.N.H.P., F.G.-L., and V.R.-M.; visualization, L.N.H.P.; supervision, F.G.-L. All authors have read and agreed to the published version of the manuscript.

Funding: This research received no external funding.

Institutional Review Board Statement: Not applicable.

Informed Consent Statement: Not applicable.

Data Availability Statement: Data are contained in the article.

Acknowledgments: Le Nam Hai Pham wants to express his gratitude to the University of South-Eastern Norway, campus Porsgrunn, for the support during his PhD period.

Conflicts of Interest: The authors declare no conflicts of interest.

References

1. Wagle, R.; Pham, L.N.H.; Tricarico, G.; Sharma, P.; Rueda, J.L.; Gonzalez-Longatt, F. Co-simulation-based optimal reactive power control in smart distribution network. *Electr. Eng.* **2024**, *106*, 2391–2405. [CrossRef]
2. Tricarico, G.; Wagle, R.; Dicorato, M.; Forte, G.; Gonzalez-Longatt, F.; Rueda, J.L. Zonal Day-Ahead Energy Market: A Modified Version of the IEEE 39-bus Test System. In Proceedings of the 2022 IEEE PES Innovative Smart Grid Technologies-Asia (ISGT Asia), Bengaluru, India, 10–13 November 2023; pp. 86–90. [CrossRef]
3. Shrestha, A.; Rajbhandari, Y.; Gonzalez-Longatt, F. Day-ahead energy-mix proportion for the secure operation of renewable energy-dominated power system. *Electr. Power Energy Syst.* **2024**, *155*, 109560. [CrossRef]
4. Tricarico, G.; Wagle, R.; Martinez, J.C.; Gonzalez-Longatt, F.; Dicorato, M.; Forte, G.; Rueda, J.L. A Co-simulation Procedure for Optimal Reactive Power Control in Active Distribution Networks. In Proceedings of the 2023 IEEE International Conference on Environment and Electrical Engineering and 2023 IEEE Industrial and Commercial Power Systems Europe (EEEIC/I&CPS Europe), Madrid, Spain, 6–9 June 2023. [CrossRef]
5. Wagle, R.; Pham, L.N.H.; Tricarico, G.; Sharma, P.; Rueda, J.L.; Gonzalez-Longatt, F. Experiences in a Cyber-Physical Co-Simulation Testbed Development for a Smart-er Distribution Network. In Proceedings of the 2023 IEEE PES Conference on Innovative Smart Grid Technologies-Middle East (ISGT Middle East), Abu Dhabi, United Arab Emirates, 12–15 March 2013; pp. 1–5. [CrossRef]
6. Urdaneta, A.J.; Restrepo, H.; Márquez, S.; Sánchez, J. Coordination of directional overcurrent relay timing using linear programming. *IEEE Trans. Power Deliv.* **1996**, *11*, 122–128. [CrossRef]
7. Chamorro, H.R.; Gomez-Diaz, E.O.; Paternina, M.R.A.; Andrade, M.A.; Barocio, E.; Rueda, J.L.; Gonzalez-Longatt, F.; Sood, V.K. Power system coherency recognition and islanding: Practical limits and future perspectives. *IET Energy Syst. Integr.* **2023**, *5*, 1–14. [CrossRef]
8. Phan, L.N.H.; Melo, A.F.S.; Gonzalez-Longatt, F. Testing a Non-directional Overcurrent Protection Relay: Power-Hardware-in-The Loop Approach. In Proceedings of the 2024 4th International Conference on Smart Grid and Renewable Energy (SGRE), Doha, Qatar, 8–10 January 2024. [CrossRef]
9. Gkavanoudis, S.I.; Tampakis, D.; Malamaki, K.N.D.; Kryonidis, G.C.; Kontis, E.O.; Oureilidis, K.O.; Maza-Ortega, J.M.; Demoulias, C.S. Protection philosophy in low short-circuit capacity distribution grids with high penetration of converter-interfaced distributed renewable energy sources. *IET Gener. Transm. Distrib.* **2020**, *14*, 4978–4988. [CrossRef]
10. Plet, C.A.; Graovac, M.; Green, T.C.; Iravani, R. Fault response of grid-connected inverter dominated networks. In Proceedings of the IEEE PES General Meeting 2010, Minneapolis, MN, USA, 25–29 July 2010. [CrossRef]
11. Aljarrah, R.; Marzooghi, H.; Yu, J.; Terzija, V. Sensitivity analysis of transient short circuit current response to the penetration level of non-synchronous generation. *Int. J. Electr. Power Energy Syst.* **2021**, *125*, 106556. [CrossRef]
12. Traupmann, A.; Kienberger, T. Test Grids for the Integration of RES—A Contribution for the European Context. *Energies* **2020**, *13*, 5431. [CrossRef]
13. Gonzalez-Longatt, F.; Adiyabazar, C.; Martinez, E.V. Setting and Testing of the Out-of-Step Protection at Mongolian Transmission System. *Energies* **2021**, *14*, 8170. [CrossRef]
14. Ding, L.; Gonzalez-Longatt, F.M.; Wall, P.; Terzija, V. Two-step spectral clustering controlled islanding algorithm. *IEEE Trans. Power Syst.* **2013**, *28*, 75–84. [CrossRef]
15. Maza-Ortega, J.M.; Zarco-Soto, F.J.; Gkavanoudis, S.; Tampakis, D.; Demoulias, C. A short communication to define the overcurrent protection system of the CIGRE European benchmark distribution networks for RES penetration studies. *Electr. Eng.* **2022**, *104*, 1331–1336. [CrossRef]
16. Pham, L.N.H.; Wagle, R.; Gonzalez-Longatt, F. Concise Definition of the Overcurrent Protection System for CIGRE European Medium Voltage Benchmark Network. In Proceedings of the 2023 IEEE PES Conference on Innovative Smart Grid Technologies-Middle East (ISGT Middle East), Abu Dhabi, United Arab Emirates, 12–15 March 2023; pp. 1–5. [CrossRef]
17. Pham, L.N.H.; Wagle, R.; Gonzalez-Longatt, F.; Acosta, M. Non-directional Overcurrent Protection Relay Testing Using Virtual Hardware-in-the-Loop Device. In *Real-Time Simulation and Hardware-in-the-Loop Testing Using Typhoon HIL*; Springer Nature Singapore: Singapore, 2023; pp. 283–337. [CrossRef]
18. Pham, L.N.H.; Gonzalez-Longatt, F. Distance Protection Relay Testing Using Virtual Hardware-in-the-Loop Device. In *Real-Time Simulation and Hardware-in-the-Loop Testing Using Typhoon HIL*; Springer Nature Singapore: Singapore, 2023; pp. 379–424. [CrossRef]
19. *IEC 60255-151:2009*; Measuring Relays and Protection Equipment—Part 151: Functional Requirements for over/under Current Protection (Revision of IEC 60255-3:1989). IEC: Geneva, Switzerland, 2009; p. 63.
20. Benmouyal, G.; Meisinger, M.; Chairnian, V.; Burnworth, J.; Elmore, W.A.; Freirich, K.; Kotos, P.A.; Leblanc, P.R.; Lerley, P.J.; McConnell, J.E.; et al. IEEE standard inverse-time characteristic equations for overcurrent relays. *IEEE Trans. Power Deliv.* **1999**, *14*, 868–872. [CrossRef]
21. Berner, W.; Basinger, S.; Driscoll, T.; Martin, K.; Poidl, J.; Throckmorton, M. Low-voltage circuit protection in hazardous locations—A comparison of codes and standards within IEC, CEC and NEC. In Proceedings of the 2009 Record of Conference Papers-Industry Applications Society 56th Annual Petroleum and Chemical Industry Conference, Anaheim, CA, USA, 14–16 September 2009. [CrossRef]

22. Leitloff, V.; Chen, H.; Chen, D.; Bonetti, A.; Xu, L.; Mohamed, A.; Byman, C. Towards a standardisation for digital inputs and outputs of protection functions in IEC 60255 series. *Prot. Control Mod. Power Syst.* **2022**, *7*, 1–11. [CrossRef]
23. Strunz, K. *Benchmark Systems for Network Integration of Renewable and Distributed Energy Resources*; CIGRE: Paris, France, 2014; ISBN 978-285-873-270-8.

engineering proceedings

MDPI

Proceeding Paper

The Analysis of Service Convergence in an Optical Access Network †

Erick Cifuentes [1,*], David Mosquera [1,*], Christian Tipantuña [1,*], Berenice Arguero [2,*] and Germán V. Arevalo [2,*]

1 Faculty of Electrical and Electronic Engineering (FIEE), Escuela Politécnica Nacional, Quito 170143, Ecuador
2 Telecommunications Engineering Career, Salesian Polytechnic University, Quito 170517, Ecuador
* Correspondence: erick.cifuentes@epn.edu.ec (E.C.); david.mosquera01@epn.edu.ec (D.M.); christian.tipantuna@epn.edu.ec (C.T.); jarguero@ups.edu.ec (B.A.); garevalo@ups.edu.ec (G.V.A.)
† Presented at the XXXII Conference on Electrical and Electronic Engineering, Quito, Ecuador, 12–15 November 2024.

Abstract: In recent years, the increasing number of internet-connected devices has exceeded the capacity of fourth-generation (4G) cellular networks, leading to the development of fifth-generation (5G) technology, designed to offer higher speeds, greater bandwidth, and lower latency. In this context, this study evaluated Universal Filtered Multi-Carrier (UFMC) and Generalized Frequency Division Multiplexing (GFDM) techniques, implementing them in a radio-over-fiber (RoF) system and a Next-Generation Radio Access Network (NG-RAN) fronthaul link, and compared the results using communication quality metrics such as bit error rate (BER). Additionally, through signal generation and processing in Matlab, the performance of UFMC and LTE signals was analyzed, confirming that simultaneous transmission over an RoF channel allows for efficient signal separation in the frequency domain, with the UFMC giving power to LTE.

Keywords: 5G; UFMC; GFDM; LTE; radio over fiber

Citation: Cifuentes, E.; Mosquera, D.; Tipantuña, C.; Arguero, B.; Arevalo, G.V. The Analysis of Service Convergence in an Optical Access Network. *Eng. Proc.* **2024**, *77*, 27. https://doi.org/10.3390/engproc2024077027

Academic Editor: Jackeline Abad

Published: 18 November 2024

1. Introduction

The rapid advancement of communication technologies has significantly transformed digital interactions, increasing the demand for faster and more efficient networks. Optical access networks play a crucial role in providing reliable, high-speed connectivity. However, as the demand grows to integrate services like LTE and 5G, the challenge of converging these services into a unified infrastructure that maximizes resource efficiency becomes evident. This convergence, while beneficial, requires overcoming technical complexities due to the different bandwidth, latency, and jitter demands that each service imposes. It is essential, therefore, that passive optical networks (PONs) are kept up-to-date to effectively support the demands of emerging technologies, particularly within the framework of the fifth generation of mobile networks [1,2].

Converging services in an optical network requires not only a robust infrastructure, but also a thorough understanding of the underlying technologies, such as LTE, 5G, and filtered multi-carrier modulation waveforms, such as the UFMC and GFDM. These waveforms, which offer superior spectral efficiency, are key to the success of the 5G network, as they enable more efficient and flexible data transmission. In addition, RoF architecture and PON networks provide the necessary technological foundations for the transmission and distribution of signals over optical access networks, facilitating the integration of multiple services on a single channel. This approach is crucial to address growing connectivity demands, enabling optical networks not only to meet current needs, but also to anticipate and adapt to future technological challenges [3].

This paper focuses on the analysis of service convergence within an optical access network, using both the simulation and practical implementation of an optical link capable of transmitting and receiving LTE and UFMC signals. Using advanced tools such as Matlab

2024a and OptSim, simulation scenarios were developed that replicate real transmission conditions in PONs. This study not only seeks to optimize bandwidth usage and improve quality of service, but also to contribute to the evolution of optical access networks toward more adaptive and efficient infrastructures. The results obtained, based on key metrics such as the BER, signal-to-noise ratio (SNR), and EVM, provide a comprehensive view of network performance, enabling a better experience for end users in an increasingly digitized environment.

2. Related Works

Table 1 presents an analysis of several recent related works. Although there have been advances in this field, these studies have not implemented private gateway monitoring in the way this article addresses.

Table 1. Comparison of related works and approaches of UFMC, LTE, and GFDM technologies.

Reference	Year	Contribution
[4]	2020	An optical access network with UFMC and LTE signals achieved −30 dBm sensitivity for the UFMC and −31 dBm for LTE, supporting up to 64 users in a radio-over-fiber configuration.
[5]	2018	An optical access network with GFDM and LTE signals achieved receiver sensitivities of −29.5 dBm for LTE and −24.5 dBm for GFDM, supporting up to 64 users in a legacy PON using IM/DD format.

3. Methodology

The methodology of this study is based on the implementation and evaluation of advanced technologies in an optical network environment using real equipment. This practical approach accurately replicates 5G network conditions, focusing on service convergence through the integration of UFMC, GFDM, and LTE technologies. The process began with the setup of an experimental infrastructure that included RoF modules and PON elements, chosen for their ability to support the high performance required by next-generation networks. To integrate the environments of signal generation, transmission, and processing, an Arbitrary Waveform Generator (AWG) was used for transmission, and a Digital Signal Analyzer (DSA) was employed for reception [6].

The interaction between the waveform generation and implementation environments is depicted in Figure 1. The transmitter block refers to the transmission stage of each system. For generating UFMC, GFDM, and LTE signals, parameters such as modulation scheme, number of blocks, and number of sub-carriers must be configured. The optical channel block illustrates the physical implementation of the optical channel, where signals enter a PON. In this step, the AWG takes the signal generated in Matlab 2024a and modulates it into an RF signal. This RF signal, combined with an intensity modulator and an optical laser, is converted into an optical signal. The optical signal then passes through the fiber optic network and a variable attenuator to adjust its power before entering a photodetector, which converts it back into an electrical signal. The receiver block shows the reception of signals that have traversed the optical channel, where demodulation of the signals is performed to calculate the BER.

Network performance analysis was performed by collecting data through multiple test iterations. These tests were conducted under different load conditions and network configurations, enabling a comprehensive assessment of the feasibility and effectiveness of UFMC-GFDM comparison and LTE-UFMC convergence in an optical network environment. The data obtained were analyzed with advanced network monitoring tools, focusing on metrics such as the BER and EVM. These metrics are essential to determine the quality of the transmitted and received signal, and their analysis allowed for the identification of optimal configurations to maximize network performance. Using real equipment instead of simulations increases the validity of the results, reflecting actual operating conditions in a converged optical network. This approach not only allows for measuring the performance of the technologies under study but also helps identify potential limitations and areas for improvement in the implementation of 5G networks [7].

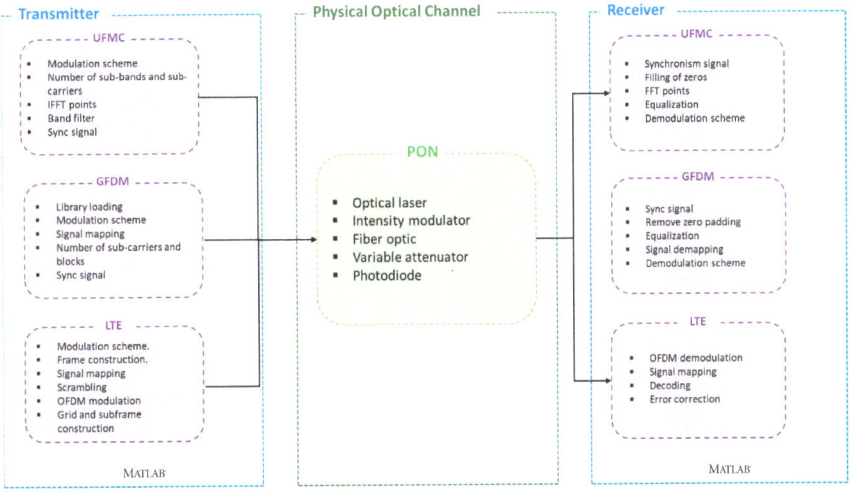

Figure 1. Block diagram modeling of the transmission, channel, and reception of the system to be implemented.

To construct the optical channel, both equipment and passive components are required. Figure 2 shows the connection diagram of these components, which form the RoF system used to transmit the UFMC and GFDM signals. In the optical network terminal (ONT) section, there is an AWG, which is loaded with the waveform generated in Matlab 2024a using Agilent Control Utility software. A continuous wave (CW) laser is also included, with its input connected to the optical intensity modulator through the first port using an FC/APC connector. A coaxial cable is then connected from the SMA port of channel 1 at one end to the RF input of the modulator at the other end. Additionally, a DC power supply is used to power the optical modulator with a voltage of 3.2 V. In the optical distribution network (ODN), a 20 km reel of G.652D optical fiber, which is manufactured by Corning in the Corning, NY, USA, is used. It has FC/APC connectors on both ends. The first connector is coupled to the output of the Mach–Zehnder-type optical modulator via an FC/APC connection, while the second is inserted into a port of the variable attenuator. The Mach-Zehnder modulator is manufactured by Thorlabs in the USA. The FC/APC end of an optical fiber patch cord is connected to the other port of the attenuator, while the FC/UPC end is inserted into the input port of the PIN photodetector. This device marks the beginning of the optical line terminal (OLT), where the transmitted signal is received and processed. To do this, one end of the coaxial cable is connected to the SMA output port of the photodetector and the other end to the SMA input port of channel 1 on the DSA. The transmitted signal can be observed in real time on the equipment screen [7,8].

For the implementation of the optical link, the parameters set in Table 2 for the various equipment used are considered. For the photodetector, its operating range must be considered to obtain accurate results. Operating above the limit of −3 dBm could damage the equipment, while working below the sensitivity of −19 dBm can result in incorrect signal recovery.

On the other hand, a key consideration for the deployment of these technologies in large-scale infrastructures is their ability to scale efficiently. In a high-demand scenario, the integration of UFMC and GFDM technologies into an extensive optical access network must account for factors such as increased data traffic, enhanced user density, and evolving service requirements. The UFMC's inherent flexibility in managing sub-carrier blocks allows for efficient scaling across larger networks, while GFDM, despite its limitations in spectral efficiency, can be optimized through techniques such as cyclic prefixing and dynamic spectrum management. Testing under simulated conditions of network stress has demonstrated that both technologies exhibit satisfactory scalability, though the UFMC tends to outperform GFDM in maintaining the BER and latency under high-load environments.

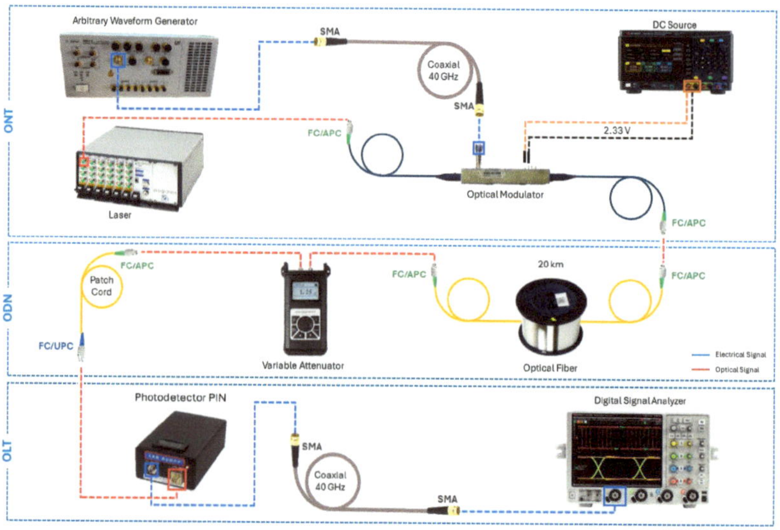

Figure 2. Optical channel equipment connection diagram.

Table 2. Description of equipment and parameter configuration in optical channel.

Equipment/Item	Brand/Model	Parameter	Value
Arbitrary Waveform Generator	Agilent N8241A	RF Data Output Signal Frequency Output Voltage	Channel 1 1 GHz 250 mV
Laser	IdPhotonics Mainframe CBMA24 and Module CoBrite MX	Type Wavelength Optical Power	N 1550 nm 16 dBm
Optical Modulator	Thorlabs LNA2124	BIAS Voltage	2.33 V
Optical Fiber	-	Type Attenuation Chromatic Dispersion	SMF G.652D 0.24 dB/km 18 ps/(nm·km)
Variable Attenuator	Gao-Voa-103	Range Interval	8–23 dB 1 dB
Photodetector	LabBuddy DSC-R402	Maximum Power Sensitivity	−3 dBm −18 dBm
Digital Signal Analyzer	KeySight DSAV164A	RF Data Input Number of Samples	Channel 1 5 GS

Finally, the methodology also considers the scalability and adaptability of the designed network, evaluating how UFMC and GFDM technologies could be integrated into larger and more complex infrastructures. This evaluation included network resiliency and flexibility tests, simulating scenarios of high demand and variability in data traffic. The results obtained offer valuable insights into the capabilities and limitations of the technologies studied, providing crucial information for future 5G network implementations and optimizations. Overall, the methodology adopted in this study provides a comprehensive and detailed view of the challenges and opportunities presented by the convergence of services in optical access networks, contributing significantly to the advancement of knowledge in this emerging field [9,10]. The code used to generate the waveforms, as well as their subsequent processing, is found in the Supplementary Material section.

4. Results and Discussion

4.1. Comparison between UFMC and GFDM

In the UFMC signal, 100 sub-bands were used, each with 100 sub-carriers, resulting in a total bit vector length of 10,000. For the GFDM signal, 128 sub-carriers were configured, with 34 used for data transmission, creating a bit vector length of 13,600. The bit vector length varies with M-QAM modulation orders. For example, with 4-QAM, the bit vector length in the UFMC is 20,000 bits, and in GFDM, it is 27,200 bits. Figure 3 shows that when using 4-QAM, the BER is acceptable up to a power of −12 dBm, with the UFMC having 10 erroneous bits and GFDM 25. Beyond this power, the BER decreases, reaching zero in both cases, ensuring high-quality transmission. The BER threshold of 10^{-3} is essential for comparing different modulation schemes (4-QAM, 16-QAM, 1024-QAM), as it reflects efficient transmission with minimal errors. While higher-order modulations typically increase the BER, maintaining a 10^{-3} BER indicates the system's ability to efficiently manage complex signals. The constellation diagrams confirm signal recovery at higher power levels without issues.

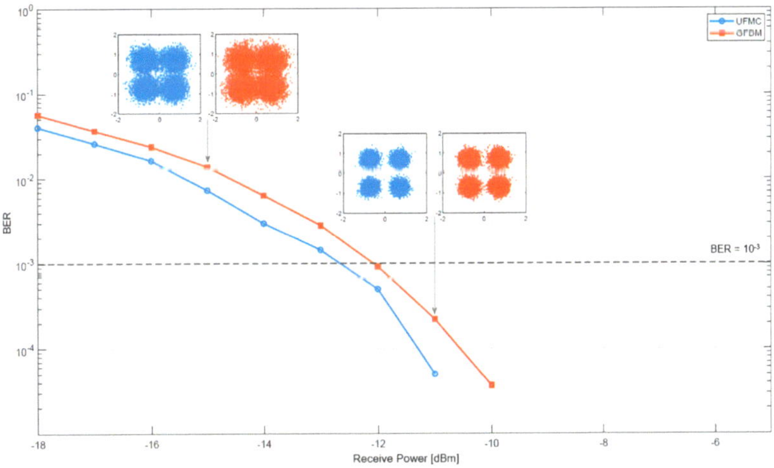

Figure 3. Rx optical power vs. bit error rate for 4-QAM.

To summarize the results obtained, Figure 4 presents a comparison of the BER curves for each modulation order in the UFMC and GFDM waveforms. It is notable that the UFMC consistently performs better than GFDM in all cases. However, the recovered signal only achieves a BER lower than 0.001 in 4-QAM and 16-QAM modulations; in other modulations, the signal experiences significant degradation, making it impossible to recover within the BER requirements. The superior performance of the UFMC is attributed to the use of filters for each sub-band, allowing each group of sub-carriers to have its own filter and corresponding equalization.

In contrast, GFDM treats all sub-carriers as a single band, which increases the likelihood of data degradation. This issue could be mitigated by using a cyclic prefix (CP), as in OFDM, although this would increase the components in the power spectral density (PSD), reducing bandwidth efficiency. For higher-order modulations, both the UFMC and GFDM exhibit high BERs, which are unacceptable in a PON environment. As the modulation order increases, the BER also increases, as more modulation states are managed, and it becomes more likely for a signal to deviate from its original position in the constellation diagram. Modulation states are closer together compared to lower orders, where it is easier to distinguish the original position of a transmitted signal.

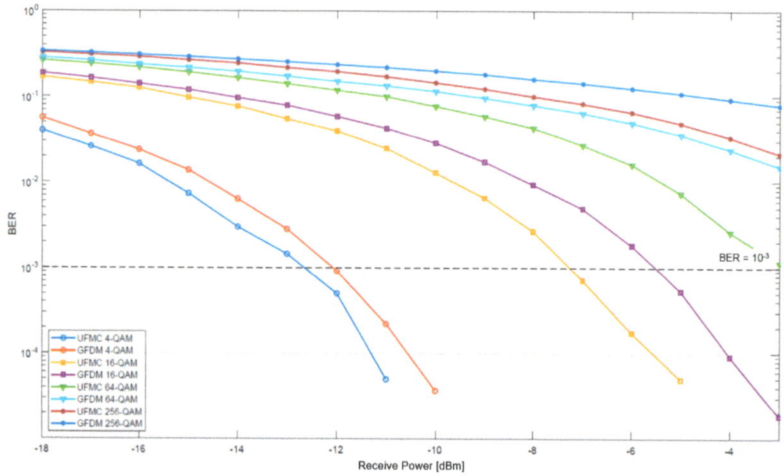

Figure 4. The BER performance of the UFMC and GFDM under different modulation schemes (4-QAM, 16-QAM, etc.). The graph illustrates the relationship between the received optical power and the BER for each scheme, demonstrating the superior performance of the UFMC at lower power levels.

4.2. Comparison between LTE and UFMC

In the evaluation of service convergence within the optical network, significant results were observed regarding optical reception power for both LTE and UFMC signals. The simulation demonstrated that the system effectively enables the transmission of both signals through the fiber optic channel, achieving a reception power of −30.987 dBm for the UFMC and −26.987 dBm for LTE, while maintaining a bit error rate (BER) within acceptable limits for service quality. These results highlight the system's capability to handle multiple high-demand services without compromising signal integrity, thereby demonstrating improved efficiency and performance in converged access networks. This optimized performance underscores the potential of optical networks to support the growing demands for data and connectivity in modern applications, paving the way for broader implementation of 5G technologies in existing networks. The results are shown in Figure 5 for LTE and in Figure 6 for the UFMC.

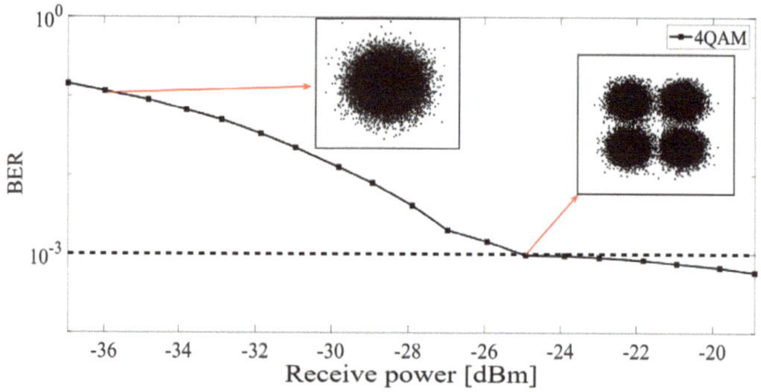

Figure 5. Rx optical power vs. bit error rate from LTE.

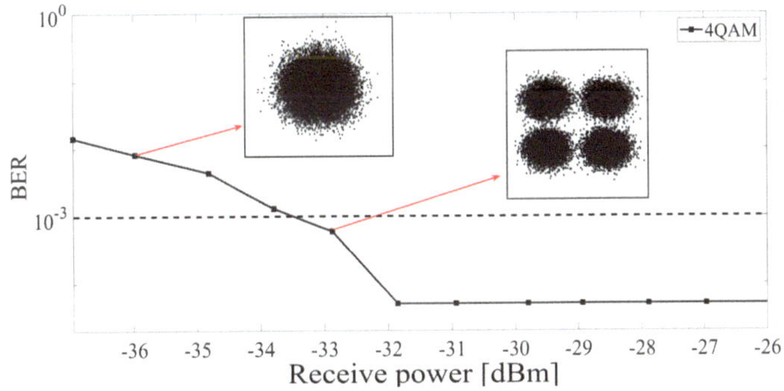

Figure 6. Rx optical power vs. bit error rate from UFMC.

In addition, while latency and throughput are commonly used metrics in networking systems to assess overall performance, they are not the primary focus in the context of radio-over-fiber (RoF) systems. RoF technologies aim to optimize the quality of signal transmission over optical media, where metrics like bit error rate (BER) and Error Vector Magnitude (EVM) provide a more accurate representation of transmission fidelity. In this study, we concentrated on these signal integrity metrics because they directly reflect the effectiveness of optical signal conversion and transmission over fiber channels. The inherent low-latency nature of the optical fiber, coupled with the high data throughput of the medium, minimizes the variability in these metrics, making them less critical for evaluating RoF systems.

4.3. Cost Analysis

While the performance metrics of the UFMC and GFDM demonstrate their technical feasibility, the cost implications for deploying these technologies in optical networks require further consideration. Key factors include the cost of upgrading infrastructure, purchasing advanced modulation and transmission equipment, and ongoing maintenance expenses, all of which must be considered in the overall analysis. For instance, the integration of additional filtering mechanisms in the UFMC could result in higher initial costs but lower operational expenses due to its efficient spectrum utilization. Conversely, GFDM may present lower upfront costs but higher operational expenses due to the complexity of signal processing.

5. Conclusions

In this work, the physical implementation of an RoF system in the context of 5G mobile networks was achieved to analyze the UFMC and GFDM waveforms using BER and EVM metrics with M-QAM modulation. Signal generation and processing were performed using Matlab, while the optical channel was set up with equipment and components available in the laboratory. The UFMC and GFDM signals were transmitted individually under 4/16/64/256-QAM modulation schemes, with the power level varied using a variable attenuator within the range of power levels acceptable for the photodetector. The results demonstrate that as optical reception power increases, the BER decreases for both waveforms. Moreover, higher modulation orders lead to greater signal degradation, reducing the rate of BER improvement. The EVM exhibits a similar behavior to the BER but tends to stabilize as power increases. Notably, in all cases, the UFMC provides better performance than GFDM in both the BER and EVM.

In the evaluation of service convergence within the optical network, significant results were observed regarding optical reception power for both LTE and UFMC signals. The simulation demonstrated that the system effectively enables the transmission of both signals through the fiber optic channel, achieving a reception power of −30.987 dBm for the UFMC and −26.987 dBm for LTE, while maintaining a bit error rate (BER) within acceptable limits for service quality. These results highlight the system's capability to handle multiple high-demand services without compromising signal integrity, thereby demonstrating improved efficiency and performance in converged access networks. This optimized performance underscores the potential of optical networks to support the growing demands for data and connectivity in modern applications, paving the way for broader implementation of 5G technologies in existing networks.

Supplementary Materials: The following supporting information can be downloaded at: https://github.com/Davosin1404/Convergencia_OptSim-Matlab, https://github.com/ErickCifuentes01/TIC-2024.git.

Author Contributions: Conceptualization, C.T., E.C. and D.M.; methodology, E.C. and D.M.; software, D.M and B.A.; validation, C.T. and B.A.; formal analysis, E.C. and D.M.; investigation, E.C. and D.M.; resources, B.A and G.V.A.; data curation, E.C. and D.M.; writing—original draft preparation, E.C. and D.M.; writing—review and editing, E.C. and D.M.; visualization, E.C. and D.M.; supervision, C.T. and B.A.; project administration, C.T., B.A., G.V.A., E.C. and D.M.; funding acquisition, E.C. and D.M. All authors have read and agreed to the published version of the manuscript.

Funding: This research received no external funding.

Institutional Review Board Statement: Not applicable.

Informed Consent Statement: Not applicable.

Data Availability Statement: The data are contained within the article and Supplementary Materials.

Conflicts of Interest: The authors declare no conflicts of interest.

References

1. Villalba, V.P.; López, K.C.P.; Mateus, J.J.R. Sistemas de radio sobre fibra y redes 5G: Perspectiva y aplicaciones. *Ingeniare* **2022**, *33*, 73–88. [CrossRef]
2. Flores Tipán, L.C.; Toscano Lucero, L.C. Sistema de Transmisión Multiportadora Mediante Multiplexación por División de Frecuencia Generalizado (GFDM). Bachelor's Thesis, Escuela Politécnica Nacional, Quito, Ecuador, 2017. Available online: http://dspace.ups.edu.ec/handle/123456789/14023 (accessed on 16 August 2024).
3. Acquah, S.; Krampah-Nkoom, A.; Adjei, M. Performance of the Candidate Modulation Waveforms for 5G Communication Systems. *Int. J. Sci. Res. Publ. (IJSRP)* **2020**, *10*, 597–644. [CrossRef]
4. Tipán, M.N.; Arguero, T.B.; Arévalo, G.V. UFMC and LTE Convergence Test in Optical Access Network. In Proceedings of the 2019 IEEE Fourth Ecuador Technical Chapters Meeting (ETCM), Guayaquil, Ecuador, 13–15 November 2019; pp. 1–4. [CrossRef]
5. Tipán, M.N.; Berenice, A.T.; Arévalo, G.V. GFDM and LTE Data Convergence Test in Optical Access Networks. In Proceedings of the 2018 IEEE Third Ecuador Technical Chapters Meeting (ETCM), Cuenca, Ecuador, 15–19 October 2018; pp. 1–4. [CrossRef]
6. Pazmiño, M.C.D. Prototipos de Redes LTE en Base a Tecnología De Radio Definida por Software e Integración de la Tecnología de Radio Sobre Fibra en Redes Móviles: Análisis de la Integración de Tecnologías de Radio Sobre Fibra en Redes Móviles: Caso de Estudio Técnica UFMC. Bachelor's Thesis, Escuela Politécnica Nacional, Quito, Ecuador, 2017. Available online: http://bibdigital.epn.edu.ec/handle/15000/23249 (accessed on 17 January 2024).
7. Aldana, A.T.; Vallejo, C. Telecomunicaciones, Convergencia y Regulación. *Rev. Econ. Inst.* **2010**, *12*, 85–101.
8. Hajlaoui, E.; Zaier, A.; Khlifi, A.; Ghodhbane, J.; Hamed, M.B.; Sbita, L. 4G and 5G technologies: A Comparative Study. In Proceedings of the 2020 5th International Conference on Advanced Technologies for Signal and Image Processing (ATSIP), Sousse, Tunisia, 2–5 September 2020; pp. 1–6.
9. Horvath, T.; Munster, P.; Oujezsky, V.; Bao, N.-H. Passive Optical Networks Progress: A Tutorial. *Electronics* **2020**, *9*, 1081. [CrossRef]
10. Abdallah, A.A.; Kassas, Z.M. Evaluation of Feedback and Feedforward Coupling of Synthetic Aperture Navigation with LTE Signals. In Proceedings of the 2019 IEEE 90th Vehicular Technology Conference (VTC2019-Fall), Honolulu, HI, USA, 22–25 September 2019; pp. 1–6. [CrossRef]

engineering proceedings

MDPI

Proceeding Paper

Comprehensive Evaluation of LoRaWAN Technology in Urban and Rural Environments of Quito †

Ricardo Mena [1,*,‡], Mario Ramos [1,‡], Luis Urquiza [1] and José D. Vega-Sánchez [2]

1 Departamentro de Electrónica, Telecomunicaciones y Redes de Información (DETRI), Escuela Politécnica Nacional, Quito 170525, Ecuador; mario.ramos@epn.edu.ec (M.R.); luis.urquiza@epn.edu.ec (L.U.)
2 Colegio de Ciencias e Ingenierías "El Politécnico", Universidad San Francisco de Quito (USFQ), Diego de Robles s/n, Quito 170157, Ecuador; dvega@usfq.edu.ec
* Correspondence: ricardo.menav@epn.edu.ec
† Presented at the XXXII Conference on Electrical and Electronic Engineering, Quito, Ecuador, 12–15 November 2024.
‡ These authors contributed equally to this work.

Abstract: The long-range wide area network (LoRaWAN) protocol is one of the most effective technologies for internet of things (IoT) applications, offering long-distance connectivity with low power consumption. This paper presents a practical approach by implementing a LoRa-based measurement prototype across urban and rural environments in the city of Quito, with the aim of assessing the performance and applicability of the technology in manifold settings. Specifically, we develop the required data collection and transmission code in the underlying network, ensuring smooth network integration. Furthermore, test environments are thoroughly characterized for numerical results, highlighting the conditions in the cities of Quito. The results obtained in both scenarios were satisfactory, allowing the comparison of the system's performance in different contexts and providing key aspects of its practical applications and effectiveness. As the main contribution, empirical data were obtained to understand how long-range low-energy connectivity behaves, providing valuable information for comparing system performance in high-altitude cities above sea level, identifying practical applications, and optimizing its use in real IoT implementations.

Keywords: LoRaWAN; prototype; Quito

Citation: Mena, R.; Ramos, M.; Urquiza, L.; Vega-Sánchez, J.D. Comprehensive Evaluation of LoRaWAN Technology in Urban and Rural Environments of Quito. *Eng. Proc.* **2024**, 77, 28. https://doi.org/10.3390/engproc2024077028

Academic Editor: Christian Tipantuña

Published: 18 November 2024

1. Introduction

Traditionally, the long-range wide area network (LoRaWAN) protocol has established itself as one of the most effective technologies for internet of things (IoT) applications, offering long-distance connectivity with low power consumption. This technology is particularly suitable for environments where other communication solutions might be unfeasible due to limited range or high power consumption. In particular, LoRaWAN enables data transmission over several kilometers, with devices featuring long battery life and low cost, making it an ideal choice for low-cost, distributed IoT networks [1]. In a LoRaWAN network, a star-type topology is usually employed where gateways act as intermediaries, transmitting messages between nodes and a central network server. Such gateways connect to the network server using internet protocol (IP) links over wired or wireless technologies. Also, nodes communicate with the gateways using LoRa- or frequency shift keying (FSK)-modulated point-to-point wireless links [2,3].

However, describing the requirements of LoRaWAN device classes is crucial because it defines the minimum capabilities needed for interoperability and operation in the network. Devices must comply with Class A functionalities, which ensures that they can operate in a basic and effective manner in the LoRaWAN network. Likewise, Classes B and C, although optional, offer additional features, such as improvements in synchronization and responsiveness, which can optimize performance in specific applications. In addition, the protocol specification details critical aspects such as the format of messages in the physical and media access control (MAC) layers, reception window management, and security methods, including encryption and integrity, ensuring that all devices can communicate securely and efficiently while complying with the necessary standards [4].

A significant amount of research has been driven by its potential in diverse areas. For example, to assess the performance measurement of LoRaWAN technology, the authors in [2] executed a performance evaluation using the specialized tool SimpleIoTSimulator, which enables the generation of traffic from tens of thousands of LoRa nodes and their direct connection to the things network (TTN) server to emulate real-world conditions. Furthermore, the authors in [5] utilized LoRa/LoRaWAN technology to connect objects to the internet reliably and securely and to study its development and application in Ecuador, emphasizing its benefits and application scenarios in simulated environments for future implementations both indoors and outdoors. Similarly, in [6], the authors presented a prototype application to detect the occupancy level of garbage containers using ultrasonic sensors based on LoRaWAN technology, aiming to optimize garbage collection in Ecuadorian cities. Lastly, in [7], the authors offered a review of various LoRa and LoRaWAN device configurations across different applications, highlighting the impact of regional regulations and the use of default settings. They investigated different deployment scenarios, ranging from indoor settings to extensive outdoor communication.

Additionally, in more recent studies such as [8,9], the future of IoT is focused on the growing number of applications communicated with LoRaWAN and complemented by the advanced processing capacity of deep learning, which can improve efficiency in the detection of anomalies, the prediction of failures, and the optimization of energy use in IoT networks.

Unlike previous studies, we present a practical approach by implementing a LoRaWAN-based measurement prototype in both urban and rural areas of Quito, using the default settings for simplicity and ensuring smooth network integration. Additionally, we examine the usability of LoRaWAN, concentrating on a comprehensive analysis of the technology's performance under various environmental conditions and its efficiency in data collection across different scenarios. In essence, we propose an experimental prototype to perform proof-of-concept tests of the LoRaWAN protocol in both urban and rural settings in Quito. This prototype enables us to assess the performance and effectiveness of LoRa in various situations, offering a practical and detailed perspective on the technology's behavior and aiding in understanding its relevance in specific urban contexts.

2. Materials and Methods

In order to carry out a proof of concept of LoRa technology in both urban and rural environments, we implement a communication prototype, as described in Figure 1. Here, we employ the structured methodology consisting of three main phases: node configuration, server deployment, and communication and monitoring establishment, as explained below.

The first phase involves configuring the LoRaWAN devices that act as nodes. These nodes are usually equipped with sensors to collect relevant data, such as temperature, humidity, or any other parameter needed for the specific application. Each node must be configured correctly to send data via the LoRa protocol. Specifically, configurations include assigning unique identifiers to each node and scheduling it to transmit data at specific intervals. Furthermore, nodes are also required to have sufficient power, either through lithium batteries or alternative power sources suitable for their operating environment [10].

The second phase focuses on configuring and integrating the necessary servers: the join server, the network server, and the application server. In particular, the join server handles the authentication and authorization of the nodes that wish to join the network. The network server manages data routing between the nodes and the application server, applying network rules, and ensuring correct packet transmission. The application server receives the processed data from the network server and stores or displays it in specific applications. Each of these servers must be configured and tested to ensure that it can handle the expected volume of data and that network security is guaranteed [11].

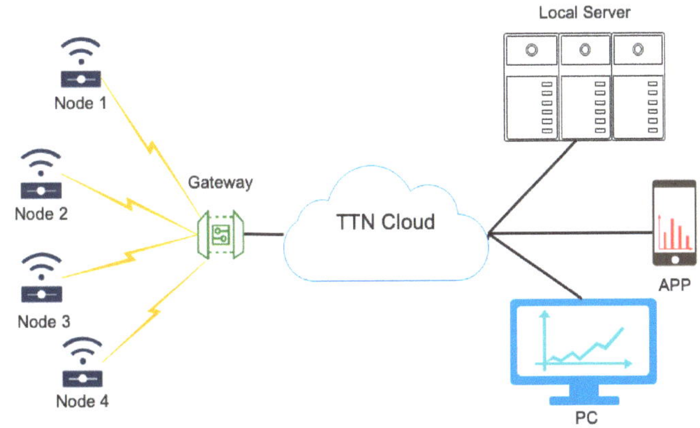

Figure 1. Structured methodology for LoRaWAN communication, based on [7].

The last phase involves establishing effective communication between the nodes and servers and implementing the necessary monitoring tools. The nodes need to be deployed in situ and checked for proper communication with the LoRaWAN gateways that relay data to the network server. It is essential to monitor the network server to ensure accurate reception and forwarding of data to the application server, avoiding any information loss. Ultimately, the application server should provide a user interface for data visualization and analysis, granting users access to the information via mobile devices or web applications. Continuous oversight of the entire infrastructure will allow for real-time problem detection and resolution, thereby guaranteeing the robustness and reliability of the LoRaWAN prototype [12].

2.1. Requirements

In this section, we thoroughly detail the essential hardware needed for the suggested prototype. This involves an in-depth analysis of each required component, as presented in Table 1. The choice of hardware is crucial, as it significantly impacts the system's performance and dependability across different conditions. Our goal is to lay a robust groundwork for the prototype's implementation and future assessment by carefully specifying the components and their characteristics.

Table 1. Required hardware for the proposed LoRaWAN network.

Device/Characteristic		Description
Arduino		Open source hardware development platform based on a board with a microcontroller and a development environment.
MKR WAN 1310 board		Arduino compatible development board, equipped with a LoRa module for long distance communication.
MKR ENV SHIELD board		Expansion module for MKR boards that includes several environmental sensors.
Sensors	ST HTS221	Temperature and humidity sensor with high precision and low power consumption.
	ST LPS22 HB	High resolution barometric pressure sensor, ideal for environmental applications.
	VISHAY TEMT 6000	Ambient light sensor that provides an analog output proportional to light intensity.
	VISHAY VEML 6075	UV sensor that measures UVA and UVB ultraviolet radiation.
Half wavelength dipole antenna		Antenna used to improve the reception and transmission of radio frequency signals.
Li-ion type battery		Rechargeable lithium-ion battery, known for its high energy density and long life.

2.2. Implementation

In Figure 2, we illustrate the primary components of The Things Network Stack Community Edition platform, which serves as the foundation for developing and implementing the proposed network prototype. In this context, Table 2 provides a concise description of each of these components.

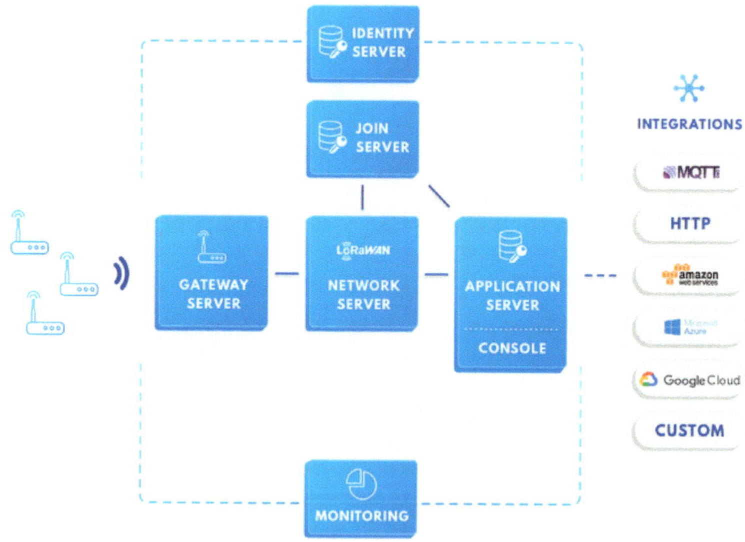

Figure 2. Architecture for a LoRaWAN network implementation [13].

Table 2. Architecture of the Proposed LoRaWAN Network Implementation [14].

Component	Description
End Nodes	These are the sensors or devices that collect data and transmit it wirelessly to the gateway. They are shown on the left side of the image, communicating with the gateway server.
Gateway Server	The gateway server receives data from the end nodes and forwards it to the network server. This acts as a bridge between the end nodes and the core network infrastructure.
Network Server	The network server is the central hub that manages the data traffic within the LoRaWAN network. It handles the communication between the gateway server and the application server. Additionally, it interacts with the join server to manage the addition of new devices.
Join Server	The join server is responsible for authenticating and managing devices as they join the network. It ensures that only authorized devices can communicate within the network.
Application Server	The application server processes the data received from the network server and makes it available for end-user applications. This server can interface with various external services and platforms for data visualization, storage, and further processing.
Identity Server	The identity server is responsible for managing the identities of devices and users within the network. It ensures secure and authorized access to the network resources.
Integrations	Various integration options are shown, including MQTT, HTTP, Amazon Web Services (AWS), Microsoft Azure, Google Cloud, and custom integrations. These integrations allow for the seamless connection of the LoRaWAN network with external platforms for enhanced data management and utilization.

2.3. Programming Data Acquisition

The operational sequence of the LoRa communication prototype starts with the inclusion of essential modules, variable definitions, and the creation of an instance of the LoRaModem class, which sets up the communication environment. The Setup function, responsible for configuration, initializes the parameters needed for sensor reading. Subsequently, the sensor data are read and converted to bytes, making it ready for transmission. This data are stored in a byte array within the payload variable.

The gateway and network server configuration are set up prior to practical deployment. The gateway is configured through files such as global_conf.json and local_conf.json, and the network server is configured to manage nodes using The Things Stack (TTS) version 3.18 platform. The node code structure is detailed in the LoRa_Tx_MonAmb.ino file, which implements the core functionality, and the secrets.h file, which contains the session keys needed for pairing with the network server. The configuration process initializes the radio module in the US915 band and uses the Over-The-Air Activation (OTAA) method to pair the node with the server. During the run cycle, the system takes readings from sensors, converts the values to byte arrays, and transmits them through the radio module.

Once the data are prepared in the payload array, the LoRa message transmission proceeds. During transmission, the reception buffer is monitored to confirm successful message sending. The received message is displayed on the serial monitor for inspection and analysis. Moreover, a function to convert floats to bytes (float2Bytes) is provided, and care is taken to ensure that configuration files such as arduino_secrets.h and global_conf.json are correctly set up and accessible for the prototype's proper operation.

2.4. Test Environment Description

Here, to carry out the proposed tests, the gateway is positioned on a three-meter high platform in a rural area of Quito, specifically in the Cocotog zone. This area is characterized by its low building density and relatively flat terrain with minimal altitude variations. These conditions favor the transmission of radio signals, especially in scenarios where there is a clear line of sight between the transmitter and the receiver, which allows the anticipation of an optimal communication range.

For the measurement campaign, in Figure 3, we illustrate a detailed view of the coverage and performance of the system as a function of distance and the specific conditions of the rural environment. Specifically, data were collected at nine points (blue markers) located at various distances from the node to the gateway (red marker).

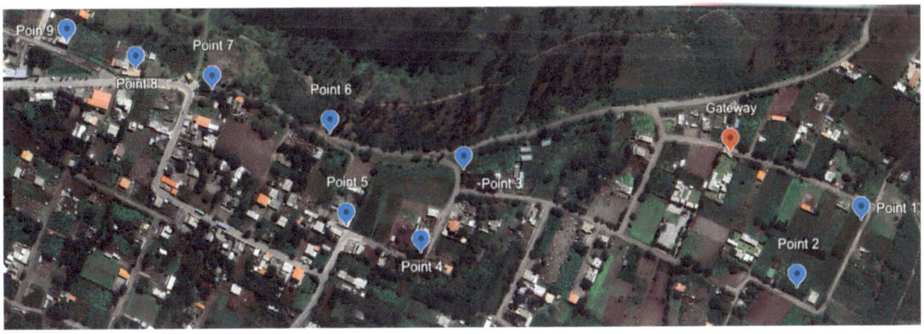

Figure 3. Test points in rural environment.

On the other hand, for the scenario in Figure 4, note that the environment is exposed to a significantly higher level of interference, which could negatively affect the range of the radio link compared to the rural environment. In particular, the gateway was installed on the terrace of a building in the La Florida zone, in the north of Quito. This location is characterized by its density of buildings and possible obstructions, which introduces a greater number of obstacles and sources of interference for the radio signal. For this setup, the tests were carried out at four different points (blue markers) within this urban area. These points were selected to assess the system's performance in more complex conditions and to measure how urban interference impacts the quality of communication and the range of the radio link.

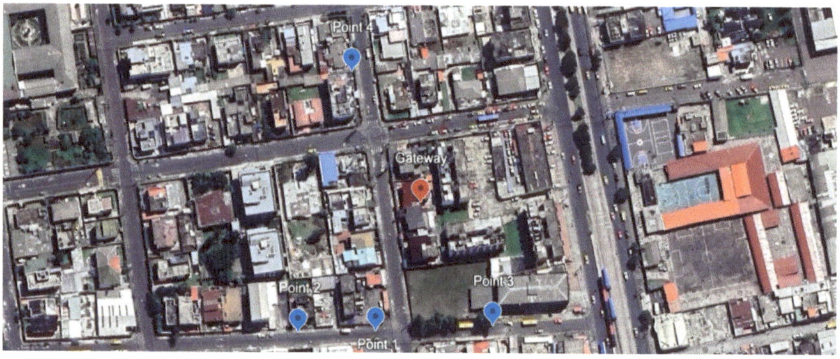

Figure 4. Test points in an urban environment.

3. Results and Discussion

In this work, the airtime of the packet within the radio link has been analyzed as an indicator of energy consumption in the node of the LoRa network. Specifically, packet airtime is defined as the duration for which the antenna actively transmits the message. Consequently, extended airtime correlates with increased energy consumption. Additionally, the signal-to-noise ratio (SNR) has been evaluated, given its utilization by the adaptive data rate (ADR) algorithm to determine the spreading factor. In summary, the operation and efficacy of the ADR mechanism are discerned through variations in these specified transmission parameters.

The LoRaWAN packet itself serves as the source for the test data, as it is the gateway that appends metadata to the packet, including the SNR, before it is forwarded to the network server. At this stage, the ADR field is processed, and the spreading factor (SF) is determined. Two distinct test environments were selected based on their unique environmental conditions, which influence the transmission characteristics of the radio link differently. In both the urban and rural test scenarios, a maximum of fifteen packets were transmitted at specific distances between the node and the gateway. Nevertheless, variations in the packet's airtime and SNR values are observed only in the packets that were successfully received.

In Figure 5, we illustrate the 3D scatter plot delineating the interrelation among the received signal strength indicator (RSSI), spreading factor (SF), and consumed airtime in milliseconds. Concerning LoRa communications, the reliable RSSI range is typically between −70 dBm and −120 dBm. Values around −70 dBm indicate a strong signal and reliable communication, while values around −100 dBm may still be acceptable, although the communication quality may start to decline. Values below −120 dBm suggest a very weak signal, commonly resulting in unreliable communication, high error rate, and connection issues.

Figure 6 exemplifies the RSSI vs. Consumed Airtime Scatterplot results grouped by SF, demonstrating that stable and effective communication is ensured. It is important to note that the variation in consumed airtime depends not only on the RSSI value, but also on the SF. Lower RSSI values (approximately between −105 dBm and −100 dBm) are associated with higher consumed airtime when the SF is higher (e.g., SF values of 9 or 10). This suggests that even if the signal is weak, communication can still be stable by increasing the SF, albeit at the cost of higher airtime. Therefore, efficient communication, especially in the trade-off between SF and RSSI, is experimentally verified, considering the specific needs of the network and the environment in which the system is deployed.

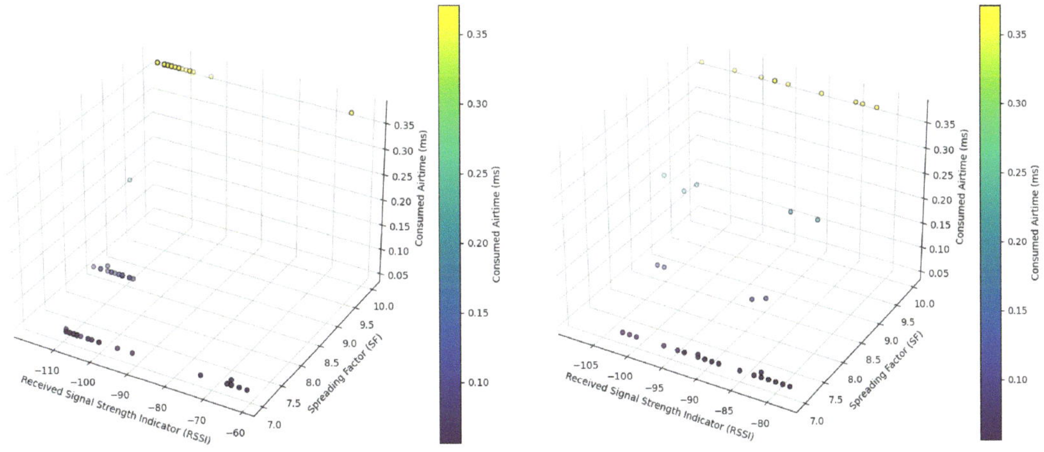

(**a**) (**b**)

Figure 5. RSSI vs. SF vs. Consumed Airtime. (**a**) Scenario rural; (**b**) Scenario urban.

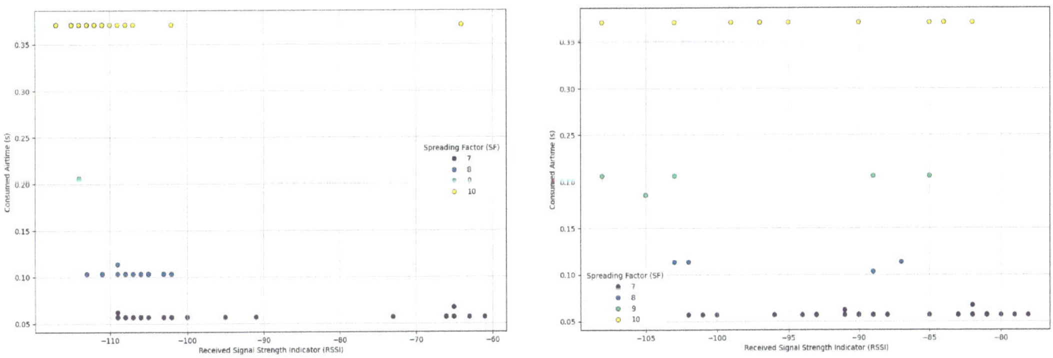

(**a**) (**b**)

Figure 6. RSSI vs. Consumed Airtime. (**a**) Scenario rural; (**b**) Scenario urban.

In both Figures 7, the expected behavior of air-time is observed as a function of SF and SNR. As SF increases, air-time also increases, which is consistent with the nature of LoRa modulation, since higher SF implies longer transmission duration. Also, SNR seems to influence air-time, with higher SNR values associated with lower air-time, reflecting the system's ability to improve transmission in cleaner signal conditions. The general alignment of the points suggests predictable and consistent behavior in both urban and rural environments.

In Figure 8, we perform a curve analysis by applying a polynomial regression to the RSSI and SNR data at different distances. We extracted distance, SNR, and RSSI data from a raw file and filtered and converted them to numerical values. A polynomial function was fitted to generate trend lines showing how RSSI and SNR vary with distance. This process was applied to both rural and urban data sets. The resulting curves illustrate the relationship between distance and RSSI, providing insight into signal behavior in different environments. The results obtained in both scenarios were satisfactory, allowing the comparison of the LoRa protocol's performance in different contexts, confirming the system's robustness against environmental variations.

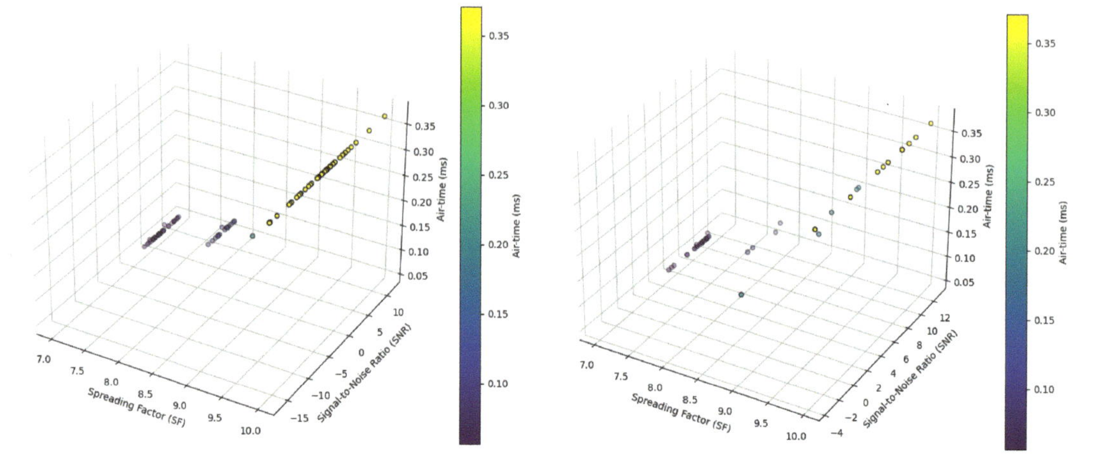

(**a**) (**b**)

Figure 7. SF vs. SNR vs. Consumed Airtime. (**a**) Scenario rural; (**b**) Scenario urban.

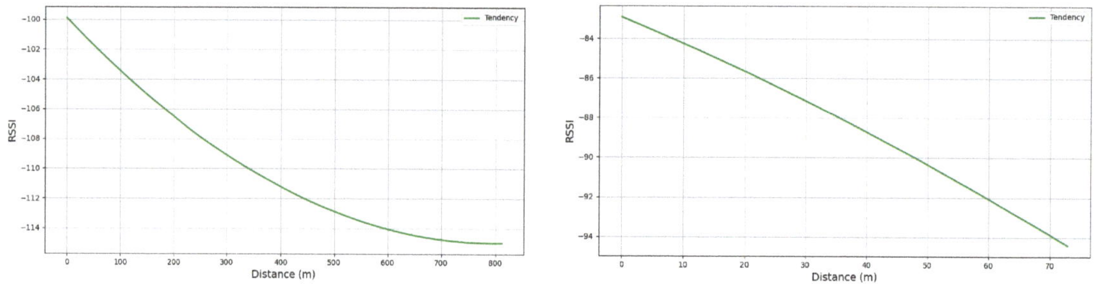

(**a**) (**b**)

Figure 8. RSSI as a function of Distance. (**a**) Scenario rural; (**b**) Scenario urban.

The results showed that in urban areas with tall buildings, the transmission signal had a shorter range due to interference but still remained stable over distances of up to 2 km. In rural areas without as many obstacles the signal reached much further, up to 5 km. This shows that the technology is effective in collecting data in different environments, but performance can vary depending on geographic features and building density.

A simple use case of environmental monitoring in Quito, a city with both urban and rural areas. The LoRaWAN-based communication prototype was deployed to measure air quality in different zones. Sensors were placed on nodes connected to a LoRa network to transmit data such as temperature, humidity, and pollution levels to a central server.

4. Conclusions

A prototyping environment for LoRaWAN solutions was created using Arduino MKR WAN 1310 for nodes, Raspberry Pi 3 model B+ for gateways, and The Things Stack for the network server, all with free software and hardware. The Raspberry Pi provided an economical and functional gateway solution, while The Things Stack supported the integration and development of applications with node data through its graphical interface and LoRaWAN 1.0.3 features.

This experiment confirmed that the LoRa protocol maintains stable communication in both rural and urban environments, even with weaker signals, ensuring reliable data transmission over long distances. Its robustness allows it to handle interference efficiently, underlining its suitability for various applications. The protocol is highly adaptable, adjusting performance based on environmental conditions and demonstrating versatility in different scenarios.

Experiments have shown that a smaller SF for short-range transmissions decreases packet airtime and battery consumption. In addition, our results indicated that packet loss increases with distance in both urban and rural settings. In urban environments, minor distance variations, for example, from 78.91 m to 85.6 m, can result in complete communication failure between the node and the gateway.

This paper lays the groundwork for future studies in network optimization, prediction models, interference, robustness analysis, scenario comparison, and practical applications. Future work will focus on: (a) Optimizing LoRaWAN node configuration to enhance coverage and reduce low reception areas. This includes examining how vegetation, topography, and buildings affect signal quality. Additionally, assess transmission power and coding techniques to improve network performance for rural applications in environmental monitoring, precision agriculture, and resource management; and (b) using machine learning algorithms to build more precise models to forecast signal attenuation due to distance and environmental variables, thus improving predictions of coverage and signal quality.

Author Contributions: Conceptualization, M.R. and L.U.; methodology, R.M. and M.R.; Investigation, R.M. and M.R.; software, M.R.; validation, M.R. and L.U.; formal analysis, R.M. and L.U.; investigation, M.R., R.M. and L.U.; Writing—Original Draft Preparation, R.M. and M.R.; Writing—Review and Editing, L.U. and J.D.V.-S.; Visualization, J.D.V.-S.; Supervision, L.U. and J.D.V.-S. All authors have read and agreed to the published version of the manuscript.

Funding: This research was funded by project PII-DETRI-2024-02, "Análisis de desempeño de comunicaciones 6G asistidas por superficies inteligentes reconfigurables o antenas fluidas"

Institutional Review Board Statement: Not applicable.

Informed Consent Statement: Not applicable.

Data Availability Statement: Data is contained within the article.

Conflicts of Interest: The authors declare no conflicts of interest

References

1. Nolan, K.E.; Guibene, W.; Kelly, M.Y. An evaluation of low power wide area network technologies for the Internet of Things. In Proceedings of the 2016 International Wireless Communications and Mobile Computing Conference (IWCMC), Paphos, Cyprus, 5–9 September 2016; pp. 439–444.
2. Lavric, A.; Petrariu, A.I.; Coca, E.; Popa, V. LoRaWAN analysis from a high-density Internet of Things perspective. In Proceedings of the 2020 International Conference on Development and Application Systems (DAS), Suceava, Romania, 21–23 May 2020; pp. 94–97.
3. Ebi, C.; Schaltegger, F.; Rüst, A.; Blumensaat, F. Synchronous LoRa mesh network to monitor processes in underground infrastructure. *IEEE Access* **2019**, *7*, 57663–57677. [CrossRef]
4. Jouhari, M.; Saeed, N.; Alouini, M.S.; Amhoud, E.M. A survey on scalable LoRaWAN for massive IoT: Recent advances, potentials, and challenges. *IEEE Commun. Surv. Tutor.* **2023**, *25*, 1841–1876. [CrossRef]
5. Pinto-Erazo, A.M.; Suárez-Zambrano, L.E.; Mediavilla-Valverde, M.M.; Andrade-Guevara, R.E. Introductory analysis of LoRa/LoRaWAN technology in Ecuador. In Communication, Smart Technologies and Innovation for Society: Proceedings of CITIS 2021; Springer: Singapore, 2022; pp. 547–557.
6. Valle Paillacho, B.R. Desarrollo de Un Prototipo de Aplicación Para Detectar el Nivel de Ocupación de Contenedores de Basura de Gran Capacidad Utilizando Tecnología Lorawan. Bachelor's Thesis, Escuela Politécnica Nacional, Quito, Ecuador, 2022.
7. Khutsoane, O.; Isong, B.; Abu-Mahfouz, A.M. IoT devices and applications based on LoRa/LoRaWAN. In Proceedings of the IECON 2017—43rd Annual Conference of the IEEE Industrial Electronics Society, Beijing, China, 29 October–1 November 2017; pp. 6107–6112.
8. Chinnasamy, P.; Sathya, K.S.; Jebamani, B.J.A.; Nithyasri, A.; Fowjiya, S. Deep Learning: Algorithms, techniques, and applications—A systematic survey. In *Deep Learning Research Applications for Natural Language Processing*; IGI Global: Hershey, PA, USA, 2023; pp. 1–17.
9. Arulananth, T.; Kuppusamy, P.; Ayyasamy, R.K.; Alhashmi, S.M.; Mahalakshmi, M.; Vasanth, K.; Chinnasamy, P. Semantic segmentation of urban environments: Leveraging U-Net deep learning model for cityscape image analysis. *PLoS ONE* **2024**, *19*, e0300767. [CrossRef] [PubMed]

10. Prauzek, M.; Konecny, J.; Borova, M.; Janosova, K.; Hlavica, J.; Musilek, P. Energy harvesting sources, storage devices and system topologies for environmental wireless sensor networks: A review. *Sensors* **2018**, *18*, 2446. [CrossRef] [PubMed]
11. Košt'ál, K.; Helebrandt, P.; Belluš, M.; Ries, M.; Kotuliak, I. Management and monitoring of IoT devices using blockchain. *Sensors* **2019**, *19*, 856. [CrossRef] [PubMed]
12. Hou, L.; Zhao, S.; Xiong, X.; Zheng, K.; Chatzimisios, P.; Hossain, M.S.; Xiang, W. Internet of things cloud: Architecture and implementation. *IEEE Commun. Mag.* **2016**, *54*, 32–39. [CrossRef]
13. CatSensors. Tecnología LoRa y LoRaWAN. 2024. Available online: https://www.catsensors.com/es/lorawan/tecnologia-lora-y-lorawan (accessed on 22 August 2024).
14. Zhou, Q.; Zheng, K.; Hou, L.; Xing, J.; Xu, R. Design and Implementation of Open LoRa for IoT. *IEEE Access* **2019**, *7*, 100649–100657. [CrossRef]

engineering proceedings

MDPI

Proceeding Paper

Detection of Non-Technical Losses in Special Customers with Telemetering, Based on Artificial Intelligence †

José Luis Llagua Arévalo * and Patricio Antonio Pesántez Sarmiento *

Department of Electrical and Electronic Engineering, National Polytechnic School, Quito 170525, Ecuador
* Correspondence: jose.llagua@epn.edu.ec (J.L.L.A.); patricio.pesantez@epn.edu.ec (P.A.P.S.)

† Presented at the XXXII Conference on Electrical and Electronic Engineering, Quito, Ecuador, 12–15 November 2024.

Abstract: The Ecuadorian electricity sector, until April 2024, presented losses of 15.64% (6.6% technical and 9.04% non-technical), so it is important to detect the areas that potentially sub-register energy in order to reduce Non-Technical Losses (NTLs). The "Empresa Eléctrica de Ambato Sociedad Anónima" (EEASA), as a distribution company, has, to reduce NTLs, incorporated many smart meters in special clients, generating a large amount of data that are stored. This historical information is analyzed to detect anomalous consumption that is not easily recognized and is a significant part of the NTLs. The use of machine learning with appropriate clustering techniques and deep learning neural networks work together to detect abnormal curves that record lower readings than the real energy consumption. The developed methodology uses three k-means validation indices to classify daily energy curves based on the days of the week and holidays that present similar behaviors in terms of energy consumption. The developed algorithm groups similar consumption patterns as input data sets for learning, testing, and validating the densely connected classification neural network, allowing for the identification of daily curves described by customers. The results obtained from the system detected customers who sub-register energy. It is worth mentioning that this methodology is replicable for distribution companies that store historical consumption data with Advanced Measurement Infrastructure (AMI) systems.

Keywords: energy under registration; fraudulent curves; validation indices; neural networks; non-technical losses

Citation: Llagua Arévalo, J.L.; Pesántez Sarmiento, P.A. Detection of Non-Technical Losses in Special Customers with Telemetering, Based on Artificial Intelligence. *Eng. Proc.* **2024**, *77*, 29. https://doi.org/10.3390/engproc2024077029

Academic Editor: Walter Vargas

Published: 18 November 2024

1. Introduction

In a society that is highly dependent on the availability, efficiency, and reliability of electricity, it is essential for the electric companies in the sector to have good management of the production and distribution of energy. One of the major problems that concerns this industry is electrical losses, which occur mostly in the distribution segment [1]. In the transportation of energy, losses are the difference between the electricity that enters the network and that which is delivered for final consumption, and they reflect the level of efficiency of the transmission and distribution infrastructure. Being able to reduce losses is essential to increasing the efficiency of energy distribution, and, in many cases, it can even help improve the financial sustainability of distribution companies. In Ecuador, at a general level, in recent years, in the electric sector, the level of losses has substantially improved, dropping from 22% in 2006 to 13.06% in 2021 [2]; however, recently, it has increased to 15.64% (6.6% technical and 9.04% non-technical). The advances in the identification of NTLs have been undeniable recently, and identification models have been established with the information to detect patterns or find areas where there are high percentages of NTLs, as well as methodologies based on energy balances at the level of distribution transformers and medium voltage branches. Smart Grids have evolved to the level of incorporating techniques based on artificial intelligence to solve real problems, such as NTLs in distributors, demand prediction, and adaptive protections, among others. These systems generate a large amount of data and cybersecurity plays an important role. The purpose of this article is to develop a methodology to detect non-technical losses using readings of energy consumption every 10 min, use various classification indexes that guarantee the formation of groups, and finally implement artificial intelligence techniques for the development of pattern recognition of fraudulent and non-fraudulent curves of industrial clients owned by EEASA. Compared to the latest advances in NTL methodologies, in which the Multiversal Recurrent Algorithm with Multiple Repetitions (MV-REMR), published in November 2023, stands out [3], the lack of information on energy consumption replaced by zeros has, in this research, been solved through the use of statistical interpolation methods. The records obtained by the AMI, with sampling every 10 min, generated a matrix of 105,264 per special client for two years; to facilitate information management, the reduction to an hourly demand was carried out, and this allowed us to improve the classification of the clients using clustering techniques supported by three k-means validation indices, complying with the class balance. Densely Connected Neural Networks (DenseNet) have a better response in non-discretized systems compared to the Convolutional Neural Networks (CNN) used in MV-REMR, allowing for a better view of the consumption pattern of special clients. Likewise, the use of AMIs, due to their high reliability in data acquisition, communication, and storage, makes the proposed methodology more effective.

2. Methodological Proposal

The methodology consists of the evaluation of historical daily curves in a two-year time interval for the identification of non-technical losses through deep learning of the classification neural network. This method will help us to classify curves with anomalies and typical curves based on learning from the historical data downloaded from the distribution company's customers, identifying patterns considered with the underreporting of customers that could be suspected of energy fraud. Figure 1 summarizes this methodology, showing three clear stages for the development of this work.

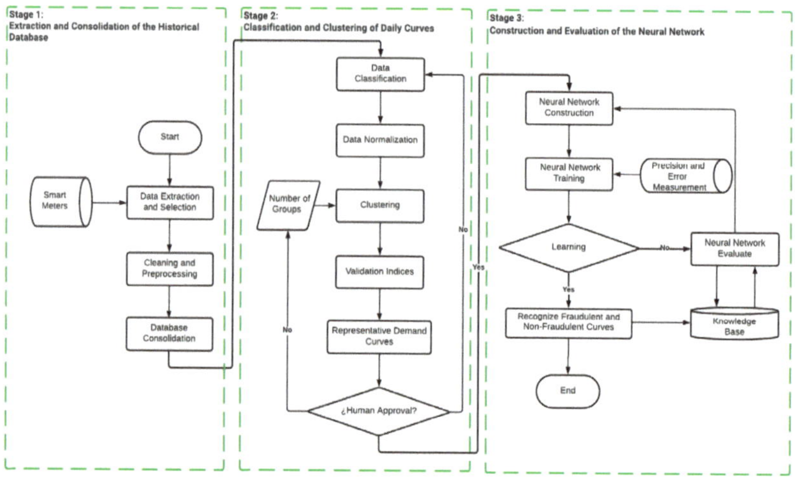

Figure 1. Methodology flowchart.

2.1. Database Extraction and Consolidation

From the historical data of the special customers (industrial and commercial) that have telemetering for the period from 31 May 2020 to 31 May 2022, a data matrix will be created for which a pre-selection of the most representative customers will be made according to their consumption characteristics and metering history.

2.1.1. Extraction and Selection

In the selection, customers have determining details, such as the characteristic of "possessing Telemetry" and "Total kW Consumption", which will be important at the time of selecting 100 of the 297 customers that meet the conditions of having historical data for greater than two years and for whom there has not been a reason for a change in the telemetering process during the designated time interval. With the stratified sample method, the number of individuals from each zone to be extracted to form the customer matrix is determined, as shown in Table 1, which specifies the number of customers chosen from each zone and the number from which the historical data will be extracted.

Table 1. Stratified sample of customers.

Zone	Description	Customers	Sample
1	Ambato	142	60
2	Pelileo	43	10
3	Pillaro	13	3
4	Baños	11	5
5	Patate	8	2
7	Pastaza	43	8
8	Palora	2	2
10	Quero	14	3
11	Tena	19	5
12	Archidona	2	2
TOTAL		297	100

The platform "Telemetering Systems for special customers EASYmetering AMI Solutions", used by the distributor, has several clusters where they store readings of voltage [V], current [A], power [kW], active energy [kWh], and reactive energy [kVArh] in intervals of 10 min; this results in 144 data entries for each day.

2.1.2. Cleaning and Preprocessing

It is important to mention that not all the records are complete due to failures in the communication between the AMI and the server; also, not all the columns contain the same format. The work is meticulous client-by-client since each file had its own particularities which did not allow for treating all the extracted files in a general way. Instead, we followed a small procedure to eliminate inconsistencies, which is detailed as follows: 1. In the debugging phase, missing values are detected and replaced by means of interpolation techniques; 2. The interpolation considers all the points and correlates the complete behavior of the set used; and 3. Cases where there was no data had the particularity that they were followed by Not a Number (NaN) values, showing that there were connection failures with the server and the AMI meter, so the data were interpolated for the interval with these values.

2.1.3. Consolidation of the Complete Data

With the customer data processed and purified, which contained the 105,264 pieces of data on energy delivered every 10 min during the 731 days of the two years, we proceeded to consolidate the database of the 100 customers. Keeping their names assigned by the zone of their location and the place they occupied in the ranking of energy consumption, the consolidated data are reduced to 17,544 rows to obtain consumption values for each hour, as shown in the green circles of the Figure 2, where the first columns have the date and time of the reading of the energy delivered and then, in chronological order, the demand values expressed in kW that each client had, thus forming the history of the two years of AMI readings.

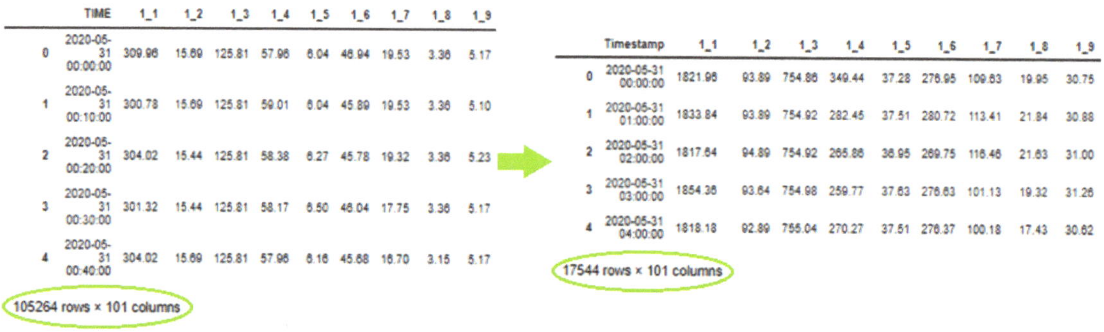

	TIME	1_1	1_2	1_3	1_4	1_5	1_6	1_7	1_8	1_9
0	2020-05-31 00:00:00	309.96	15.69	125.81	57.96	6.04	46.94	19.53	3.36	5.17
1	2020-05-31 00:10:00	300.78	15.69	125.81	59.01	6.04	45.89	19.53	3.36	5.10
2	2020-05-31 00:20:00	304.02	15.44	125.81	58.38	6.27	45.78	19.32	3.36	5.23
3	2020-05-31 00:30:00	301.32	15.44	125.81	58.17	6.50	46.04	17.75	3.36	5.17
4	2020-05-31 00:40:00	304.02	15.69	125.81	57.96	6.16	45.68	16.70	3.15	5.17

105264 rows × 101 columns

	Timestamp	1_1	1_2	1_3	1_4	1_5	1_6	1_7	1_8	1_9
0	2020-05-31 00:00:00	1821.96	93.89	754.86	349.44	37.28	276.95	109.63	19.95	30.75
1	2020-05-31 01:00:00	1833.84	93.89	754.92	282.45	37.51	280.72	113.41	21.84	30.88
2	2020-05-31 02:00:00	1817.64	94.89	754.92	265.86	36.95	269.75	116.46	21.63	31.00
3	2020-05-31 03:00:00	1854.36	93.64	754.98	259.77	37.63	276.63	101.13	19.32	31.26
4	2020-05-31 04:00:00	1818.18	92.89	755.04	270.27	37.51	276.37	100.18	17.43	30.62

17544 rows × 101 columns

Figure 2. Variabilit.

2.2. *Classification and Clustering of the Daily Curves*

2.2.1. Data Reduction and Sorting

To obtain a more effective data reduction without losing important information in the energy records, all energy values within each hour were summed, thereby obtaining a reduction from "105,264 rows × 100 columns" to "17,544 rows × 100 columns".

The classification of the data was performed with prior knowledge of the way in which EEASA's special customers record their energy consumption, knowing beforehand that most industrial sector workers work typical working days between Monday and Friday, that, on weekends, few industrial customers work fully, and that holidays are a very unfavorable option to work. The behavior of the daily curves generated on the basis of the energy consumption supplied is similar in most of the special customers, clearly affecting their shape on weekdays; for more information see [4].

Table 2 shows a summary of the classification performed, detailing the days of each group, the data within each group, and the matrix formed for the grouping with the k-means indexes.

Table 2. Classification data frames.

Ranking	Days	Data Frame
Group from Monday to Friday	500	12,000 × 100
Weekend group	173	4152 × 100
Holiday group	58	1392 × 100

2.2.2. Data Normalization

The minimum–maximum normalization methodology is used—see [5]. Figure 3 shows the curves described with the normalized data for the Group from Monday to Friday corresponding to the normalized consumptions represented in different colors for each of the 100 customers, where all values are within the interval 0 and 1.

2.2.3. Clustering

This sub-process conglomerates the daily curves into compact groups with distinct and significant properties from the rest of the groups. In this work, the clustering capability of the k-means algorithm and its variations are investigated; see [6] to understand the heuristic iterative procedure to cluster the Representative Demand Patterns into k groups. Where the procedure is fulfilled, choose centroids of k groups (randomly chosen from the set of the Daily Curves), then each Daily Curve is integrated into the nearest group (according to the optimization function), and, finally, the centroids are recalculated by averaging the Daily Curves of its members. The process is repeated until the centroids of the cluster are stable—see [7].

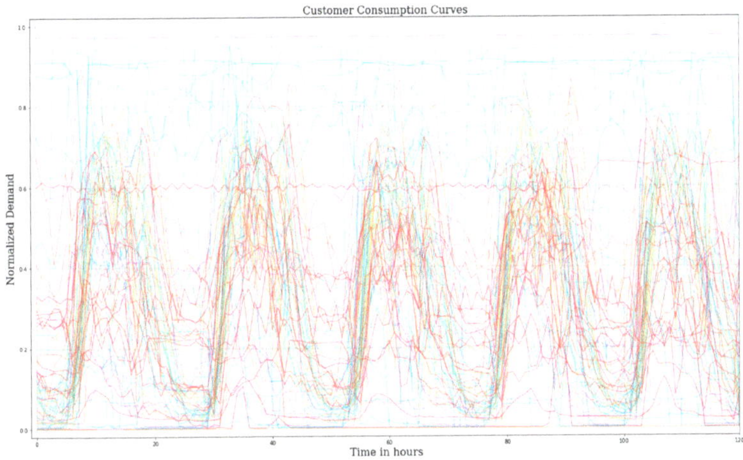

Figure 3. Demand.

2.2.4. Validation Indexes

The indexes used are Euclidean k-means, Dynamic Time Warping (DTW), Barycenter Averaging k-means (DBA), and Soft-DTW k-means, which employ evaluation coefficients such as DTW, Within Cluster Sum of Saqueis (WCSS), and Silhouette Coefficient (SC)—see [8].

In Figure 4, the red curve represents the centroids of each cluster, and the black ones represent the clients that make up the group; the value of "*n*" gives the number of clients that are in each group. The visual analysis of each group with the location of the centroids with respect to the curves that make up the cluster and the experience in the behavior of the curves of the industrial clients are important factors for the definition of the value of k and the most suitable classification method for each group.

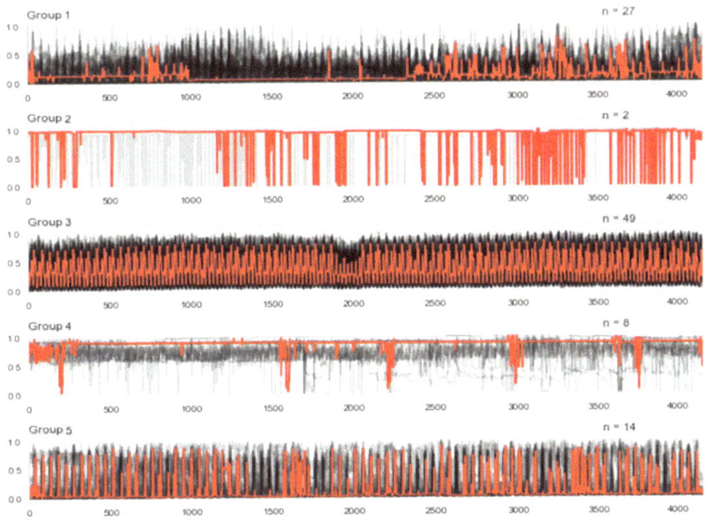

Figure 4. Grouping using the Soft-DTW k-means index for a k = 5, represented the centroid curves in red color.

2.2.5. Definition of Groups

In the visual evaluation process, it is important to carefully observe the graphs described in each evaluation index to determine the method that best describes the group, taking into account the position of the centroids calculated and fully defined in red and the behavior of the curves of the customers with respect to the centroid within each group. Figure 5 shows how they were classified for a k = 5 and the value of 'n' with the number of customers found in each group. It is important to mention that the ideal k value for this project is 5, since 'k' values equal to 4, 6, and 7 do not show homogeneity in the clusters. For more details about evaluation indexes, see [9].

It is important to emphasize that, for the previously classified groups (Monday through Friday, holidays, and weekends), the same methodology is applied, but the behavior of the evaluation indexes with respect to their daily curves is different. Any index of the three that were evaluated can be chosen, observing the best grouping, the number of individuals in the group, and the way the centroid describes the complete group.

2.2.6. Conformation of Groups

With the method and the value of k defined, the assigned values for the grouping and the calculated centroids are determined; with an array of the assignments to each group presented, the centroids calculated for each hour of energy consumption are also normalized. With a small array of rows and columns, each customer can be assigned to the group to which they belong, as shown in Figure 5, where the first column shows the customer, followed by the normalized data of the daily curves, and the last column shows the number of the group to which the customer was assigned. With the values assigned in the Data Frame, the five groups are created with the given assignments—see [10].

	0	1	...	1390	1391	Groups
1_1	0.675224	0.677938	...	0.568357	0.581090	0
1_2	0.748290	0.758575	...	0.103997	0.102846	3
1_3	0.967538	0.967640	...	0.918640	0.918844	3
1_4	0.055345	0.018177	...	0.507596	0.418882	4
1_5	0.116467	0.054153	...	0.018108	0.019235	3
...	...	...	...	...	...	...
11_4	0.019700	0.025805	...	0.195339	0.208935	0
11_5	0.013710	0.007192	...	0.103012	0.074843	4
12_1	0.088134	0.101428	...	0.168554	0.155096	4
12_2	0.170968	0.178710	...	0.272258	0.208710	4
12_3	0.603521	0.601006	...	0.610855	0.611693	4

100 rows × 1393 columns

Figure 5. Grouping assigned values.

2.2.7. Creation of Fraudulent Customers

With the lack of data on daily curves that present abnormal consumptions by the distributor, patterns that under-register energy are created to simulate typical frauds that may occur in special customers of the distributor. Therefore, for the creation of these patterns, the aim is to simulate this type of behavior with a set of random variables, simulating the different types of under-registration that can be created by using malicious activities and directly affecting the distributor and increasing the non-technical losses of the same—see [11]. Therefore, three types of frauds are created, and are described as follows: Type 1 (T1): Constant Proportional Decrease in Time; Type 2 (T2): Proportional Decrease in

Time Windows with Time Band at Maximum Consumption (Peak Hours); and Type 3 (T3): Proportional Decrease in Time Windows with Indistinct and Random Time Band.

To create the fraudulent curves, the parameters detailed in Table 3 were used, in which the values are specified to create the synthetic base that contains the curves with underregistration. With the use of statistical metrics, such as standard deviation, the minimum and maximum values of variability of the data of each group in the daily consumption curves are found.

Table 3. Creation of fraudulent curves.

	Standard Deviation		**Fraudulent Curves**		
	MIN	**MAX**	**Min. % of Admissible Variation**	**% Min. of Variation Chosen**	**Max. % Variation Chosen**
Group Monday to Friday	0.172888139	0.303566006	30.35%	35.00%	85.00%
Group Weekend	0.228457341	0.32703843	32.70%	35.00%	85.00%
Group Holidays	0.210337241	0.344457664	34.45%	35.00%	85.00%

The percentages of decrease, which are between 35% and 85%, create fraudulent curves; these are randomly chosen values and form curves with abnormal consumptions for each type of fraud (in blue color), with respect to the normal energy consumption curves (in orange color), as shown in Figure 6. Each chart has a label that shows the customer code plotted, the type of fraud performed, and the percentage that has been decreased to the normal consumption curve.

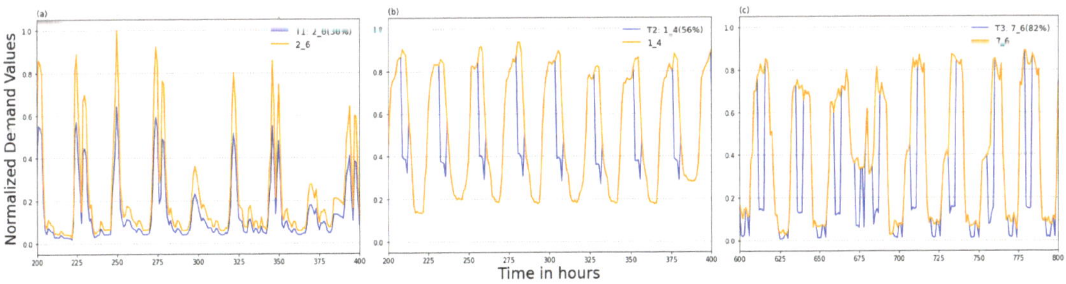

Figure 6. Normal and fraudulent consumption curves with percentage decrease. (**a**) Type 1 with 36% of customer 6 in zone 2. (**b**) Type 2 with 56% of customer 4 in zone 1 and (**c**) Type 3 with 82% of customer 6 in zone 7.

2.3. Construction of the Neural Network

The neural network model to be built is shown in Figure 7, where can see the input, hidden and output layer (rectangles blue color), the main hidden layer is constituted for three hidden layers (rectangles red color) with eighty nine, ninety, eighty nine neurons respectively and the black arrows show the density connection into NN; It is one of the three models to be built for each group is shown. The hyperparameters are parameters external to the model itself, set by the neural network programmer, for example, the selection of the activation function to use or the batch size used in training. The model parameters are internal to the neural network, for example, the weights of the neurons. Stating that the hyperparameters are the adjustable parameters used to control the training process of the model, it must be specified that the values set for each model are different since each data set has different row dimensions.

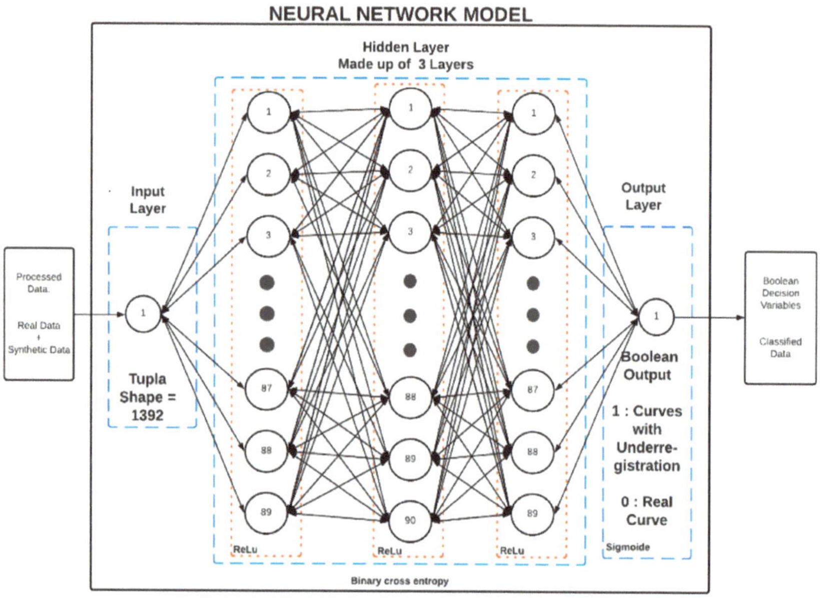

Figure 7. Model network design for the holiday group.

2.3.1. Construction of the Neural Network

For the construction of the neural network, the free software KNIME Analytics Platform 4.7.1 is used which, despite its high performance in the handling of large data sets, must be linked to the installation of the "Python Deep Learning" package with the Keras and Tensor Flow 2 libraries, which are compatible with Python 3.6.13, which comes by default in the installation of Conda in its latest version (Conda 4.12.0). Figure 8 shows the linked software so that the libraries can be used for the construction of the neural network.

2.3.2. Configuration of the Neural Network

For the configuration of the neural network model, the following nodes must be configured: Keras Input Layers, Keras Dense Layers, Keras Network Learner, Keras Network Executor, and a node for the visualization of results which will be Line Plot (local) after forming the complete data (real data and fraudulent data of the three types). The hyperparameters shown in Figure 8 exist in order to form the densely connected neural network that will be trained and tested to check if it is learning and generating knowledge in the identification and classification of curves with sub-register with a value of 1 and normal curves with values of zero. For more information on this, see the thesis work [10], where the process followed in the reading, classification (training, validation, and test data), learning, validation, and testing of the formed groups are shown in detail.

The neural network model and the data execution nodes will be shown configured in the work area, as shown in Figure 9, where all the nodes are executed and in green color, showing the correct procedure in data loading and filtering, configuration, training, validation, and testing of the neural network model. The yellow nodes represented the data conditioning, the magenta nodes show the tag assignment, the brown nodes present the neural network architecture and finally the green nodes evince the trained network model.

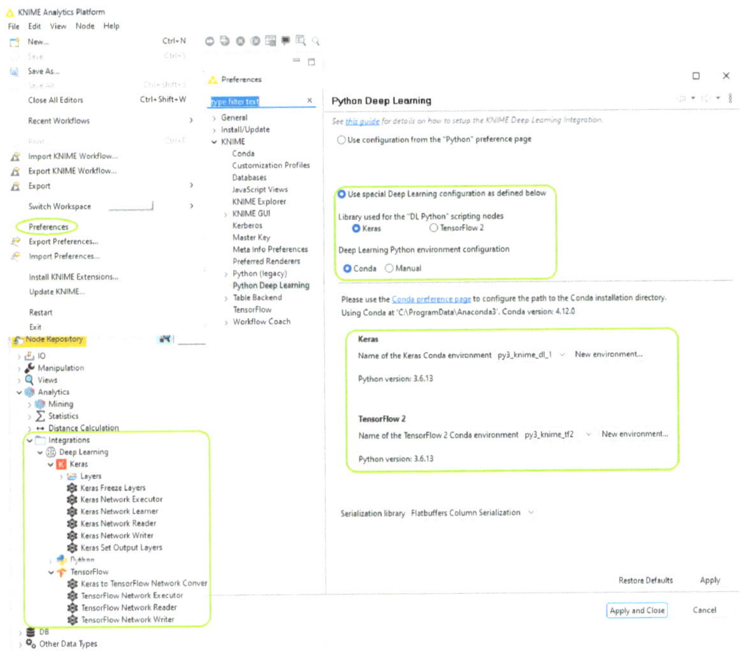

Figure 8. KNIME—Python link and deep learning libraries.

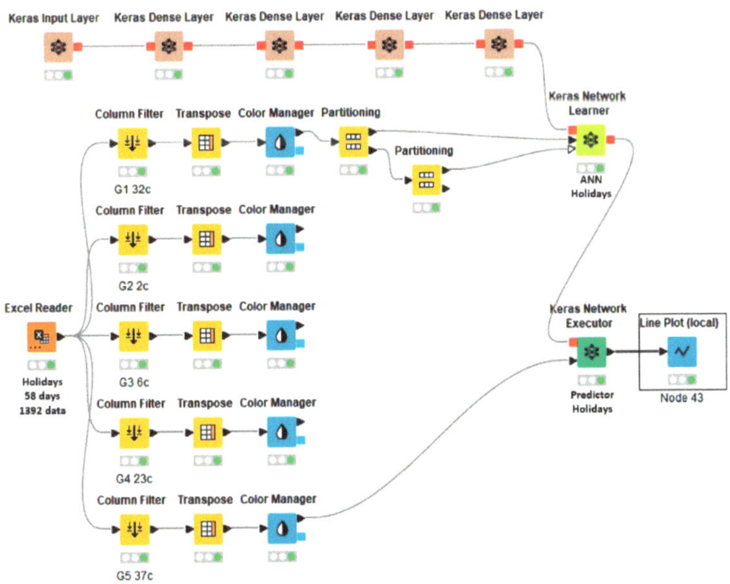

Figure 9. Completed neural network in the working environment.

This process of building the neural network model must be performed for each classification group (Monday to Friday, weekends, and holidays), so, in the end, three neural network models will have to be trained, see [10].

3. Results

The results are focused on the prediction performed by the neural network to classify real and fraudulent curves, so they will focus directly on the accuracy and loss margins of the trained neural network.

3.1. Accuracy

In the option *View: Learning Monitor* of the *Keras Network Layer*, the *Accuracy* of the training can be observed, which indicates the model accuracy and how the model is learning with the *Batches* configured during the construction of the network, as shown in Figure 10. In Figure 10 it can also be seen that the smoothed curve of the training data (in red color) is reaching the values of the smoothed curve of the validation data (in blue color), with small and acceptable error margins within the training. This confirms that Adam's optimization function and the activation functions configured in the hidden layers work together for the learning of the neural network. Also, these results can be verified between the curves of the training data (in pink color) and the validation data (in lilac color), as the values change abruptly at the beginning and the others describe the step function of the activation function *Binary cross entropy* as they pass the *Batch* configured during the learning process of the model, adjusting until the necessary knowledge base for the neural network is achieved.

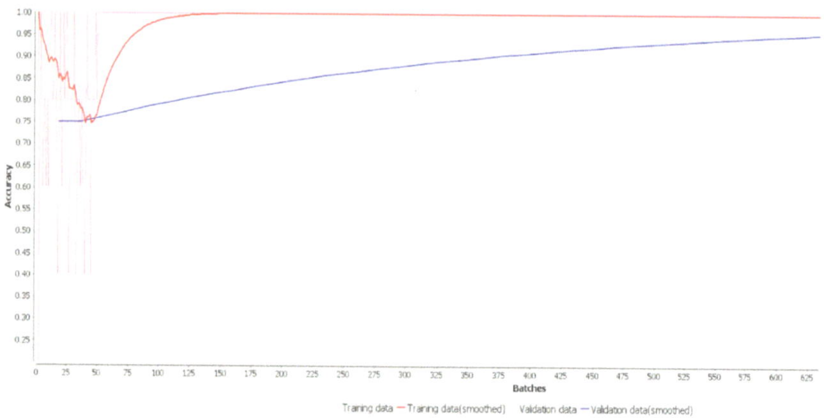

Figure 10. Accuracy curves of the neural network.

3.2. Losses

In the Loss tab, within the same View: Learning Monitor, it can be observed how the values of the error calculated by the Binary cross entropy function, used for the classification of the curve patterns, make the values desirable since the error tends toward zero and the curves converge as the Batch and Epoch are advancing, as shown in Figure 11. Here, it can also be seen that the red curve (Smoothed Training Data) and the blue curve (Smoothed Validation Data) generate increasingly similar values, reaching a minimum distance between the predicted values and the desired values. Between the purple curve (Validation Data) and the pink curve (Training Data), we can see the abrupt changes that the network model undergoes, showing how the configured hyperparameters make the model predict and learn the expected responses. When the accuracy and error curves follow the same trend, it is an indicator that it is a good model.

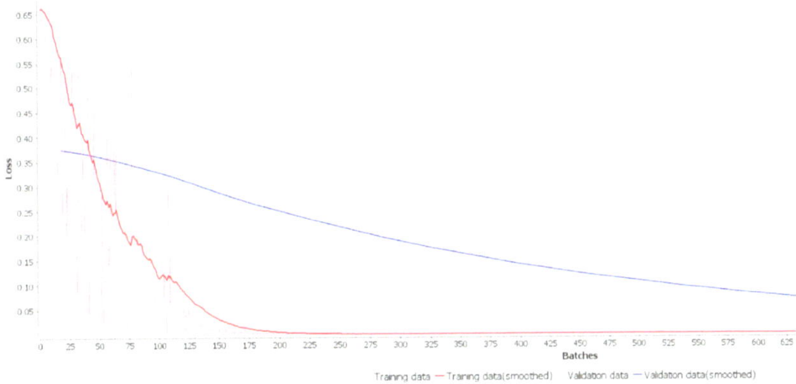

Figure 11. Losses curves of the neural network.

3.3. Tests

With the trained neural network, tests are performed with data from 20 new customers that are completely unknown to the trained network, where the network analyzes 2 months of historical data from 1 June to 31 July 2023. This period has the particularity of not containing holidays, making 61 days to be analyzed, and the results will be shown with the *Line Plot Node*.

The results of the prediction accuracy in red line of the weekend data evaluated on the neural network (18 days) are shown in Figure 12.It can be said that customers 1_2, 1_4, 1_5, 1_8, 3_1, and 7_2, present anomalous consumptions because the red dots have values close to Boolean value one, being possible customers that are performing to register abnormal consumptions or under-registering energy.

Finally, we analyze the data for the group from Monday to Friday (43 days), and we can say that customers 1_4, 1_8, and 7_2, repeat this pattern of anomalous consumption or under-registration of energy, as shown in Figure 13, where we can clearly observe that the high Boole values of these customers are approximated to one, corresponding to abnormal energy consumption.

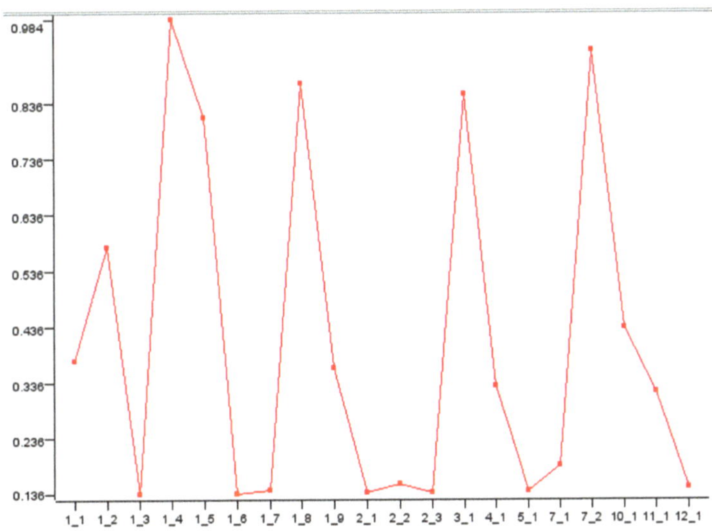

Figure 12. Weekend neural network results.

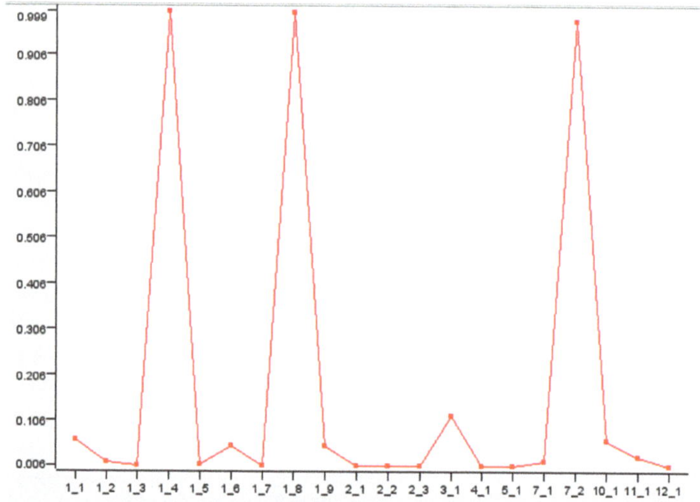

Figure 13. Results of the neural network from Monday to Friday.

With the results given, it can be affirmed that clients 1_4, 1_8, and 7_2 are candidates to be inspected for abnormal consumption. Previously to the culmination of this article, EEASA indicated that, after an inspection was carried out, one of the clients is a frequent recipient of fines for this particularity and the other two presented technical reports of corrective maintenance for problems in their machinery, causing the low energy consumption that they regularly register during their regular operation.

4. Conclusions

In the development of the neural network model, the training process is important, since the hyperparameters must be correctly chosen and configured (structure and topology of the neural network: number of layers, number of neurons in each layer, activation functions, etc.; at the level of the learning algorithm: the Epoch, the Batch Size, the Learning Rate, the Momentum, etc.). The hyperparameters (structure and topology of the neural network: number of layers, number of neurons in each layer, activation functions, etc.; at the level of the learning algorithm: Epoch, Batch Size, Learning Rate, Momentum, etc.) are important because they fulfill their function at each stage, being an essential step to achieve an acceptable neural network model, so that it fulfills the objective for which it was designed.

In the training process, if the expected results are not achieved, the hyperparameters must be changed, such as varying the number of neurons in the hidden layers in accordance with the number of variables entered, as well as choosing appropriate activation functions. This is important since one of their main functions is to model the nonlinearity of the data and these are transcendental decisions in the training of the model and since they are in charge of activating or not activating these neurons in each training cycle.

Loss and Accuracy functions should be analyzed and monitored during the model training process because the shape and trend allow us to know if the network is learning.

Overfitting (overfitting or overtraining) and underfitting (undertraining or underfitting) are references that make the neural network model unable to capture the underlying trends of the data, i.e., when generalizing the model, it does not fit with the knowledge it is expected to acquire, causing the Accuracy and Loss curves to diverge and the network model to fail to fulfill the purpose for which it was created.

Author Contributions: Conceptualization, J.L.L.A. and P.A.P.S.; methodology, J.L.L.A. and P.A.P.S.; software, J.L.L.A.; validation, J.L.L.A.; formal analysis, J.L.L.A. and P.A.P.S.; investigation, J.L.L.A.; resources, J.L.L.A.; data curation, J.L.L.A.; writing—original draft preparation, J.L.L.A. writing—review and editing, J.L.L.A.; visualization, J.L.L.A.; supervision, P.A.P.S.; project administration, J.L.L.A.; funding acquisition, J.L.L.A. All authors have read and agreed to the published version of the manuscript.

Funding: This research received no external funding.

Institutional Review Board Statement: Not applicable.

Informed Consent Statement: Not applicable.

Data Availability Statement: The data can be obtained from the corresponding author on request.

Acknowledgments: To the director of this degree work, Patricio Antonio Pesántez Sarmiento, who collaborated tirelessly in the development of this degree project; to the National Polytechnic School, for the opportunity provided for professional development through its master's program; and finally, to my wife, children, and family, who are always present.

Conflicts of Interest: The authors declare no conflicts of interest.

References

1. Estupiñan, J.J.; Giral, D.A.; Martínez, F. *Implementation of Algorithms Based on Support Vector Machines (SVM) for Electrical Systems: Subject Review*; Francisco José de Caldas District University: Bogotá, Colombia, 2016.
2. Bermeo, J.C.; Arguello, G.A.; Cepeda, J.C. *Annual and Multiannual Statistics of the Ecuadorian Electrical Sector*. Agency for the Regulation and Control of Energy and Non-Renewable Natural Resources. 2022, p. 30. Available online: https://www.controlrecursosyenergia.gob.ec/wp-content/uploads/downloads/2022/04/Estadistica2021.pdf (accessed on 30 July 2022).
3. Berghout, T.; Benbouzid, M.; Ferrag, M.A. Multiverse Recurrent Expansion With Multiple Repeats: A Representation Learning Algorithm for Electricity Theft Detection in Smart Grids. *IEEE Trans. Smart Grid* **2023**, *14*, 4693–4703. [CrossRef]
4. Monedero, Í.; Biscarri, F.; León, C.; Biscarri, J.; Millán, R. MIDAS: Detection of non-technical losses in electrical consumption using neural networks and statistical techniques. In *Computational Science and Its Applications–ICCSA 2006*; Springer: Berlin/Heidelberg, Germany, 2006.
5. Chicco, G.; Napoli, R.; Piglione, F. Comparisons Among Clustering Techniques for Electricity Customer Classification. *IEEE Trans. Power Syst.* **2016**, *21*, 933–940. [CrossRef]
6. Angelos, E.W.S.D.; Saavedra, O.R.; Cortés, O.A.C.; de Souza, A.N. Detection and Identification of Abnormalities in Customer Consumptions in Power Distribution Systems. *IEEE Trans. Power Deliv.* **2011**, *26*, 2436–2442. [CrossRef]
7. Rossoni, A. Estimation of Technical and Commercial Losses: Methods Based on Load Flow and State Estimator. Master's Thesis, Federal University of Rio Grande do Sul, Porto Alegre, Brazil, 2014.
8. Monedero, I.; Biscarri, F.; León, C.; Guerrero, J.I.; Biscarri, J.; Millán, R. Detection of Frauds and other Non-Technical Losses in a Power Utility using Pearson Coefficient, Bayesian Networks and Decision Trees. *Int. J. Electr. Power Energy Syst.* **2012**, *34*, 90–98. [CrossRef]
9. Jonathan, M.M.B.; Antonio, D.C. Customer Segmentation Using Electricity Smart Meter Data. Master's Thesis, Higher School of Engineering and Technology, International University La Rioja, La Rioja, Spain, 2022.
10. Llagua, J.L. *Detection of Non-Technical Losses in Special Customers with Telemetering, Based on Artificial Intelligence with Application in EEASA*; National Polytechnic School: Quito, Ecuador, 2023.
11. Pau, G.C. *Design and Implementation of a Classifier Using Neural Networks for a 3D Industrial Inspection System*; Polytechnic University of Valencia: València, Spain, 2019.

engineering proceedings

MDPI

Proceeding Paper

Analysis of Conventional Direct Detection and Coherent Optical Receivers in Optical Access Networks †

Johanna Berenice Arguero Tello [1,*], Milton N. Tipán [1], Germán V. Arévalo [1] and Christian Tipantuña [2]

1 Carrera de Ingeniería en Telecomunicaciones, Universidad Politécnica Salesiana, Quito 170702, Ecuador
2 Departamento de Electrónica, Telecomunicaciones y Redes de Información, Escuela Politécnica Nacional, Quito 170525, Ecuador
* Correspondence: jarguero@ups.edu.ec

† Presented at the XXXII Conference on Electrical and Electronic Engineering, Quito, Ecuador, 29 November–1 December 2023.

Abstract: This study evaluated the use of GFDM transmission in passive optical networks (PONs) by comparing the performance of coherent and non-coherent optical receivers using OptSim 2023.12sp2 and Matlab 2018b ®. The study concentrated on transmitting 10 Gb/s radio frequency signals over optical fiber, emphasizing the significance of high-speed fronthaul links for 5G networks. The findings demonstrated that coherent detection markedly enhances receiver sensitivity by approximately 3 dB compared to direct detection, thereby augmenting the capacity of optical fronthaul networks despite the elevated cost. Additionally, the study recommended investigating pre- and post-compensation techniques to mitigate signal dispersion in optical fibers for further performance optimization.

Keywords: IM-DD; coherent optical receivers; PON; optical access network

Citation: Arguero Tello, J.B.; Tipán, M.N.; Arévalo, G.V.; Tipantuña, C. Analysis of Conventional Direct Detection and Coherent Optical Receivers in Optical Access Networks. *Eng. Proc.* **2024**, *77*, 30. https://doi.org/10.3390/engproc2024077030

Academic Editor: Jackeline Abad

Published: 18 November 2024

1. Introduction

The growing demand for higher data transmission speeds presents considerable challenges for network operators and service providers. Next-generation networks provide a crucial solution to address these challenges, particularly in complex scenarios involving advanced mobility, large user bases, and rigorous data requirements. Optical access networks are particularly well-suited for emerging applications such as mixed reality, augmented reality, holographic communication, the Internet of Things (*IoT*), and digital sensing. These applications necessitate high-capacity, low-latency, and high-reliability networks to ensure the delivery of quality experiences and adherence to the requisite Quality of Service (*QoS*) levels [1,2].

PONs have undergone significant advancements to meet the escalating bandwidth requirements and user numbers. Conventional direct detection (*IM-DD)* methodologies are being augmented by sophisticated techniques to augment network capacity. One such advancement is the G.989.1-Next-Generation Passive Optical Network 2 (*NG-PON2*) standard, which supports data transfer rates of up to 100 Gbps or more [3,4]. To attain these speeds, coherent detection—a technology heretofore employed in long-haul networks for its superior performance—is now being adapted for PONs. Coherent detection offers markedly enhanced receiver sensitivity in comparison to IM-DD, rendering it an optimal choice for fulfilling the performance expectations of contemporary optical networks [5,6]. The latest advances in digital signal processing (*DSP*) have further enabled coherent detection-based PON systems to support high-speed optical access networks with data rates exceeding 100 Gbps [7].

This study examines the potential applications of PON technology in the fronthaul link of a 5G mobile network, with a particular focus on comparing two optical transmission schemes: the two optical transmission schemes under consideration are intensity modulation with direct detection (*IM-DD*) and Coherent Transmission-Detection. Through co-simulation with OptSim 2023.12sp2 and Matlab2018b®, the performance of these schemes was evaluated using GFDM signals modulated with 4QAM and 16QAM over 20 km of single-mode fiber at a rate of 10 Gb/s [8].

The study's objective is to enhance the fronthaul link's capacity by considering the type of optical detection, modulation formats, and digital post-processing. The findings demonstrate that coherent transmission markedly enhances receiver sensitivity by approximately 3 dB compared to direct detection, indicating a considerable expansion in network capacity. However, the high cost of implementing coherent detection remains a significant drawback. Furthermore, the study investigates pre- and post-compensation techniques for dispersion in optical fibers.

Using BER and EVM analysis, this paper will compare receiver sensitivity between coherent and direct detection. Section 2 reviews previous work, Section 3 provides a theoretical analysis of detection methods, Section 4 discusses system architecture and co-simulation scenarios, and Section 5 presents the co-simulation results and summarizes the findings.

2. Related Works

The enhancement of optical system capacities represents a pivotal challenge driven by the expanding demand for data and the necessity for integration with existing technologies. The evolution of PONs aims to enhance their performance, with the deployment of coherent detection techniques offering notable advantages over direct detection. This approach has the potential to boost capacity and sensitivity. OOK (on–off keying) modulation is a prevalent choice in coherent detection systems due to its simplicity and efficacy. It facilitates implementation and compatibility with direct detection techniques, reducing system complexity. Moreover, its capacity to adapt to noise and turbulence conditions renders it a promising candidate for enhancing BER in PON-WDM networks. A modified OOK system was demonstrated to enhance BER performance and electrical signal-to-noise ratios (*SNRs*) in ID/MM systems. At a BER of 10^{-4}, a 2.2 dB and 4.2 dB improvement in SNR was observed between an optimized and an adaptive detection system, respectively. Moreover, a reduction in the power penalty of between 2.0 and 4.0 dB was observed at a BER of 10–12 under conditions of weak turbulence [9]. Another study [10] presented a 64 QAM phase-modulated system capable of transmitting at 168 Gbps over 170 km. This system demonstrated improvements in receiver sensitivity and a reduction in BER through coherent detection.

Building upon this line of inquiry, the work of [11] continues to serve as a foundational reference, illustrating a receiver sensitivity of −53 dBm at 311 Mb/s with an optical link budget of 50 dB through real-time data processing. The optical power/loss budget represents a pivotal aspect, necessitating the incorporation of intensity modulation/direct detection (*IM-DD*) solutions to attain a 29 dB budget, thereby ensuring compatibility with existing fiber-optic infrastructures [12]. Furthermore, the intricate nature of PON systems operating at 100G or above necessitates the incorporation of sophisticated components, including photon detectors and transimpedance amplifiers, which could potentially lead to increased costs and complexity. The study [13] presented a 100 Gb/s UD-WDM coherent PON system for fronthaul in 5G networks, improving receiver sensitivity to −26 dBm. These advances reflect continued progress towards faster and more efficient optical networks designed to meet the growing demand for data.

Consequently, recent research has demonstrated that, following a propagation distance of 120 km, a received coherent OFDM signal's radio frequency (*RF*) spectrum undergoes expansion due to chromatic dispersion and polarization mode dispersion (*PMD*) in single-mode fiber. OFDM technology is employed to mitigate these effects, offering a more cost-effective alternative to traditional OOK systems. Furthermore, it has been observed that by increasing the laser power from −16 dBm to 0 dBm, the Q factor improves significantly, and the BER is reduced, indicating an improved system performance [14]. However, the authors do not mention the receiver sensitivity or if they are close to the values recommended by GPON standards.

3. Background

3.1. Direct Detection

Due to its simplicity and cost-effectiveness, direct detection is the prevailing technique in contemporary PONs. In this configuration, the optical signal is converted directly into an electrical signal by a photodetector, typically an inversely polarized photodiode. This generates a photocurrent that is proportional to the received optical power. The resulting electrical signal is proportional to the intensity of the modulated optical signal. However, this technique has significant limitations, as it only captures the optical signal intensity, omitting phase information, which limits the data transmission capability and reduces the spectral efficiency. Furthermore, the electrical signal is susceptible to various forms of noise, including thermal, shot, and photodetection noise, which can impair signal quality. Moreover, the encoding of the signal in a single polarization results in a reduction in both the spectral and power efficiency of the system [5,15].

The signal strength constrains the functionality of these devices, as they are only capable of detecting and demodulating the information present within the optical signal strength. Encoding the signal in a single polarization reduces both the spectral efficiency and the power efficiency of the system, and information regarding the signal phase is lost. The optical signal $E(t)$, as described by Equation (1), is converted into an electrical signal $i(t)$ at the output. This transformation results in the conversion of an optical power pulse into an electrical current pulse, as illustrated in Figure 1. This conversion enables the information to be processed and recovered. However, only the intensity information is retained, resulting in the loss of the phase of the original signal.

$$E(t) = A(t)e^{j\theta(t)}e^{j\omega(t)} \quad (1)$$

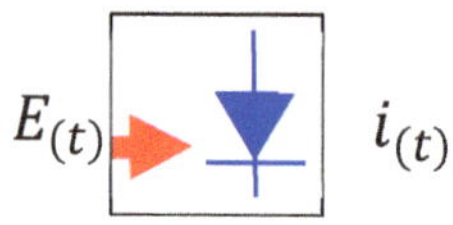

Figure 1. Scheme of direct detection on the photodetector.

The received optical signal is represented by three parameters: the amplitude, given by A; the phase, which is expressed as $e^{j\theta(t)}$, and the angular frequency, which is represented by $e^{j\omega(t)}$. The direct detection response is the electrical signal at the output of the photodetector $i(t)$,which is determined by Equation (2). This Equation shows that the electrical current is proportional to the optical field power $P(t)$ entering the receiver. The proportionality constant is related to the responsivity of the photodetector (R) and the thermal noise components generated in the photodetector process (θn). Consequently, the phase information of the transmitted signal is irretrievably lost [16,17].

$$i(t) = RP(t) + \theta n(t) \quad (2)$$

3.2. Coherent Detector

The coherent detection system utilized in optical networks is more intricate and costly than direct detection. It comprises a set of high-sensitivity photodiodes, a local oscillator, and various high-performance optical and electronic components. The coherent receiver's advanced design enables the complete extraction of information from the received signal, including amplitude, phase, and frequency. This markedly enhances spectral efficiency and detection capability.

One of the most notable advantages of coherent detection is an increase in receiver sensitivity of up to 20 dB, as documented in the reference [18]. The extent of this increase is contingent upon the modulation format utilized and the multiplexing mechanisms employed. For instance, sophisticated modulation formats, such as quadrature amplitude modulation (*QAM*) or phase shift keying (*PSK*), can considerably benefit from the capabilities of coherent detection [5,19]. Furthermore, coherent detection mitigates the impact of nonlinear effects within the optical fiber. This is made possible by the complex photodiode system's nonlinear response, which enables more effective management and correction of nonlinear distortions that can impact data transmission at high speeds and over long distances. Mitigating nonlinear effects results in cleaner and more efficient transmission, enabling a longer range and superior signal quality.

Coherent detection also enables a more efficient utilization of the available spectrum in terms of spectral efficiency. The ability to demodulate and detect signals with greater accuracy allows for using narrower channels, thereby increasing the amount of data that can be transmitted per unit of bandwidth. This is particularly significant in high-capacity applications, where bandwidth is a valuable and scarce resource.

Figure 2 illustrates the schematic of the coherent detector, which incorporates a 90° hybrid coupler. This component combines a 2 × 2 coupler with a 90° phase shifter, thereby enabling the combination of the input signal and the local oscillator signal. Because of this combination, two current signals, designated as $i_1(t)$ and $i_2(t)$, are generated. These signals are ideally modeled as a nonlinear response with an intermediate frequency (*IF*). The IF is the difference between the input signal and the local oscillator frequency. This relationship is expressed in Equation (3), where ω_s represents the frequency of the received optical signal and ω_{LO} corresponds to the frequency of the local oscillator [16].

$$\omega_{IF} = \omega_s - \omega_{LO} \tag{3}$$

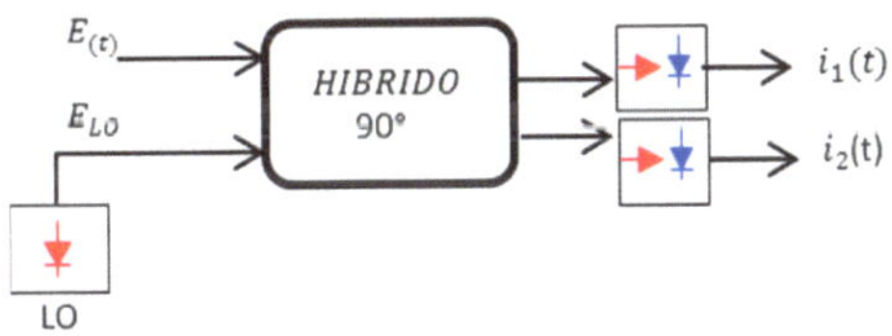

Figure 2. Homodyne coherent detector scheme with a 90° hybrid coupler.

Figure 2 shows the complex envelope of the received signal $E(t)$ and the signal E_{LO}. In Equations (6) and (7), P is the optical power of the input signal, P_{LO} is the optical power of the local oscillator (*LO*), θ_{ns} contains the modulated phase information, and θ_{LO} represents the phase noise generated by the devices.

$$E(t) = Re\{E * e^{j\omega t} * \hat{E}\} \tag{4}$$

$$E_{LO}(t) = Re\{E_{LO} * e^{j\omega t} * \hat{E_{LO}}\} \tag{5}$$

$$E(t) = \sqrt{P} * e^{j(\theta+\theta_{ns})} \tag{6}$$

$$E_{LO}(t) = \sqrt{P_{LO}} * e^{j\omega_{IF}} * e^{j\theta_{LO}} \tag{7}$$

At the output of the 90° hybrid, two current signals are obtained as described in Equation (8), where $\Delta\theta$ is the difference between the noise θ_{ns} and θ_{LO}.

$$i_{12(t)} = \frac{R}{2}\left\{P + P_{LO} \pm 2\sqrt{P * P_{LO}}\sin(\omega_{IF}t + \Delta\theta + \theta) * \hat{E} * \hat{E_{LO}}\right\} \tag{8}$$

where $(P + P_{LO})$ represents the interference between the received signal and the LO reference signal, which can be canceled by electronic domain processing or by balanced coherent detection. Balanced detection results in a current signal described by Equation (10). Balanced detection suppresses the DC component, so the current signal depends on the amplitude and phase of the signal.

$$i_{DE} = i_1(t) - i_2(t) \tag{9}$$

$$i_{DE} = 2R\sqrt{P * P_{LO}}\sin(\omega_{IF}t + \Delta\theta + \theta) * \hat{E} * \hat{E_{LO}} \tag{10}$$

On the other hand, to cancel the phase noise, Equation (11) [20] can be used for the two current signals obtained at the output of the 90° hybrid. This operation improves the signal power.

$$I_o(t) = {i_1}^2(t) + {i_2}^2(t) \tag{11}$$

4. System Design and Implementation

4.1. System Implementation

Figure 3 shows the block diagram of the transmission system, where the dotted blocks represent the Matlab 2018b® tool, while the solid line represents the Optsim 2023.12sp2 simulation tool within the same environment [21].

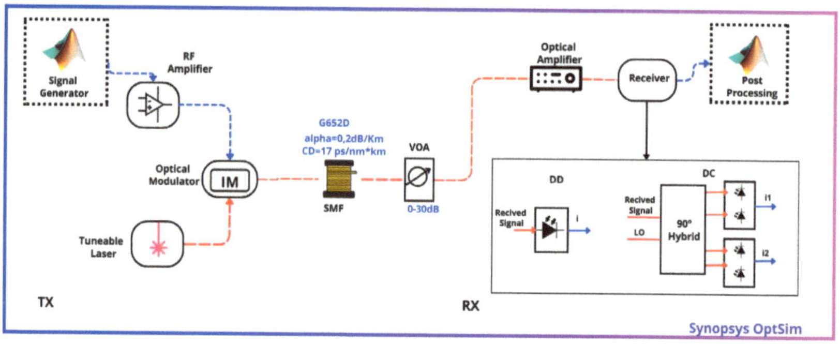

Figure 3. Access network co-simulation configuration with ID-/MM-IDDC SetUp.

The optical transmitter comprises a co-simulation block in Matlab 2018b®, where, in the GFDM signal for the fronthaul link of the 5G network, it is generated through the co-simulation interface with OptSim 2023.12sp2. A vector comprising 10,100 random symbols is created, which is then mapped into a digital modulator to produce a vector $d = [d_0, d_1, \ldots\ldots, d_n\ ]$. The 4-QAM and 16-QAM symbols have been optimized for BER of 10^{-3}. The analysis considers an additive white Gaussian noise (*AWGN*) channel, assuming a uniform error probability among the bits. A continuous wave (*CW*) laser source at 1550 nm, with a pulse width of 10 MHz, serves as the optical source, interfacing with the optical distribution network (*ODN*) and functioning as the transmitter. The minimum optical power required for injection is 6 dBm, corresponding to Class C.

The amplified signal was modulated using an intensity modulation and direct or coherent detection (*IM-DD*/*IM-CD*) system. Correctly selecting the bias voltage for the Mach–Zehnder modulator (*MZM*) is essential to ensure its optimal performance. An appropriate bias voltage is critical to avoid signal clipping during the intensity modulation process. In this case, a bias voltage of 2.5V was determined to obtain an adequate BER value, less than 10^{-3} [22].

After modulation, the optical signal was transmitted over a 20 km single-mode fiber (*SMF*) G652D. This fiber has an attenuation of 0.25 dB/km, a dispersion not exceeding 17.0 ps/(nm km), and a polarization mode dispersion (*PMD*) less than or equal to 0.1 ps/$\sqrt{}$km, including the nonlinear effects of the fiber. A variable optical attenuator

(*VOA*) was incorporated into the optical link to simulate variations in the received optical power, especially as the number of users increases. This attenuator allowed the attenuation to vary between 10 dB and 30 dB, corresponding to a received power range of −13 dBm to −30 dBm. This attenuation range is critical for determining BER and EVM under different operating conditions.

The direct detector is configured with a PIN photodiode that exhibits a sensitivity of 0.8 A/W, a dark current of 0.1 nA, and a bandwidth of 10 GHz. In the case of the homodyne coherent detector configuration, an external continuous wave (*CW*) mode laser is employed as the local oscillator (*LO*), in conjunction with a 90° optical hybrid. In this configuration, the local oscillator (LO) and the received carrier are set to a zero-phase shift; that is, the LO laser frequency is tuned to the same frequency as the transmitter, at 193.41 terahertz (*THz*), with an optical power of 6-decibel milliwatts (*dBm*) and a pulse width of 10 megahertz (*MHz*), matching the transmitter values. The coherent detector with a 90° hybrid comprises four PIN photodiodes, each with specifications of 0.8 A/W, 0.1 nA, and 10 GHz, in conjunction with an ideal phase conjugator that mixes the received optical signal with the local oscillator signal with an offset of 90°.

The electrical signal recovered in both detection schemes is processed in Matlab 2018b®, where demodulation is performed to recover the transmitted bits. In this process, the signal power is normalized, the offset voltage is eliminated, Butterworth low-pass filters are applied, and an equalizer is incorporated for the GFMD signal. This results in an improvement in the quality of the recovered signal and the optimization of system performance. Moreover, the two currents obtained at the output of the coherent detector are combined into a single current using Equation (11) before proceeding with the post-processing above.

4.2. Limitations of the System

While a reduction in the optical power necessary to achieve specific BER levels by the GPON standard has been attained, several limitations warrant consideration. Primarily, the system's intricacy is significantly greater than that of the direct detection, necessitating sophisticated equipment such as local, stable lasers and specialized modulators, thereby increasing operational complexity. Furthermore, the financial burden associated with coherent detection is considerable, largely due to the necessity for specialized equipment and DSP, which can also result in latency and increased power consumption. Local laser phase stability and noise have the potential to negatively impact signal quality, while sensitivity to fluctuations in signal polarization introduces an additional challenge [22]. Coherent detection systems necessitate precise frequency alignment between the received signal and the local laser to prevent misinterpretation. Compatibility with existing technologies can be complex and costly, and although DSP helps mitigate chromatic dispersion and other nonlinear effects, these extreme conditions still present significant challenges. These limitations underscore the necessity for developing technical and economical solutions to optimize coherent detection performance in practical applications.

5. Performance Evaluation

As illustrated in Figure 4a, both detection techniques demonstrate an impressive ability to achieve a remarkably low BER of 10^{-4}. However, coherent detection exhibits a distinct advantage in requiring approximately 3 dB less received power to achieve a BER of 10^{-3} ompared to direct detection, as evidenced by a leftward shift in receiver sensitivity. This advantage can be attributed to the ability of the coherent receiver to combine the received signal with the local oscillator power in a 90° hybrid detector, complemented by more sophisticated signal processing. The enhanced sensitivity attained with coherent detection is particularly pertinent in high-speed optical communication applications, where it is crucial to sustain consistent performance under high attenuation conditions. However, to reinforce and contextualize these findings, it is vital to juxtapose them with recent studies that have showcased improvements in receiver sensitivity through advanced techniques, such as digital dispersion compensation and the deployment of high-efficiency modulators. These emerging technologies could offer further insights to optimize coherent detection in high-demand environments.

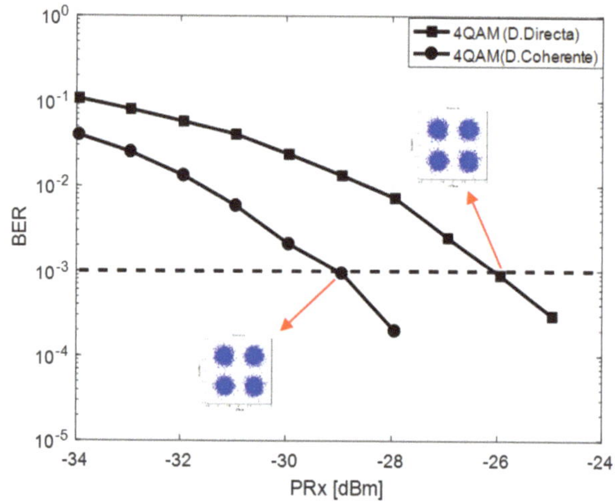

(a)

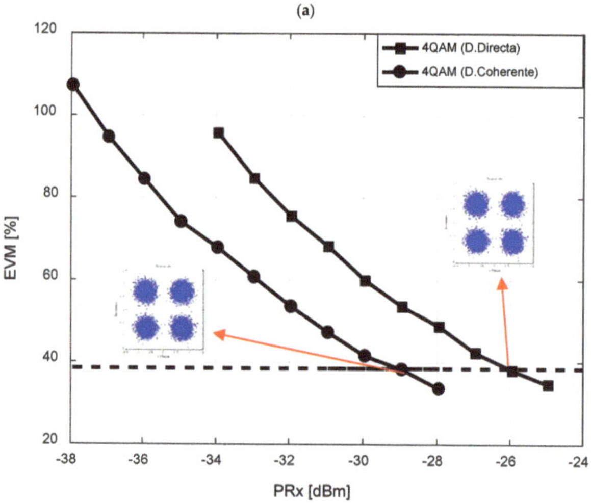

(b)

Figure 4. (**a**) Characteristic curves of BER vs. received power. (**b**) Characteristic curves of EVM vs. received power IM-DD/IM-CD for 4 QAM.

The network performance was evaluated by EVM, for four QAM, where coherent detection requires a minimum receiver power of -29 dBm to achieve an EVM of 28%, corresponding to a BER of 10^{-3} indicated by the dotted line in Figure 4a. This represents a 3 dB improvement over direct detection, which requires at least −26 dBm at the receiver to achieve the same EVM threshold in Figure 4b. The results highlight the significant advantage of coherent detection regarding signal sensitivity and accuracy. The ability of coherent detection to operate with higher signal attenuation without compromising transmission quality is essential for optimizing performance in high-speed, long-haul optical communications systems.

A comparative analysis of BER and EVM curves as a function of received power was conducted for the 16-QAM modulated signal. The findings are presented in Figure 5a. The results indicate that the highest receiver sensitivity is achieved when a coherent detector is employed. To achieve a BER of 10^{-3}, the receiver sensitivity with coherent detection is approximately −23 dBm, whereas direct detection necessitates a received power of −19 dBm. The 4 dB discrepancy between these methods, as illustrated in Figure 5a, highlights the superiority of coherent detection. Moreover, the examination of the EVM curve regarding received power, illustrated in Figure 5b, demonstrates comparable results. The superior performance of the coherent detector, which requires less received power to maintain a BER of 10^{-3}, is confirmed. It is crucial to acknowledge that the penalty associated with employing a non-coherent receiver compared to a coherent one is approximately 3 dB. This implies that a non-coherent receiver necessitates 3 dB more power to attain the same BER. This underscores the significant benefit of coherent detection in systems where received efficiency is of paramount importance.

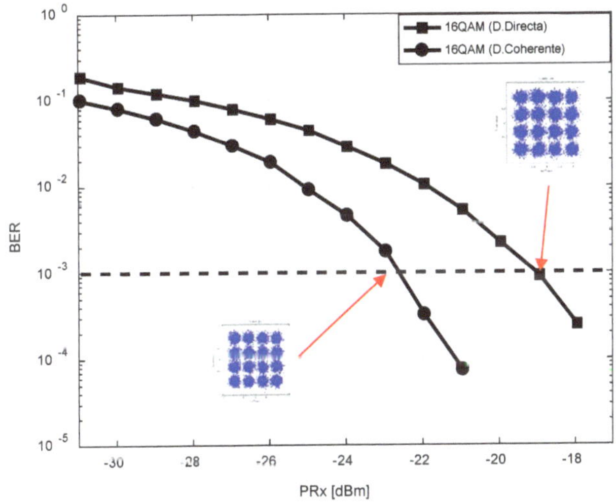

(**a**)

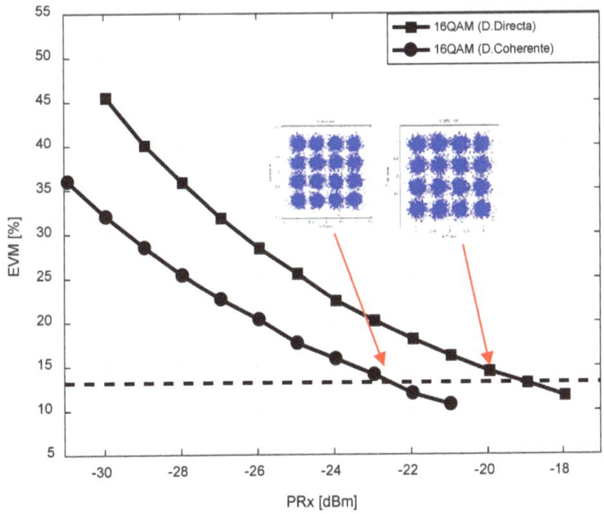

(**b**)

Figure 5. (**a**) Characteristic curves of BER vs. received power (**b**) Characteristic curves of EVM vs. received power; IM-DD/IM-CD for 16 QAM.

The analysis additionally encompasses the 5G optical fronthaul link, wherein enhanced performance was discerned by deploying a coherent detector. The system achieved the targeted BER of 10^{-3} with a reduced received power of approximately −29 dBm, which met the specified performance criteria. In this analysis, it is assumed that there is a loss of 0.2 dB per kilometer over 20 km for the GFDM signal. These findings reinforce the superiority of the proposed coherent receiver, particularly in the context of 4QAM and 16QAM schemes, concerning the requisite received signal power. The performance of a 5G fronthaul optical link was evaluated using coherent detectors, demonstrating a notable enhancement in comparison to non-coherent detectors. The results show that a BER of 10^{-3} was attained with an approximate received power of −29 dBm, which exceeded the performance requirements established by the ITU-T G.948.2 standard, which is −26 dBm using an OOK signal in this standard.

6. Conclusions

In conclusion, this study comprehensively compares direct and coherent detection schemes in PONs, demonstrating that coherent detection exhibits superior performance for 4QAM and 16QAM signals. The results demonstrated that coherent detection exhibited approximately 3 dB higher sensitivity than direct detection, consistent with the ITU-T G.948.2 PON specifications. The implementation of GFDM modulation and signal post-processing has been demonstrated to enhance the spectral efficiency and system capacity compared to traditional OOK systems. Notwithstanding the elevated complexity and cost associated with coherent technology, its capacity to maintain a BER of less than 10^{-4} and achieve acceptable EVM levels underscores its effectiveness in advanced optical communications. Moreover, while 4QAM requires less power and is less susceptible to distortion, 16QAM offers twice the information capacity, and coherent detection achieves a more favorable sensitivity penalty compared to direct detection. In conclusion, the results demonstrate the robustness and efficiency of coherent detection in high-speed optical communications, making it the preferred choice for next-generation fronthaul networks, provided that cost and complexity issues can be effectively addressed.

Author Contributions: Conceptualization, J.B.A.T.; methodology, J.B.A.T., M.N.T. and G.V.A.; validation, G.V.A. and M.N.T.; formal analysis, J.B.A.T.; investigation, J.B.A.T.; resources, J.B.A.T.; data curation, J.B.A.T.; writing—original draft preparation, J.B.A.T.; writing—review and editing, C.T.; visualization, G.V.A.; supervision, G.V.A. and M.N.T.; project administration, J.B.A.T.; funding acquisition, G.V.A. All authors have read and agreed to the published version of the manuscript.

Funding: This work has been supported by the Carrera de Telecomunicciones—Universidad Polittécnica Salesiana Sede Quito project 030-01-2024-01-30.

Data Availability Statement: Data sharing is not applicable.

Conflicts of Interest: The authors declare no conflict of interest.

References

1. Wong, E.; Mondal, S.; Ruan, L. Machine learning enhanced next-generation optical access networks—Challenges and emerging solutions [Invited Tutorial]. *J. Opt. Commun. Netw.* **2023**, *15*, A49–A62. [CrossRef]
2. Shahi, B.; Jha, D.; Dushyanth, N.D.; Suresh, D. Recent advances in optical network technologies and its role in telecommunication industry. *Int. J. Res.-Granthaalayah* **2017**, *5*, 15–22. [CrossRef]
3. Abejide, A.E.; Duarte, P.; Patel, R.; Pandey, S.; Kota, M.; Pinho, C.M.R.; Novo, C.C.; Popoola, J.J.; Alimi, I.A.; Lima, M.; et al. PON-FTTX Architecture and Bandwidth Analysis for Future Broadband Communications. Available online: www.intechopen.com (accessed on 27 July 2024).
4. Abbas, H.S.; Gregory, M.A. The next generation of passive optical networks: A review. *J. Netw. Comput. Appl.* **2016**, *67*, 53–74. [CrossRef]
5. Welch, D. Digital Subcarrier Multiplexing: Enabling Software-Configurable Optical Networks. *J. Light. Technol.* **2023**, *41*, 1175–1191. [CrossRef]
6. Xu, T. Digital Signal Processing for Optical Communications and Networks I: Linear Compensation. *arXiv* **2017**, arXiv:1705.05284.
7. Zhao, J.; Liu, Y.; Xu, T. Advanced DSP for coherent optical fiber communication. *Appl. Sci.* **2019**, *9*, 4192. [CrossRef]

8. Ali, F.; Wadday, A. *Study the Performance Evaluation of Radio over Optical Fiber System with Optical OFDM Using Different Modulation Techniques*; European Alliance for Innovation: Bratislava, Slovakia, 2020. [CrossRef]
9. Elsayed, E.E.; Yousif, B.B. Performance evaluation and enhancement of the modified OOK based IM/DD techniques for hybrid fiber/FSO communication over WDM-PON systems. *Opt. Quantum Electron.* **2020**, *52*, 385. [CrossRef]
10. Feng, N.; Ma, M.; Zhang, Y.; Tan, X.; Li, Z.; Li, S. Key Technologies for a Beyond-100G Next-Generation Passive Optical Network. *Photonics* **2023**, *10*, 1128. [CrossRef]
11. Smolorz, S.; Gottwald, E.; Rohde, H.; Smith, D.; Poustie, A. Demonstration of a Coherent UDWDM-PON with Real-Time Processing. In Proceedings of the 2011 Optical Fiber Communication Conference and Exposition and the National Fiber Optic Engineers Conference, Los Angeles, CA, USA, 6–10 March 2011.
12. ECOC. *ECOC 2016 42nd European Conference on Optical Communication: Düsseldorf, September 18–22, 2016*; VDE Verlag: Berlin, Germany, 2016.
13. Mohammed, N.A.; Mansoor, R.; Abed, A.K.; Abd, H.J. Performance Analysis of WDM Coherent Optical OFDM Systems. In Proceedings of the 2022 2nd International Conference on Advances in Engineering Science and Technology (AEST), Babil, Iraq, 24–25 October 2022; pp. 385–388. [CrossRef]
14. Iglesias, P.J.R. Nuevos Esquemas de Recepción Coherente Multipuerto Para Comunicaciones Ópticas. PhD Thesis, Universidad De Málaga, Málaga, Spain, 2015.
15. Roudas, I. Coherent Optical Communication Systems. In *WDM Systems and Networks*; Springer: New York, NY, USA, 2012; pp. 373–417. [CrossRef]
16. Ip, E.; Pak, A.; Lau, T.; Barros, D.J.F.; Kahn, J.M. Coherent detection in optical fiber systems. *Opt. Express* **2008**, *16*, 753–791. [CrossRef] [PubMed]
17. Folland, T.G.; Marshall, O.P.; Beere, H.E.; Ritchie, D.A.; Chakraborty, S. Coherent detection of THz laser signals in optical fiber systems. *Opt. Express* **2017**, *25*, 25566. [CrossRef] [PubMed]
18. Brandt-Pearce, M.; Noshad, M. Optical transmission. In *Academic Press Library in Mobile and Wireless Communications: Transmission Techniques for Digital Communications*; Elsevier: Amsterdam, The Netherlands, 2016; pp. 661–687. [CrossRef]
19. Chen, X.; Yao, J. Wavelength reuse in an RoF link based on CS-DSB, coherent detection and DSP. *IEEE Photonics Technol. Lett.* **2017**, *29*, 975–978. [CrossRef]
20. Arguero Tello, J.B.; Berenice, J. Escuela Superior Politécnica de Chimborazo, Análisis Comparativo del Desempeño de Receptores Ópticos Coherentes y no Coherentes para GFDM en redes PON(2018). Available online: http://dspace.espoch.edu.ec/handle/123456789/8917 (accessed on 27 July 2024).
21. Dávila, M.; Camacho, J.; Arguero, B.; Arévalo, G.V.; Tipantuña, C. Analysis of the Optimal Bias Voltage for an Optical Fronthauling Modulator in a RoF System. In Proceedings of the 2023 IEEE 7th Ecuador Technical Chapters Meeting (ECTM 2023), Ambato, Ecuador, 10–13 October 2023; Institute of Electrical and Electronics Engineers Inc.: Piscataway, NJ, USA, 2023. [CrossRef]
22. El-Nahal, F.I. Special Issue on Coherent Optical Communications. *Photonics* **2023**, *10*, 683. [CrossRef]

engineering proceedings

MDPI

Proceeding Paper

Design and Construction of Inductive Compensation for Extra-High-Voltage Transmission Line Models of Physical Laboratory of Electric Power Systems †

Anderson Anrrango Delgado, Anghelo Navarrete Cumbal and Jesús Játiva Ibarra *

Escuela Politécnica Nacional, Quito 170143, Ecuador; anderson.anrrango@epn.edu.ec (A.A.D.); anghelo.navarrete@epn.edu.ec (A.N.C.)
* Correspondence: jesus.jativa@epn.edu.ec

† Presented at the XXXII Conference on Electrical and Electronic Engineering, Quito, Ecuador, 12–15 November 2024.

Abstract: The implementation of inductive compensation in the two-scale models of the extra-high-voltage transmission line, without and with transposition, of the Physical Laboratory of Electric Power Systems (PLEPS) is presented. As they have voltages above the normal operating range, for a situation like that of the real 500 kV Coca Codo Sinclair–El Inga lines in Ecuador, the problem was solved by incorporating parallel inductive reactors located at the ends of the lines. The fixed bus compensation used was carried out by means of three-phase inductors in star connection with neutral-to-ground voltage and built-in iron cores with pitches of 50, 75 and 100% of the design capacity.

Keywords: inductive compensation; transmission line scaled models; design and construction

Citation: Anrrango Delgado, A.; Navarrete Cumbal, A.; Játiva Ibarra, J. Design and Construction of Inductive Compensation for Extra-High-Voltage Transmission Line Models of Physical Laboratory of Electric Power Systems. *Eng. Proc.* **2024**, *77*, 31. https://doi.org/10.3390/engproc2024077031

Academic Editor: Walter Vargas

Published: 18 November 2024

1. Introduction

Extra-high-voltage (EHV) transmission lines tend to have overvoltages due to the capacitive effect. To maintain voltage levels within the safety range, one of the solutions used in transmission systems is the injection of reactive power [1]. Transmission lines operating with light or no loads may experience phenomena such as the Ferranti effect. This phenomenon manifests itself when the voltage at the receiving end of the line exceeds the voltage at the sending end, due to the current injected by the capacitance distributed along the line, which causes a voltage increase in the phase with the voltage of the sending end. This increase builds up to the receiving end, resulting in a higher voltage under low- or no-load conditions [2].

To address these challenges, inductive compensation is presented as an effective solution to counteract the adverse effects of line capacitance, absorbing reactive power to keep the receiving-end voltage magnitude close to the rated voltage [3]. This work focuses on the design and construction of scaled inductive compensators, specifically for a model of the Coca Codo Sinclair–San Rafael–El Inga transmission lines, developed at the PLEPS. The steps involved in the process are detailed, ranging from the dimensioning of the compensating power to the operational tests of the system. The work provides a comprehensive view of the calculation of electrical parameters, selection of materials and experimental procedures necessary to validate the technical specifications established during the design.

One of the best ways to understand and learn the electromagnetic phenomena that occur in an electrical system is through a complete physical modeling of its three-phase components, within its size limitations, since it considers asymmetry, electromagnetic couplings and nonlinearities of its elements, as well as the imbalances of its variables, among others. The following statement is relevant in this context: "The complexity of modern power systems is increasing, and thus is requiring of both power engineers and systems operators a corresponding greater knowledge base upon which to draw form to understand the phenomena occurring in the system, from a systems point of view. This increased understanding cannot be achieved solely through digital modeling but must be complemented by physical modeling" [4].

In the 1970s and 1980s, electromagnetic phenomena were studied in TNA; however, with the development of software, it came to replace it, leaving a void in the research and development of electrical engineering. Some teaching equipment suppliers have three-phase components with many limitations, such as mutual impedance modeling and unloaded and loaded taps on transformers with voltage and angle control. Therefore, a main objective of the Department of Electrical and Electronic Engineering of the National Polytechnic School is to continue developing a physical laboratory based on a scaled power system with four energy sources, two transmission corridors and load centers with residential, commercial and industrial load types as well as a LVDAC-EMS computerized data acquisition system, as shown in Figure 1.

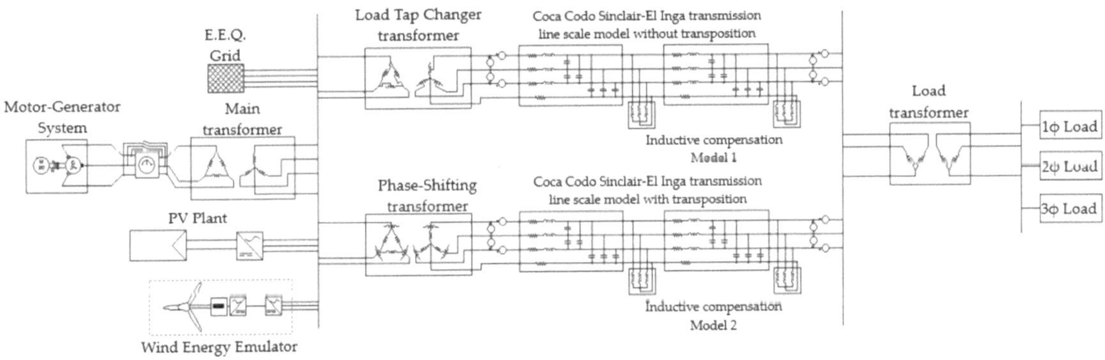

Figure 1. Diagram of the Physical Laboratory of Electrical Power Systems.

Through this research, we seek to demonstrate how the implementation of inductive compensators can mitigate the effects of the Ferranti effect and optimize the performance of transmission lines in a controlled environment with a laboratory voltage level, providing a practical and detailed approach to inductive compensation in EHV transmission systems, such as 500 kV Coca Codo Sinclair–El Inga lines in Ecuador [5].

2. Description of Scale Models of 500 kV Transmission Lines

Inductive compensation is usually placed at the beginning and end of high-voltage transmission lines. The specifications of the actual transmission system and the scale models of the laboratory are presented in Table 1.

Table 1. Specifications of actual transmission line and scale models.

Description	Actual System	Model 1 *	Model 2 **
Base power	1000 MVA	5 kVA	5 kVA
Base voltage	500 kV	220 V	220 V
Base impedance	250 Ω	9.68 Ω	9.68 Ω
Frequency	60 Hz	60 Hz	60 Hz
Resistance	0.015324 Ω/km	0.06708 Ω	0.17947 Ω
Inductive reactance	0.332239 Ω/km	1.55696 Ω	1.52105 Ω
Parallel admittance	4.9857 μS/km	18,835.762 μS	19,049.285 μS

*: Without transposition, **: with transposition.

The scale models were built to represent the Coca Codo Sinclair–San Rafael–El Inga transmission system, which covers approximately 126 km at 500 kV, Figure 2. This system was specifically chosen due to its critical role in the National Interconnected System, as it interconnects the largest hydroelectric plant in Ecuador. One of its complexities is the installation of inductive reactors of 30 MVAr per line at both ends. These reactors are essential for controlling voltage levels and minimizing reactive power imbalances.

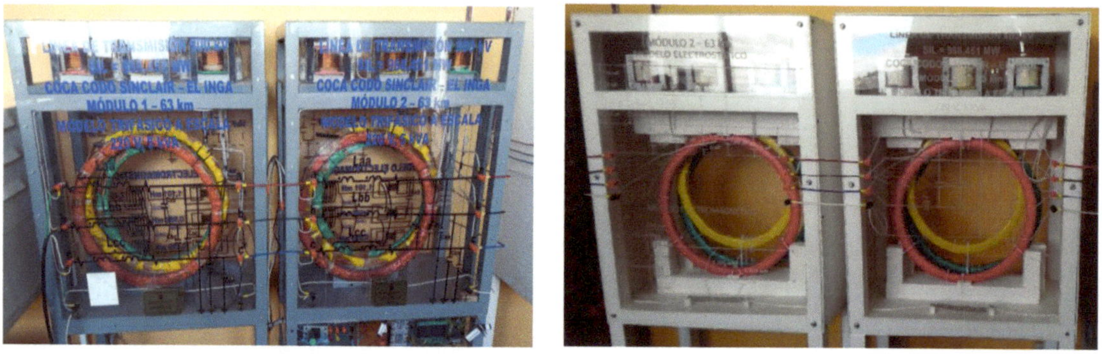

(a) (b)

Figure 2. Scale models of the Coca Codo Sinclair–San Rafael–El Inga transmission lines. (**a**) Model 1 [6]; (**b**) Model 2 [6], Lii and Lij are self and mutual inductances.

The decision to focus on this system was also due to the need to model real conditions in the transmission of large powers.

To obtain the transmission line models, scaling factors are determined considering the voltage and base power of the real system and PLEPS. To calculate the values in per unit of the actual transmission line, the following base quantities are considered:

$$V_{actual\ base} = 500\ \text{kV and } S_{actual\ base} = 1000\ \text{MVA},$$

while the base quantities compatible with the PLEPS are as follows:

$$V_{model\ base} = 220\ \text{V and } S_{model\ base} = 5\ \text{kVA}.$$

Based on the above relationships, the scaling factors of the transmission line scale model are presented [6,7].

$$Z_{model}\ [\Omega] = \frac{Z_{model\ base}\ [\Omega]}{Z_{actual\ base}\ [\Omega]} Z_{actual}\ [\Omega]$$

$$Z_{model}\ [\Omega] = \frac{(V_{model\ base})^2}{S_{model\ base}} \cdot \frac{S_{actual\ base}}{(V_{actual\ base})^2} \cdot Z_{actual}\ [\Omega]$$

$$Y_{model}\ \left[\Omega^{-1}\right] = \frac{Y_{model\ base}\left[\Omega^{-1}\right]}{Y_{actual\ base}\left[\Omega^{-1}\right]} Y_{real}\ \left[\Omega^{-1}\right]$$

$$Y_{model}\ \left[\Omega^{-1}\right] = \frac{S_{model\ base}}{(V_{model\ base})^2} \cdot \frac{(V_{actual\ base})^2}{S_{actual\ base}} \cdot Y_{actual}\ \left[\Omega^{-1}\right]$$

3. Design and Construction of the Inductive Compensation Coils

The design of inductive compensation coils involves a series of key steps that ensure their effective operation within the scaled transmission models in a laboratory environment.

The flowchart in Figure 3 illustrates the logical sequence followed in the design process from the acquisition of initial parameters to the core construction.

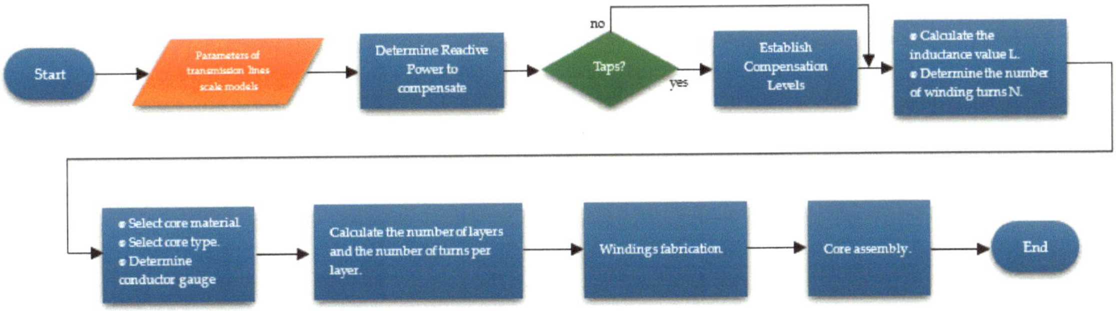

Figure 3. Design flowchart for inductive compensation winding construction.

Determining the reactive power for compensation is one of the crucial steps in the design of inductive compensation coils. This calculation is approached through two methodologies: using the power of the inductive reactors installed in the actual transmission system Coca Codo Sinclair San Rafael–El Inga and considering the parameters of the scale model of the transmission line. Both methodologies allow compensation adjustments to be made according to the specific characteristics of the power transmitted to the load.

The first approach involves carrying out the process of scaling up the power of the inductive reactors installed in the 500 kV transmission system to the bases of the PLEPS (220 V and 5 kVA) using Equation (1).

$$S_{model} = S_{real}\left(\frac{S_{model\ base}}{S_{actual\ base}}\right) \quad (1)$$

The second approach uses the PLEPS transmission line scale model, as the calculation of the compensation power is based on the electrical parameters of the scale transmission line considering a compensation factor k_c of 70% to avoid the resonance effect of the parallel branches of the line [1,8].

$$k_c = \frac{B_L}{B_C} \quad (2)$$

The calculation of the compensating power is performed per phase according to Equation (3), where $Q_{c/phase}$ is the compensating power per phase, V_{fn} is the nominal neutral-phase voltage of the transmission line, Z_{comp} is the compensating impedance, B_C is the parallel capacitive susceptibility of the transmission line in µS/km, and ℓ is the length of the transmission line in km, Equation (4).

$$Q_{c/phase} = \frac{V_{fn}^2}{Z_{comp}}[\text{VAr}] \quad (3)$$

$$Z_{comp} = \frac{1}{\frac{B_c}{2} * \ell * k_c}[\Omega] \quad (4)$$

The total capacity of the inductive compensation modules is equal to the sum of the compensation power of each phase, Equation (5).

$$Q_c = 3Q_{c/phase}\ [\text{VAr}] \quad (5)$$

In this work, we decided to design and build three-phase modules made up of three inductors with a silicon steel core, which reflects the operation of an inductive compensation reactor in parallel that is installed in the EHV transmission lines. Steel-core inductor banks allow for reactive power absorption, counteracting the capacitance effects of laboratory transmission line models. Considering the constructive layout of the transmission line scale models, these compensators are placed at the end of each module.

Three levels have been established: 100%, 75% and 50% of the compensation power determined with the methodologies described above. This step, indicated in Figure 3, allows for controlling reactive power according to the needs of the transmission line.

The next step in the flowchart is to calculate the inductance and number of turns. To achieve this, compensating power is used in Equation (6). And with the inductance obtained, the number of turns of the windings of each inductor is calculated with Equation (7) [9].

$$Q_{c/phase} = \frac{V_{fn}^2}{2\pi f L}[\text{VAr}] \tag{6}$$

$$L = \frac{4\pi A_c N^2}{\frac{MPL}{\mu_m}} \times 10^{-7}\ [\text{H}] \tag{7}$$

The core of the inductors is constructed using standardized Type EI sheets made of silicon steel, as shown in Figure 3. The EI-type laminate allows easy assembly of the cores, ensuring uniform distribution of magnetic flux and reduction in eddy current losses. At this step of the flowchart, it is necessary to determine the gauge of the conductor. The maximum current flowing through the winding is calculated with Equation (8) and with the use of manufacturers' tables, the appropriate conductor gauge can be chosen.

$$I = \frac{S}{V}\ [\text{A}] \tag{8}$$

The calculations obtained in the design are presented in Table 2.

Table 2. Summary of inductive compensator design calculations.

Design Parameter	Symbol	Case 1	Case 2
Single-phase compensation power (100%, 75%, 50%)	$Q_{c/phase}$	50, 37.5, 25 VAr	53.78, 40.34, 26.89 VAr
Core material	-	Silicon steel	Silicon steel
Lamination type	-	EI-125	EI-125
Inductance (100%, 75%, 50%)	L	0.8557, 1.1408, 1.7113 H	0.7957, 1.0609, 1.5914 H
Number of turns (100%, 75%, 50%)	N	304, 351, 430	293, 338, 413
Conductor gauge	-	22 AWG	22 AWG

With the calculations of the design values, the necessary materials for the manufacture of both the windings and the core are obtained. As illustrated in Figure 3, for the construction of the inductors, the wire windings are made in the formworks; for this, a machine with a turn counter is used that guarantees compliance with the number of turns specified in Table 2. For this, it is necessary to determine the number of layers (N_{layers}) and the number of turns (N_b) per layer with Equations (9)–(11), which are also indicated in the flowchart. In these equations, h_b is the height of the coil, G is the internal height of the EI sheets, C is the thickness of the form that isolates the winding from the steel core, f_s is a safety factor, and d_c is the diameter of the conductor. Considering the number of turns of the last compensation level N, Table 3 presents constructive characteristics of the windings.

$$N_b = \frac{h_b}{f_s d_c} \tag{9}$$

$$h_b = G - 2C \tag{10}$$

$$N_{layers} = \frac{N}{N_b} \tag{11}$$

It should be noted that derivations must be made for the winding depending on the number of turns of each compensation level. Once the windings of each of the inductors are available, the core is built by superimposing one sheet after another, completing the process described in Figure 3. If the sheets are placed properly, the edges are aligned correctly; otherwise, it could cause loss of inductance. There are many techniques for the construction of the core, but the most common and the one used in this study was the method that consists of alternating the position of the "E" and the "I"; in this way, maximum permeability is guaranteed.

Table 3. Winding design data.

Construction Parameter	Symbol	Unit	Case 1	Case 2
Total number of turns	N	turns	430	413
Internal height of EI sheets	G	cm	4.8	4.8
Formwork thickness	C	cm	0.1	0.1
Conductor diameter	d_c	cm	0.0644	0.0644
Number of turns per layer	N_b	turns/layer	68	68
Number of layers	N_{layers}	layers	7	6

Figure 4 shows the windings constructed for an inductive compensation module for the transposed line model of the PLEPS.

Figure 4. Single-phase inductor bank for an inductive compensation module: 53 VAr, 220 V, 1.5914 H, 1.0609 H and 0.7957 H, respectively.

4. Results

The tests, performed on the inductive compensation modules, are Ferranti effect test, and operational tests with/without loads with each level of compensation, hysteresis curves and saturation curves.

4.1. Hysteresis Curve and Saturation Curve

To validate the design of the compensators, an experimental test was carried out that consisted of assembling an electronic circuit to obtain the hysteresis loop of each of the single-phase inductors built. On the other hand, according to IEC 60076-6 [10], the magnetic characteristic of an inductive reactor is the relationship between the flow links and the current. Since flux bonds cannot be measured directly, it is possible to obtain the saturation curve through an indirect method that consists of obtaining instantaneous measurements of voltage and current at nominal frequency.

The curves obtained from the performance of these tests are those that are presented in Figure 5.

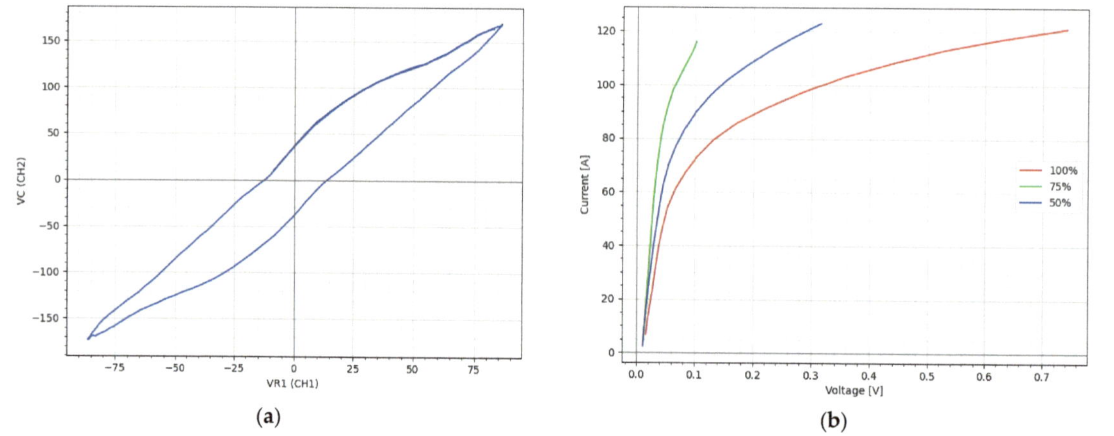

Figure 5. Curves obtained for a single-phase inductor. (**a**) Hysteresis curve; (**b**) saturation curve for each compensation level.

4.2. Ferranti Effect

In principle, the no-load and no-compensation test is carried out, which allows the establishment of fundamental reference parameters to demonstrate the operation of the transmission line model before connecting the load and the inductive compensators. In this test, it is possible to observe the Ferranti effect that occurs due to the capacities distributed in the modules of each scale transmission line model. The results of this test are presented in Table 4.

Table 4. Test results without load and without compensation.

	Model 1		**Model 2**	
	Vab	**Vcb**	**Vab**	**Vcb**
Sending bus	217.5 V	217.9 V	220.9 V	221.8 V
Reception bus	222.2 V	221.7 V	224.5 V	226.8 V
Percentage increase	2.16%	1.65%	1.63%	2.25%

In Model 1, it can be observed that the voltage increase between phases A and B is 2.16%, which represents an elevation of 4.6 V in the receiving bus. In Model 2 of the extra-high-voltage line, the capacitive effect causes an increase in the receiving terminals, with 2.25% between phases C and B representing an increase of 5 V.

4.3. No-Load Function Test and with Each Level of Compensation

To analyze the operation of the transmission line models of the laboratory with the inductive compensators connected, measurements were obtained at the sending and receiving terminals of each line model. The results of the test with compensators connected without loads are presented in Table 5.

Considering the tests carried out on the transmission line models without loads attached, Figure 6 shows how the voltage increases in the reception bus vary with the different levels of compensation. In the two initially uncompensated line models, the reception bus shows voltage increases between 3 and 5 V.

Table 5. Test results with no-load inductive compensation.

Compensation Level		Model 1		Model 2	
		Vab	Vcb	Vab	Vcb
50%	Sending bus	219.5 V	219.2 V	223.1 V	224.8 V
	Reception bus	222.7 V	222.5 V	224.2 V	225.9 V
	Percentage increase	1.46%	1.51%	0.493%	0.489%
75%	Sending bus	219.5 V	219.2 V	220.5 V	222.8 V
	Reception bus	222.3 V	222.1 V	221.2 V	223.8 V
	Percentage increase	1.27%	1.32%	0.317%	0.449%
100%	Sending bus	219.5 V	219.2 V	221.1 V	223.7 V
	Reception bus	221.3 V	220.9 V	221.2 V	224.5 V
	Percentage increase	0.82%	0.78%	0.045%	0.358%

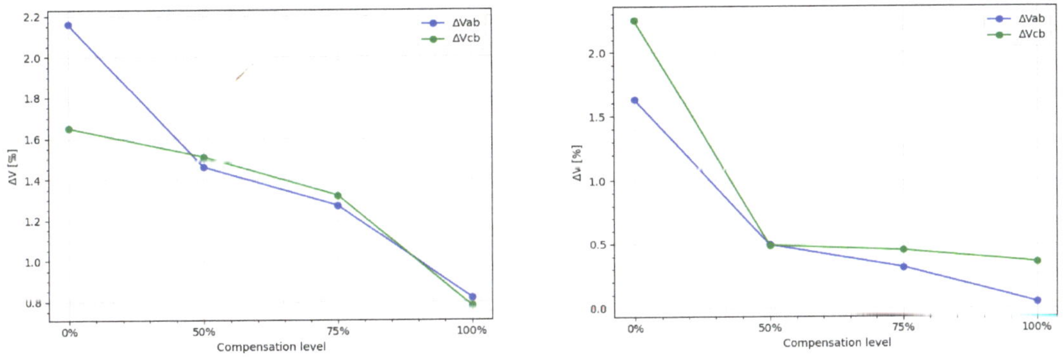

Figure 6. Percentage increase in voltage at the receiving bar for each level of no-load compensation in transmission line-scale models. (**a**) Model 1; (**b**) Model 2.

4.4. Load Test Operation with Each Level of Compensation

The execution of this test allows us to evaluate the behavior of the inductive compensators in the state of operation with loads. At this stage, interactions of the laboratory's transmission line models and reactive compensation to handle the energy demand are observed. In addition, it provides the final validation of the inductive compensation modules, ensuring efficient behavior with all connected components. The results of the test with the compensators connected with loads are presented in Table 6.

Table 6. Test results with 100% inductive and load compensation.

Compensation Level		Model 1		Model 2	
		Vab	Vcb	Vab	Vcb
100%	Sending bus	220.8 V	221.3 V	219.6 V	223.8 V
	Reception bus	221.4 V	221.8 V	219.8 V	224.1 V
	Percentage increase	0.27%	0.23%	0.091%	0.134%

Considering the tests performed on the load and unloaded transmission line models with 100% offset connected, Figure 7 shows the percentage increases for Vab and Vcb in Model 1 and Model 2, with and without loads.

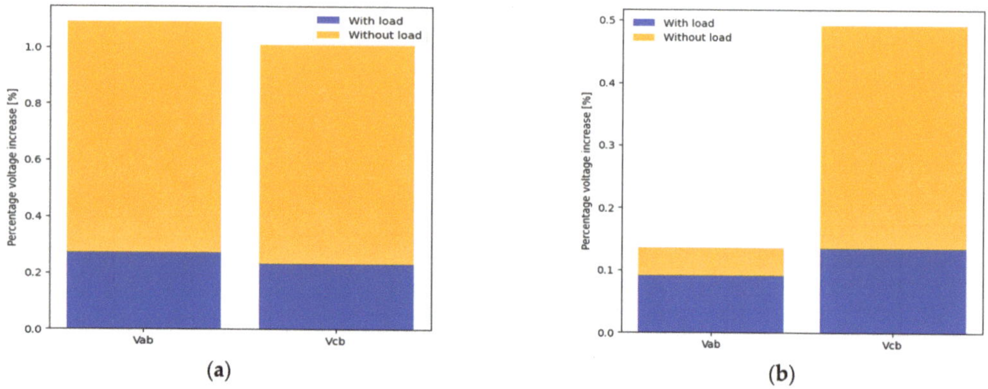

Figure 7. Percentage voltage increase in the reception bus with 100% compensation with and without load in transmission line scale models. (**a**) Model 1; (**b**) Model 2.

The graph corresponding to Model 1 shows the variations in the percentage increase for Vab and Vcb both under load (blue bar) and non-load (orange bar) conditions. The figure reveals that the percentage increase in Vab is greater than in Vcb for both conditions, with significantly higher values in the no-load test compared to the no-load test.

On the other hand, a similar trend is observed in Model 2, although with smaller magnitudes compared to Model 1. This test provides a clear visualization of how load conditions affect percentage increments in both models with their respective inductive compensation connected at 100%.

5. Conclusions

The implementation of the compensation modules proved to be effective in reducing overvoltage at the ends of the transmission line models, which translates into a significant improvement in the quality of electrical variables in the PLEPS.

The effectiveness of the proposed design was verified, where the greater the number of turns of the inductors, the lower the reactive compensating power. The highest number of turns corresponded to 50% compensation with 413 turns, and for 75%, 338 turns were recorded. On the other hand, for 100% compensation, a winding of only 293 turns was required, which corresponded to inductance values of 1.5914 H, 1.0609 H and 0.7957 H, respectively.

The inductive compensation implemented in the PLEPS will serve to carry out research and development in steady-state, dynamic, and transient UHV transmission systems at the undergraduate and graduate levels of electrical engineering.

Author Contributions: Conceptualization, J.J.I.; methodology, A.N.C. and A.A.D.; software, A.N.C.; validation, A.A.D. and A.N.C.; formal analysis, J.J.I.; investigation, A.A.D. and A.N.C.; resources, A.A.D. and A.N.C.; data curation, J.J.I.; writing—original draft preparation, A.A.D.; writing—review and editing, A.A.D. and J.J.I.; supervision, J.J.I.; project administration, J.J.I.; funding acquisition, A.A.D. and A.N.C. All authors have read and agreed to the published version of the manuscript.

Funding: This research received no external funding.

Institutional Review Board Statement: Not applicable.

Informed Consent Statement: Not applicable.

Data Availability Statement: Data are contained within the article.

Conflicts of Interest: The authors declare no conflicts of interest.

References

1. Kundur, P. *Power System Stability and Control*; McGraw-Hill: New York, NY, USA, 1993.
2. Foqha, T.; Alsadi, S.; Refaat, S.S.; Abdulmawjood, K. Experimental Validation of a Mitigation Method of Ferranti Effect in Transmission Line. *IEEE Access* **2023**, *11*, 15878–15895. [CrossRef]
3. Mondal, D.; Chakrabarti, A.; Sengupta, A. *Power System Small Signal Stability Analysis and Control*, 2nd ed.; Elsevier: London, UK; Academic Press: Cambridge, MA, USA, 2020.
4. Domijan, A. *Overall Conceptual Development, Planning, and Design Aspects of an Electric Power System Laboratory with an Energy Management System Control Center*; The University of Texas at Arlington: Arlington, TX, USA, 1986.
5. *Plan Maestro de Electrificación 2013-2022, Perspectiva y Expansión del Sistema Eléctrico Ecuatoriano*; Consejo Nacional de Electricidad CONELEC: Quito, Ecuador, 2013.
6. Pagalo, D.; Quintana, E. *Diseño y Construcción de un Transformador Trifásico para Control de Ángulo y un Modelo a Escala de Línea de Transmisión de 500 kV para el Laboratorio de Sistemas Eléctricos de Potencia*; Trabajo de Titulación, Escuela Politécnica Nacional: Quito, Ecuador, 2019.
7. Ramos, M.C.; Tupiza, S.R. *Diseño y Construcción de un Modelo a Escala de la Línea De Transmisión de 500 kV Coca Codo Sinclair-El Inga para el Laboratorio de Sistemas Eléctricos de Potencia*; Trabajo de Titulación, Escuela Politécnica Nacional: Quito, Ecuador, 2018.
8. Yonggao, Z.; Wanqing, L.; Yanli, G. Research on reactive compensation in ultra-high voltage power grid using shunt reactors. In Proceedings of the 2016 IEEE 11th Conference on Industrial Electronics and Applications (ICIEA), Hefei, China, 5–7 June 2016.
9. McLyman, C.W.T. *Transformer and Inductor Design Handbook*, 4th ed.; CRC Press: Boca Raton, FL, USA, 2011.
10. *IEC 60076-6*; Power Transformers—Part 6: Reactors. Available online: https://www.saiglobal.com/PDFTemp/Previews/OSH/iec/iec60000/60000/iec60076-6%7Bed1.0%7Db.pdf (accessed on 30 September 2024).

engineering proceedings

MDPI

Proceeding Paper

Machine Learning G-Code Optimization †

Héctor Lasluisa-Naranjo [1,*], David Rivas-Lalaleo [2], Joaquín Vaquero-López [1] and Christian Cruz-Moposita [3]

1 International Doctoral School, Universidad Rey Juan Carlos, 28933 Madrid, Spain; joaquin.vaquero@urjc.es
2 Department of Electronics and Telecommunications, Universidad de las Fuerzas Armadas ESPE, Sangolquí 171103, Ecuador; drrivas@espe.edu.ec
3 DIDAE Aeronautical Research and Development Department, Ejército Ecuatoriano, 170403 Quito, Ecuador; crcruzm@ejercitoecuatoriano.mil.ec
* Correspondence: h.lasluisa.2022@alumnos.urjc.es

† Presented at the XXXII Conference on Electrical and Electronic Engineering, Quito, Ecuador, 12–15 November 2024.

Abstract: G-codes are essential in CNC systems, providing crucial instructions for controlling machine parameters and operations in manufacturing, including 3D printing. They may contain errors affecting product quality and increasing resource consumption. This research applies the K-means machine learning clustering algorithm to optimize G-code parameters such as extruder and heated bed temperature, deposition speed, and flow control. The objective is to reduce manufacturing time and material usage while maintaining surface quality. A line-by-line analysis and rewriting of the G-code resulted in an average 24.36% reduction in time and 5% in material use, with minimal impact on quality, validated with the Taguchi method.

Keywords: G-code; machine learning; 3D printing

Citation: Lasluisa-Naranjo, H.; Rivas-Lalaleo, D.; Vaquero-López, J.; Cruz-Moposita, C. Machine Learning G-Code Optimization. *Eng. Proc.* **1970**, *77*, 32. https://doi.org/10.3390/engproc2024077032

Academic Editor: Christian Tipantuña

Published: 18 November 2024

1. Introduction

G-codes, which are fundamental to computerized numerical control (CNC) systems, play a critical role in both machining and 3D printing. These commands provide instructions to control the motion and actions of CNC machines during manufacturing [1]. In addition, they play a key role in 3D printing, where the trajectory is controlled by these codes, allowing the quality of printed products to be evaluated through image analysis and result comparisons [2]. Understanding G-codes is essential for CNC programming, as they dictate parameters such as cutting feed, depth, and speed, and they are the foundation of modern manufacturing processes [3].

In 3D printing, G-codes control critical parameters such as trajectories, motion speed, fill patterns, material flow, and temperature [4]. Problems with G-codes originate at the CAD design stage, where NURBS-based programs offer higher quality finishes than those based on parametric curves and polygonal approximations [5]. G-code errors can also be introduced during the conversion of 3D models to STL format, creating gaps, holes, and open loops in cross sections that affect manufacturing quality [6]. In addition, another source of G-code errors comes from the wide variety of 3D printing equipment brands and slicing software, each with its own idiosyncrasies and standards [7].

Machine learning (ML), a branch of artificial intelligence (AI), is increasingly being used to optimize G-code. Neural networks efficiently transform numerically controlled G-codes [8]. AI algorithms optimize G-codes in CNC machines, reducing paths and improving processes (e.g., PCB drilling) [9]. These advances promise greater efficiency and accuracy in programming and machining. The best practices for debugging G-codes include statistical models and ML to detect malicious edits without access to original models, thus ensuring the quality of 3D-printed parts [10].

This study proposes the optimization of G-codes in 3D printing using the K-means algorithm, which is widely used in unsupervised clustering to evaluate parameter variation among objects [11]. In the context of 3D printing, K-means is applied to the modular partitioning of FDM 3D printers to improve performance, functionality, and product cost by generalizing modules [12].

In addition, K-means is used to predict uncertain material parameters when updating finite element models in 3D printing by optimizing an objective function based on dynamic characteristics to determine material properties [13]. The algorithm also improves the efficiency of 3D printing processes by determining optimal initial centroids, reducing iterations, and execution time [14]. It is also used to select interpolation points in the ISDF method, reducing computational cost and memory usage [15].

2. Methodology

This section describes the method used to optimize 3D printing parameters, specifically to reduce printing time and material consumption without compromising the quality of the printed model. This was achieved using the K-means algorithm, a machine learning technique that clusters data into specific groups. In the field of artificial intelligence, K-means is known for its ability to efficiently identify patterns and structures in large data sets [16]. This algorithm is particularly suitable for the present study due to its computational efficiency, its ability to identify patterns and trends in pressure parameters, and its versatility in addressing clustering problems in various contexts [17].

2.1. Experimental Model Selection

The 3Dbenchy model [18] was selected for the development of the 3D printing tests. This popular benchmark is widely used in the maker community as a standard for evaluating printer performance. The 3Dbenchy model presents multiple printing challenges, including changes in layer height, speed, and temperature, making it ideal for testing parameter optimization. Using the 3Dbenchy allows for a comprehensive evaluation of how the optimized parameters affect print quality, build time, and material consumption. Figure 1 shows the 3Dbenchy CAD model.

Figure 1. Three-dimensionalbenchy CAD model.

2.2. Slicing Software

To generate the input G-code file and evaluate the effectiveness and applicability of the algorithm, three different slicers were employed: Cura, PrusaSlicer, and IdeaMaker. Each slicer possesses distinctive features, path generation algorithms, and configurations, as illustrated in Table 1, allowing for a comprehensive and diverse comparison of the features.

Table 1. Main features of slicing software.

	Cura	Prusa Slicer	Ideamaker
Developer	Ultimaker	Alessandro Ranellucci	Raise 3D
Algorithm	Arachne Engine	Slic3r	Pathfinder
Principle	Voronoi Diagram	Planar Intersection, tessellación	Algorithms A star Dijkstra
Disadvantage	Unnecessary non-extrusion movements and longer printing time.	False retraction movements and other non-essential movements.	Low trajectory optimization

2.3. Coding

The proposed solution begins with the reading of a G-code file, which is provided by the slicing software. The file contains specific codes, including M104, M109, M140, and M190 for temperature control, G1 for speed control, and M221 for flow control. In the ML process, the K-means algorithm employs the input parameters to cluster the G-code data and calculate new values (centroids). The infill density is fixed at 10%, and the infill speed is fixed at 80 mm/s. These new values are then replaced with the relevant lines of code, resulting in the creation of an optimized G-code file. For more detailed information, please refer to Figure 2.

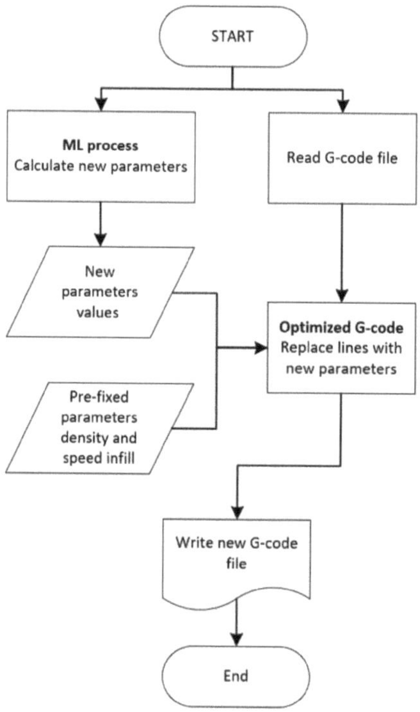

Figure 2. G-code optimizer flowchart.

2.4. Data Preprocessing

For the machine learning process, the data from successful 3D prints were collected, which included essential parameters such as layer height, print speed, first layer speed, perimeter speed, extruder temperature, bed temperature, and flow rate. These data points are presented in Table 2 for reference.

Table 2. Three-dimensional printing input parameters.

Parameter	Measures				
Layer height (mm)	0.2	0.2	0.3	0.1	0.3
Print speed (mm/s)	50	60	45	55	65
First layer speed (mm/s)	20	30	25	15	35
Perimeter speed (mm/s)	40	50	45	35	55
Extruder Temp (°C)	200	210	195	205	220
Bed Temp (°C)	60	65	55	60	70
Flow rate (%)	100	95	105	90	110

The data were organized into a dictionary and subsequently converted into a DataFrame using the pandas library, allowing for convenient manipulation and analysis.

2.4.1. Library Imports

NumPy (imported as np) and pandas (imported as pd) were imported for data manipulation and mathematical operations, and KMeans from sklearn cluster was imported to apply the K-means clustering algorithm for data grouping.

```
import numpy as np
import pandas as pd
from sklearn.cluster import KMeans
```

2.4.2. Data Definition

Dictionary-named data were defined, where each key represents a 3D printing parameter, and each value is a list of corresponding measurements. Parameters include layer height, print speed, first layer speed, perimeter speed, extruder temperature, bed temperature, and flow rate.

data = {'layer_height': [0.2, 0.2, 0.3, 0.1, 0.25],'print_speed': [50, 60, 45, 55, 65],'first_layer_speed': [20, 30, 25, 15, 35],'perimeter_speed': [40, 50, 45, 35, 55],'extruder_temp': [200, 210, 195, 205, 220],'bed_temp': [60, 65, 55, 60, 70],'flow_rate': [100, 95, 105, 90, 110]}.

2.4.3. DataFrame Creation

A DataFrame df is created using pandas from the data dictionary, organizing the data into a tabular format where each column corresponds to a parameter and each row to an observation.

```
df = pd.DataFrame(data)
```

2.5. K-Means Model Training

2.5.1. K-Means Algorithm Initialization

An instance of the K-means clustering algorithm from the sklearn library is created with *n*_clusters = 5, specifying that the data should be grouped into 5 clusters. This parameter can be adjusted based on the analysis goals.

```
kmeans = KMeans(n_clusters = 5)
```

2.5.2. Model Fitting

The fit() method was used to train the K-means model on the data contained in the DataFrame df. During this process, the algorithm clusters the data based on the features provided.

```
kmeans.fit(df)
```

2.5.3. Assigning Cluster Labels

After fitting the model, the labels_ attribute of the K-means object contain the cluster labels for each observation in the DataFrame. These labels indicate the cluster to which each row belongs. A new column, 'cluster', was added to the DataFrame to store these cluster labels.

```
df['cluster'] = kmeans.labels_
```

2.6. Optimal Parameter Selection

2.6.1. Identifying the Most Frequent Cluster

df['cluster'].value_counts() counted the occurrences of each cluster label in the 'cluster' column of the DataFrame. The value_counts() method returned a pandas Series with counts in descending order. idxmax() retrieved the index of the maximum value in the Series, which corresponds to the most frequent cluster label. This result was stored in the variable optimal_cluster, indicating the most common cluster.

```
optimal_cluster = df['cluster'].value_counts().idxmax()
```

2.6.2. Calculating Average Parameters for the Optimal Cluster

- df[df['cluster'] == optimal_cluster] filters the DataFrame to select only the rows belonging to the most frequent cluster (optimal_cluster).
- drop(['cluster'], axis = 1) removes the 'cluster' column from the filtered rows as it is no longer needed for the average parameter calculation.
- mean() computes the mean of each column in the filtered DataFrame, resulting in a pandas Series with average values for each parameter.
- to_dict() converts this Series into a dictionary where the keys are parameter names and the values are the computed averages. The result is stored in the variable optimized_params.

```
optimized_params = df[df['cluster'] == optimal_cluster].drop(['cluster'], axis = 1).
mean().to_dict()
```

Additionally, the printing parameters were set to an infill speed of 80 mm/s and an infill density of 10%.

```
infill_speed = 80
infill_density = 10
```

2.7. G-Code Optimization

2.7.1. Import the Regular Expression Module

The re module was imported, which provides functions for working with regular expressions in Python. This step is useful for performing search and replace operations in the text.

```
import re
```

2.7.2. Define the Function

The function optimize_gcode was defined to take four arguments: input_file_path (path to the input G-code file), output_file_path (path to the optimized G-code file), params (a dictionary containing printing parameters), infill_speed (speed of infill), and infill_density (density of infill).

```
def optimize_gcode(input_file_path, output_file_path, params, infill_speed, infill_density):
```

2.7.3. Read the Input File

The input file was opened in read mode ('r') and all lines were read into a list called lines.

with open(input_file_path, 'r') as file:
lines = file.readlines()

2.7.4. Initialize the List of Optimized Lines

An empty list was created called optimized_lines to store the modified lines of G-code.
optimized_lines = []

2.7.5. Process Each Line to Apply Optimizations

Each line in the input G-code file was iterated to apply the required modifications based on the content and parameters provided.

for line in lines:

2.7.6. Update Layer Height Comment

If the line started with; (indicating a comment) and contained 'Layer height', the comment was updated to reflect the layer height specified in params.

if line.startswith(';') and 'Layer height' in line:
line = f';Layer height: {params["layer_height"]}\n'

2.7.7. Adjust Print Speed

For lines containing G1, F (indicating speed), and not starting with ;, replaced the speed value (F) with the speed specified in params, converted from mm/s to mm/min.

if 'G1' in line and 'F' in line and not ';' in line.split('G1')[0]:
line = re.sub(r'(F)(\d+)', lambda m: f'F{int(params["print_speed"] * 60)}', line)

2.7.8. Adjust First Layer Speed

This step is similar to print speed adjustment but applied to lines containing G1, Z (indicating layer height), and F, modifying the speed for the first layer.

if 'G1' in line and 'Z' in line and 'F' in line and not ';' in line.split('G1')[0]:
line = re.sub(r'(F)(\d+)', lambda m: f'F{int(params["first_layer_speed"] * 60)}', line)

2.7.9. Adjust Perimeter Speeds

We updated the speeds for outer and inner wall types, replacing F with the perimeter speed specified in params.

if ';TYPE:WALL-OUTER' in line or ';TYPE:WALL-INNER' in line:
line = re.sub(r'(F)(\d+)', lambda m: f'F{int(params["perimeter_speed"] * 60)}', line)

2.7.10. Adjust Infill Speed

For lines containing;TYPE:FILL, the speed F was replaced with the specified infill_speed.

elif ';TYPE:FILL' in line:
line = re.sub(r'(F)(\d+)', lambda m: f'F{int(infill_speed * 60)}', line)

2.7.11. Modify Extruder and Bed Temperatures

The extruder and bed temperatures were adjusted by updating lines containing M104, M109, M140, or M190 with the temperatures specified in params.

if 'M104' in line or 'M109' in line:

line = f'M104 S{int(params["extruder_temp"])}\n' if 'M104' in line else f'M109 S{int (params["extruder_temp"])}\n'

if 'M140' in line or 'M190' in line:

line = f'M140 S{int(params["bed_temp"])}\n' if 'M140' in line else f'M190 S{int(params ["bed_temp"])}\n'

2.7.12. Adjust Material Flow Rate

For lines containing G1 and E (indicating material flow), the flow value E is adjusted based on the specified flow_rate in params.

if 'G1' in line and 'E' in line and not ';' in line.split('G1')[0]:

line = re.sub(r'(E)(\d+\.?\d*)', lambda m: f'E{float(m.group(2)) * params["flow_rate"] / 100}', line)

2.7.13. Update Infill Density

Lines containing M221were modified to set the infill density using the provided infill_density. The function optimized_lines.append(line) guaranteed that all lines, both modified and unmodified, are incorporated into the final optimized G-code file.

if 'M221' in line:

line = f'M221 S{int(infill_density)}\n'

optimized_lines.append(line)

2.7.14. Write to Output File

Finally, the output file was opened in write mode ('w') and all the optimized lines were written to this file.

with open(output_file_path, 'w') as file:

file.writelines(optimized_lines)

2.8. Deployment of Optimization Function

The paths for the original and optimized G-code files were defined. The optimize_gcode function was then called with the optimized parameters, including the values set for infill speed and density. This function generated the optimized G-code file.

input_file_path = r"file_path"
output_file_path = r"file_path"
optimize_gcode(input_file_path, output_file_path, optimized_params, infill_speed, infill_density)

The input_file_path and output_file_path variables were assigned the respective paths to the source and destination G-code files. The prefix r indicates that these strings were raw literals, ensuring that backslashes were handled correctly.

This function call executed optimize_gcode, passing:

- input_file_path: the location of the G-code file to be optimized.
- output_file_path: the destination path for the optimized G-code file.
- optimized_params: a dictionary of parameters including layer height, print speed, first layer speed, perimeter speed, extruder temperature, bed temperature, and material flow rate.
- infill_speed: the speed of infill printing, in millimeters per second.
- infill_density: the percentage of infill density to be applied.

In summary, this method enables the optimization of 3D printing parameters through the use of the K-means algorithm, which adjusts the resulting G-code to enhance both the efficiency and quality of the printing process.

3. Results and Discussion

With the three new optimized G-code files, the 3D benchy model was 3D printed in PLA material, as shown in Figure 3.

Figure 3. Models printed with optimized G-code based on slicers (a) Cura, (b) Prusa Slicer, (c) Idea Maker.

Despite having undergone optimization, the models displayed a multitude of characteristics and surface details, which can be attributed primarily to the slicer software utilized. Table 3 presents a comparative analysis of the features between the optimized models and those that have not undergone optimization. The results of the percentage variation in each case are highlighted.

Table 3. Comparison of traditional and optimized 3D printing performance measures.

		Time	Material	Surface Finish	Accuracy
CURA	Original	48 m 14 s	3724 mm	15 µm	99.41%
	Optimized	39 m 33 s	3538 mm	14.8 µm	99.39%
	% variation	**−18%**	**−5%**	**−1%**	**0.02%**
PRUSASLICER	Original	1 h 22 m 21 s	4297 mm	16.5 µm	99.4%
	Optimized	51 m 56 s	4082 mm	16.3 µm	99.38%
	% variation	**−36.90%**	**−5%**	**−1%**	**−0.01%**
IDEAMAKER	Original	52 m 48 s	3908 mm	18 µm	99.13%
	Optimized	43 m 12 s	3712 mm	17.7 µm	99.12%
	% variation	**−18.18%**	**−5.02%**	**−1.5%**	**−0.01%**

To assess the efficacy of the G-code optimization script, the Taguchi statistical method, a widely utilized approach in experimental design, will be employed. This method will be implemented through an experimental framework, with four levels selected for each factor, based on both the recommendations for optimal FDM printing parameters for PLA [19] and the technical specifications of the Artillery Hornet printer [20]. The values of this configuration are presented in Table 4.

Table 4. Input parameters.

Layer Height (mm)	Extruder Temp (°C)	Bed Temp (°C)	Print Speed (mm/s)	Flow Rate (%)
0.1	195	55	45	90
0.15	200	60	50	95
0.2	205	65	55	100
0.3	210	70	60	105

Results

The XLSTAT software employs the provided data to assign an L16 orthogonal matrix for the experimental design, thereby generating 16 potential parameter combinations.

A test object is produced for each combination, and the resulting output variables are subsequently measured, as illustrated in the results presented in Table 5.

Table 5. Combinations of input parameters and measured printing variables.

Experiment	Layer Height (mm)	Extruder Temp (°C)	Bed Temp (°C)	Print Speed (mm/s)	Flow Rate (%)	Fabrication Time Variation (%)	Material Consumption Variation (%)	Accuracy Variation (%)
1	0.1	195	55	45	90	−12.5	−2.5	−0.5
2	0.1	200	60	50	95	−15	−1.5	−1
3	0.1	205	65	55	100	−17.75	−3	−0.75
4	0.1	210	70	60	105	−20.5	−4.25	−1.5
5	0.15	195	60	55	105	−19	−3.75	−1.25
6	0.15	200	55	60	100	−21.5	−2.25	−1.75
7	0.15	205	70	45	95	−23.25	−4.5	−0.85
8	0.15	210	65	50	90	−22	−3.5	−1
9	0.2	195	65	60	95	**−24**	**−5**	**−0.01**
10	0.2	200	70	55	90	−21	−3	−1.1
11	0.2	205	55	50	105	−19.5	−2.5	−1.4
12	0.2	210	60	45	100	−16.75	−2.75	−1.2
13	0.3	195	70	50	100	−14	−1.75	−1.3
14	0.3	200	65	45	105	−10	−1.25	−1.5
15	0.3	205	60	60	90	−7.5	0.5	−1.75
16	0.3	210	55	55	95	−5	1.25	−1.9

As illustrated in Figure 4, the extruder temperature (195 °C) in experiment 9 is notably lower than in the other experiments, which may have contributed to the prevention of over-extrusion. Conversely, the bed temperature (65 °C) is within the typical range, facilitating optimal adhesion of the initial layers without warping the test object. The layer height (0.2 mm) represents an intermediate value, and in conjunction with the printing speed (60 mm/s), it facilitated a reduction in fabrication time without a significant impact on dimensional accuracy in this experiment.

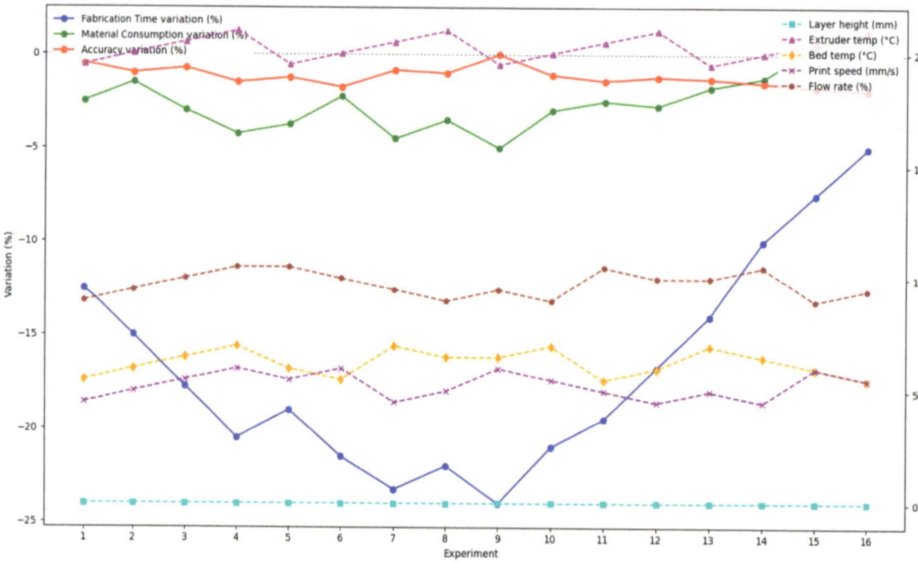

Figure 4. Fabrication time, material consumption, and accuracy variation.

A level of linearity consistent with this manufacturing technology can be observed, which facilitates the identification of general trends and the primary influences of input parameters on output variables. The Taguchi method is designed to enhance efficiency in the search for optimal configurations by limiting the number of experiments required. In this context, the linearity of the data allows for practical inferences to be drawn without the need for an excessive number of trials. A fully nonlinear approach would necessitate a significantly greater number of experiments, which would unnecessarily complicate the subsequent analysis.

4. Conclusions

The analysis using the Taguchi method demonstrated that optimizing 3D printing parameters using the developed code is effective in enhancing manufacturing process efficiency. The optimal combination of parameters successfully reduced print time by an average of 24.36% across all three cases and decreased material consumption by 5%, without compromising the surface quality and accuracy of the printed object.

These findings confirm that, despite the advanced algorithms employed by current slicers (such as Cura with Voronoi diagrams, PrusaSlicer with plane intersection and tessellation, and IdeaMaker with A* and Dijkstra algorithms), there is significant room for improvement through the application of machine learning techniques and robust analysis like the Taguchi method. This additional optimization addresses inherent weaknesses in slicer algorithms, achieving a more efficient and higher quality 3D printing process.

This study provides a robust methodology for parameter optimization in 3D printing, demonstrating that combining advanced machine learning techniques with experimental design can lead to substantial improvements in manufacturing performance and final product quality.

Author Contributions: Conceptualization, H.L.-N. and D.R.-L.; methodology, J.V.-L. and C.C.-M.; software, C.C.-M.; validation, H.L.-N. and J.V.-L.; formal analysis, H.L.-N.; investigation, H.L.-N. and C.C.-M.; resources, H.L.-N.; data curation, D.R.-L.; writing—original draft preparation, H.L.-N.; writing—review and editing, J.V.-L.; visualization, C.C.-M.; supervision, D.R.-L.; project administration, H.L.-N.; funding acquisition, H.L.-N. All authors have read and agreed to the published version of the manuscript.

Funding: This research received no external funding.

Institutional Review Board Statement: This study did not require ethical approval.

Informed Consent Statement: Not applicable.

Data Availability Statement: The data presented in this paper were extracted from practical experiments, are not stored in databases and do not violate any privacy policy.

Conflicts of Interest: The authors declare no conflict of interest.

References

1. Daskalogrigorakis, G.; Kirakosian, S.; Marinakis, A.; Nikolidakis, V.; Pateraki, I.; Antoniadis, A.; Mania, K. G-Code Machina: A Serious Game for G-code and CNC Machine Operation Training. In Proceedings of the 2021 IEEE Global Engineering Education Conference (EDUCON), Vienna, Austria, 21–23 April 2021. [CrossRef]
2. Bae, H.C. G-code 3D Method and Apparatus for Managing 3D Printing Using G-code. 2019.
3. Negi, P.K.; Ram, M.; Yadav, O.P. CNC Machine Programming Codes (G-Codes and M-codes). In *Basics of CNC Programming*, 1st ed.; Taylor & Francis: New Dehli, India, 2018. [CrossRef]
4. John, P.; Komma, V.R.; Bhore, S.P. Development of MATLAB Code for Tool Path Data Extraction from the G Code of the Fused Filament Fabrication (FFF) Parts. *Eng. Res. Express* **2023**, *5*, 025018. [CrossRef]
5. Pascual, A.; Ortega, N.; Plaza, S.; Holgado, I.; Arrizubieta, J.I. A RE Methodology to Achieve Accurate Polygon Models and NURBS Surfaces by Applying Different Data Processing Techniques. *Metals* **2020**, *10*, 1508. [CrossRef]
6. Husiev, O.; Nikiforova, T. Research of the Converting Stages for the Volume Model of the Product into the Control Code for a 3D Printer in the Context of Automated Construction of 3D Printing Technology. *Ukr. J. Constr. Archit.* **2022**, *2312*, 250822. [CrossRef]
7. Eva, S.-C.; Sover, A.; Ermolai, V. The Impact of the G-code Flavour Selection in FFF. In Proceedings of the 4th International Conference. Business Meets Technology, Ansbach, Germany, 7–9 July 2022. [CrossRef]

8. Xu, H.; Wu, J.; Qin, Y.; Lin, X.; Song, H.; Zhu, W. Method for Converting Numerically Controlled G-Codes. Patent WO2007078025A1, 12 July 2007.
9. Aciu, R.-M.; Ciocarlie, H. G-code Optimization Algorithm and Its Application on Printed Circuit Board Drilling. In Proceedings of the 2014 IEEE 9th IEEE International Symposium on Applied Computational Intelligence and Informatics (SACI), Timisoara, Romania, 15–17 May 2014. [CrossRef]
10. Beckwith, C.; Naicker, H.S.; Mehta, S.; Udupa, V.R.; Nim, N.T.; Gadre, V.; Pearce, H.; Mac, G.; Gupta, N. Needle in a Haystack: Detecting Subtle Malicious Edits to Additive Manufacturing G-Code Files. *IEEE Embed. Syst. Lett.* **2022**, *14*, 111–114. [CrossRef]
11. Li, X.; Tan, H. K-Means Algorithm Based on Initial Cluster Center Optimization. In *Cyber Security Intelligence and Analytics*; Lecture Notes in Computer Science; Springer: Cham, Switzerland, 2020; Volume 12337, pp. 477–485. [CrossRef]
12. You, Y.; Liu, Z.; Liu, Y.; Huang, Y.; Huang, Q. K-Means Module Division Method of FDM 3D Printer-Based Function–Behavior–Structure Mapping. *Appl. Sci.* **2023**, *13*, 7453. [CrossRef]
13. Mathur, M.B. K-means Optimizer: An Efficient Optimization Algorithm for Predicting the Uncertain Material Parameters in Real Structures. In Proceedings of the 5th International Conference on Numerical Modelling in Engineering, Ghent, Belgium, 23–24 August 2022; Lecture Notes in Electrical Engineering. Springer: Singapore, 2023; Volume 833, pp. 95–106. [CrossRef]
14. Qin, X.; Li, J.; Hu, W.; Yang, J. Machine Learning K-Means Clustering Algorithm for Interpolative Separable Density Fitting to Accelerate Hybrid Functional Calculations with Numerical Atomic Orbitals. *J. Phys. Chem. A* **2020**, *124*, 7397–7407. [CrossRef] [PubMed]
15. Vinué, G.; Simó, A.; Alemany, S. The K-means Algorithm for 3D Shapes with an Application to Apparel Design. *Adv. Data Anal. Classif.* **2016**, *10*, 297–312. [CrossRef]
16. Fitri, M.; Wardhani, A.R.; Purnomowati, W.; Vitianingsih, A.V.; Maukar, A.L.; Puspitarini, E.W. Potential Customer Analysis Using K-means with Elbow Method. *JIKO (J. Inform. Dan Komput.)* **2023**, *7*, 911. [CrossRef]
17. Géron, A. *Hands-On Machine Learning with Scikit-Learn, Keras, and TensorFlow*, 2nd ed.; O'Reilly Media: Sebastopol, CA, USA, 2019.
18. 3DBenchy. 3DBenchy Brochure. 2015. Available online: http://www.3dbenchy.com/wp-content/uploads/2015/04/3DBenchy_Broschure_3DBenchy.com_.pdf (accessed on 31 July 2024).
19. Wang, Y.; Tang, M.; Zhou, Z.; Wang, C. Study on the temperature field distribution of the interface and process parameter optimization for material extrusion of PLA. *Polym. Eng. Sci.* **2024**, *64*, 718–732. [CrossRef]
20. Artillery. (n.d.). Manual de la Impresora 3D Sidewinder X1, Hornet V4. Manuals Plus. Available online: https://manuals.plus/es/artillery-hornet/sidewinder-x1-3d-printer-v4-manual (accessed on 31 July 2024).

engineering proceedings

MDPI

Proceeding Paper

Application of Multi-Agent Systems: Modeling of Interacting Tanks †

Silvana Gamboa * and Jackeline Abad Torres

Department of Automation and Industrial Control, National Polytechnic School, Quito 17-01-2759, Ecuador; jackeline.abad@epn.edu.ec
* Correspondence: silvana.gamboa@epn.edu.ec

† Presented at the XXXII Conference on Electrical and Electronic Engineering, Quito, Ecuador, 12–15 November 2024.

Abstract: This paper presents a comprehensive approach to modeling interacting tanks as a multi-agent system. The primary goal is to develop a model that considers the dynamics of each agent and their interconnection so that the behavior of the whole system can be inferred from their coupling via graph theory, spectral graph theory and control systems. Given the tools used to model the system, not only is the proposed model scalable to n agents/tanks, but it also considers any configuration among them.

Keywords: modeling; interconnected tanks; multi-agent systems; linear systems

Citation: Gamboa, S.; Abad Torres, J. Application of Multi-Agent Systems: Modeling of Interacting Tanks. *Eng. Proc.* **2024**, *77*, 33. https://doi.org/10.3390/engproc2024077033

Academic Editor: Pablo Proaño

Published: 20 November 2024

1. Introduction

A collection of components known as agents communicate with one another to accomplish a shared goal in multi-agent systems (MASs) [1–5]. MAS applications are numerous and expanding all the time. Numerous applications can be found here, including data traffic in computer science; industrial plant control with actuator saturation [6]; multi vehicle coordination [7]; emergency response operations involving drones and mobile robots [8]; and smart grids or micro-grids [9–11], among others. In all of these applications, the agents are interconnected and share resources, and their individual behavior affects the collective response. Because of MASs' complexity, these systems remain an open challenge in terms of modeling, design, development, and coordination. As a result, a number of works have focused on studying their structure, evolution, and control in diverse fields [3–5,12].

Despite the fact that interactive tank models have been studied in great detail [13–15] because of their wide range of applications in power generation, chemical processes, food and beverage production and storage, water supply and transportation, and oil and gas refining and storage, most of the studies have focused on their control. Further, the few that considered a MAS approach concentrated on computational techniques for its estimation, simulation, and control [16–18].

In contrast, we describe the dynamics of this system as two parts: an individual behavior with their individual inputs/outputs (Section 2) and the coupling (interaction) among them as well as external inputs and outputs to form the whole system, all based on engineering and physical principles. In order to make the model easily usable with any number of interconnected agents (connected in any configuration), we apply MAS analysis tools such as spectral graph theory and graph theory (Sections 3 and 4.1). Further, control theory serves to understand the behavior of the whole system based on each agent's response (Sections 3.3 and 4.2). Finally, we show illustrative examples (Section 5).

2. Background

To model the fluid flow rate and pressure in a system of interconnected reservoirs or tanks, one considers the following variables: (1) fluid height in the reservoir/tank

h; (2) fluid resistance R, defined as the opposition to the fluid flow in the pipes, i.e., a restriction/valve on the pipe or a load/drain valve; and (3) the coefficient of flow storage in the reservoir (capacitance of the fluid) C, which depends on the area of the reservoir and the density of the fluid [19] (see Figure 1).

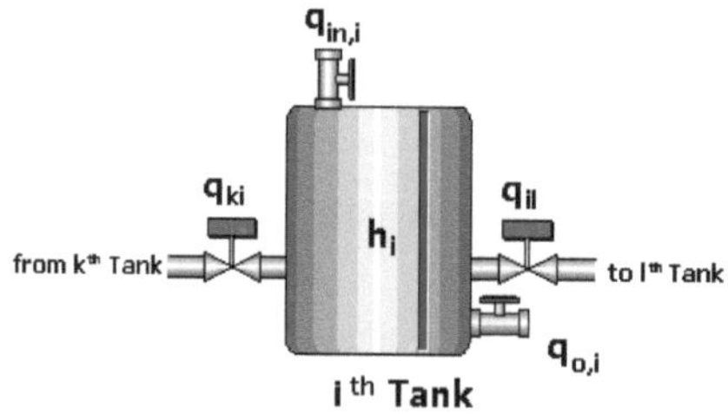

Figure 1. Summary of variables in the ith tank.

For now, system uncertainties and external disturbances are not considered for modeling. According to the principle of conservation of material [19], in the ith tank/reservoir, the relationship between fluid level h_i and flow rate q_i is

$$C_i \frac{dh_i}{dt} = q_i \tag{1}$$

Further, the total flow rate, i.e., the summation of all flow rates related to the ith tank in an n-tank system, could be written as

$$q_i = q_{in,i} - q_{o,i} - \sum_{j,\{i,j\}\text{interact}} q_{ij} \tag{2}$$

where $q_{in,i}$ and $q_{o,i}$ are the flow rates for the source and drain of the fluid, respectively, and q_{ij} is the exchanged flow rate between the tanks j that interact with i. Also, the flow rate on the drain $q_{o,i}$ could be approximated to the equation that describes the flow rate through a valve of constant $R_{o,i}$:

$$q_{o,i} = \frac{1}{R_{o,i}} h_i. \tag{3}$$

3. Modeling as First-Order Coupled System

3.1. Modeling Interacting Tank Without Inputs and Outputs

If the tank does not have external input/output, i.e., a source and drain, and is interacting with other tanks, then $q_{in,i} = 0$ and $q_{o,i} = 0$ in (2). Further, the exchanged flow rate between tank i and j is $q_{ij} = \frac{1}{R_{ij}}(h_i - h_j)\ \forall i, j > 0$, where R_{ij} is the flow resistance in the pipe (or valve) between tank i and j. Consequently, the level of tank i is given by

$$\dot{h}_i = -\frac{h_i}{C_i}\left(\sum_{j,\{i,j\}\text{interact}} \frac{1}{R_{ij}}\right) + \frac{1}{C_i}\left(\sum_{j,\{i,j\}\text{interact}} \frac{h_j}{R_{ij}}\right), \tag{4}$$

where "$\{i,j\}$interact" indicates that there is an interaction between tank i and j.

Considering that the reservoirs/tanks are interconnected through pipes and valves, and hence flows q_{ij}, one can specify the coupling between the tanks using a directed weighted graph $G = (V, E)$, where $V = \{i, \ldots, n\}$ is the vertex set that represents each tank, and $E = \{(i,j) : w_{ij}\}$ is the edge set that represents the interconnecting pipes/valves, with weights given by $w_{ij} = \frac{1}{R_{ij}C_i}$ [20]. Further, the dynamic model of the whole system of interacting tanks is given by

$$\dot{\vec{h}} = -\mathcal{L}\vec{h}, \tag{5}$$

where $\vec{h} = [h_1, \ldots, h_n]^T$ is the state vector containing the tanks' heights, and $\mathcal{L}$ is the Laplacian matrix associated with graph G, defined as [21]

$$\mathcal{L}_{ij} = \begin{cases} -\frac{1}{C_i R_{ij}} & \text{if } i \neq j, \{i,j\} \in E, \\ -\sum\limits_{j,\{i,j\} \in E, i \neq j} \mathcal{L}_{ij} & \text{if } i = j \\ 0 & \text{if } \{i,j\} \notin E \end{cases}, \tag{6}$$

for all $i, j = \{1, 2, 3, \ldots, n\}$. It should be noted that in this case, the Laplacian matrix $\mathcal{L}$ has non-positive off-diagonal elements, and its diagonal elements are equal to the negative sum of its row elements. Therefore, it has a left eigenvector equal to an all-ones vector ($\vec{1}$) associated with a zero eigenvalue, i.e., $\mathcal{L}\vec{1} = 0\vec{1}$ [21].

3.2. Modeling Interacting Tank with Inputs and Outputs

Considering that the system has external sources, which forcibly take out or inject fluid through local control valves that allow the tanks to interact or drain the tanks, the dynamic model of the system of interacting tanks is given by

$$\dot{h}_i = -\frac{h_i}{C_i}\left(\frac{1}{R_{o,i}} + \sum_{j,\{i,j\}\text{interact}} \frac{1}{R_{ij}}\right) + \frac{1}{C_i}\left(\sum_{j,\{i,j\}\text{interact}} \frac{h_j}{R_{ij}}\right) - \frac{1}{C_i}\left(\sum_{\{i,j\}\text{interact}} q_{i,j}\right) + \frac{q_{in,i}}{C_i}, \tag{7}$$

In general, one can consider a system of interacting tanks with both external inputs/outputs (flow rates) and local input and outputs to have the form

$$\dot{\vec{h}} = -\mathcal{L}\vec{h} + E\vec{q} + B\vec{u}, \tag{8}$$

where $\mathcal{L}$ is a Laplacian matrix, $\vec{u} = [q_{in,1}, \ldots, q_{in,p}]^T$ is the external input vector, $\vec{q} = [q_{i,j}, \ldots, q_{p,q}]^T$ is the local input vector, and the local input matrix is

$$E_{i,m} = \begin{cases} \frac{1}{C_i} & \text{if there is control valve } m \text{ between tanks i and j} \\ -\frac{1}{C_j} & \text{if there is control valve } m \text{ between tanks j and i,} \\ 0 & \text{otherwise} \end{cases}$$

for all $m = 1, \ldots, M$ and $i = 1, \ldots, n$, where M and n are the numbers of control valves and tanks, respectively. And the input matrix is

$$B_{i,m} = \begin{cases} \frac{1}{C_i} & \text{if there is an external input } m \text{ to the tank } i \\ 0 & \text{otherwise} \end{cases},$$

for all $i = 1, \ldots, n$ and $m = 1, .., p$, where p is the number of external inputs. Further, $\mathcal{L}$ in (8) is the Laplacian matrix of the system's graph. In this case, the graph associated with the system can include self-edges; that is,

$$\mathcal{L}_{ij} = \begin{cases} -\frac{1}{C_i R_{ij}} & \text{if } i \neq j, \{i,j\} \in E, \\ \frac{1}{C_i R_{o,i}} - \sum\limits_{j,\{i,j\} \in E, i \neq j} \mathcal{L}_{ij} & \text{if } i = j \\ 0 & \text{if } \{i,j\} \notin E \end{cases}, \tag{9}$$

for all $i, j = \{1, 2, 3, \ldots, n\}$, where $\frac{1}{C_i R_{o,i}}$ is a self-edge in the graph G, caused by the drain in a tank i. It should be noted that in this case, the Laplacian matrix $\mathcal{L}$ has non-positive off-diagonal elements, and its diagonal elements are equal to the negative sum of its row elements plus the weight of a self-edge, i.e., it is non-negative. In this case, the matrix $-\mathcal{L}$ is a Metzler matrix, which is commonly found in positive linear dynamical systems [22,23].

3.3. Interconnected Characteristics of the Coupled Model

One of the advantages of representing these systems as a MAS is that Equation (5) works for any interconnection among the tanks. Further, applying the same tools used in networked control systems and MAS [24], one can obtain the following result.

A system of interconnected tanks without inputs and outputs modeled as (4) or (5) reaches consensus if and only if the graph G associated with it is connected, i.e., there is a path between any two vertices in the graph. Further, the consensus height of fluid is $h_i = \sum_i w_i h_i(0)\ \forall i = 1, \ldots, n$ (weighted average of the initial tanks' levels), where $h_i(0)$ is the initial height of tank i, and w_i is the ith entry of the left eigenvector associated with the eigenvalue $\lambda_1 = 0$ of the Laplacian matrix $\mathcal{L}$ [25]. Moreover, if all the tanks have the same capacitance ($C_i = C \forall i = 1, \ldots, n$), then the tanks' heights converge to $\frac{1}{n}\sum_i h_i(0)$.

When considering the dynamics with local inputs/outputs q_{ij} (8), the graph representing the coupling may not be connected. However, the local input q_{ij} can locally control the tank level and also connect all the tanks in the systems (through the control valves available in the infrastructure). In this case, the exchange flow rate is $q_{ij} = -\frac{1}{R_{p_{i,j}}}(h_i - h_j)$, where $R_{p_{i,j}}$ is the flow resistance through the valve between tanks i and j, which can be considered static gain in a simple control algorithm. Further, the values of $\frac{1}{R_{p_{i,j}}} \neq 0$ generate new edges in the graph G. From the control and graph theory perspective, these new edges must connect the graph, in order to achieve a consensus in the system. Therefore, Equation (8) of the system becomes

$$\dot{\vec{h}} = -\tilde{\mathcal{L}}\vec{h} + B\vec{u}, \tag{10}$$

where $\tilde{\mathcal{L}}$ is the Laplacian matrix of the new graph, and hence has all the properties of any Laplacian matrix.

Moreover, if the tanks have drains, then all the eigenvalues, γ_i, of the Laplacian $\tilde{\mathcal{L}}$ have negative real parts. Hence, the natural response, i.e., $\vec{u} = \vec{0}$, of the dynamics in (10) is $\vec{h}(t) = \sum_{i=1}^{n} e^{\gamma_i t} v_i w_i^T \vec{h}(0)$, whose steady-state value is zero, that is, the dynamics converge to $\lim_{t\to\infty} h_i(t) = 0 \forall i = 1, \ldots, n$.

4. Modeling as Second-Order Coupled System

4.1. Model Development

In order to define the total flow rate q_i as a state variable, expression (2) is derived:

$$\dot{q}_i = \dot{q}_{in,i} - \frac{1}{R_{oi}}\dot{h}_i - \sum_{j,\{i,j\}\text{interact}} \dot{q}_{ij}. \tag{11}$$

Moreover, using the analysis of the pressure difference at the ends of the pipes that interconnect tanks i and j [19], it is possible to describe the derivative of the flow rate q_{ij} through that pipe as a general expression for the interconnection between tanks i and j: $\dot{q}_{ij} = a_{ij}\frac{\rho}{L_{ij}} g(h_i - h_j)$, where a_{ij} and L_{ij} are the cross-section and the length of the pipe between tanks i and j, ρ is the liquid density, g is the gravity constant, and h_i is the fluid height/level in tank i. Therefore, the total derivative of the flow rate exchanged by tank i to the rest of the system can be expressed through the expression

$$\sum_{j,\{i,j\}\text{interact}} \dot{q}_{ij} = \rho g \sum_{j,\{i,j\}\text{interact}} \frac{a_{pij}}{L_{ij}}(h_i - h_j), \tag{12}$$

Furthermore, in taking into account (1) and (3), (11) becomes

$$\dot{q}_i = -\frac{1}{R_{o,i}}\frac{1}{C_i} q_i - \rho g \sum_{j,\{i,j\}\text{interact}} \frac{a_{pij}}{L_{ij}}(h_i - h_j) + \dot{q}_{in,i}, \tag{13}$$

As in Section 3, a weighted graph $G = (V, E)$ represents the coupling between the tanks, where $V = \{i, \ldots, n\}$ is the vertex set that represents each tank, and $E = \{(i,j) : w_{ij}\}$ is the edge set that represents the interconnecting pipes, with weights given by $w_{ij} = w_{ji} =$

$\frac{a_{pij}}{L_{ij}}$. Therefore, the dynamics of the system of interacting tanks can be expressed by a second-order model as

$$\begin{bmatrix} \dot{\vec{h}} \\ \dot{\vec{q}} \end{bmatrix} = \begin{bmatrix} 0_{n,n} & \widetilde{C} \\ -\rho g \mathcal{L}_K & -\widetilde{C}\widetilde{R}_o \end{bmatrix} \begin{bmatrix} \vec{h} \\ \vec{q} \end{bmatrix} + \begin{bmatrix} 0_{n,n} \\ \mathsf{I}_{n,n} \end{bmatrix} \vec{u} \tag{14}$$

where $0_{n,n}$ and $\mathsf{I}_{n,n}$ are a zero matrix and the identity matrix, respectively, $\widetilde{C} = \text{diag}(C_1^{-1}, \ldots, C_n^{-1})$ is the matrix whose diagonal corresponds to the inverse of each coefficient of flow storage in every reservoir, $\widetilde{R}_o = \text{diag}(R_{o,1}^{-1}, \ldots, R_{o,n}^{-1})$ is the matrix whose diagonal corresponds to the inverse of the fluid resistances of every reservoir, $\vec{u} = [\dot{q}_{in,1}, \ \ldots, \dot{q}_{in,n}]^T$, and $\mathcal{L}_K$ is the Laplacian matrix

$$\mathcal{L}_{K_{ij}} = \begin{cases} -\frac{a_{pij}}{L_{ij}} & \text{if } i \neq j, \{i,j\} \in E, \\ -\sum\limits_{j,\{i,j\} \in E, i \neq j} \mathcal{L}_{K_{ij}} & \text{if } i = j \\ 0 & \text{if } \{i,j\} \notin E \end{cases}, \tag{15}$$

for all $i, j = \{1, 2, 3, \ldots, n\}$.

4.2. Interconnected Characteristics in Tank System Behavior

The state matrix of the dynamics in (14) has the following characteristic equation:

$$\left| \lambda^2 + \lambda \widetilde{C}\widetilde{R}_o + \rho g \mathcal{L}_K \widetilde{C} \right| = 0, \tag{16}$$

where λ denotes the eigenvalues of the state matrix. Therefore, the eigenvalues λ must satisfy the equation $\lambda^2 + \lambda \frac{1}{C_i R_{o,i}} + \rho g \frac{1}{C_i} \gamma_i = 0$, where γ_i is an eigenvalue of $\mathcal{L}_K$. Also, given that n eigenvalues of the Laplacian matrix $\mathcal{L}_K$ are $0 = \gamma_1 \leq \gamma_2 \leq \ldots \leq \gamma_n$ [21], the state matrix has at least two real eigenvalues: $\lambda_1 = 0$ and $\lambda_2 = -\frac{1}{CR_o}$. Further, the remaining $2n - 2$ eigenvalues have negative real parts.

Additionally, if G is connected, then the dominant mode of the dynamics in (14) is e^{0t}, of which its response is maintained over time while the effect of the other modes disappears as $t \to \infty$ since the associated eigenvalues have a negative real part. In this regard, we can analyze the behavior of this mode as time grows and whhile considering $\dot{q}_{ii} = 0$, i.e., the natural response.

The natural response of (14) is given by

$$\begin{bmatrix} \vec{h}(t) \\ \vec{q}(t) \end{bmatrix} = \sum_{i=1}^{2n} \left(e^{\lambda_i t} v_i w_i^T \right) \begin{bmatrix} \vec{h}(0) \\ \vec{q}(0) \end{bmatrix}, \tag{17}$$

where v_i and w_i^T are the right and left eigenvectors associated with the eigenvalue λ_i. Since the eigenvalues $\lambda_i \forall i = 2, \ldots, 2n$ have a negative real part and $\lambda_1 = 0$, the steady state natural response of the system is

$$\lim_{t \to \infty} \begin{bmatrix} \vec{h}(t) \\ \vec{q}(t) \end{bmatrix} = v_1 w_1^T \begin{bmatrix} \vec{h}(0) \\ \vec{q}(0) \end{bmatrix}, \tag{18}$$

where $v_1^T = \begin{bmatrix} \vec{1}^T & \vec{0}^T \end{bmatrix}$ and $w_i^T = \frac{1}{n} \begin{bmatrix} \vec{1}^T & \widetilde{R}_o^{-1} \vec{1}^T \end{bmatrix}$ are the right and left eigenvectors associated with the eigenvalue $\lambda_1 = 0$. Therefore, both the fluid height and the flow rate converge to

$$\begin{aligned} \lim_{t \to \infty} h_i(t) &= \frac{1}{n} \sum_{i=1}^{n} h_i(0) + \frac{1}{n} \sum_{i=1}^{n} R_{o,i} q_i(0) \\ \lim_{t \to \infty} q_i(t) &= 0 \end{aligned} \tag{19}$$

Moreover, if the initial flow rate is zero, the height in each tank converges to $\frac{1}{n} \sum_{i=1}^{n} h_i(0)$, which is the same as in the first-order model.

5. Illustrative Examples

First, let us model the system of three tanks shown in Figure 2 (left) using first-order dynamics.

Figure 2. System of three interacting tanks without inputs/outputs (**left**) and its associated graph (**right**).

The graph associated with this system has three vertices and edges, as shown in Figure 2 (right). Then, the directions and the weights of the edges depend on the direction of the fluid flow, capacitance of each tank, and resistance of the pipes. For instance, the flows $q_{12} = -q_{21} = \frac{1}{R_{12}}(h_1 - h_2)$ affect the dynamics of tank 1 and 2 and are represented as the directions and weights of the edges between vertices 1 and 2. Therefore, the state matrix of the model (5) is given by

$$\mathcal{L} = \begin{bmatrix} \frac{1}{R_{12}C_1} & -\frac{1}{R_{12}C_1} & 0 \\ -\frac{1}{R_{12}C_2} & (\frac{1}{R_{12}C_2} + \frac{1}{R_{23}C_2}) & \frac{1}{R_{23}C_2} \\ 0 & -\frac{1}{R_{23}C_3} & \frac{1}{R_{23}C_3} \end{bmatrix},$$

which is the negative of the Laplacian matrix of the graph shown in Figure 2 (right). The natural response of the three-tank dynamics is shown in Figure 3. One observes that the tanks' height reaches consensus with the weighted average of the initial heights, while the flow rate in each tank becomes zero.

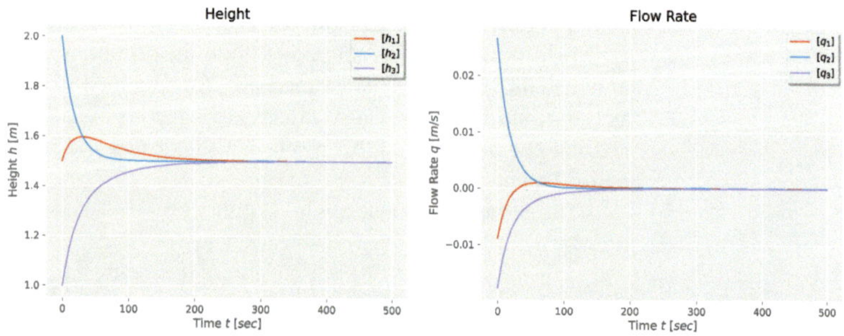

Figure 3. Height and flow rate for a first-order model of a three-tank system.

Meanwhile, using second-order dynamics, (14), where $\mathcal{L}_K$ is

$$\mathcal{L}_K = \begin{bmatrix} \frac{a_{p12}}{L_{12}} & -\frac{a_{p12}}{L_{12}} & 0 \\ -\frac{a_{p12}}{L_{12}} & (\frac{a_{p12}}{L_{12}} + \frac{a_{p23}}{L_{23}}) & -\frac{a_{p23}}{L_{23}} \\ 0 & -\frac{a_{p23}}{L_{23}} & \frac{a_{p23}}{L_{23}} \end{bmatrix}$$

the natural response of the system is faster than its first-order counterpart, and it has an oscillatory behavior due to the presence of complex modes in system (14), as shown in Figure 4. It is observed that the tanks' height reaches consensus with a value that depends of the initial heights, and the flow rate in each tank becomes zero, which means that exchanged flow rate through the pipes between tanks will be zero too.

Now, consider a five-tank system as shown in Figure 5, where there are control valves between tanks 2 and 3 and between 3 and 4, i.e., the dynamics corresponding to tanks 2, 3, and 4 have local inputs/outputs. Further, tanks 1, 3, and 5 have external inputs and tanks1, 2, 4, and 5 have an output flow.

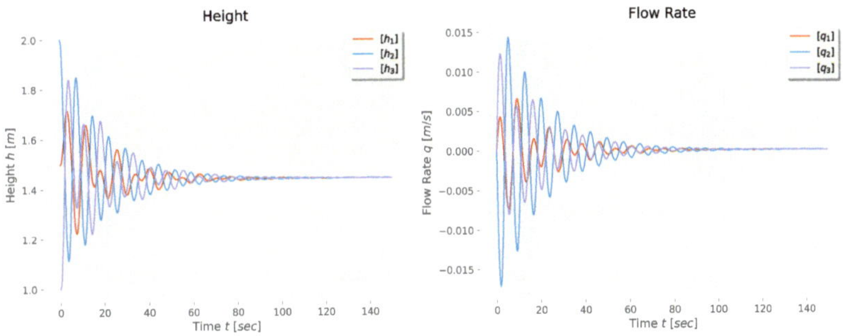

Figure 4. Height and flow rate for a second-order model of a three-tank system.

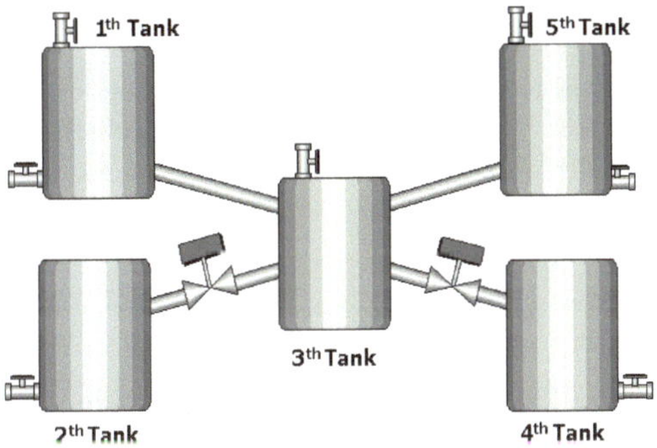

Figure 5. System of five interacting tanks.

The graph associated with this system is shown in Figure 6 (right), which includes self-edges in all the vertices. Moreover, the graph is not connected since vertices 2 and 4 are isolated; however, these vertices, together with vertex 3, have local inputs. The system dynamics are

$$\dot{\vec{h}} = -\mathcal{L}\vec{h} + \begin{bmatrix} 0 & 0 \\ \frac{1}{C_2} & 0 \\ -\frac{1}{C_3} & -\frac{1}{C_3} \\ 0 & \frac{1}{C_4} \\ 0 & 0 \end{bmatrix} \begin{bmatrix} q_{23} \\ q_{43} \end{bmatrix} + \begin{bmatrix} \frac{1}{C_1} & 0 & 0 \\ 0 & 0 & 0 \\ 0 & \frac{1}{C_3} & 0 \\ 0 & 0 & 0 \\ 0 & 0 & \frac{1}{C_5} \end{bmatrix} \begin{bmatrix} q_{in,1} \\ q_{in,3} \\ q_{in,5} \end{bmatrix} \tag{20}$$

where $\vec{h} = \begin{bmatrix} h_1 & \dots & h_5 \end{bmatrix}^T$ and $\mathcal{L} = \begin{bmatrix} \frac{1}{C_1}\left(\frac{1}{R_{o1}} + \frac{1}{R_{13}}\right) & 0 & -\frac{1}{C_1}\frac{1}{R_{13}} & 0 & 0 \\ 0 & \frac{1}{C_2}\frac{1}{R_{o,2}} & 0 & 0 & 0 \\ -\frac{1}{C_3}\frac{1}{R_{13}} & 0 & \frac{1}{C_3}\left(\frac{1}{R_{35}} + \frac{1}{R_{13}}\right) & 0 & -\frac{1}{C_3}\frac{1}{R_{35}} \\ 0 & 0 & 0 & \frac{1}{C_4}\frac{1}{R_{o,4}} & 0 \\ 0 & 0 & -\frac{1}{C_5}\frac{1}{R_{35}} & 0 & \frac{1}{C_5}\left(\frac{1}{R_{o,5}} + \frac{1}{R_{35}}\right) \end{bmatrix}$.

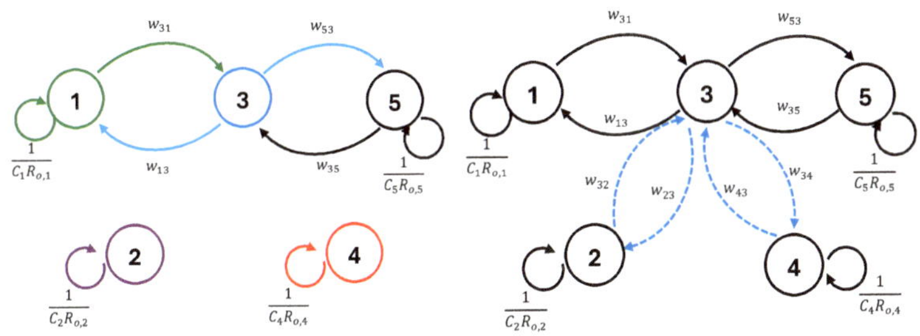

Figure 6. Graphs of the system with five interconnected tanks with closed control valves (**left**) and open control valves (**right**) between tanks 2, 3, and 4.

Once, the valves between tanks 2 and 3 and tanks 3 and 4 are open, the dynamics of the system are given by (10), as shown in Figure 6 (left), with Laplacian matrix

$$\tilde{\mathcal{L}} = \begin{bmatrix} \frac{1}{C_1}\left(\frac{1}{R_{o1}} + \frac{1}{R_{13}}\right) & 0 & -\frac{1}{C_1}\frac{1}{R_{13}} & 0 & 0 \\ 0 & \frac{1}{C_2}\left(\frac{1}{R_{o,2}} + \frac{1}{R_{p2,3}}\right) & -\frac{1}{C_2}\frac{1}{R_{p2,3}} & 0 & 0 \\ -\frac{1}{C_3}\frac{1}{R_{13}} & -\frac{1}{C_3}\frac{1}{R_{p2,3}} & \frac{1}{C_3}\left(\frac{1}{R_{35}} + \frac{1}{R_{13}} + \frac{1}{R_{p2,3}} + \frac{1}{R_{p3,4}}\right) & -\frac{1}{C_3}\frac{1}{R_{p3,4}} & -\frac{1}{C_3}\frac{1}{R_{35}} \\ 0 & 0 & -\frac{1}{R_{p3,4}} & \frac{1}{C_4}\left(\frac{1}{R_{o,4}} + \frac{1}{R_{p3,4}}\right) & 0 \\ 0 & 0 & -\frac{1}{C_5}\frac{1}{R_{35}} & 0 & \frac{1}{C_5}\left(\frac{1}{R_{o,5}} + \frac{1}{R_{35}}\right) \end{bmatrix}$$

For the second-order model, the dynamics of the five tanks given by (14) uses

$$\mathcal{L}_K = \begin{bmatrix} \frac{a_{p13}}{L_{13}} & 0 & -\frac{a_{p13}}{L_{13}} & 0 & 0 \\ 0 & \frac{a_{p23}}{L_{23}} & -\frac{a_{p23}}{L_{23}} & 0 & 0 \\ -\frac{a_{p13}}{L_{13}} & -\frac{a_{p23}}{L_{23}} & \left(\frac{a_{p13}}{L_{13}} + \frac{a_{p23}}{L_{23}} + \frac{a_{p34}}{L_{34}} + \frac{a_{p35}}{L_{35}}\right) & -\frac{a_{p34}}{L_{34}} & -\frac{a_{p35}}{L_{35}} \\ 0 & 0 & -\frac{a_{p34}}{L_{34}} & \frac{a_{p34}}{L_{34}} & 0 \\ 0 & 0 & -\frac{a_{p35}}{L_{35}} & 0 & \frac{a_{p34}}{L_{34}} \end{bmatrix}$$

Finally, the natural response of the five-tank system for a first-order and second-order model are shown in Figures 7 and 8. Again, one observes that the tanks' height and flow rate reach consensus: the heights converge to the weighted average of the initial heights, while the flow rates converge to zero.

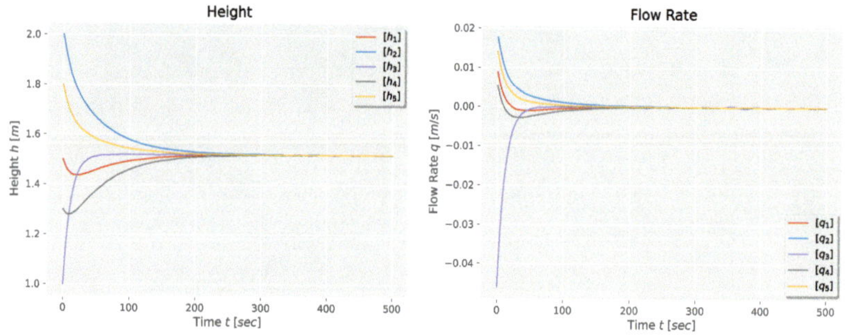

Figure 7. Height and flow rate for a first-order model of a five-tank system in star configuration.

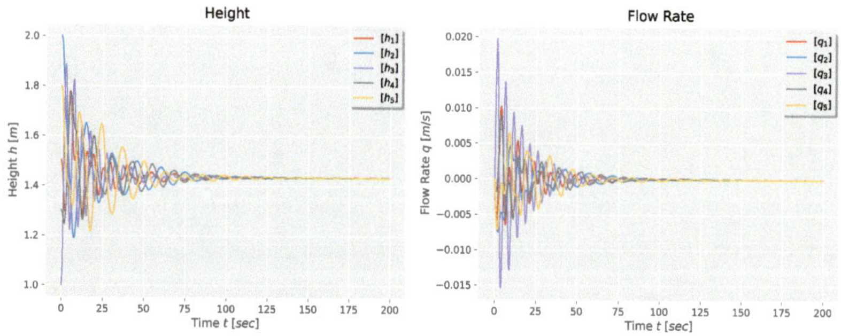

Figure 8. Height and flow rate for a second-order model of a five-tank system in star configuration.

6. Conclusions

The interacting tank model described in this study was modeled as a coupled/interconnected system that can employ any configuration for the interconnection between the tanks and is scalable to n agents. Furthermore, it employs analytical techniques to support the modeling, including graph and spectral graph theory, and control theory to analyze and characterize the behavior of individual agents as well as the system as a whole. Using this MAS approach, the system is viewed as a collection of independent agents cooperating to achieve a shared goal, namely the convergence of heights, which acts as a stepping stone to apply control algorithms designed for cyber-physical systems to this type of system.

Author Contributions: Conceptualization and methodology, S.G. and J.A.T.; software and validation, S.Gamboa; formal analysis, S.G. and J.A.T.; investigation, S.G. and J.A.T.; resources, S.G. and J.A.T.; data curation, S.G. and J.A.T.; writing—original draft preparation, S.G.; writing—review and editing, J.A.T.; visualization, S.G. and J.A.T.; supervision, J.A.T.; project administration, S.G. All authors have read and agreed to the published version of the manuscript.

Funding: This research received no external funding.

Institutional Review Board Statement: Not applicable.

Informed Consent Statement: Not applicable.

Data Availability Statement: http://emps.exeter.ac.uk/engineering/research/cws/resources/benchmarks (accessed on 31 August 2024).

Acknowledgments: The authors thank the Department of Automation and Industrial Control of the National Polytechnic School for providing the time for the development of this work.

Conflicts of Interest: The authors declare no conflict of interest.

References

1. Uhrmacher, A.M.; Weyns, D. *Multi-Agent Systems: Simulation and Applications*; CRC Press: Boca Raton, FL, USA, 2009.
2. Busoniu, L.; Babuska, R.; De Schutter, B. A comprehensive survey of multiagent reinforcement learning. *IEEE Trans. Syst. Man Cybern. Part C (Appl. Rev.)* **2008**, *38*, 156–172. [CrossRef]
3. Rizk, Y.; Awad, M.; Tunstel, E.W. Decision Making in Multiagent Systems: A Survey. *IEEE Trans. Cogn. Dev. Syst.* **2018**, *10*, 514–529. [CrossRef]
4. Oprea, M. Applications of multi-agent systems. In *Information Technology*; Springer: Cham, Switzerland, 2004; pp. 239–270.
5. Sycara, K.P. Multiagent systems. *AI Mag.* **1998**, *19*, 79.
6. Wang, B.; Wang, J.; Zhang, B.; Li, X. Global Cooperative Control Framework for Multiagent Systems Subject to Actuator Saturation With Industrial Applications. *IEEE Trans. Syst. Man Cybern. Syst.* **2017**, *47*, 1270–1283. [CrossRef]
7. Wang, B.; Wang, J.; Zhang, B.; Chen, W.; Zhang, Z. Leader–Follower Consensus of Multivehicle Wirelessly Networked Uncertain Systems Subject to Nonlinear Dynamics and Actuator Fault. *IEEE Trans. Autom. Sci. Eng.* **2018**, *15*, 492–505. [CrossRef]
8. Nunavath, V.; Prinz, A. Visualization of exchanged information with dynamic networks: A case study of fire emergency search and rescue operation. In Proceedings of the 2017 IEEE 7th International Advance Computing Conference (IACC), Hyderabad, India, 5–7 January 2017; pp. 281–286.
9. Merabet, G.H.; Essaaidi, M.; Talei, H.; Abid, M.R.; Khalil, N.; Madkour, M.; Benhaddou, D. Applications of multi-agent systems in smart grids: A survey. In Proceedings of the 2014 International Conference on Multimedia Computing and Systems (ICMCS), Marrakech, Morocco, 14–16 April 2014; pp. 1088–1094.
10. Maldonado, D.; Cepeda, J.; Gamboa, S.; Torres, J.A.; Cruz, P.J. Optimum frequency response with decentralized drivers in power system. *IET Gener. Transm. Distrib.* **2023**, *17*, 1985–1998. [CrossRef]
11. Kuo, M.T.; Lu, S.D. Design and implementation of real-time intelligent control and structure based on multi-agent systems in microgrids. *Energies* **2013**, *6*, 6045–6059. [CrossRef]
12. Leite, J.; Omicini, A.; Sterling, L.; Torroni, P. *Declarative Agent Languages and Technologies: First International Workshop, DALT 2003, Melbourne, Australia, July 15, 2003, Revised Selected and Invited Papers*; Springer: Cham, Switzerland, 2004; Volume 2990.
13. Gatzke, E.P.; Meadows, E.S.; Wang, C.; Doyle Iii, F.J. Model based control of a four-tank system. *Comput. Chem. Eng.* **2000**, *24*, 1503–1509. [CrossRef]
14. Naik, R.B.; Kanagalakshmi, S. Mathematical modelling and controller design for interacting hybrid two tank system (IHTTS). In Proceedings of the 2020 Fourth International Conference on Inventive Systems and Control (ICISC), Coimbatore, India, 8–10 January 2020; pp. 297–303.
15. Balakrishnaa, A.; Arun, N. Liquid level control of interacting coupled spherical tank system using PI and Fuzzy Pi Controller. In Proceedings of the 2022 3rd International Conference for Emerging Technology (INCET), Belgaum, India, 27–29 May 2022; pp. 1–5.
16. Hajebi, S.; Barrett, S.; Clarke, A.; Clarke, S. Multi-agent simulation to support water distribution network partitioning. In Proceedings of the 27th European Simulation and Modelling Conference, Alesund, Norway, 27–30 May 2013.
17. Kordestani, M.; Safavi, A.A.; Sharafi, N.; Saif, M. Novel multiagent model-predictive control performance indices for monitoring of a large-scale distributed water system. *IEEE Syst. J.* **2016**, *12*, 1286–1294. [CrossRef]
18. Dötsch, F.; Denzinger, J.; Kasinger, H.; Bauer, B. Decentralized real-time control of water distribution networks using self-organizing multi-agent systems. In Proceedings of the 2010 Fourth IEEE International Conference on Self-Adaptive and Self-Organizing Systems, Budapest, Hungary, 27 September–1 October 2010; pp. 223–232.

19. Wellstead, P.E. *Introduction to Physical System Modelling*; Academic Press: London, UK, 2005; Volume 4.
20. Diestel, R. *Graph Theory*; (print edition); Springer: New York, NY, USA, 2024.
21. Chung, F.R. *Spectral Graph Theory*; American Mathematical Soc.: Providence, RI, USA, 1997; Volume 92.
22. Duan, X.; Jafarpour, S.; Bullo, F. Graph-theoretic stability conditions for Metzler matrices and monotone systems. *SIAM J. Control Optim.* **2021**, *59*, 3447–3471. [CrossRef]
23. Cvetković, A. Stabilizing the Metzler matrices with applications to dynamical systems. *Calcolo* **2020**, *57*, 1. [CrossRef]
24. Oh, K.K.; Park, M.C.; Ahn, H.S. A survey of multi-agent formation control. *Automatica* **2015**, *53*, 424–440. [CrossRef]
25. Pirani, M.; Sundaram, S. Spectral properties of the grounded Laplacian matrix with applications to consensus in the presence of stubborn agents. In Proceedings of the 2014 American Control Conference, Portland, OR, USA, 4–6 June 2014; pp. 2160–2165.

MDPI AG
Grosspeteranlage 5
4052 Basel
Switzerland
Tel.: +41 61 683 77 34

Engineering Proceedings Editorial Office
E-mail: engproc@mdpi.com
www.mdpi.com/journal/engproc

www.ingramcontent.com/pod-product-compliance
Lightning Source LLC
Chambersburg PA
CBHW041426210926
51850CB00013B/3361